Lecture Notes in Physics

Editors

Patrick B. Byrne
Armagh Observatory
Armagh BT61 9DG, Northern Ireland

Dermott J. Mullan
Bartol Research Foundation, University of Delaware
Newark, DE 19716, USA

ISBN 978-3-662-13884-7 ISBN 978-3-540-47025-0 (eBook)
DOI 10.1007/978-3-540-47025-0

Originally published by Springer-Verlag Berlin Heidelberg New York in 1992
Softcover reprint of the hardcover 1st edition 1992

Typesetting: Camera ready by author

58/3140-543210 - Printed on acid-free paper

Preface

The idea for a colloquium dedicated to surface structures on active late-type stars, and taking place during the Armagh Observatory bicentenary year, was first discussed during the period leading up to IAU Coll. No. 104, *Solar & Stellar Flares.* Preliminary discussions with members of the cool star research community met with an enthusiastic reception and, after some initial uncertainty over the precise format of the meeting, a workshop catering for 60 - 100 people was decided upon. The experience of the meeting itself proved that this was a good choice, allowing for a great deal of interaction between the participants. Although a good portion of this interaction took place outside the meeting hall, we have tried to present as much as possible of the discussions as part of this volume. It is the hope of the editors that readers will find the result a useful addition to the papers themselves.

The editors, as co-chairpersons of the colloquium, wish to take the opportunity to thank our fellow members of the Scientific Organising Committee: J.G. Doyle, V.P. Grinin, J.L. Linsky, B.W. Lites, P. Maltby, M. Rodonó, A. Skumanich and H. Spruit for the work they did in preparing the programme. We are also grateful to the chairpersons of the various sessions: M. de Groot, J.B. Gurman, J.L. Linsky, M. Rodonó, A. Skumanich, H.C. Spruit.

No meeting is a matter of scientific organisation alone, and so our thanks go to the members of the Local Organising Committee: M. de Groot (Chairperson), A.D. Andrews, B. Bates, C.J. Butler, P.B. Byrne, D.J. Clement, J.G. Doyle, T. Duncan, S. Grew and A.E. Kingston, who worked so hard for over a year before the meeting took place. During the meeting invaluable assistence was given by Mrs. M. Cherry, Miss A. Brannigan and Mr. R. Scott.

Special thanks are due to the our sponsors: the Department of Education of Northern Ireland, our principal sponsor, and the Northern Ireland Tourist Board, who provided transport for delegates within Northern Ireland and a magnificent barbeque. With regard to the latter event we also acknowledge the contribution made by the Old Bushmills Distillery, a contribution which will be remembered by many of the delegates! Our thanks also go to Aer Lingus, Fane Valley Co-op Ltd, Mr. Henry Emerson, Short Brothers PLC, the Trustee Savings Bank, Annavale Construction, Castles' Travel and the Drumsill Hotel. On the day after the meeting itself a substantial number of the delegates went on bus tours to the Giant's Causeway and North Antrim coast, where we were provided with a reception by the Moyle District Council, while another contingent visited the Boyne Valley megalithic monuments and early Christian ruins and were the guests at lunch of the Meath County Council. During the tours expert commentary was provided by P. Doughty and R. Warner of the Ulster Museum. Our thanks to them.

We would also like to record our special thanks to the headmaster of the Royal School, Armagh, who gave us the use of his historic premises to house the meeting. A mention should also be made of the exceptional cooperation of the Irish weather which broke several records during the week of the meeting! Last but not least we, as editors, would like to thank Miss A. Brannigan of Armagh Observatory for her help in preparing the manuscript of this book.

During the final session of the colloquium the speaker (D.J. Mullan) paid the following tribute to Dr. E.J. Öpik, a former member of the staff of the Armagh Observatory. We felt that it was worth recording for posterity. "It is said that President Kennedy, on the occasion of a Brains' Trust meeting in the White House, remarked to the assembled dignitaries that there had not been so much talent gathered in the White House since Thomas Jefferson had dined there alone. In a similar vein, with no disrespect to present company, I would venture to say that on the occasion of this colloquium, Armagh Observatory has not seen such a gathering of talent since Dr Ernst J. Öpik worked in his office alone."

P.B. Byrne
D.J. Mullan
Armagh Observatory
July 1991

Contents

Invited and Orally Presented Papers

LIST OF ATTENDEES

(Numbers in bracketts refer to the group photograph)

ANDREWS, D.A. (51)
Armagh Observatory, College Hill, Armagh BT61 9DG, N. Ireland

ANTONOPOULOU, E. (19)
Department of Physics, Panepistimiopolis GR-15783, Athens, Greece

BATES, B. (38)
Department of Pure & Applied Physics, The Queen's University, Belfast BT7 1NN, N. Ireland

BELL, S. (54)
University Observatory, Buchanan Gardens, St Andrews, Fife KY16 9LZ, Scotland

BROMAGE, G.E. (29)
Rutherford Appleton Laboratory, Chilton, Didcot, Oxon OX11 0QX, England

BROWN, S.
Department of Applied Mathematics, The University of Sydney, NSW 2006, Australia

BUDDING, E. (24)
Department of Astronomy, University of Manchester, Oxford Road, Manchester M13 9PL

BUTLER, C.J. (5)
Armagh Observatory, College Hill, Armagh BT61 9DG, N. Ireland

BYRNE, P.B. (4)
Armagh Observatory, College Hill, Armagh BT61 9DG, N. Ireland

CHENG, C.C. (15)
Code 4177, Naval Reserach Laboratory, 4555 Overlook Ave, Washington DC 20375-5000, USA

CIARAVELLA, A.
Osservatorio Astronomico, Palazzo dei Norhanni, Palermo, Italy

COLLIER CAMERON, A. (41)
School of Mathematical & Physical Science, University of Sussex, Brighton BN1 9QH, England

CUTISPOTO, G. (6)
Osservatoiro Astrofisico, V.le A. Doria 6, I-95125 Catania, Italy

DADONAS, V. (10)
Astronomical Observatory, Vilnius University, 232009 Vilnius, Ciurlionio 29, Lithuania, USSR

DAMIANI, F.
Osservatorio Astronomico, Palazzo dei Norhanni, Palermo, Italy

DOYLE, J.G. (56)
Armagh Observatory, College Hill, Armagh BT61 9DG, N. Ireland

EATON, J.A. (49)
Suite 265, 330 Tenth Avenue North, Nashville TN 37203-3401, USA

ELLIOTT, I. (3)
Dunsink Observatory, Dublin 15, Ireland

FALCHI, A.
Arcetri Astrophysical Observatory, I-50125 Firenze, Italy

FOING, B.H. (7)
Institut d'Astrophysique Spatiale, B.P. 10, F-91371 Verrieres-le-Buisson, France

GALLEGO, J. (34)
Armagh Observatory, College Hill, Armagh BT61 9DG, N. Ireland
GIAMPAPA, M. (62)
National Solar Observatory, 950 N Cherry Avenue, PO Box 26732, Tucson AZ 85726-6732, USA
DE GROOT, M. (60)
Armagh Observatory, College Hill, Armagh BT61 9DG, N. Ireland
GURMAN, J.B. (52)
NASA Goddard Space Flight Center, 7474 Greenway Center Dr, Suite 500, Greenbelt MD 20770, USA
HATZES, A. (40)
Astronomy Department, University of Texas at Austin, Austin TX 78746, USA
HOUDEBINE, E.R. (27)
Armagh Observatory, College Hill, Armagh BT61 9DG, N. Ireland
HAYRAPETYAN, V.S. (9)
Byurakan Astrophysical Observatory, Byurakan, Armenia 378433, USSR
INNIS, J. (35)
School of Physics & Space Research, University of Birmingham, Birmingham B15 2TT, England
ISRAELIAN, G. (53)
Armagh Observatory, College Hill, Armagh BT61 9DG, N. Ireland
KATSOVA, M. (37)
Sternberg State Astronomical Institute, Moscow State University, 119899 Moscow, USSR
KELLETT, B.J. (58)
Rutherford Appleton Laboratory, Chilton, Didcot, Oxon OX11 0QX, England
KINGSTON, A.E.
Dept. of Applied Mathematics & Theoretical Physics, Queen's University, Belfast BT7 1NN, N Ireland
KJURKCHIEVA, D.P.
Department of Physics, Higher Pedagogical Institute, Shoumen, Bulgaria
KURSTER, M. (43)
Max-Planck Institut für Extraterrestriche Physik, D-8046 Garching bei Munchen, Germany
LANDSTREET, J.D.
Department of Astronomy, University of Western Ontario, London, Ontario N6A 3K7, Canada
LANZA, A.F. (32)
Instituto di Astronomia, Viale A.Doria 6, I-95125 Catania, Italy
LANZAFAME, A. (25)
Armagh Observatory, College Hill, Armagh BT61 9DG, N. Ireland
LEAN, J.
Code 4165, Naval Research Laboratory, Washington D.C. 20315, USA
LINNELL, A.P. (33)
Dept of Physics and Astronomy, Michigan State University, E. Lansing, MI 48824, USA
LINSKY, J.L. (36)
Joint Institute for Laboratory Astrophysics, University of Colorado, Boulder CO 80309, USA
MACERONI, C.
Osservatorio Astronomico di Roma, Viale del Parco Mellini 84, I-001136, Roma, Italy

MALTBY, P. (8)
Institute of Theoretical Astrophysics, PO Box 1029, 0315 Oslo 3, Norway
MATHIOUDAKIS, M. (18)
Armagh Observatory, College Hill, Armagh BT61 9DG, N. Ireland
MICELA, G.
Osservatorio Astronomico di Palermo, Piazza del Parlamento 90, Italy
MIRZOYAN, A.L. (61)
Byurakan Astrophysical Observatory, 378433 Byurakan, Armenia, USSR
MONSIGNORI-FOSSI, B.
Arcetri Astrophysical Observatory, Largo E. Fermi 5, Firenze, Italy
MULLAN, D.J. (1)
Bartol Research Institute, University of Delaware, Newark DE 19716, USA
MURPHY, M.
Armagh Observatory, College Hill, Armagh BT61 9DG, N.Ireland
NEFF. J.E. (48)
Code 681, NASA/GSFC, Greenbelt MD 20771, USA
NHA, I-S. (59)
Yonsei University Observatory, Seoul 120-749, Korea
OLAH, K. (11)
Konkoly Observatory, 1525 Budapest, PO Box 67, Hungary
O'SULLIVAN, C. (21)
Armagh Observatory, College Hill, Armagh BT61 9DG, N. Ireland
PALLAVICINI, R. (23)
Osservatorio Astrofisico di Arcetri, Largo Fermi 5, 50125 Firenze, Italy
PANAGI, P.M. (57)
Armagh Observatory, College Hill, Armagh BT61 9DG, N.Ireland
PANOV, K.P. (20)
National Astronomical Observatory, Lenin Blvd 72, Sofia 1784, Bulgaria
PARSAMIAN, E. (22)
Byurakan Astrophysical Observatory, Yerevan, Armenia, 378433 USSR
PASQUINI, L. (39)
European Southern Observatory, Casilla 19001, Santiago 19, Chile
PERES, G.
Osservatorio Astrofisico di Catania, Viale Andrea Doria 6, I-95125 Catania, Italy
PETTERSON, B.R. (16)
Institute of Theoretical Astrophysics, University of Oslo, PO Box 1029, N-0315 Oslo 3, Norway
POLETTO, G.
Osservatorio di Arcetri, Largo Fermi 5 - 50125, Firenze, Italy
RANDICH, S.
Dipartimento di Astronomia, University of Florence, Largo E. Fermi 5, 50125 Firenze, Italy
RIPODAS, P. (28)
Instituto de Astrofisica de Canarias, La Laguna, Tenerife E-38200, Spain

RODONO, M.
Institute of Astronomy, University of Catania, Viale A. Doria 6 95125, Catania, Italy
SCIORTINO, S.
Osservatorio Astronomico di Palermo, Piazza del Parlamento, Italy
SKUMANICH, A. (45)
High Altitude Observatory/NCAR, PO Box 3000, Boulder, Colorado, USA
SPRUIT, H.C.
Max Planck Astro, Schwarzschild-Str. 1, D-8046 Garching, Germany
STAUDE, J. (46)
Sonnen Observatorium Einsteinturm, DDR-1561 Potsdam, Germany
STRASSMEIER, K.G. (44)
Institute for Astronomy, University of Vienna, Turkenschanzstr 17, A-1180 Vienna, Austria
TAGLIAFERRI, G. (26)
Exosat-Sae ESTEC Keplerlaan 1, 21725C Noordwijk, The Netherlands
TONER, C.G. (14)
Institute for Astronomy, University of Hawaii, 2680 Woodlawn Drive, Honolulu HI 96822, USA
TRIBIOLI, F.
Arcetri Astrophysical Observatory, Largo E Fermi 5, Firenze, Italy
TSIKOUDI, V. (2)
Physics Department, University of Ioannina, 45110 Ioannina, Greece
VAN 'T VEER, F.
Institut d'Astrophysique, 98 bis Boulevard Arago, Paris 14, France
YOUNG, A. (30)
Astronomy Department, San Diego State University, San Diego, CA 92182, USA
ZAJTSEVA, G.V.
Sternberg State University, Southern Station, Nauchny 334413, Crimea, USSR
ZEILIK, M. (50)
Department of Physics & Astronomy, University of New Mexico, 800 Yale Blvd NE, Albuquerque NM 87131, USA

Invited

and

Orally Presented

Papers

SURFACE INHOMOGENEITIES ON LATE-TYPE STARS

P.B. Byrne
Armagh Observatory
Armagh BT61 9DG
Northern Ireland

1. Introduction

This Colloquium has been assembled to consider the current status of our knowledge of surface structures on stars of the latest spectral types, especially those which are "chromospherically active". We shall be examining solar magnetic structures, i.e. active regions, spots, plages, etc. as paradigms for the stellar features. We have accordingly invited speakers from appropriate areas of expertise in both the fields of stellar and solar research to examine this comparison and its validity. The purpose of this introduction is to give an overview of, and to make preliminary comments on, some of the basic observational phenomena.

2. Starspots

The most striking observable characteristic of late-type stellar surface structures occurs in two groups of stars, *viz.* the BY Dra and RS CVn variables. Both these groups of objects vary in broadband visible light with periods in the range ~0.5 - 10 day. The variations are periodic in the short-term, *i.e.* over 10 - 100 cycles, but their amplitudes, phases, mean brightnesses and detailed light curve shapes change on longer timescales (see Figure 1).

The generally accepted model for this behaviour is one of large-scale, dark, photospheric spots, analogous to sunspots, which grow, decay and move with time in stellar longitude with respect to one another.

2.1 BY Dra stars

BY Dra stars are K- or M-type dwarfs. They are all characterized in their optical spectra by the hydrogen Balmer-alpha line in emission. Thus they all possess dense, active chromospheres and are also active Flare Stars. These indications of non-radiative heating of their outer atmospheres are echoed in strong transition region (UV) and coronal (X-ray) emission. BY Dra stars may be binary or isolated objects and the majority have periods of $\leq$ 5 days.

2.2 RS CVn stars

RS CVn stars are all binaries, in which one component is generally main sequence G/early-K and its companion is a K subgiant or giant. Since the generally more active KIV component is post-Main Sequence, their binary nature is thought to be fundamental in preserving rapid rotation in RS CVn's. Like the BY Dra stars, the RS CVn's are strong XUV emitters.

3. Photometric mapping of surface features

Much of the invited material of this Colloquium will be concerned with examining methods used for mapping the inhomogeneous surface brightness of these objects. It is worthwhile therefore to remind ourselves of some of the most striking results which need to be addressed. In this section we will use the recent results on the optical light curve of the RS CVn system, II Peg, to illustrate some of the extreme behaviour which needs to be explained.

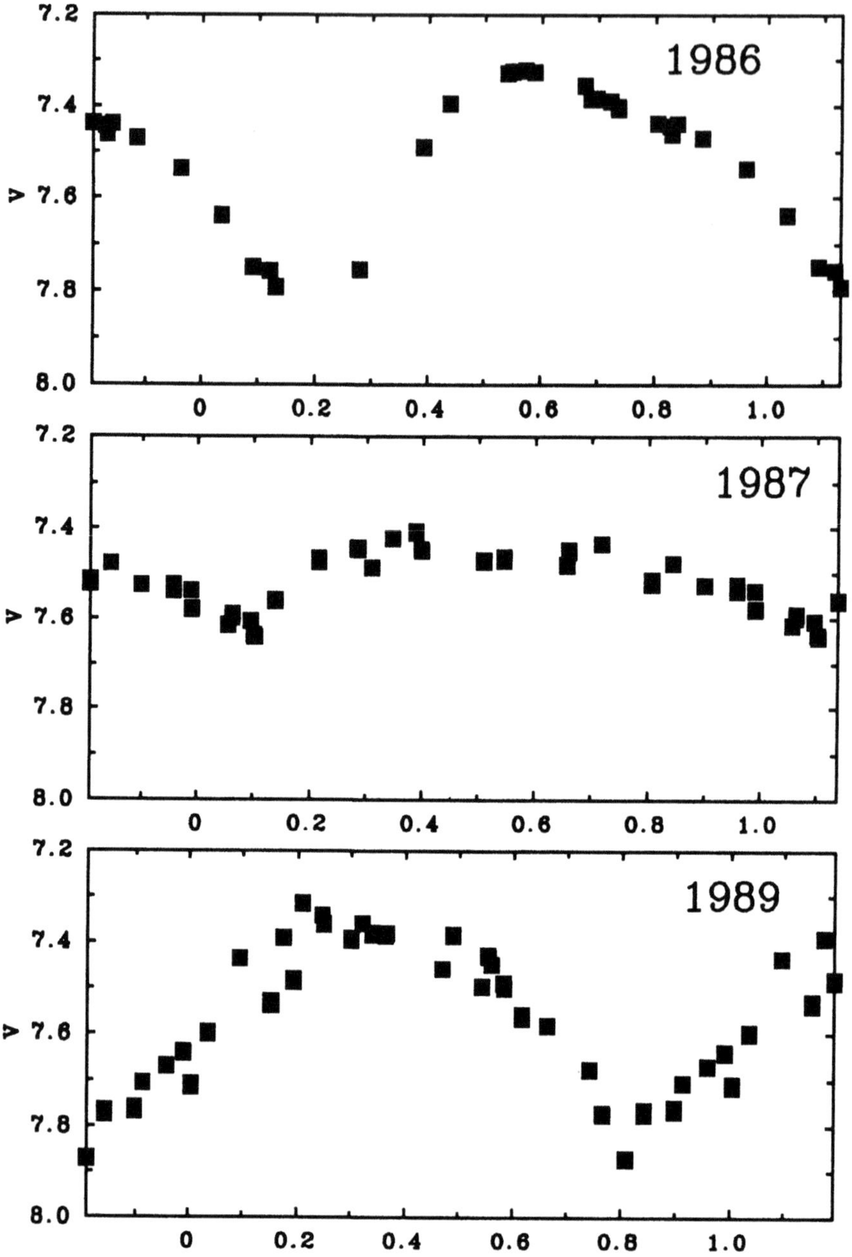

Figure 1: Optical (V) light curves of the 6.7 day period RS CVn star, II Peg, at three different epochs during the years 1986 - 1989, illustrating the changes in the light curve shape from season-to-season.

3.1 The 1986 light curve of II Peg

Figure 2 shows the V and $(V\text{-}I)_{KC}$ light curves of II Peg taken from Byrne and Marang (1987) and Cutispoto *et al.* (1987). The following gross features should be noted. First the amplitude of the V light curve is at least 0.5 mag. Second, there is a significant variation in the $(V\text{-}I)_{KC}$ colour in the sense that the star is reddest when it is faintest. These features of the light curve are consistent with a large contrast between the hemispheres of II Peg in terms of mean surface brightness and temperature and in the sense that the faintest hemisphere is also the coolest.

We can estimate the minimum areal coverage by a cool spot needed to account for the light curve as follows. We will assume a perfectly black spot of circular cross section whose centre is carried across the observer's line-of-sight by the rotation of the star. Such a solution is minimal

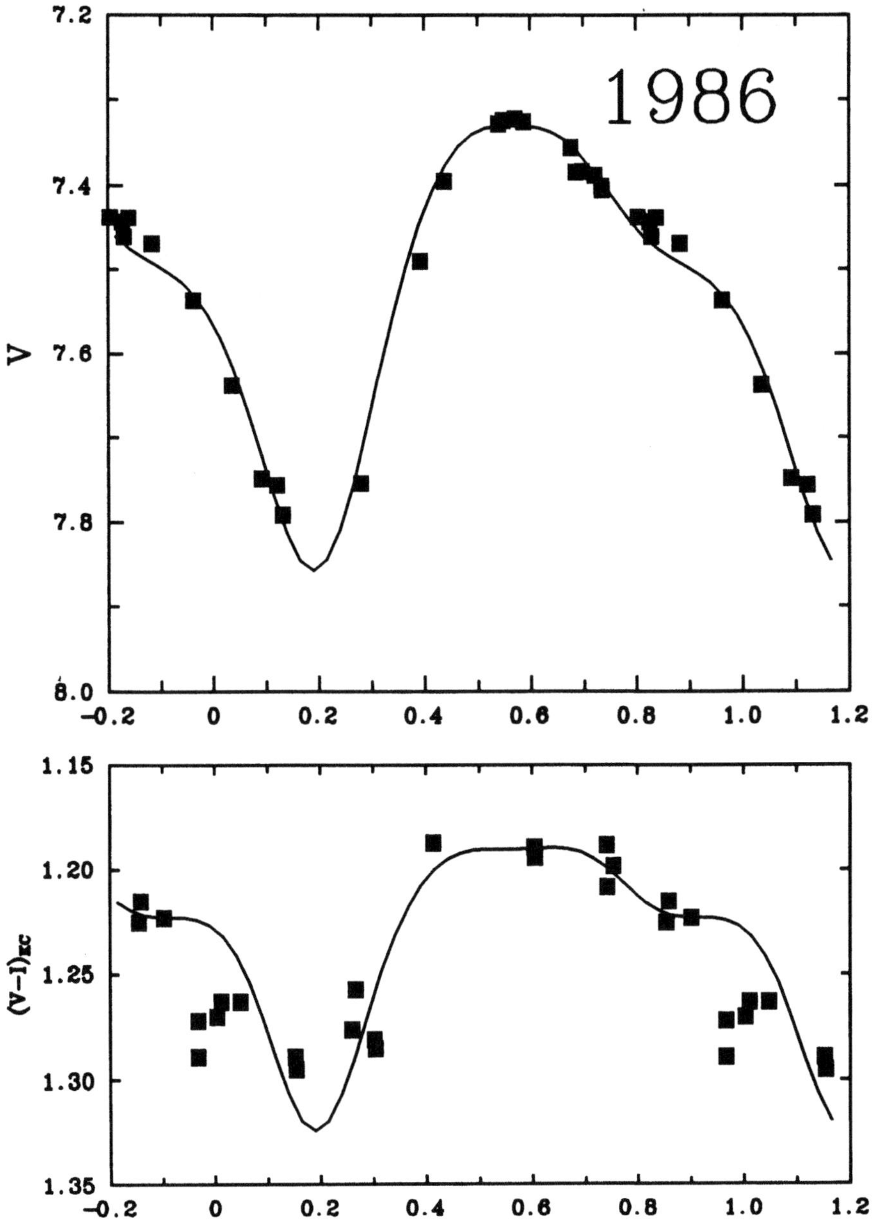

Figure 2: Multicolour broadband optical lightcurves for II Peg in 1986 with model fits.

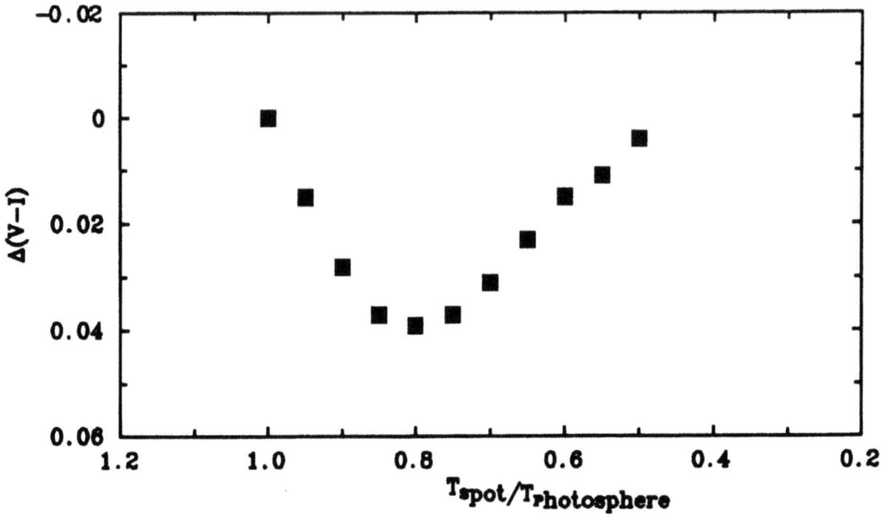

Figure 3: Variation of $(V\text{-}I)_{KC}$ colour with difference in spot and photospheric temperature

in area because there is maximum contrast between the spot and the unspotted hemisphere, because it transits the observer's meridian in the least limb-darkened portion of the disk and the circular cross section also ensures that the spot removes the light of the least limb-darkened portion of the disk. Such a solution for the 1986 light curve of II Peg is shown in Figure 2 superimposed on the data points. It requires that ~18% of the darker hemisphere be occupied by such an idealised spot (Byrne and Marang, 1987).

In fact the darker hemisphere, while dark, must contribute appreciably to the optical/IR light since colour changes are observed. A perfectly black spot would contribute no light and so would have no effect on the colour of the star. An attempt to model the colour variation led Byrne and Marang to conclude that a more realistic spot, ~ 1000°K cooler than the unspotted photosphere, but still crossing the observer's line-of-sight, would occupy ~24% of the visible hemisphere at light minimum.

There is another sense, however, in which the above solution is a minimum. Implicit in the above analysis is the assumption that light maximum corresponds to the brightness of an entirely unspotted hemisphere. Not only is this assumption unjustified but there is observational evidence that the unspotted star is at least 0.12 mag brighter in V. Chugainov (1976) observed II Peg to have a maximum of V=7.18, at least 0.12 brighter than our observed maximum. Furthermore, his light curve has no interval of constant brightness, as might be expected from an "immaculate" hemisphere. Thus, the total spottedness of II Peg in 1986 may exceed that quoted above but this additional component, because of its distribution in longitude, does not modulate the light of the star and so its distribution cannot be deduced from photometric data alone. To first approximation this depression of the light maximum from its "historical" maximum would suggest that an additional ~10 - 12% of the entire surface area of the star is spotted but we conclude nothing about its distribution, except that it be uniformly distributed in longitude at about the 1 - 2% level.

In passing, we note that this uncertainty in the placing of the "unspotted" level is the origin of polar spots in photometric models. This arises because, within the formalism of the commonly-used two-circular spot model (see *e.g.* Rodono *et al.* . 1986), it can be convenient to represent the depression of the light at all phases by a spot overlapping the pole.

Thus, from considering photometric data alone, we can demand of spot theories of late-type active stars that they be able to account for spot areas of up to $\sim 25\%$ of the stellar surface.

3.2 Spot temperatures

We have made explicit reference above to the fact that a perfectly black (0°K) spot contributes no light and therefore provides no colour variation. In fact, owing to differential limb-darkening effects, a small amplitude colour variation is seen even with a perfectly black spot (Poe and Eaton, 1985). This effect is small, however, compared to that recorded in light variations such as that in Figure 2.

Various models of colour variation with spot temperature have been made ranging from black body approximations to using the measured spectral distributions of very cool stars. There is, in the literature a general concurrence on the matter of temperature contrast between unspotted photosphere and mean spot temperature. These values range over $\sim 900°$ - 1200°K. Caution needs to be exercised in accepting these values when they are based on a single colour index. Byrne and Marang (1987) demonstrated that the effect of a spot of typical size on the popular colour index, $(V\text{-}I)_{KC}$, undergoes a maximum just in this range of temperature difference (Figure 3). Thus there is the tendency for a spot solution to drive to a temperature in this very range of values.

4. Doppler Imaging of surface features

A second powerful technique for investigating the distribution of surface inhomogeneities is that of Doppler Imaging, *i.e.* using the Doppler component of temperature- or magnetically-sensitive spectral line profiles in these rapidly rotating stars to locate and track individual surface features. Later speakers will deal with the techniques of Doppler Imaging and Zeeman Doppler Imaging in some detail but I would like to sound a cautionary note on the subject of spots found at the poles of stars.

In a Doppler Imaging line profile the location of a given feature in longitude is determined from its displacement from line centre. Latitude is determined, rather less well in general, from the radial velocity amplitude of such features. It is apparent that a feature located at the pole and symmetric about it is stationary with time at line centre. Therefore its detection relies on a knowledge of the difference between the shape of the line core in the absence of these gross surface inhomogeneities and that with the inhomogeneity present.

A widely used method of determining the shape of the reference profile is to spin up the profile of a slowly rotating star of similar spectral type (see, for instance, Vilhu *et al.* 1991). The danger in this method lies in the assumption of similar line formation in the two types of star, one chromospherically inactive, the other highly active. Lines, normally considered as photospheric in non-active cool stars (*e.g.* CaI λ4226 Å, NaI D_1 and D_2), have been found to have increasingly strong emission reversals in their cores in more rapidly rotating stars. It seems quite possible that Doppler mapping lines could be similarly affected in the most rapidly rotating objects. Unfortunately, such filling in of the line core would mimic a stationary feature in the Doppler Image map and produce an apparent "Polar Spot". Thus, as in the photometric imaging, caution needs to be exercised in treating uncertainties in the "zero-point" reference level of the signal from the unblemished star.

5. Mapping chromospheric/coronal features

Since photospheric sunspots are magnetically generated and since it is generally accepted

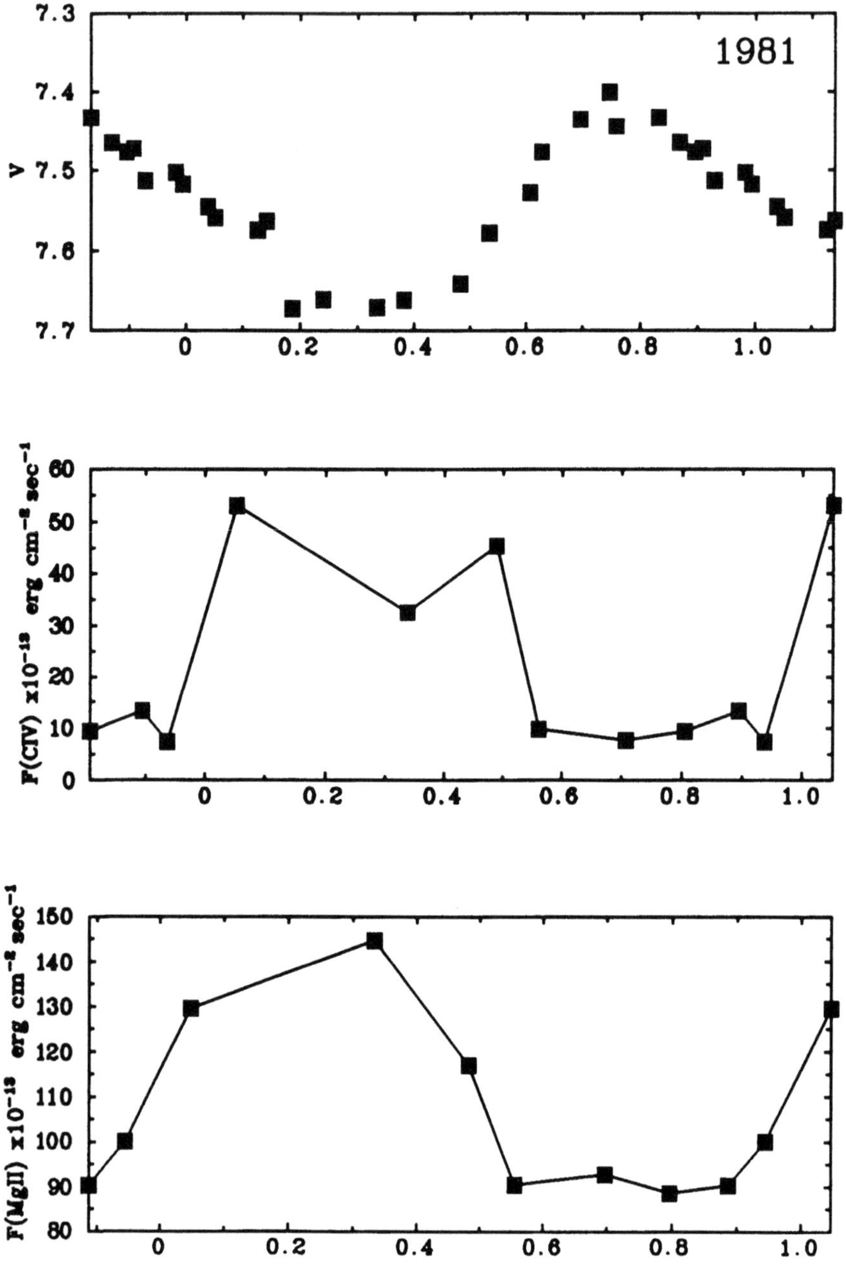

Figure 4: V band optical light curve for II Peg in 1981 and contemporaneous variations in MgII hk (chromospheric) and CIV (transition region) line emission.

that the chromospheres and coronae of stars are magnetically heated, it is a reasonable assumption that there should be some kind of positive correlation between starspots and excess chromospheric/coronal emission. Considerable effort has been expended in trying to establish the existence of such a correlation and its precise nature. There is no question that late-type

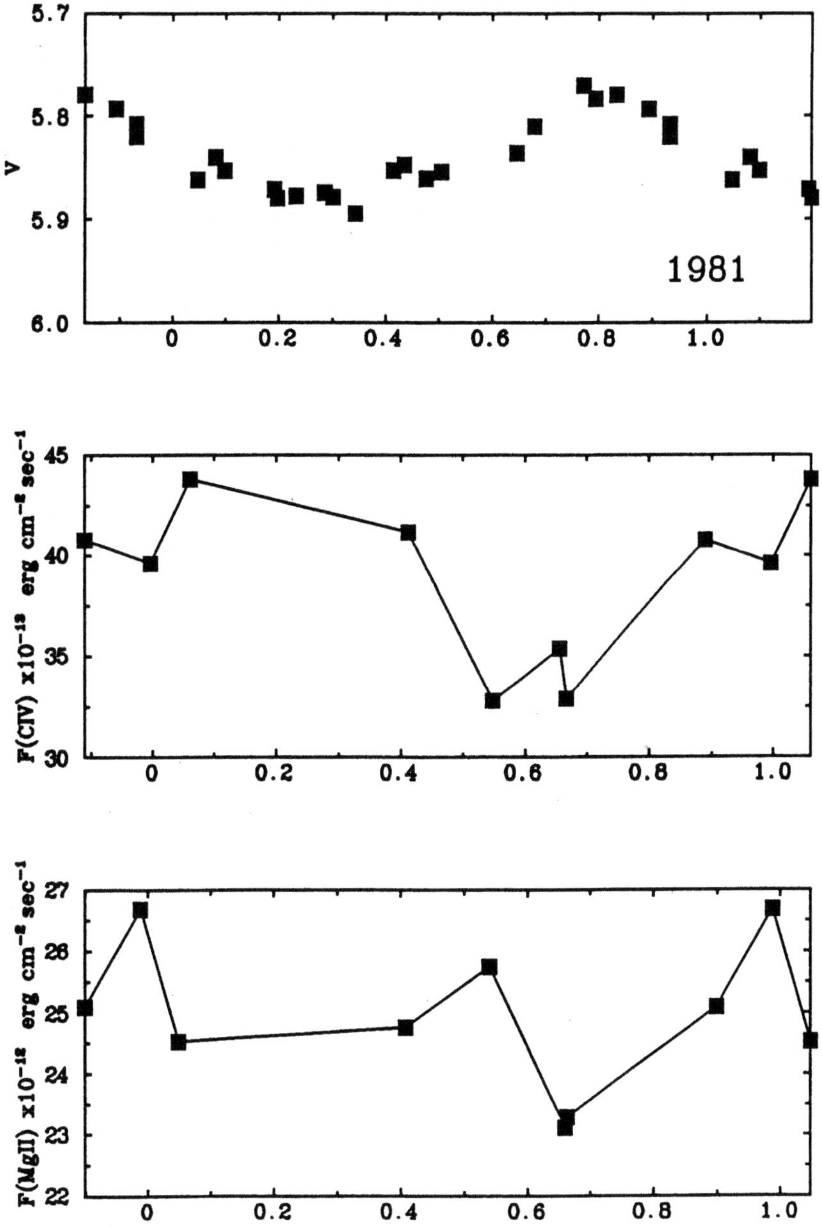

Figure 5: V band optical light curve for V711 Tau in 1981 and contemporaneous variations in MgII hk (chromospheric) and CIV (transition region) line emission.

stars of the RS CVn and BY Dra classes are among the "chromospherically" most active objects as measured by suitable diagnostic emissions, such as Balmer/Ca HK and mid-UV line emission, as well as X-ray global flux. However, attempts to establish more precise correlations have met with mixed success.

5.1 Transition region/starspot correlation

A much-quoted result in this area of study is the beautiful anticorrelation seen between transition region line emission and visible light in II Peg during 1981 (Figure 4) (Rodono *et al.* 1987). Here, three measurements at light minimum (maximum spottedness) showed a five- to ten-fold increase in transition region line intensity over those made at light maximum. The three measurements were separated by ~4 days and then 2 days which made a flare origin seem unlikely (Byrne *et al.* 1987). Subsequent work (Doyle *et al.* 1989a) pointed out the flare rates seen in this part of the ultraviolet spectrum made it more likely than previously thought that the three separate enhancements might be flares (see below).

This conclusion is borne out by both subsequent studies of II Peg itself and of other comparable RS CVn and BY Dra systems (Butler *et al.* 1987, Andrews *et al.* 1988, Doyle *et al.* 1989b, Byrne *et al.* 1991). Figure 5 shows the same kind of data as Figure 4 but for the star V711 Tau in 1981. Clearly a comparable variation to that on II Peg is not found. This behaviour is now known to be more typical of the behaviour of spotted star's transition regions.

5.2 Chromosphere/starspot correlation

The correlation between starspot positions and enhanced chromospheric emission is much more convincing. Doyle *et al.* (1989b) and Byrne *et al.* (this proceedings) include measurements of the chromospheric MgII hk emission line fluxes. Clear evidence of rotational modulation is seen with a good anticorrelation between it and visible light intensity.

6. Stellar flares

Although this colloquium does not set out to address the question of stellar flares directly, it is difficult to discuss any aspect of late-type stellar activity without some reference to the occurrence of flares and how they might have a common origin with surface inhomogeneities.

6.1 Energies of stellar flares

Just as the starspot observations outlined above suggest a solar analogy for sunspots, but on a very much enlarged scale, so stellar flares have many properties in common with their solar counterparts, but, again, on a larger scale. To illustrate this point we again turn to examples from the RS CVn star, II Peg.

Doyle *et al.* (1989a) first drew attention to the frequency and large energy content of flares on II Peg as seen in transition region line emission. Their observations suggested typical integrated flare energies in line losses from the lower transition region ($4.2 \leq \log T_e \leq 5.5$) $\sim 3\,10^{35}$ erg. Furthermore the rate of flaring was such that one such flare could be recorded once per ~10-20 hr. Such flares are orders of magnitude larger than typical solar flares and would indicate that flare energies scale by similar factors to spot area ($\sim 10^3$).

The above observations were made in transition region lines only and so energies in photospheric (optical continuum), chromospheric (UV/optical emission lines) and coronal (X-ray) radiation were unknown. An intensive campaign on II Peg, involving X-ray (GINGA), UV(IUE), ground-based spectroscopy and photometry was mounted over almost three days in August 1989 to attempt to detect the bolometric behaviour of such flares. Two X-ray/optical flares were recorded, including observations of strong broadband optical radiation (Byrne *et al.* 1990; Doyle *et al.* 1991; Avgoloupis *et al.* 1991)).

The better observed of these was that of 17 August (Figure 6). In broadband optical light the flare's integrated energy was $\sim 10^{36}$ erg, with a comparable amount of energy emitted in soft X-rays (1 - 10 keV). The X-ray spectrum was well fitted by a thermal brehmstrahlung model with an maximum observed temperature, well into the decay phase (*cf.* Figure 6), of $6.6\,10^{7}$°K.

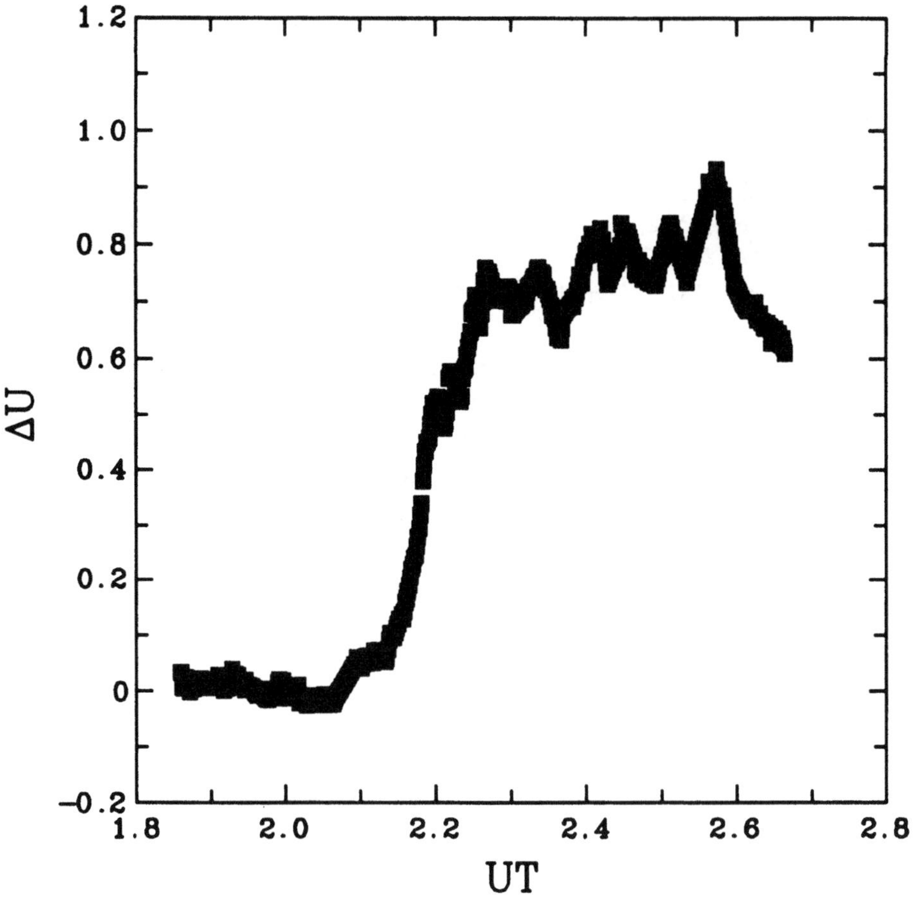

Figure 6: The optical light curve of the 17 August flare on II Peg.

The implied peak temperature would be in the region of 8-10 10^7°K.

It is clear, therefore, that these spotted RS CVn stars are capable of generating flares many orders of magnitude greater in energy than the Sun and probably beyond the upper end of the power and temperature scale of the BY Dra stars. This has led some authors to speculate on whether the solar paradigm is valid in this regime or whether additional effects become important due to their binary nature (see, for instance, Doyle *et al.* 1989a).

7. Conclusions

The study of surface inhomogeneities in late-type stars relies heavily on the solar paradigm for interpretation of the various phenomena observed. In general this model seems to serve well. However, there are results from the active late-type star field which have no solar counterpart. The most striking of these are polar spots. We have shown, however, that the case for spots at the poles is not yet fully proven. In this colloquium we will address these issues and look forward, particularly in the presence of some of our solar physics colleagues, to a detailed examination of the issues involved.

References

Andrews, A.D., Rodono, M., Linsky, J.L., Brown, A., Butler, C.J., Catalano, S., Scaltriti, F., Busso, M., Il-Seong Nha, Oh, J.Y., Henry, M.C.D., Hopkins, J.L., Landis, H.J., Engelbrektson, S.: AA, 204, 177

Avgoloupis, S., Doyle, J.G., Mavridis, L.N., Seiradakis,J.H., Mathioudakis, M.: 1991, this volume

Butler, C.J., Doyle, J.G., Andrews, A.D., Byrne, P.B., Linsky, J.L., Bornmann, P.L., Rodono, M., Pazzani, V., Simon, T.: AA, 174, 139

Byrne, P.B., Marang, F.: 1987, Irish AJ, 18, 84

Byrne, P.B., Doyle, J.G., Brown, A., Linsky, J.L., Rodono, M.: 1987, AA, 180, 172

Byrne, P.B., Doyle, J.G., Tsuru, T., Avgoloupis, S., Mavridis, L.N., Seiradakis, J.H., Kilkenny, D.W., Laing, J.D., Marang, F.: 1990, Irish AJ, 19, 150

Byrne, P.B., Agnew, D., Cutispoto, G., Kilkenny, D., Neff, J.E., Panagi, P.M.: 1991, this volume

Chugainov, P.F.: 1976, Izv. Krymskoi Astron. Obs. 54, 89

Cutispoto, G., Leto, G., Pagano, I., Santagati, G., Ventura, R.: 1987, IBVS, No. 3034

Doyle, J.G., Byrne, P.B., van den Oord, G.H.J.: 1989a, AA, 224, 153

Doyle, J.G., Butler, C.J., Byrne, P.B., Rodono, M., Swank, J., Fowles, W.: 1989b, AA, 223, 219

Doyle, J.G., Kellett, B.J., Avgoloupis, S., Byrne, P.B., Tsuru, T., Bromage, G.E., Seiradakis, J.H., Pye, J.P., Barstow, M.A., McHardy, I.M., Makashima, K., Mavridis, L.N.: 1991, MNRAS, 248, 503

Poe, C.H., Eaton, J.A.:1985, ApJ, 289, 644

Rodono, M., Cutispoto, G., Pazzani, V., Catalano, S., Byrne, P.B., Doyle, J.G., Butler, C.J., Andrews, A.D., Blanco, C., Marilli, E., Linsky, J.L., Scaltriti, F., Busso, M., Cellino, A., Hopkins, J.L., Okazaki, A., Hayashi, S.S., Zeilik, M., Helston, R., Henson, G., Smith, P., Simon, T.:1986, AA, 165, 135

Rodono, M., Byrne, P.B., Neff, J.E., Linsky, J.L., Simon, T., Butler, C.J., Catalano, S., Cutispoto, G., Doyle, J.G., Andrews, A.D., Gibson, D.: 1987, AA, 176, 267

Vilhu, O., Gustafsson, B., Walter, F.M.: 1991, AA, 241, 167

Discussion: Byrne

Zeilik You mentioned that we should examine how well the solar analogy "holds up". I have the impression that it does so only selectively - that we are for too constrained by the Sun. For active stars, we have a *very* different regime of physical parameters.

Byrne I disagree. It is difficult, in fact, to avoid the solar analogy. The late-type active stars have chromospheres and coronae whose physical parameters are not very different from those of Active Regions on the Sun. Such estimates as we have of electron densities and emission measures of stellar active regions overlap with the upper end of the range of solar values.

Mullan Is there observational evidence of changes in bolometric magnitude as spots come and go? Petterson's U-to-K photometry should be repeated at different epochs to determine whether total output from star varies.

Byrne Some work has been done in this area by Doyle, Butler and Mathioudakis who have recently recorded simultaneous U-to-L photometry of a large number of late-type active stars. This work is in press in AA

Eaton How is the question of energy balance required by giant spots any different from the question of energy transfer in contact binaries? Perhaps we have a way in the contact binaries to study the same process.

Byrne I believe that there are fundamental differences between the contact and non-contact systems in this respect. Most of the "conventional" RS CVn systems are fully detached and, beyond synchronization, there are no large-scale gravitational effects on their atmospheric structure. It seems to me better to pursue the question of the nature of the spots by observing the detached, longer-period RS CVn's wherein the additional complicating factors associated with contact are not present.

Giampapa The lack of modulation in high excitation lines at times of large spot coverage may be a manifestation of even larger (perhaps uniform) coverage of plage-like or loop regions. This should be associated with an enhanced over-all level of emission on e.g. II Peg. Is this what you observed?

Byrne No. In fact the reverse is true. During the 1986 season, when there was the greatest degree of contrast yet observed between the stellar hemispheres, the global emission in the high-excitation lines was depressed.

Giampapa I think the question of the uniqueness of models of single-spots which persist for decades should be addressed during this colloquium. Can, in fact, a variety of evolving spot surface distributions give rise to about the same light curve over many years?

Byrne I agree that this question should be addressed. The question of long-lived single spots is, of course, closely related to the question of active longitudes. If spots tend to occur at preferred longitudes, for dynamical or other reasons, the evolving spot groups can mimic single long-lived spots.

Innis Following question of Giampapa and response of Byrne concerning the presence/absence of the correlation of spot wave with UV line flux in II Peg, can geometric effects (eg spot latitude, angle of inclination of rotation axis of star) explain how the correlation (between spot wave and UV line flux) comes and goes?

Byrne No, I don't believe so. When we are measuring lines such as C IV or N V, these are formed in the coronal approximation and so are optically thin. As such they do not suffer from "limb-darkening" type effects and so would be insensitive to variations of latitude.

Foing The observed longterm stability of spots, rather than being totally an artifact of the method of imaging them from photometry, should not be over-interpreted. It may correspond only to the recurrent emergence of fields in active regions, even if individual spots lifetime may be significantly shorter.

Byrne Yes, I agree fully, as I said in my reply to Giampapa.

Foing The lack of exact temporal correlation between photometric light curves and emission line variation, as well as solar observations showing large magnetic patches, or emission faculae far from spot areas, suggest non spatial coexistence of these different degrees of activity. How are we to put these observations in a physical framework?

Byrne The relationship between optical spots and chromospheric plage emission is obviously complicated in both the stars and Sun. We look forward to hearing the reports of our invited speakers working in solar physics in this area so that a comparison may be made with the stellar case.

(l to r) J. Lean looks away while E. Budding catches up on local news. K. Olah and F. van't Veer have more serious things to discuss. S. Bell and A. Linnell are also deep into it!

MODELS OF STARSPOTS AND RESULTS

Joel A. Eaton
Center of Excellence in Information Systems
Tennessee State University, Nashville, TN 37203 USA

1. Introduction

In the landmark paper signalling the beginning of today's interest in spotted stars, Hall (1972) proposed a model analogous to solar activity for understanding the properties of RS Canum Venaticorum and stars with similar properties. While many of the mechanisms he advanced have fallen quietly by the wayside, a surprisingly large number of them are with us today providing the backbone for our present understanding of highly spotted stars. Specifically, we may mention the concept of the very large spots we use to calculate such stars' light curves, the differential rotation that carries these spots around the star at slightly different rates at different times, and the hope that we may actually detect magnetic cycles in these stars analogous to the solar cycle.

During the past five years we have seen the development of the systematic approach to studying spotted stars. This has resulted in a burgeoning body of information about chromospherically active, spotted stars that promises to define the relationship between magnetic activity and spottedness in a way never before possible. In my opinion, the most significant recent advances have been (1) applying standardized models to large bodies of data and (2) combining results of photometry and spectroscopy to define the locations and properties of starspots. Let me review these points, first discussing developments in light-curve analysis, then considering the impact of some of the recent results and their uncertainties on our understanding of spotted stars.

2. Trends in light-curve solutions

There have recently been three major developments in analysis of spotted stars' light curves: the analysis of spotted systems with the Wilson-Devinney code, the systematic approach to fitting circular spots on stars taken by Budding and Zeilik, for example, and the inelegant, simple-minded yet extremely efficient technique of fitting combinations of an idealized light curve of a single spot to many observed light curves.

2.1 Application of the Wilson-Devinney model to spotted stars

Eaton and Hall (1979) demonstrated convincingly that the light curve of RS CVn can be fitted quite accurately with a spot model and that, indeed, a spot model is required by the data. The distribution of spots they used, confined to a belt centered on the equator in analogy to the Sun, has fallen from favor, and there are reasons to believe the spots on RS CVn are not simply a collection of small, solar-type spots. Yet, the conclusion remains; in 1964 the F star was eclipsing a distribution of spots spread unevenly over the K subgiant's disc. Furthermore, at least those spots eclipsed at secondary minimum had to be within about 35 degrees of the equator to be covered.

Two groups have now included spots in the Wilson-Devinney code (Wilson and Devinney 1971) and used it to solve actual light curves of spotted stars. Kang and Wilson (1989) presented solutions of light curves of RS CVn, AR Lac, and V711 Tau in which the spots' locations and other attributes were adjusted automatically by least squares. Zhai and Chen have devised an independent procedure for adjusting the spots in the Wilson-Devinney program and have obtained spot solutions for RT CrB (Zhai and Chen 1989) and WY Cnc (Xie et al. 1989).

Kang and Wilson demonstrated convincingly that such a procedure could place spots reliably and impersonally on the star and converge to a reasonable fit of the data. Furthermore, they were able to fit the eclipses of the spotted component with the same two spots that fit the distortion wave. The Chinese group have obtained similar results for RT CrB (P = 5.1 days) and WY Cnc (P = 0.83 days). The former was found to have one large spot and a smaller one, the large spot extending from a pole to past the equator. WY Cnc, on the other hand, was found to have two moderately sized spots (radii between 13 and 33 degrees) at two epochs.

We cannot stress enough the importance of analyzing **eclipses** of spots for determining their detailed properties. To illustrate this, I should like to draw your attention to the light curve of RS CVn in 1984 (Eaton and Poe 1985). The data were obtained at Kitt Peak National Observatory with the late, lamented No. 4 16-inch telescope. They are shown in Figure 1, where we can see the huge distortion wave at this epoch. Significant aspects of these curves for spot models are (1) the shape of secondary minimum, in which the eclipse of a large spot causes pronounced distortion, (2) the shapes of the color curves for (B-V) and (V-I), which reflect the temperatures of the spots.

We have analyzed V, (B-V), and (V-I) with an improved version of the computer program developed by Eaton and Hall (1979), which allows two barrel-shaped spots bounded by circles

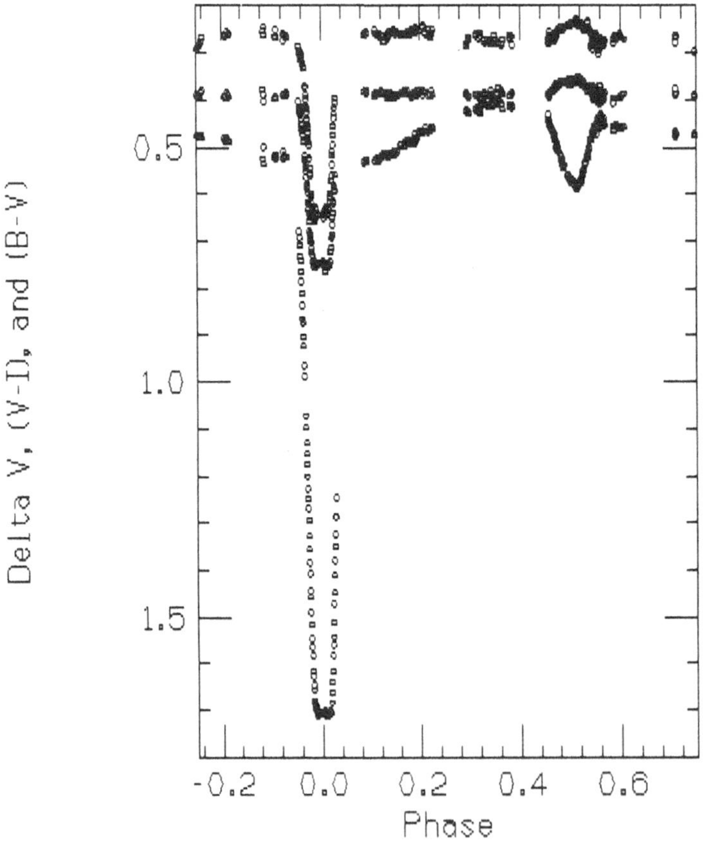

Figure 1: BVI$_c$ photometry for RS CVn at epoch 1984.7. The 0.11-mag wave in the visual light curve would be 0.30 mag if light of the F star were removed, comparable to the huge amplitude found by Doyle et al. (1988) in II Peg.

Table 1:
Elements for RS CVn at 1984.7

$q = 1.0$ $r_1 = 0.117$ $r_2 = 0.235$ $i = 87.0$
$T_1 = 6650\,K$ $T_2 = 4700 \pm 10\,K$ $A_{Bol} = 0.5$ $g = 0.32$
$x_{1B} = 0.70$ $x_{1V} = 0.55$ $x_{1I} = 0.35$
$x_{2B} = 0.90$ $x_{2V} = 0.71$ $x_{2I} = 0.45$

Properties of the Spots

	Latitudes	Width in Longitude	Central Longitude	Fractional Coverage	T(spot)
Spot 1	-16 to +16	166	180	0.795	3450 K
Spot 2	0 to +24	50	335	1.000	3100 K

Note: Latitudes and longitudes are in degrees. All elements were fixed except the temperatures and spot parameters.

of latitude and longitude to be put on the star. This spot model can have two spots of different temperature, and this property is actually needed to fit the data for 1984. Figure 2 shows the secondary minimum and illustrates well what the eclipse of a spot does to an ordinarily clean light curve. The slope from second to third contact in the color curves is sensitive to spot temperature. Also, we have represented the dominant spot (seen at phases 0.75–0.25) as a belt extending 166 degrees in longitude, but this shape is not significant. Poe represented it well in a non-eclipsing solution with a circular spot of radius 47 degrees at 45 degrees latitude (or 36 degrees on the equator).

2.2 Use of two-spot models to follow spots

There have been many studies of active stars that do not eclipse using two-spot models (e.g., Vogt 1981b; Dorren and Guinan 1982; Poe and Eaton 1985; Olah et al. 1985; Rodono et al. 1986). Most, however, have not adjusted the spot parameters impersonally and have not considered the uncertainties of the fits in detail. Budding and Zeilik (1987), however, have developed a particularly elegant version of this approach, fitting two spots to observations by least squares with careful consideration of the determinacy of the solution and the possibility of finding local minima in χ^2. At favorable epochs this makes it possible to determine spots' latitudes and radii as well as their longitudes. Their program has now been applied to a number of short-period RS CVn systems (P < 1 day; Hall 1976) by Zeilik and his collaborators. SV Cam was analyzed by Zeilik et al. (1988), RT And by Zeilik et al. (1989), BH Vir by Zeilik et al. (1990a), and WY Cnc by Zeilik et al. (1990b). Lazaro (1990) has recently applied the model to the growing body of data for II Peg.

These studies provide objective measurements of spot longitudes, latitudes, and radii at very many epochs, which allows the distribution of spots to be characterized statistically. Zeilik et al. find that the spots on short-period RS CVn binaries are generally found in the sectors of longitude centered on 90 and 270 degrees and that they tend to have relatively high latitudes, at least in RT And and BH Vir. Furthermore, they find the spots on these stars are relatively small. The longitudes derived by Lazaro for II Peg seem to be more randomly distributed, as

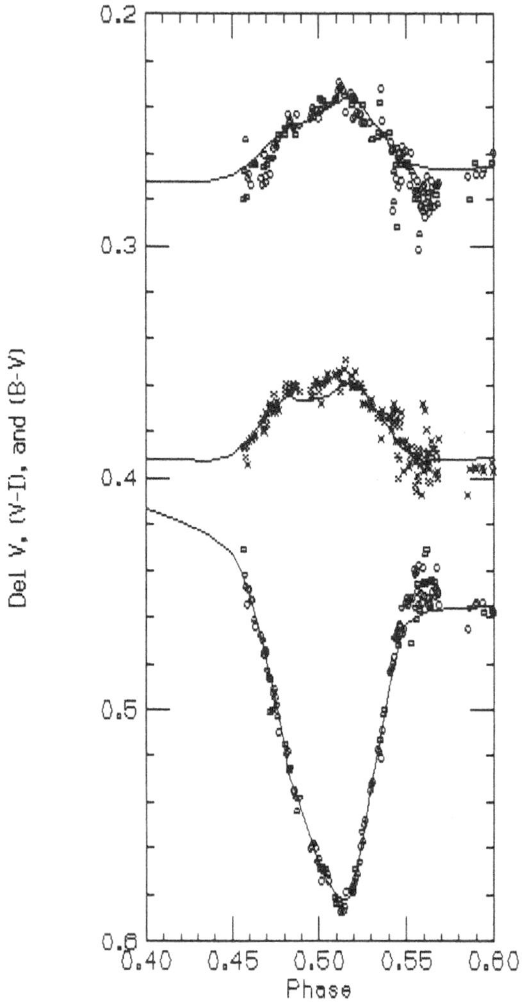

Figure 2: RS CVn at 1984.7: the fits of secondary minimum. Eclipse of a spot centered on the disc at phase 0.57 gives the pronounced distortion of the eclipse. Since the phases are well determined from primary minimum, the displacement of this eclipse is caused solely by mottling of the eclipsed star. For this epoch the slopes of the two color curves are sensitive to spot temperature and lead to a value near 3100 K.

might be expected for this non-eclipsing system with relatively larger spots.

2.3 A light-curve fitting technique

Most of our studies of spots to date have looked at a few epochs only, providing mere fleeting glimpses of the surface of a star. This makes it impossible to follow the development and change of individual spots reliably. A third approach to spot modelling which overcomes these limitations has been pioneered by Hall, Henry, and Sowell (1989), who reasoned that a single spot transiting a star's disc will depress its light by an amount calculable from the lower branch of a sine curve. When the spot is out of sight, they assume, there is no light loss. This gives a particularly simple form to the light loss, thus allowing one or two spots to be added together with appropriate phase shifts and amplitudes to fit an observed light curve. The parameters describing the spots – two phases (longitudes), two amplitudes, and a zero point – are found by changing the parameters to minimize χ^2, the various parameters being changed in succession until an acceptable fit is achieved. There are reasons to criticise this approach. It adds the effects of spots in magnitude space, which assumes the small-amplitude limit, and it eschews all pretense of determining the spots' latitudes. On the other hand, it determines the **longitudes** of spots accurately and quickly, and it provides a measurement of the strength of the spot in the light curve. Furthermore, its efficiency allows the photometry for a given year to be divided into different subgroupings which can be fitted separately, thereby making it possible to study the development and changes of individual spots in the large bodies of data now pouring in from today's automatic photometric telescopes (APT's). This latter property is the true strength of the method.

Results of this approach are indeed impressive. Hall, Henry, and Sowell (1989) were able to identify 7-9 individual spots in V478 Lyr (G8 V, P = 2.13 days) which lived between 3.5 months and 2 years. A study of HR 7428 = V1817 Cyg (K0 III, P = 108 days) by Hall et al. (1990) reveals four spots with lifetimes of about two years each.

3. Active topics

Let us now turn our attention to four topics relating to the interpretation and reality of the results of spot models – the possibility that spots emerge or are found only at preferred stellar longitudes, the latitudes of spots, the interpretation of spot temperatures, and the possibility of a significant facular contribution to the unspotted photosphere of the star.

3.1 The Possibility of Active Longitudes

In the Sun spots tend to emerge within very restricted zones of **latitude** which change position with the magnetic cycle. The idea that spots are somehow produced in restricted active zones of **longitude** has been hanging with us for as long as we have considered spotted stars. Losh (1939) proposed it for the Sun, although more recent evidence that spots are generally confined to active longitudes has not been convincing. Coronal holes in the Sun do exhibit a long-lived sector structure, which may persist from one spot cycle to the next, so spottedness in magnetically active stars could well be expected to have the same behavior. The idea of active longitudes was used by Eaton and Hall (1979) to explain the persistence of the light-curve wave in RS CVn in the context of a spot model featuring very many small solar-type spots. Such spots have much shorter lifetimes on the Sun than the decade-long persistence of the wave, so new spots would necessarily emerge at just the right longitudes for the wave to persist. It now appears spots are generally **much** larger than sunspots and live much longer (Hall and Busby 1990), so we no longer have the same rationale for active longitudes. Nevertheless, spot solutions for many years of photometry of certain stars have continued to reveal spots congregated at certain longitudes. Zeilik et al. find a concentration of spots near longitudes 90 and 270 degrees in short-period eclipsing RS CVn systems (see Section IIb), while Hall, Henry, and Sowell (1989) have argued for a **possible** structure rotated 90 degrees from this in V478 Lyr, a main-sequence star with a somewhat longer period of 2.2 days. Other claims of active longitudes have been advanced by Olah and Hall (1988) for HK Lac and by Olah et al. (1989) for σ Gem.

Just what do the spot solutions tell us about active longitudes? The data in hand are **not** conclusive. In II Peg, Lazaro has found spots that are typically 20 degrees in radius and which would thus live about five years (Hall and Busby 1990). These spots have time to migrate in

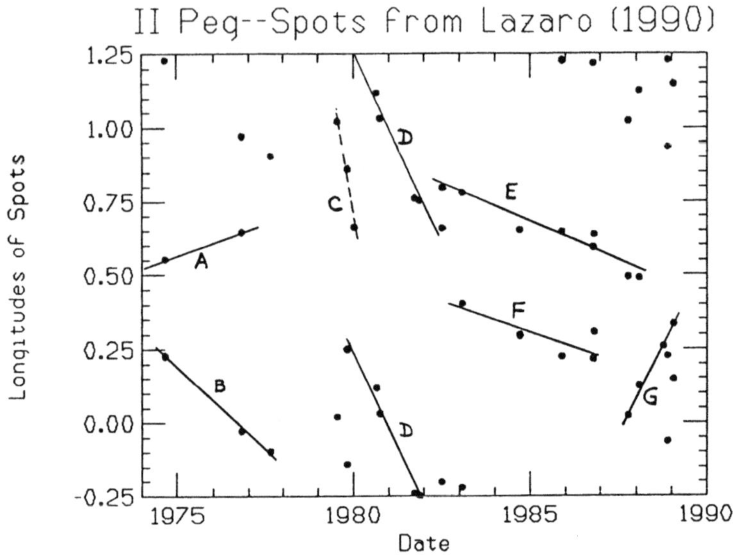

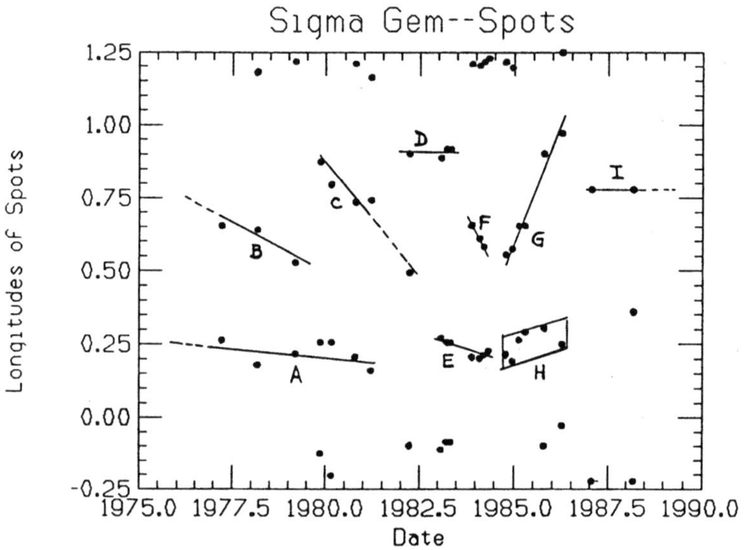

Figure 3: Migration diagrams for II Peg and σ Gem. Data for II Peg have been taken from the solutions of Lazaro (1990); those for σ Gem from Strassmeier et al. (1988) and Olah et al. (1989). Shown are possible identifications of individual spots, based on the migration seen as well on as the sizes derived in the spot solutions.

longitude from their place of origin through differential rotation. Spots on the short-period systems analyzed by Zeilik and collaborators are much smaller, in both the angular and linear senses. The median spot sizes in SV Cam, RT And, and BH Vir are 12, 10, and 12 degrees, respectively, implying lifetimes of 1-3 months. Unless the migration rates for these spots are especially high, we are seeing them very near their point of origin. Their concentration near longitudes 90 and 270 degrees argues that there **are** active longitudes at which spots preferentially emerge. Data for the few non-eclipsing systems observed extensively enough for an analysis contradict this conclusion. In their study of V478 Lyr, Hall, Henry, and Sowell (1990) identified the emergence of four of the ~ 8 spots detected, and these emerged in the sectors centered on longitudes 0 and 180 degrees, 90 degrees from the sectors in the short-period systems. In V1817 Cyg, Hall et al. (1990) observed three spots that formed during the period covered, one near phase 0.25, two near phase 0.0. We can extend this analysis by attempting to identify individual spots in the models for II Peg (Lazaro 1990) and σ Gem (Strassmeier et al. 1988; Olah et al. 1989). These are shown in Figure 3, where I have used continuity of the migration curve and stated sizes of the spots to identify likely spots in the jumble of spot longitudes. We can see roughly seven spots in II Peg and ten in σ Gem. The A and B spots were in existence when the observations began. Of the others, six originate in sectors centered on 0 or 180 degrees, while seven originate in sectors centered on 90 or 270 degrees. Clearly these other stars do not have the same sector structure found in SV Cam, RT And, BH Vir, and XY UMa. Are there systematic errors introduced by not calculating directly the effects of eclipses of the spots in these stars? Although Zeilik et al. have considered this question very carefully, there is some evidence such errors do occur. Xie et al. (1989) have fitted the short-period system WY Cnc at two epochs with the Wilson-Devinney model, finding four spots, one of which was small enough (13 degrees) to be near its point of origin. This spot was at longitude 358 degrees, facing the other star.

I must conclude that the present evidence favors a random distribution of spots in longitude, but that a final resolution of the question must await consideration of data for further stars and, especially, the analysis of more short-period RS CVn systems with the Wilson-Devinney method.

3.2 The Latitudes of Starspots

Latitudes are found in a number of ways, although their determination from photometry is difficult. If the spot's shape is reliably known, a latitude can be determined from a rotational light curve, as is being done most reliably by Zeilik and collaborators. A latitude can also be found from eclipses of spots, from Doppler images of spotted stars, and in an ancillary way from studies of active regions. Recent results have suggested very different placements of spots. As we have seen, the secondary eclipses of RS CVn often require large spots to be within 35 degrees of the equator. Studies to date with Doppler images have all found spots covering the star's pole or very near it. This was shown most convincingly for V711 Tau, a star whose 32-degree inclination makes the pole readily visible, by Vogt and Penrod (1983) and Vogt (1988). More recently in HD 26337 (EI Eri), Strassmeier (1990) has found a large polar spot and several small ones at lower latitudes in Doppler images. A study of active regions on HK Aqr by Young et al. (1990) suggests these, too, are confined to the poles. On the other hand, Gray (1989) has argued active regions in the typical, **slowly** rotating G-K giants are found in a belt centered on the equator. Likewise, the reconciliation of a light curve of UX Ari with high-dispersion spectra (Noah, Bopp, and Fekel 1987) does not require a polar spot.

Another powerful tool for identifying spots at different latitudes is the migration curve of the sort shown in Figure 3. Individual spots migrate at different rates because of the difference in rotation of their different, fixed latitudes (e.g., Hall and Henry 1990). Thus in II Peg, σ Gem, V478 Lyr, RT Lac, V1817 Cyg, HK Lac, etc. we know there are spots on the star at **different** latitudes.

The photometric studies of BH Vir (Zeilik et al. 1990a) and RT And (Zeilik et al. 1989) found latitudes near 50 degrees in 13 of 14 determinate cases. SV Cam (Zeilik et al. 1988), in contrast, was found to have spots extending from the equator to 70 degrees, and II Peg (Lazaro 1990) also was found to have spots extending up to 70 degrees. How accurate are these latitudes? Does the constant latitude of spot emergence seen in BH Vir and RT And represent the same phenomenon seen in the Sun? There are two ways to approach these questions, namely comparing latitudes derived from light curves and Doppler images and comparing the derived latitudes of spots identified in migration diagrams. Strassmeier (1990) compared spots from Doppler images and from photometry for EI Eri, finding the **longitudes** of the individual small spots may agree but that the **latitudes** in his light-curve solution were systematically lower than those from the line profiles and that the spots were bigger. This is not necessarily a definitive comparison, since the individual spots were too small for much light variation. A more favorable comparison, however, comes from V711 Tau in which the light-curve solutions of Rodono et al. (1986) were very similar to the Doppler images of Vogt and Penrod (1983). The spot latitudes for II Peg derived by Lazaro agree very poorly with the spot migration seen in Figure 3. This indicates the latitudes are simply incorrect, or that our assumption about migration at constant latitude is wrong. We cannot, unfortunately, test the constancy of spot latitudes in BH Vir and RT And, since the light curves analyzed were obtained so infrequently that no one spot could possibly be observed migrating. This is a problem also in SV Cam for which the most complete coverage, by Cellino, Scaltriti, and Busso (1985) consisting of four light curves at ~ monthly intervals, cannot reliably distinguish between a one-month migration period and a short spot lifetime.

3.3 Temperatures of Starspots

Spot temperatures have now been derived from multi-color photometry for more than a decade. The effect studied is simple; since actual starspots are cooler than the unspotted photosphere, possibly by a considerable amount, the spot will contribute much more light at long wavelengths than at shorter ones. The spot shows up as variation in a color, generally (V-I), (V-R), or (B-R), in phase with the visual light variation. The star appears redder when it becomes fainter. Table 2 is a compilation of spot temperatures measured with multi-color photometry. Most of these determinations are rather old, possibly reflecting the extirpation of stellar photometry from Kitt Peak National Observatory and the lack of precise photometry in the near infrared from APT's. However, we note that measurements for two short-period RS CVn binaries have now been derived by Zeilik's group. The results found by various investigators are remarkably similar, and a typical spot in the subgiant component of an RS CVn binary seems to be about 1100 K cooler than the surrounding photosphere and to have a temperature around 3300-3800 K on the scale for dwarf stars.

Just what do these spot temperatures represent? This question was addressed by Poe and Eaton (1985) and by Dorren (1987). The answer goes a long way toward clarifying the nature of starspots in chromospherically active stars. Dorren showed that a spot consisting of a penumbra and an umbra in solar proportions and with temperatures proportioned to the surrounding photosphere as in the Sun would give light and color variations just like those of a spot with a single, low, umbral temperature (see, also, Byrne 1987). Thus, much of the visual light variation could come from a very cool, relatively small umbra, while the spot's light would be emitted by the warm penumbra. This complication was mooted previously by Rucinski (1981). Simulations with the computer program of Eaton and Poe (1985) indicate this effect persists for spots consisting of roughly equal areas of black umbra and warm penumbra, but it disappears if the relative amounts of umbra and penumbra are reversed. In that case, the temperature derived from photometry would be a few tens of degrees higher than the umbral temperature in the case tested.

Spectroscopy has shown convincingly that the spots must be much cooler than the surrounding photospheres. Infrared spectra detect features of TiO and VO, (Ramsey and Nations 1980;

Table 2
Spot Temperatures

Star	Spectrum	P_{Orb}(dy)	Epoch	T_{spot}	T_{star} - T_{spot}	Source
RS CVn	K0 IV	4.798	1984.7	3100 + 3400 K	1600 + 1300 K	1
RT And	F8 V	0.629	1982.8	5180 K	1070 K	2
			1987.9	5040 K	1200 K	2
BH Vir	G0 V	0.817	1984.3	...	1300-1400 K	3
			1986.5	4900 K	1100 K	4
SV Cam	G2-3 V	0.593	1984.1	3780 K	1550 K	12
WY Cnc	G5 V	0.829	1989.3	3640 K	1900 K	12
II Peg	K2-3 IV-V	6.724	1977.6	3420 K	910-1200 K	5,7
			1979.7	3330 K	1020 K	5
			1980.7	3620 K	730 K	5
			1986.7	3700 K	950 K	13
HK Lac	K0 III	24.428	1980.7	$\leq$3520 K	$\geq$1080 K	5
				$\leq$3840 K	$\geq$950 K	7
HR 7275	K1 III-IV	28.59	1980.7	3400 K	1200 K	5
IM Peg	K2 II-III	24.649	1980.7	3520 K	920 K	5
λ And	G8 III-IV	20.521	1980.7	3730 K	1050 K	5
UX Ari	K0 IV	6.438	1980.7	3360 K	1420 K	5
σ Gem	K1 III	19.604	1980.7	3870 K	570 K	5
LX Per	K0 IV	8.038	1980.7	4050 K	730 K	5
VY Ari	G9 V	13.198	1984.7	3500 K	800 K	6
FK Com	G3 III	2.40	1979.3	...	600-1000 K	8
			1983.2	...	600 K	11
SZ Psc	K1 IV	3.966	1976.9	3500 K	1200 K	9
V711 Tau	K1 IV	2.838	1977.7		1800 K	10

References: 1 = this paper, 2 = Zeilik et al. (1989), 3 = Scaltriti et al. (1985), 4 = Zeilik et al. (1990a), 5 = Poe and Eaton (1985), 6 = Eaton and Poe (1986), 7 = Vogt (1981b), 8 = Rucinski (1981), 9 = Eaton and Hall (1979), 10 = Dorren et al. (1981), 11 = Holtzman and Nations (1984), 12 = Cellino et al. (1985), 13 = Byrne (1987), 14 = Zeilik et al. (1990b).

Vogt 1981a; Huenemoerder et al. 1989a,b), molecules that can only exist in significant concentrations in stellar atmospheres cooler than about 3500 K. In these studies the relative visibility of TiO bands at 7100 Å and 8860 Å suggests spots with spectral types M3–M6, or 2700-3600 K. These spectroscopic spot temperatures agree reasonably well with the photometric ones. Thus the light we see from spots cannot be diluted light of the warm penumbrae, as it would be if the spots had the same structure as sunspots, but instead must be **emitted mostly by the umbrae**.

3.4 Contamination by Faculae?

Dorren and Guinan (1990) have argued that observations of V711 Tau over the past decade have detected a magnetic cycle through which changes have occurred in a global facular contribution to the active star's continuum. This effect is now thought to make the Sun more luminous when it is more active, but it has not seemed to apply to the light variations of highly spotted stars (e.g., Radick et al. 1988). This is an important consideration toward our understanding the energy balance in stellar convective zones. We know that colors of chromospherically active stars can be abnormal. Spectra seem to be flatter than those of other stars (Fekel, Moffett, and Henry 1986), and infrared excesses are detected in many objects (e.g., Arevalo and Lazaro 1990) which are probably not contributed by spots (Spruit and Weiss 1986; Soderblom 1989).

This is essentially unexplored territory in that we have traditionally ignored the question of background spottedness and spectral peculiarity in obtaining spot solutions. Rodono et al. (1986), following Vogt, assumed all the light loss was caused by two spots and got high spot latitudes as a result. Eaton and Hall (1979) and Poe and Eaton (1986) assumed a variable background of spots. Lazaro (1990) has found it much easier to fit light curves with a variable zero point, as well. So it may well be that the surface brightness of the unspotted photosphere changes with magnetic activity. Hall, Henry, and Sowell (1989) suggested such a change in V478 Lyr.

The evidence for significant facular changes in V711 Tau, however, is weak. The change in chromospheric line emission detected was a mere 50% and was poorly correlated with the 0.1 mag change in mean visual brightness. Further, the changes in the V band can be accounted for rather well with spots. Since 1976 the star has never been brighter than $V = 5.63$ (Dorren, Guinan, and Wacker 1986; Bartolini et al. 1983; Henry 1990), which may represent the unspotted brightness of the star. TiO observations by Ramsey and Nations (1980) indicate weak TiO at $V = 5.70$ and noticeable TiO at $V = 5.90$. The Doppler images of Vogt and Penrod give a general depression of 0.15 mag in the light of the spotted star, or about 0.11 mag for the system (binary plus $V = 8.83$ background star). This leads to an unspotted magnitude of $V = 5.70$ in late 1981, when the spottedness and chromospheric emission were near their maximum level, exactly the same continuum level as maximum light in 1979. Thus, much of the increase in chromospheric emission in the early 1980's was **not** accompanied by a change in the unspotted magnitude. More recent evidence is contradictory. We can identify a possible 5-10% brightening since 1979, i.e., from $V = 5.70$ to $V = 5.63$ in late 1984. Doppler images for 1984 and 1985 by Vogt (1988) suggest the dominant polar spot may have been present even at $V = 5.63$, and this would require a huge (~0.2 mag) increase in the facular contribution between 1979–81 and 1984, just when Dorren and Guinan found the chromospheric line emission constant to 15%.

A second approach to this question of facular contribution comes from W UMa binaries. Hα spectra show the more massive components of cooler systems, which are known to have at least some large spots (e.g., Linnell 1987) and may be **very** highly spotted (Mullan 1975; Eaton 1986), are more active than their companions (Barden 1985). Furthermore, hydrostatic equilibrium should force the surfaces of these objects to be at roughly the same temperature (see Kahler 1989) without the effects of magnetic fields. In those systems cool enough to be spotted, the more massive, presumably more active components have lower surface brightnesses than their companions. This has often been taken to indicate a global temperature difference, but this explanation requires the stars' colors to change too much with phase, even in the optical (Eaton 1986). That the more massive, more active components are never brighter than their companions argues against a large facular contribution.

Acknowledgements: I would like to thank my two colleagues Douglas S. Hall and Gregory W. Henry for helpful discussions about observations of spotted stars and their interpretation.

References

Arevalo, M.J., Lazaro, C. 1990, AJ, 99, 986.

Barden, S.C. 1985, ApJ, 295, 162.

Bartolini, C, et al. 1983, AA, 117, 149.

Budding, E., Zeilik, M. 1987, ApJ, 319, 827.

Byrne, P.B. 1987, in Cool Stars, Stellar Systems, and the Sun, p. 491.

Celino, A., Scaltriti, F., Busso, M. 1985, AA, 144, 315.

Dorren, J.D. 1987, ApJ, 320, 756.

Dorren, J.D., Guinan, E.F. 1982, ApJ, 252, 296.

Dorren, J.D., Guinan, E.F. 1990, ApJ, 348, 703.

Dorren, J.D., Guinan, E.F., Wacker, S.W. 1986, in New Insights in Astrophysics, ESA SP-263, p. 201.

Dorren, J.D., Siah, M.J., Guinan, E.F., McCook, G.P. 1981, AJ, 86, 572.

Doyle, J.G. 1989, AA, 214, 258.

Doyle, J.G., Butler, C.J., Morrison, L.V., Gibbs, P. 1988, AA, 192, 275.

Eaton, J.A., Hall, D.S. 1979, ApJ, 227, 907.

Eaton, J.A., Poe, C.H. 1985, BAAS, 16, 914.

Eaton, J.A., Poe, C.H. 1986, IBVS, No. 2846.

Endal, A.S., Sofia, S., Twigg, L.W. 1985, ApJ, 290, 748.

Fekel, F.C., Moffett, T.J., Henry, G.W. 1986, ApJS, 60, 551.

Gray, D.F. 1989, PASP, 101, 1126.

Hall, D.S. 1972, PASP, 84, 323.

Hall, D.S. 1976, IAU Coll. No. 29, Multiple Periodic Variable Stars, Part I, ed. W.S. Fitch (Dordrecht: Reidel), p. 287.

Hall, D.S., Busby, M.R. 1990, in Active Close Binaries, ed. C. Ibanoglu and I. Yavuz (Kluwer: Dordrecht), in press.

Hall, D.S., Gessner, S.E., Lines, H.C., Lines, R.D. 1990, AJ, submitted.

Hall, D.S., Henry, G.W. 1990, in Active Close Binaries, ed. C. Ibanoglu and I. Yavuz (Kluwer: Dordrecht), in press.

Hall, D.S., Henry, G.W., Sowell, J.R. 1989, AJ, 99, 396.

Henry, G.W. 1990, private comm.

Holtzman, J.A., Nations, H.L. 1984, AJ, 89, 391.

Huenemoerder, D.P., Ramsey, L.W., Buzasi, D.L. 1989a, AJ, 98, 2264.

Huenemoerder, D.P., Buzasi, D.L., Ramsey, L.W. 1989b, AJ, 98, 1398.

Kahler, H. 1989, AA, 209, 67.

Kang, Y.W., Wilson, R.E. 1989, AJ, 97, 848.

Lazaro, C. 1990, AJ, submitted.

Linnell, A.P. 1987, ApJ, 316, 389.

Losh, H.M. 1939, Publ. Obs. Univ. Michigan, 7, 127.

Mullan, D.J. 1975, ApJ, 198, 563.

Noah, P.V., Bopp, B.W., Fekel, F. 1987, in Cool Stars, Stellar Systems, and the Sun, p. 506.

Olah, K., et al. 1985, Ap&SS, 108, 137.

Olah, K., and Hall, D.S. 1988, Comm. Konkoly Obs., 10, 121.

Olah, K., Panov, K.P., Pettersen, B.R., Valtaoja, E., and Valtaoja, L. 1989, AA, 218, 192.

Poe, C.H., Eaton, J.A. 1985, ApJ, 289, 644.

Radick, R.R., Lockwood, G.W., Skiff, B.A., Baliunas, S.L. 1988, BAAS, 20, 995.

Ramsey, L.W., Nations, H.L. 1980, ApJ, 239, L121.

Rodono, M., et al. 1986, AA, 165, 135.

Rucinski, S.M. 1981, AA, 104, 260.

Scaltriti, F., Celino, A., Busso, M. 1985, AA, 149, 11.

Soderblom, D.R. 1989, ApJ, 342, 823.

Spruit, H.C., Weiss, A. 1986, AA, 166, 167.

Strassmeier, K.G. 1990, ApJ, 348, 682.

Strassmeier, K.G., et al. 1988, AA, 192, 135.

Vogt, S.S. 1981a, ApJ, 247, 975.

Vogt, S.S. 1981b, ApJ, 250, 327.

Vogt, S.S. 1988, in The Impact of Very High S/N Spectroscopy on Stellar Physics, ed. G. Cayrel de Strobel and M. Spite, p. 253.

Vogt, S.S., Penrod, G.D. 1983, PASP, 95, 565.

Vogt, S.S., Penrod, G.D., Hatzes, A.P. 1987, ApJ, 321, 496.

Wilson, R.E., Devinney, E.J. 1971, ApJ, 166, 605.

Xie, X., Zhai, D., Chen., H., Li, Y. 1989, Chinese AA, 9, 97.

Young, A., Skumanich, A., MacGregor, K.B., and Temple, S. 1990, ApJ, 349, 608.

Zeilik, M., Cox, D.A., De Blashi, C., Rhodes, M., Budding, E. 1989, ApJ, 345, 991.

Zeilik, M., Cox, D.A., Ledlow, M.J., Rhodes, M., Heckert, P.A. Budding, E. 1990b, ApJ, in press.

Zeilik, M., De Blashi, C., Rhodes, M., Budding, E. 1988, ApJ, 332, 293.

Zeilik, M., Ledlow, M., Rhodes, M., Arevalo, M.J., Budding, E. 1990a, ApJ, 354, 352.

Zhai, D., and Chen., H. 1989, Chinese AA, 13, 9.

INFORMATION LIMIT OPTIMIZATION: STARSPOT SOLUTIONS FOR ECLIPSING RS CVn SYSTEMS

Michael Zeilik
Institute for Astrophysics, University of New Mexico,
Albuquerque, NM87131, USA

Edwin Budding
Carter Observatory, Wellington, New Zealand

1. Introduction

To probe the magnetic activity of late-type stars at many levels requires coordinated, multiwavelength observations - an admirable goal, difficult to achieve! For some systems that have attracted early attention, we have broadband photometric observations that often cover decades in the past. The time base spanned in archival observations provides a unique resource to sample the variability of the magnetic activity, which can provide information on activity cycles. Hence, a combination of archival and contemporary data of high quality opens a path to determine the phenomenological nature of short-term and long-term trends.

Because late-type stars are cool but not cold, the bulk of their radiation is emitted from the photosphere at optical and near-infrared wavelengths. The ability to detect active, spotted regions requires a measurable contrast compared to the photospheric background. With even modest telescopes (apertures less than 100 cm), modern detectors (such as CCD cameras) can achieve S/N ratios of 100 or better for broad-band differential photometry with time resolutions of a few tens of seconds. So we can obtain the quality observations needed to tell us about the location (longitude and latitude), sizes, and temperatures from the photometric distortion waves in the light curves of late-type systems. We then need a reliable, objective means to extract this information.

2. The information limit approach to starspot modelling

Photometric modelling of distortion waves (Rodonó, 1986) has received blows of justifiable criticisms, which I sense has too much undermined the actual usefulness of the technique to provide information that complements that obtained from spectroscopy. And photometric models are the *only* way to extract information from historic light curves. Hence, photometric starspot modelling should be used appropriately rather than rudely ignored.

The main thrust of the problem is usually stated as a "lack of uniqueness" in the photometric solutions of even high-quality data. This opinion developed around solutions to *non-eclipsing* systems, for which the orbital inclinations (or even periods) were not known that well - or not at all! In these cases, latitude, inclination, and spot size are all correlated variables - only the longitudes have a clear characterization. In contrast, especially for the RS CVn stars, far fewer solutions have been attempted for eclipsing systems, for which the orbital inclination is known within a degree or so and for which the eclipses provide fiducial markers for longitude. Hence, eclipsing systems provide a much more constrained modelling problem; the drawback is that we need to separate eclipse and proximity effects to reveal the distortion waves. We have done so by modifying Budding's 16-parameter, optimizing curve fitter (Budding, 1973; Budding and Najim, 1980) to generate the model curves for the eclipsing system. These models use Roche lobes for the distorted stars, and include both the reflection and ellipticity effects. The difference between a theoretical curve and the observed one defines the observed distortion wave. Another curve fitter, corresponding to the case of one or two dark, circular spots generates an optimized solution to these residuals. The results provide the basic starspot parameters: longitude, latitude, effective radius, and temperature - if the quality of the data is sufficient. To infer a temperature requires observations at two wavelengths; we have found that simultaneous V and R or V and I work well.

Our programs (Budding and Zeilik, 1987) contain an optimizer routine that uses a variety of methods to search efficiently for a minimum in χ^2 hyperspace; the two main ones are a vector search and a parabolic interpolation (Banks, 1989). The χ^2 routine contains the particular fitting function appropriate to generate a representation of the eclipsing binary light curve or of the distortion wave from starspots (Budding, 1977). Budding's algorithm for circular, dark spots is used by many other spot modellers, and a χ^2 minimization is standard practice. What sets our procedure apart is our information limit analysis, which provides an objective evaluation of the results from the optimized parameters, in the sense of the information content of the data.

As detailed in Budding and Najim (1980), our programs evaluate the curvature Hessian in the vicinity of the optimal solution in χ^2 hyperspace. (The curvature Hessian is the matrix of the second-order derivatives of χ^2 with respect to the various parameters whose values are to be optimized.) We also calculate the error matrix, which is the inverse of the curvature Hessian. For mathematically-determinate solutions, the curvature Hessian will have a positive eigenvalue for each optimized parameter. And the error matrix then provides a correlated error for each well-determined parameter. Hence, we go beyond simply minimizing χ^2; we discover, given the quality of the data, which parameters have strictly determined values and what the ranges in those values are. We approach the information limit by increasing the number of parameters until the Hessian becomes nonpositive definite; we then know that we have exceeded the information inherent in the data. To try to evaluate any more parameters is to ask the data to deliver more information that they contain. We then have a clear way to evaluate the quality of archival light curves, so that we can extract from them in an objective way only as much information as they have to offer.

3. Application to eclipsing, chromospherically-active systems

The *Catalog of Chromospherically Active Binary Stars* (Strassmeier et al. 1988) contains 16 systems with orbital periods of less than one day. Of these, 12 are classified as RS CVn systems and 9 are eclipsing. We have concentrated on observations and modelling of the short-period, eclipsing RS CVn stars because their rapid rotation provides a unique laboratory to understand the magnetic activity of cool, sunlike stars - especially to see if "cycles" actually exist. To do so over an extended time requires both new observations on a systematic basis and an analysis of archival data. We have completed work so far on SV Cam (Zeilik et al. 1988), RT And (Zeilik et al. 1989), BH Vir (Zeilik et al. 1990a), and WY Cnc (Zeilik et al. 1990b). We will attack CG Cyg and UV Psc next. The CCD camera at Capilla Peak Observatory in New Mexico is well suited for multichannel photometry of systems for which a suitable comparison star appears in the 3 x 5 arcmin field. These include CG Cyg, RT And, XY UMa, and BH Vir.

Our main results to date, which apply to the hotter star in these systems, are: (1) The primary activity signature are two active longitude belts (ALBs) at the quadrature points; one belt is more active than the other at any given time. This result agrees with the analysis of Nelson and Zeilik (1990), who found for a large sample of RS CVn stars (mostly non-eclipsing) that one or two spot groups were sufficient to account for the distortion waves. (2) The secondary activity signature is that the active regions appear most often at high latitudes (near 45°); they resemble the polar active regions inferred from both Doppler Imaging (Vogt and Penrod, 1983) and photometric modelling (Rodonó et al. 1986) for HR1099 = V711 Tau, a non-eclipsing RS CVn system. We have also found that the spotted regions cover large areas of the primary, active stars (radii are typically 10°) and have temperatures about 1500 K below that of the photosphere with a variation of only a few hundred degrees Kelvin.

Although based on a small number of systems, we expect that these results will hold up as the sample is expanded because they are based on decades of time and tens of thousands of points of data. The great puzzle to us so far is the lack of a clear and forceful indicator of activity cycles. Perhaps the data are too often gapped for long times. Or, the cycles may occur on time scales much longer than decades or much shorter than months, both of which are unlikely given the characteristics of the solar magnetic cycles. Or, we may have to examine the starspot parameters in a different way to find the cycle signature.

Acknowledgement: This research was supported in part by NSF grant AST-8903174.

References

Budding, E.: 1973, Ap&SS, 22, 87

Budding, E.: 1977, Ap&SS, 48, 287

Budding, E., Najim, N.N.: 1980, Ap&SS, 72, 369

Budding, E., Zeilik, M.: 1987, ApJ, 319, 827

Nelson, E., Zeilik, M.: 1980, ApJ, 349, 163

Rodonó, M.: 1986, in *Cool Stars, Stellar Systems and the Sun,* ed. M. Zeilik, D. Gibson (Heidelberg: Springer-Verlag), p. 470

Rodonó, M. et al.: 1986, A&A, 165, 135

Strassmeier, K.G., Hall, D.S., Zeilik, M., Nelson, E., Eker, Z., Fekel, F.C.: 1988, A&AS, 72, 291 Vogt, S.S., Penrod, G.D.: 1983, PASP, 95, 565

Zeilik, M., Cox, D.A., De Blasi, C., Rhodes, M., Budding, E.: 1989, ApJ, 345, 991

Zeilik, M., Ledlow, M.J., Rhodes, M., Arevalo, M.J., Budding, E.: 1990a, ApJ, 345, 352

Zeilik, M., Cox, D.A., Ledlow, M.J., Rhodes, M., Heckert, P.A., Budding, E.: 1990b, ApJ, 363, in press.

Discussion: Zeilik

Cameron In your blind tests of the modelling procedure, did you examine the systematic errors in spot latitude, caused by non-circular spots in the input model?

Zeilik We made a rough "elongated" spot by placing two spots at the same latitude, just in contact with each other. Within the correlated errors, the optimized latitude was the same.

Cameron Given the size of the error bars on your spot latitudes, how confident are you about the presence/absence of active latitudes on these stars?

Zeilik The latitude errors bars range from 5° to 45°. Hence, we cannot identify "active latitudes" on these stars. Basically, we do not resolve a "butterfly" diagram as for the Sun; we can only say with confidence if the spots are near the equator or at high latitudes.

MAJOR CHANGES IN POLAR STARSPOTS ON V833 Tau

K. Oláh[1] and B.R. Pettersen[2]

[1]Konkoly Observatory, Hungarian Academy of Sciences
P.O. Box 67, H-1525 Budapest XII, Hungary

[2]Institute of Theoretical Astrophysics, University of Oslo,
P.O. Box 1029 Blindern, N-0315 Oslo 3, Norway

1. Introduction

V833 Tau = Gliese 171.2A = BD +26°730 is a late type (dK5e) single lined spectroscopic binary. Long term variations over 60 years were discovered by Hartmann et al. (1981). A detailed photometric and spectroscopic history of the binary is given in Griffin et al. (1985). Pettersen (1989) estimated the photometric period to be 1.85 day from the rotational modulation of the brightness of V833 Tau.

2. Observations

We have carried out starspot photometry of V833 Tau in 1983, 1987, 1988, 1989, 1990. At Konkoly Observatory, the 50 cm telescope was equipped with an integrating photometer and UBV filters, and the 1 m telescope with a UBV(RI)$_C$ photon counting photometer. At McDonald Observatory, data were taken with the 76 cm and 91 cm telescopes through UBVR$_C$ filters.

We used as comparison star C1 = BD+26°729 = HD 283749 and as a check star C2 = BD +27°683 = HD 283772.

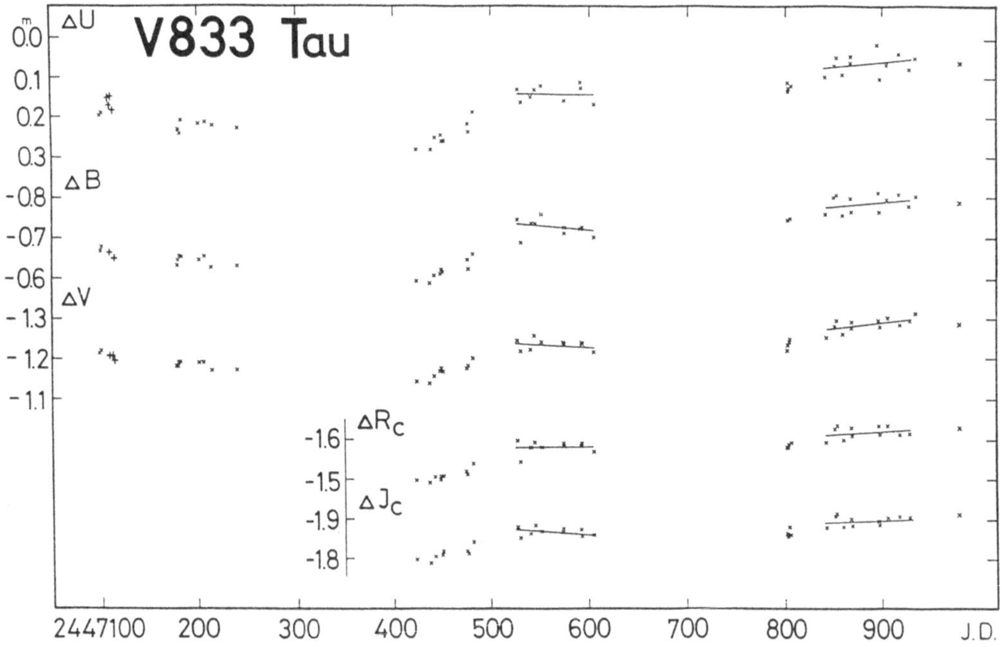

Figure 1. Photometry of V833 Tau between 1987 - 1990 showing both long term changes and effects of rotational modulation.

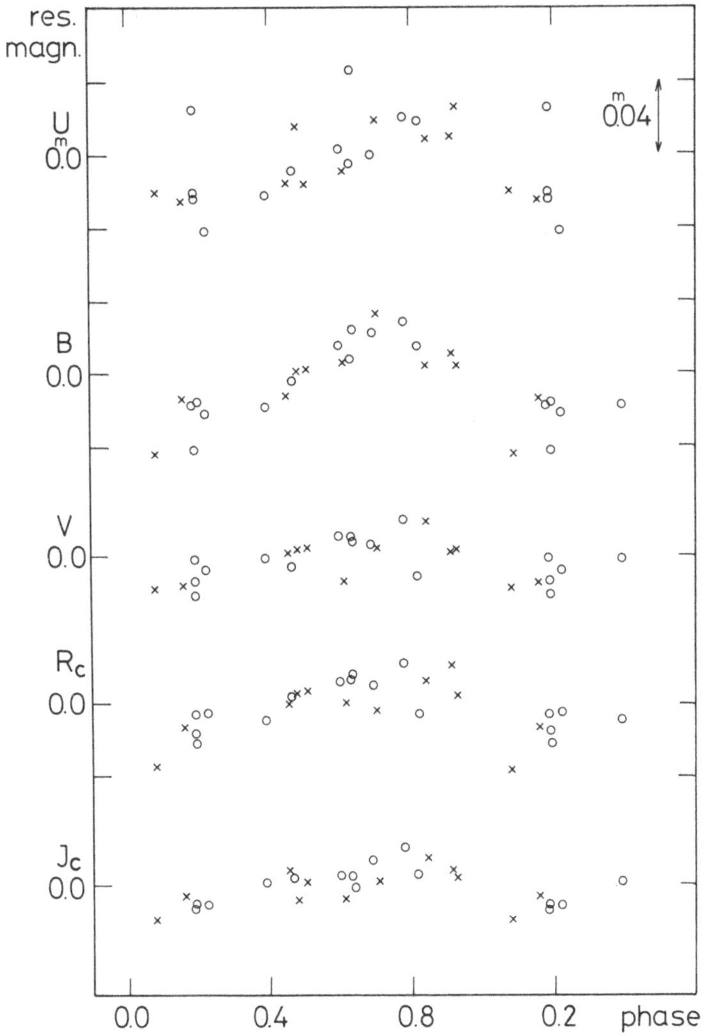

Figure 2. UBV(RI)$_C$ changes in 1989 (*circles*) and 1989/90 (*crosses*) due to rotational modulation by starspots, with $P_{rot} = 1.7975$ day.

3. Results and discussion

Our 1987 - 1990 observations are shown in Figure 1. The data do not contradict the long term trend discovered by Hartmann et al. (1981). Slow changes on timescales of months are found in all filters. The amplitude reaches 0.22 mag in the U filter and towards longer wavelengths this amplitude is smaller: 0.20 in B, 0.15 in V, 0.12 in R_C and 0.11 in I_C bands. Our measurements show the star being faintest in the autumn of 1988 and brightest in early 1990.

The rotational modulation of V833 Tau was first observed in 1983. At that time the overall light level of the star was constant, all variability originated from the rotational modulation. The slow changes in 1987 - 1990 make it difficult to detect the low amplitude rotational modulation. In 1987 no rotational modulation was detected, but the overall brightness was slowly decreasing.

In the autumn of 1988 the star was fainter than at the end of the previous season but started to brighten quite fast until the beginning of 1989. The two subsets of data in early 1989 and 1989/90 (see Figure 1) seem to distribute data points around almost constant average values, however, and the rotational modulation can be analysed. The light curve displayed in Figure 2 was constructed by taking residual values from the fitted brightness levels, in all the five colours.

We found $P_{rot} = 1.79$ day for the 1983 data. The 1988 - 1990 data provide a longer time base, yielding $P_{rot} = 1.7975$ day.

Observations from 1983 and 1987 - 1990, combined with a very low inclination ($\approx 20°$), demonstrate the existence of the polar spot on V833 Tau. The suppressed brightness level and a very small amplitude of the rotational modulation (if any) suggest a dominating polar spot in 1987-88. Towards the end of 1988 the area of this spotted region began to decrease and the star brightened somewhat. In early 1989 the spotted region covered an area slightly asymmetric around the pole, thus producing a low amplitude rotational modulation in the brightness. A similar state seems to apply to the 1989/90 season. Asymmetric light curves at suppressed light levels observed at these seasons can be equally well described by one elongated spot which partly covers the polar region or by two spots, one on the pole and one in a lower latitude.

The rotational modulation observed in 1983 was caused by a very similar total spot coverage to that six years later. The symmetric light curve at suppressed brightness level suggests one circular spotted area which covers partly the polar region.

Our estimates for the rotational period of V833 Tau (1.79 day in 1983 and 1.797 day in 1988-1990) are longer than $P_{orb} = 1.788011$ day obtained by Griffin et al. (1985). If this minute difference proves to be real, it would support a solar type differential rotation with longer rotation periods away from the equator. The low inclination of $\approx 20°$ for V833 Tau means that the polar region weights heavily in the photometric lightcurve.

Acknowledgements

This collaboration bas received funding support from the Cultural Exchange Program between Hungary and Norway. K.O. expresses sincere thanks to all of her colleagues, especially to Dr. J. Jurcsik, who helped in obtaining the observational data at Konkoly Observatory.

References

Griffin, R.F., Gunn, J.E., Zimmermann, B.A., Griffin, K.E.M.: 1985, AJ, 90, 609

Hartmann, L., Bopp, B.W., Dussault, M., Noah, R.V., Klimke, A.: 1981, ApJ, 249, 662

Pettersen, B.R.: 1989, AA, 209, 279

Discussion: Pettersen

Landstreet I am not sure how far your data force one to assume polar spotting. If the star has strong and variable equatorial spotting, this would probably redistribute the total stellar luminosity into polar regions, with corresponding brightness variations.

Pettersen Of course there is the usual ambiguity between a homogeneous distribution of equatorial spots and a spotted area covering the polar region of the star. The very low inclination value ($\approx 20°$) and the suppressed light level of ≈ 0.3 mag imply very large spot areas if positioned in the equatorial region. If a polar spot is employed to account for the initial suppression of light from an assumed immacualate level, then only a small spot extension towards lower latitudes is needed to produce the observed rotational modulation.

MODELLING STELLAR PHOTOSPHERIC SPOTS USING SPECTROSCOPY

A. Collier Cameron
Astronomy Centre, University of Sussex,
Falmer, Brighton BN1 9QH, England.

1. Introduction

The most easily-observable manifestation of the solar dynamo is the sunspot cycle. The spatial distribution of spots, in two well-defined active belts which drift towards the equator during the course of the stellar magnetic activity cycle, provides important observational constraints on solar dynamo models. Until a decade or so ago, it seemed unlikely that such detailed effects could ever be observed on the surfaces of other stars. Since then, however, advances in solid-state detectors, computing power and methods of solving large-scale inversion problems, have provided us with the tools needed for such observations. We are now in a position to study the evolution of starspot distributions on active, rapidly-rotating stars with a surface resolution comparable with that of the ancient Chinese naked-eye sunspot records.

We know from analyses of visual and near-IR photometry that the stellar surface features responsible for the rotational variability of active late-type stars are in general cooler than the surrounding photosphere (Vogt, 1981; Eaton 1990). Multi-colour lightcurves of spotted stars yield useful information about the total surface areas and average temperatures of the spots, but do not constrain their locations on the stellar surface very well.

High resolution spectroscopy offers us a more powerful way of detecting localised features on stellar surfaces. Deutsch (1958, 1970) recognised that migrating distortions in the rotationally-broadened photospheric line profiles of peculiar A stars betrayed the positions of localised abundance anomalies on the stellar surface. Falk and Wehlau (1974) attempted to map the distribution of equivalent widths on the stellar surface using harmonic analysis methods, but with a simple Gaussian form assumed for the line and instrumental profiles. Khokhlova (1975) and Khokhlova and Rjabchikova (1975) adopted an approach which allowed a more rigorous treatment of the line formation, by working the problem forward from a set of trial surface EW distributions to obtain synthesised datasets. The mismatch between the observed and computed data can then be quantified in terms of the observational errors, usually in the form of the χ^2 statistic. The dependence of this statistic on each pixel in the image can then be calculated, and the image adjusted accordingly. This process is repeated until a satisfactory fit is achieved.

The main problem with this approach is that the data are inevitably noisy and incomplete. Under these circumstances the problem of inverting the profiles to recover a surface map becomes ill-conditioned: any of a very large number of possible surface images with varying degrees of unresolved complexity can give a satisfactory fit to the data. The most common solution to the uniqueness problem is to impose an extra constraint on the image in the form of a regularising function. This means that we seek that image which is in some sense the smoothest or the simplest of all the "feasible" images that fit the observations satisfactorily. Goncharsky e al. (1977, 1982) were the first to apply this method to the Ap star problem, using a smoothness criterion developed by Tikhonov (1963) as their regularising function.

The history of spectral surface imaging on late-type stars has followed a pattern very similar to the development of surface abundance mapping on Ap stars. Vogt & Penrod (1983) used a trial-and-error profile fitting method very similar to that used in the mid-70s by Khokhlova and co-workers, to derive the first maps of the starspot distribution on the active subgiant in the

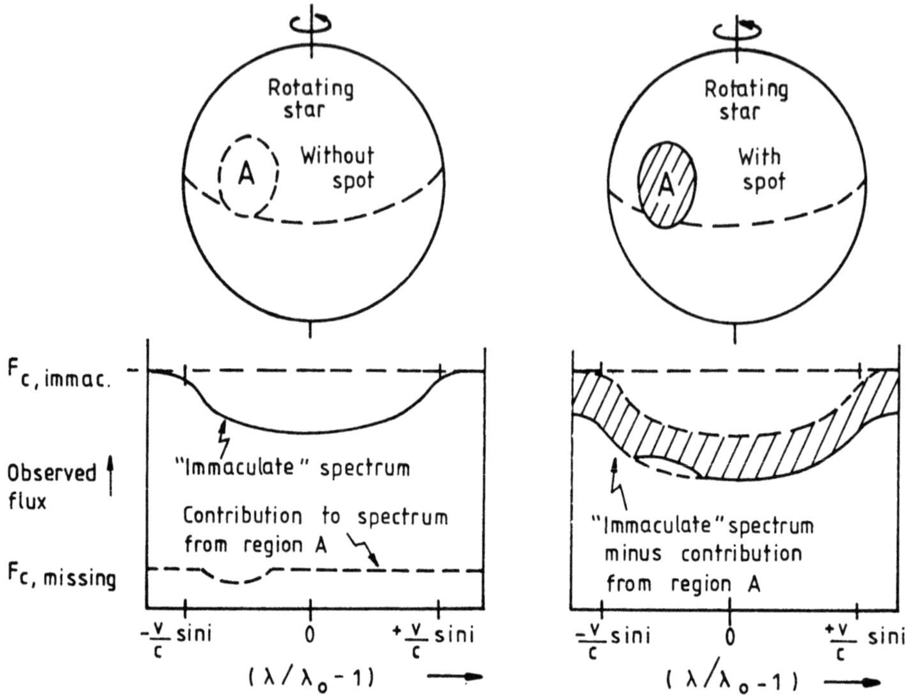

Figure 1. A schematic picture of the formation of a bright "bump" in the rotationally-broadened spectrum of a star with a dark starspot. λ_0 is the central wavelength of the line.

RS CVn binary HR 1099. This paper was followed by a more sophisticated treatment by Vogt, Penrod & Hatzes (1987) in which a method very similar to that of Goncharsky e al. was used to maximise the entropy of the stellar surface image on the $\chi^2 = M$ surface in image space.

Since then, several different groups have used large scale inversion techniques to produce maps of the spot distributions on active stars, using time-series spectroscopy. Two recent papers, by Vogt (1988) and by Piskunov, Tuominen & Vilhu (1990), review the workings, merits, shortcomings of the inversion techniques currently in use, and present some of the first stellar surface maps derived from observations of several stars.

2. Observational data constraints

2.1 Starspot bumps in line profiles

The classic signature of a dark starspot in a rotationally-broadened absorption line profile is a bright "bump" in the profile. This bump appears at a velocity shift from line centre corresponding to the projected distance of the spot from the stellar rotation axis (Figure 1). Why does the bump appear bright? The answer is intuitively apparent when we consider what would happen to the observed residual profile if we could cut out part of the stellar surface at the temperature of the photosphere, throw it away and calculate the resulting residual line profile.

The spectrum at each point in the missing patch (region A in Figure 1) contains a narrow absorption feature displaced from line centre by rotation. The flux we measure at the velocity shift of the spot from line centre has decreased, but it has not decreased as much as it has at other wavelengths, where we have thrown away continuum rather than line flux. Even if we add in the much fainter spectrum of a cool starspot over the missing patch, the observed flux will be

$$R_{\text{pert}}(\lambda) = \frac{F_{\text{immac}}(\lambda) - F_{\text{missing}}(\lambda)}{F_{\text{c,immac}} - F_{\text{c,missing}}}. \tag{1}$$

Here $F_{\text{c,immac}}$ and $F_{\text{immac}}(\lambda)$ represent the observed continuum flux level and the observed monochromatic flux at wavelength λ, integrated over the unspotted, rotating stellar disc. The differences between the spotted and unspotted flux contributions from region A alone, are $F_{\text{c,missing}}$ and $F_{\text{missing}}(\lambda)$ respectively.

The equivalent width of the absorption feature in the the spectrum of the "missing" light from region A is

$$W_{\text{missing}} = \int_{-\infty}^{\infty} \left[1 - \frac{F_{\text{missing}}(\lambda)}{F_{\text{c,missing}}}\right] d\lambda. \tag{2}$$

The equivalent width of the bump in the normalised spectrum is obtained simply by re-scaling W_{missing} to the new continuum level:

$$W_{\text{bump}} = \frac{F_{\text{c,missing}}}{F_{\text{c,immac}} - F_{\text{c,missing}}} W_{\text{missing}}. \tag{3}$$

The *amplitude* of the bump at any wavelength λ can be expressed as a fraction of the observed continuum flux:

$$R_{\text{bump}} = \frac{F_{\text{c,missing}} - F_{\text{missing}}(\lambda)}{F_{\text{c,immac}} - F_{\text{c,missing}}}. \tag{4}$$

Since F_{missing} incorporates the convolution of the local missing flux profiles over the rotational Doppler profile of the spot, the only missing ingredient is the instrumental profile. Once R_{bump} has been convolved with the instrumental profile we are in a position to start worrying about the minimum detectable spot size.

2.2 Detector noise and surface resolution

The size of the smallest feature that can be resolved on the stellar surface is limited by

1. the ratio of the widths (full width at half maximum depth) of the intrinsic, non-rotating mapping line profile and the stellar rotation profile;
2. the drift in wavelength of the spot bump as the star rotates during the exposure; and
3. the ratio of the starspot bump amplitude to the RMS scatter in the measured spectral fluxes.

When a starspot lies on the observer's meridian, the ratio of its Doppler profile width to that of the star is roughly equal to r, the ratio of the spot radius to the stellar radius. If x is defined as the ratio of the mapping line FWHM to the stellar Doppler profile width, then in order to resolve the spot we require $r \geq x$.

During the course of an exposure, the bump due to a spot on the stellar equator drifts through the line profile from blue to red at a rate

$$\dot{\lambda} = \frac{2\pi}{P_{\text{rot}}} \frac{\lambda_0 v \sin i}{c}. \tag{5}$$

The effective width of the stellar rotation profile (defined here as the ratio of the observed line EW to the central depth of the pure rotation profile) is roughly $\Delta\lambda_{\text{rot}} \simeq \pi\lambda_0(v \sin i)/2c$. The blurring of the spot feature during an exposure of duration Δt can be expressed in units of the rotation profile width, giving $\tau \equiv \dot{\lambda}\Delta t/\Delta\lambda_{\text{rot}} \simeq 4\Delta t/P_{\text{rot}}$.

If the central residual intensity $R_{\rm line}(\lambda_0)$ of the photospheric mapping line profile is assumed constant over the stellar disc, we can write the EW of the mapping line as $W_{\rm obs} \simeq (1 - R_{\rm line}(\lambda_0))\Delta\lambda_{\rm rot}x$. The observed equivalent width of a photospheric absorption line is a continuum intensity-weighted average of the local equivalent width over the projected stellar disc, and does not change with increasing $v_e \sin i$. The central residual intensity $R_{\rm obs}(\lambda_0)$ of the observed, rotationally-broadened profile is thus given by

$$\begin{aligned}(1 - R_{\rm obs}(\lambda_0)) &\simeq \frac{W_{\rm obs}}{\Delta\lambda_{\rm rot}(1+x^2)^{1/2}} \\ &\simeq (1 - R_{\rm line}(\lambda_0))\frac{x}{(1+x^2)^{1/2}}. \end{aligned} \tag{6}$$

A completely dark, circular spot located near the centre of the stellar disc will give a bump with approximate equivalent width

$$W_{\rm bump} \simeq \frac{r^2}{(1-r^2)} W_{\rm obs}. \tag{7}$$

The central bump amplitude is then given as a fraction of the continuum flux by $R_{\rm bump} \simeq W_{\rm bump}/\Delta\lambda_{\rm bump}$. The FWHM of the bump is obtained from the convolution of the rotational profile of the spot with the intrinsic profile of the mapping line and the effects of blurring during the exposure, so that $\Delta\lambda_{\rm bump} \simeq \Delta\lambda_{\rm rot}(r^2 + x^2 + \tau^2 + y^2)^{1/2}$. This gives

$$R_{\rm bump} \simeq (1 - R_{\rm line}(\lambda_0))\frac{x}{(r^2+x^2+\tau^2+y^2)^{1/2}}\frac{r^2}{(1-r^2)}, \tag{8}$$

where y is the ratio of the instrumental profile width to the stellar rotation profile width.

For pure photon statistics, negligible sky background, a perfect flat field and readout noise of $N_{\rm R}$ photo-electrons, a single spectral bin from a CCD spectrum yields $N_{\rm det} \pm \sigma$ photo-electrons, with $\sigma^2 = N_{\rm R}^2 + N_{\rm det}$. If the mapping line profile is oversampled by the spectrograph and detector, the signal-to-noise ratio (S:N) can be improved by binning until the effective bin size samples the mapping line profile at the Nyquist frequency. This means that the ratio u of the effective bin width to the stellar rotation profile width should be $u \simeq x/2$. After binning, the RMS error in the normalised profile at line centre due to readout noise and photon statistics is

$$\sigma_{\rm norm} = \frac{(\dot{N}_\lambda u \Delta\lambda_{\rm rot}\Delta t(1 - R_{\rm obs}) + mN_R^2)^{1/2}}{\dot{N}_\lambda u \Delta\lambda_{\rm rot}\Delta t}. \tag{9}$$

Here m is the number of detector pixels per spectral bin and $\dot{N}_\lambda$ is the photon detection rate per unit wavelength in the continuum. For a successful spot detection, at a level s times the RMS error, the bump amplitude must satisfy

$$R_{\rm bump} \geq s\sigma_{\rm norm}. \tag{10}$$

In reality, the surface resolution for stars with low $v_e \sin i$ values will be limited by the instrumental profile width and (more importantly) by microturbulence, which gives a lower limit of roughly 0.15Å on the FWHM of any photospheric absorption line. Also, our ability to detect spots on rapidly-rotating stars is restricted because we cannot use lines with equivalent widths greater than about 0.25Å. The cores of stronger lines are formed in the chromosphere and will vary in strength according to the local chromospheric heating rate.

These two restrictions mean that the best lines to use for Doppler imaging lie in a fairly narrow range of equivalent widths. Since x is effectively fixed by these constraints, the optimal

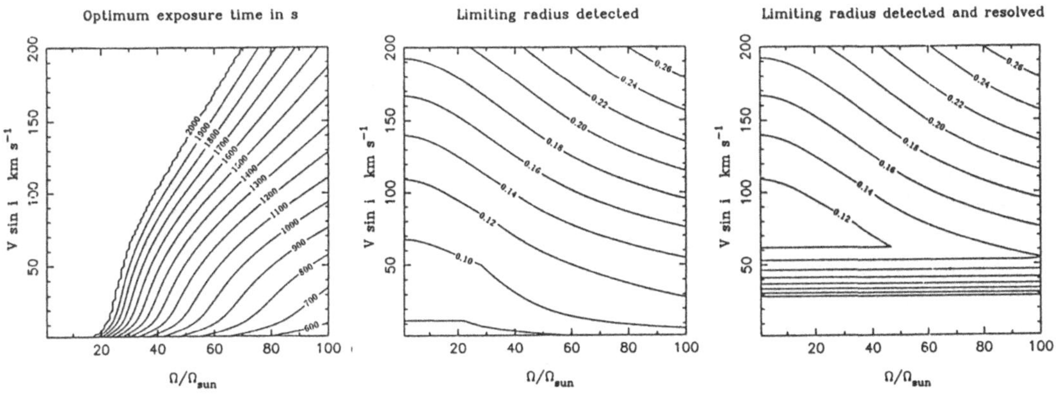

Figure 2. Optimum exposure times and minimum detectable and resolvable spot radii as functions of $\Omega/\Omega_\odot$ and $v_e \sin i$.

observing strategy is determined by the requirement that the smallest detectable spot should not drift through more than its own radius during the exposure. We thus set $\tau \simeq r$, bin the spectrum with $2u = \max(y, x)$ and determine the value of r needed for a marginal detection.

In some stars with very long rotation periods, it will not be possible to expose for long enough to take full advantage of the time available for the spot to drift through its own width. This is because exposure times are ultimately limited by detector saturation and cosmic ray hits. In such a situation the best strategy is to expose for the longest possible time, bin the spectrum with $2u = \max(y, x)$ and determine the value of r that gives a marginal detection. The weakest features detected in this case should be fully-resolved. The S:N can also be improved by co-adding spectra taken at the same phase.

The optimum exposure times, and the resulting threshold radii above which a spot gives a detectable and/or a resolved bump, are shown as contour plots in Figure 2 as functions of stellar rotation rate and $v_e \sin i$. These plots are for a model situation where $\dot{N}_\lambda = 1130\,\mathrm{ct\ s^{-1}\ Å^{-1}}$, spectrograph resolving power $\lambda/\Delta\lambda = 40000$, $\lambda_0 = 6431$Å, line FWHM $= 0.25$Å, $R_{\rm line} = 0.2$ and the maximum exposure time is 2000s. These numbers are typical of those obtained for an early K star with V=7.0, using the UCL echelle spectrograph on the 3.9m Anglo-Australian telescope.

The differences between Figures 2b and c emphasise that at low values of $v_e \sin i$, unresolved features may give detectable distortions in the line profile even when the mapping line width is comparable with the rotation profile width. This effect has been observed by Toner & Gray (1988), Gray (1988) and Toner (1990), who used line bisector methods to detect unresolved starspot bumps in the spectrum of the active G8 dwarf ξ Boo A, which has $v_e \sin i \simeq 10\,\mathrm{km\,sec^{-1}}$.

3. The image reconstruction procedure

We now turn to the problem of using the information contained in a time-series of distorted line profiles, to recover an image of the stellar surface brightness distribution. In practice, the stellar surface grid and the fluxes in the spectral bins comprising the dataset are treated as vectors. Each possible image is represented by a single point in a vector space with as many dimensions as there are pixels in the image. The set of "feasible" images that produce adequate fits to the data within a specified level of significance (using a measure of discrepancy such as

χ^2) occupy a closed region of the image space. This region is bounded by a surface on which the discrepancy measure is constant (e.g. $\chi^2 = M$, where M is the number of measurements in the dataset). If $\boldsymbol{f}$ is an image vector containing the image pixel values, the regularising function $R(\boldsymbol{f})$ must be a convex function if it is to have a unique maximum on the boundary of the feasible region in image space.

For each image vector $\boldsymbol{f}$, a synthetic data vector $\boldsymbol{D}$ can be calculated. In spectral surface imaging, $\boldsymbol{D}$ contains a set of line profiles measured at different stellar rotation phases. The image-data transformation involves calculating the local continuum intensity and line profile for every pixel in the stellar image, and its Doppler-shifted contribution to the observed spectrum at each rotation phase. The goodness of the fit between the observed data $\boldsymbol{F}$ (with observational errors $\boldsymbol{\sigma}$) and the synthetic data $\boldsymbol{D}$ is measured by the statistic

$$\chi^2(\boldsymbol{f}) = \sum_{k=1}^{M} \left(\frac{F_k - D_k(\boldsymbol{f})}{\sigma_k} \right) \tag{11}$$

The image reconstruction problem then involves finding an extremum of the function $Q(\boldsymbol{f}) = R(\boldsymbol{f}) - \lambda\chi^2(\boldsymbol{f})$. The value of the Lagrange multiplier λ is chosen so that $\chi^2(\boldsymbol{f}) \simeq M$, defining the outer boundary surface of the "feasible" subset of image space. We thus seek the image $\boldsymbol{f}$ for which $\partial/\partial f_j(R - \lambda\chi^2) = 0$ for all image pixels j.

For any image $\boldsymbol{f}$ the local gradients in image space of both the regularising function and the discrepancy measure are easily calculated. Standard iterative techniques such as conjugate gradients can then be used to solve the constrained maximisation problem, provided that the image-data transformation is not severely nonlinear.

3.1 The image-data transformation

Unfortunately the starspot imaging problem is not linear. First, the relationship between the value of an image pixel and its contribution to the data is in general a nonlinear function of both temperature and limb angle. Clearly it is desirable to use an accurate model of the full temperature and limb angle dependence of the continuum intensity and line profile. The computational cost of incorporating a full model atmosphere in the innermost loop of an iterative code would be prohibitive, so the most widely-used technique is to use a pre-computed grid of spectral intensities. The most sophisticated treatment of this kind that has so far been applied to real data was that of Piskunov e al. (1990). These authors used a set of model atmospheres and an LTE line-transfer code to generate a full three-dimensional lookup table with intensities computed as functions of wavelength, temperature and limb angle, convolved with the instrumental profile. Vogt e al. (1987) used a simpler approach, assuming that the spots were dark enough to contribute a negligible amount to the observed spectrum. They thus calculated a grid of profiles for the mapping line (Fe I λ 6431) as a function of limb angle, but not as a function of temperature.

A second type of nonlinearity enters the image-data transformation when (as is often the case) we do not know the observed continuum flux level in each spectrum. We then have to content ourselves with normalising the observations to a continuum level of unity and renormalising the synthetic spectra to this level at each iteration. Provided that the relationship between the image parameter and the emergent flux is not too strongly nonlinear, a local linearised treatment usually leads successfully to a solution.

3.2 The regularising function: smoothness versus entropy

Two recent stellar surface imaging papers by Rice, Wehlau & Khokhlova (1989) and by Piskunov, Tuominen and Vilhu (1990) have addressed the sticky question of which is the best regularising function to use. Goncharsky e al. (1977, 1982) used a form originally developed by

Tikhonov (1963)

$$r_{\mathrm{T}}(f) = \int\!\!\int_{\mathrm{surface}} |\boldsymbol{\nabla} f(M)|^2 \, dS(M) \tag{12}$$

which is a measure of the smoothness of the solution. The other commonly-used regularising function is the image entropy, which takes the general form

$$r_{\mathrm{ME}}(f) = -\int\!\!\int_{\mathrm{surface}} f(M) \log(f(M)) \, dS(M). \tag{13}$$

A recent review of the maximum entropy principle and its application to large-scale inversion problems in astronomy can be found in Narayan & Nityananda (1986). Gull & Skilling (1984) give demonstrations of intermediate formality which illustrate the way in which this form of regularising function eliminates correlations between different parts of an image unless they are demanded by the data, based on a more rigorous proof by Shore & Johnson (1980).

In contrast to this, the Tikhonov function induces strong correlations between the values of f at neighbouring points in the image, and so tends to produce a smoother image in the absence of strong data constraints. Various arguments have been advanced to support the use of each technique in terms of expectations about the physical processes on the surfaces of heavily-spotted stars. Fortunately, the differences between the results obtained with the two methods are usually small if the image is well-constrained by the data.

4. Choice of image parameter

A much more important issue is the choice of image parameter f that is most appropriate to the starspot imaging problem. Both the Tikhonov function and the image entropy ensure that all values of f are positive. The entropy, however, can only be defined for images which have the property of being *additive* as well. The nature of additivity can be understood by considering the quantities in two neighbouring but non-overlapping pixels a and b: the quantity f_{a+b} measured when the two pixels are combined is given by the sum of the values f_a and f_b for the two pixels considered individually. The bolometric power radiated by individual pixels (bolometric flux times pixel area) is a good example of an image parameter that is additive ($P_{a+b} = P_a + P_b$), while pixel temperature is a good example of one that is not: $T_{a+b} \neq T_a + T_b$. For the remainder of this paper I shall concentrate on the entropy, but the arguments which follow are for the most part applicable to the Tikhonov method as well.

4.1 Bolometric surface flux as image parameter

In a model where the image pixel values f_j are bolometric surface fluxes, the total power radiated over the whole image is $P = \sum_j w_j f_j$. The w_j are the pixel areas in units of the stellar surface area, so that $\sum_j w_j = 1$ and P is the stellar luminosity. Pixel j thus emits a fraction $p_j = w_j f_j / P$ of the stellar luminosity.

In some image reconstruction applications, the total flux P can be measured and used as a constraint. In stellar surface imaging applications, however, we have no information concerning the polar region in the stellar hemisphere facing away from us. Instead, we can set a "default" value m for the pixel surface brightnesses. The image entropy is defined so that it has its global maximum when all pixels in the image are set at the default level:

$$S(\boldsymbol{f}) = -\sum_{j=1}^{n} w_j f_j \log \frac{f_j}{m}. \tag{14}$$

The maximum entropy reconstruction is thus found by solving

$$w_j(\log m - \log f_j) = \lambda \frac{\partial C}{\partial f_j} \tag{15}$$

for all j. Note that if $\partial C/\partial f_j = 0$ for any pixel, the data do not constrain the value of f_j and the maximum entropy condition forces $f_j \rightarrow m$.

This kind of model allows the surface flux to vary continuously over the surface of the star, and leaves the reconstruction free to develop spots brighter than the default value, as well as dark ones if the data and the maximum entropy criterion demand them.

4.2 Starspot covering fraction as image parameter

Let us now look at the surface imaging problem in a slightly different way. If we adopt a simple two-temperature model (warm photosphere, cool spots) for the stellar surface, the fraction f_j of pixel j occupied by starspots is related to the total spotted fraction P of the stellar surface, by $P = \sum_j w_j f_j$. Note that the spotted fraction of any pixel cannot be greater than unity. In fact we are dealing with two images here: one of the spotted fraction f_j in each pixel, and one of the unspotted fraction $g_j = 1 - f_j$. This image is also additive, with $\sum_j w_j g_j = Q = 1 - P$. Pixel j thus contains a fraction $p_j = w_j f_j / P$ of the total spot area, and a fraction $q_j = w_j g_j / Q$ of the total unspotted area. If we maximise the entropy of both images simultaneously, both g_j and f_j must be positive. This restricts both f_j and g_j to lie between 0 and 1. The joint entropy function takes the form

$$S(f) = -\sum_{j=1}^{n} w_j \left[f_j \log \frac{f_j}{m} + (1 - f_j) \log \frac{(1 - f_j)}{(1 - m)} \right] \tag{16}$$

where m is the default spotted fraction and is set to a very small, positive value. In this case the f_j for the final reconstructed image are found by solving

$$w_j \left[\log \left(\frac{m}{1 - m} \right) - \log \left(\frac{f_j}{1 - f_j} \right) \right] = \lambda \frac{\partial C}{\partial f_j}. \tag{17}$$

This definition of the stellar image and its entropy forces the reconstruction to maintain a uniform photospheric brightness, and does not allow bright spots to develop. It allows only one pre-defined surface brightness for the spots, but allows spots and clean photosphere to co-exist within a single pixel.

5. Errors and artifacts

5.1 Geometrical and line formation errors

In the real world, we generally have to perform reconstructions using empirical values (or blind guesses) for stellar parameters such as the axial inclination, equatorial rotation speed, microturbulent velocity distribution, stellar effective temperature and line equivalent width. Unfortunately, errors in these parameters produce various types of artifact in the reconstructed image.

One of the most widely-practised ways of identifying such artifacts involves generating artificial datasets from known synthetic stellar images using a known set of geometric parameters and line formation physics. Errors in the physics used for the reconstruction can then be introduced under controlled conditions, and the resulting artifacts categorised.

Vogt e al. (1987) examined the effects of altering the stellar axial inclination, $v_e \sin i$, limb darkening coefficient and temperature-dependence of the mapping-line strength, while Rice e al. (1989) studied the effects of errors in the line formation physics for the Ap star abundance-mapping problem. Both groups concluded that simple errors in the geometric parameters either prevented their codes from converging at all, or else gave easily recognisable artifacts. More subtle errors such as neglecting the temperature-dependence of the mapping line strength could,

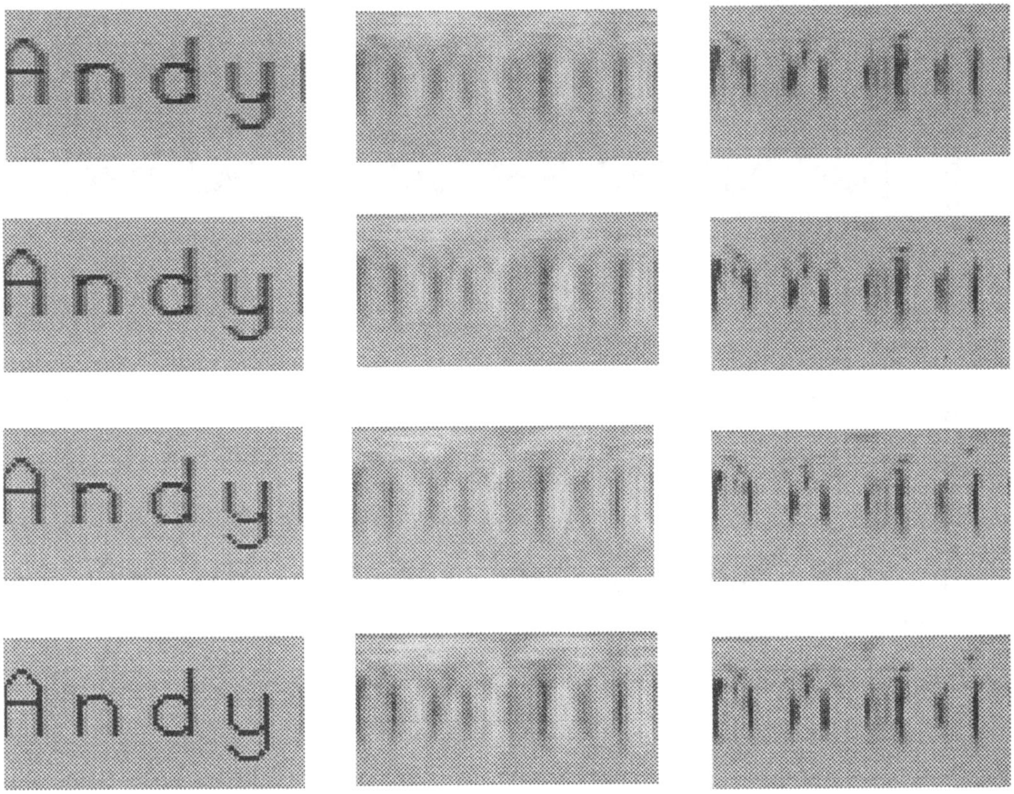

Figure 3. Rotationally-blurred initial images (*left*), with reconstructions using the bolometric (centre) and two-temperature (*right*) image definitions.

however, give reconstructions which bore little resemblance to the input map, but gave no clues to the fact that anything had gone wrong.

5.2 Surface blurring and fictitious temperatures

Another more subtle type of error results from the use of bolometric surface flux as the image parameter. It occurs when poorly-resolved surface features interact with nonlinearities in the image-data transformation. This problem has received little or no attention in the literature to date, but merits close attention. To illustrate its effects, I have calculated a series of maximum entropy reconstructions, using both bolometric surface flux and spot filling factor as the image parameter.

The differences between the results obtained using the two different forms of the entropy are clearly seen in Figure 3. The images in the left-hand column are cartesian surface maps of an artificial stellar surface brightness distribution, in which the bolometric surface brightness of the spots is 0.3 times that of the photosphere, and the spot filling-factor is unity. The rotation period and $v_e \sin i$ of the model star were set at 0.51479 day and 85 km sec^{-1} respectively. The rows in this figure have exposure times (from top to bottom) of 2000, 1600, 1200 and 800s, giving different amounts of rotational blurring and noise. Artificial datasets were generated at sixteen equally-spaced rotation phases. Their noise patterns were identical, but the noise ***amplitudes*** were calculated from the exposure times using the same instrumental parameters and photon count rate used in Section 2.2 to calculate Figure 2. The resulting line profiles for the 1600s exposure time are shown in Figure 4.

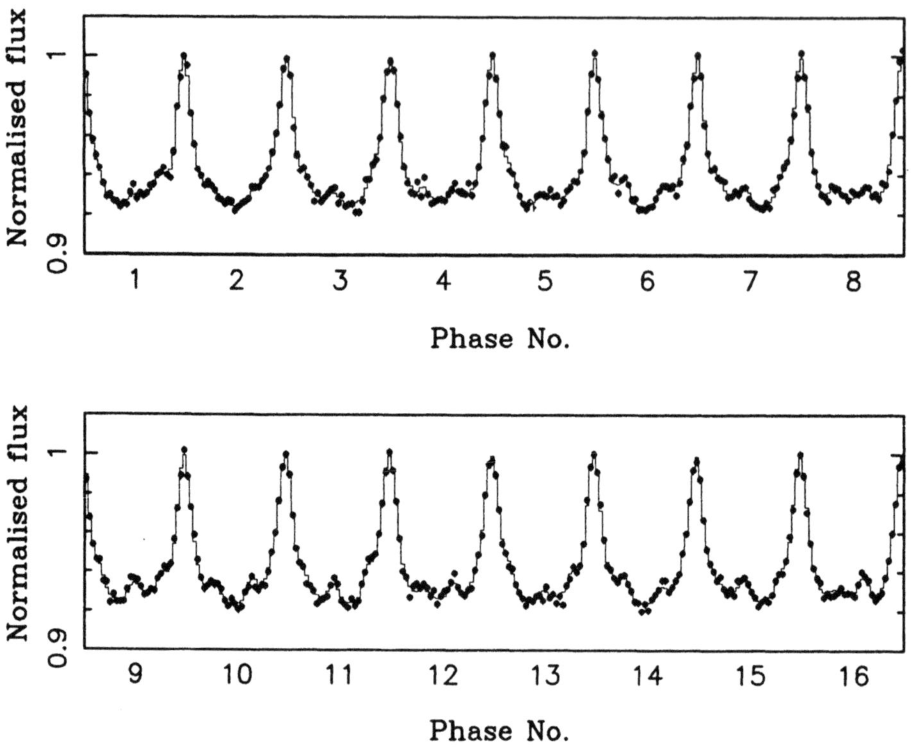

Figure 4. A series of sixteen noisy synthetic line profiles (*dots*), equally spaced in rotation phase, with model fit (*solid line*) from a two-temperature reconstruction.

The Maximum Entropy image reconstructions were carried out using the MEMSYS subroutines of Gull & Skilling (Skilling & Bryan 1984). The reconstructions obtained with bolometric flux as image parameter are in the central column of Figure 3, while those obtained with the two-temperature model appear in the right-hand column. The trade-off between the strength of spurious noise features in the reconstructions and the surface resolution is readily apparent. As suggested by Figure 2a, the optimum exposure time appears to be of order 1600s in this instance. Much more striking, however, are the bright, plage-like artifacts in the bolometric flux-based reconstructions. The same effect can also be seen in Piskunov e al. 's (1990) comparison of the MEM and Tikhonov regularisation methods. Something is clearly wrong here.

When bolometric flux is used as the image parameter, the assumption is generally made that *every pixel in the image can be assigned a unique surface temperature, on the basis of its bolometric surface flux.* This is tantamount to assuming that all surface temperature variations on the star are fully-resolved. As we have already seen, however, this is unlikely to be the case in real stellar data obtained with finite S:N ratios, mapping lines of finite width and exposure times limited by rotational smearing.

If we take the "true" image of the stellar surface and blur it, the total bolometric power missing from a spotted region should be conserved. The bolometric surface brightnesses of the spotted pixels must increase to compensate for the larger area coverage. If each image pixel is assigned a unique temperature corresponding to its bolometric flux, the final ingredient in this recipe for disaster is a line profile that varies in a non-linear fashion with bolometric surface brightness. The combined spectrum of a pair of pixels with unequal bolometric fluxes is different from the spectrum that emerges when their bolometric fluxes are averaged, and we are in trouble. Because of their large areas and high surface brightnesses, the blurred spots pollute the computed spectrum with large amounts of flux whose spectral distribution corresponds to a temperature component that does not exist on the star. The correct non-linear treatment of the temperature dependence of the line profile actually worsens the problem!

The two-temperature model is more robust in this respect. The bolometric flux of a pixel depends only on the relative amounts of spot and photosphere within its boundaries. If two pixels with unequal spot filling factors are smeared together, the total spot area is conserved and the emergent spectrum remains unchanged. Admittedly, the two-temperature model is more restrictive than the bolometric flux model. It allows only a single spot temperature, and imposes a uniform photospheric brightness distribution. Fortunately, if the spots are bright enough to contribute significantly to the observed spectrum, the mean difference between spots and photosphere can be measured independently using VRI photometry (Vogt 1981).

Bright photospheric features such as faculae which may be present on the stellar surface are specifically prevented from developing on the reconstructed image. This limitation is at least consistent with what we observe on the Sun. First, the distribution of sunspot umbral surface brightnesses is reasonably narrow at optical wavelengths (Maltby 1990). Secondly, the thermal diffusivity of the convective zone is sufficient to redistribute the heat flux blocked by solar or stellar spots over the rest of the photosphere on a short timescale (Spruit 1982,1990), thus maintaining uniform surface brightness on large spatial scales. Finally, the contrast between faculae and photosphere on the Sun is small at the longer optical wavelengths that are generally used in spectral surface imaging (Lean 1990).

The continuous temperature model may be less restrictive in this respect, but I believe that its tendency to produce bright faculae between starspots in simulated reconstructions, even when none were present in the initial image, argues against its usefulness as an unbiased means of studying stellar surface temperature distributions.

6. Concluding remarks

The success of any attempt to map a stellar surface brightness distribution depends critically on the quality of the data used to constrain the reconstruction. Great care needs to be taken with the design of the observations if the best results are to be achieved. This is because, for any given instrument and mapping line, longer exposures allow the detection of progressively smaller surface features, but rotational smearing during a long exposure degrades the surface resolution attainable. For each combination of stellar parameters $P_{\rm rot}$ and $v_e \sin i$, an optimal balance exists between signal-to-noise ratio and surface resolution on the star. This balance dictates the exposure time and the amount by which the spectrum should be binned before an image reconstruction is attempted.

The form of regularising function used in the image reconstruction does not affect the final reconstruction seriously unless the data do not constrain the solution adequately. The correct treatment of the line formation problem as a function of limb angle and temperature is of paramount importance, as is the correct choice of image parameter. Since the image-data transformation is nonlinear and some degradation of surface resolution is inevitable, spurious results will be obtained if a one-to-one relationship is assumed between pixel surface brightness and pixel temperature.

In short, spectral imaging is a very good technique for determining the *positions* of features on the stellar surface, but not so good at determining their *temperatures*. Instead, it seems more prudent to use independent techniques – such as VRI photometry – to determine the temperatures of the spots. Ideally these measurements can and should be incorporated in future surface imaging codes as additional data constraints that must be satisfied by the reconstructed image.

This work was supported by UK Science and Engineering Research Council, through the award of an Advanced Fellowship. The computational work presented in this paper was carried out at the Sussex node of the Starlink Project, which is operated by SERC. I also acknowledge the financial support provided by the meeting organisers during this conference.

References

Deutsch, A.: 1958, in Proc. IAU Symp. No. 6: *Electromagnetic Phenomena in Cosmological Physics*, ed. B. Lehnert, p. 209, Cambr. Univ. Press.

Deutsch, A.: 1970, ApJ. 159, 985.

Eaton, J.: 1990, these proceedings.

Falk, A.E., Wehlau, W.H.: 1974, ApJ. 192, 409.

Goncharsky, A.V., Stepanov, V.V., Khokhlova, V.L., Yagola, A.G.: 1977, Sov. Astron. Lett. 3, 147.

Goncharsky, A.V., Stepanov, V.V., Khokhlova, V.L., Yagola, A.G.: 1982, Sov. Astron. 26, 690.

Gray, D.F.: 1988, *Lectures on Spectral-Line Analysis: F, G, and K Stars* The Publisher, Ontario.

Gull, S.F., Skilling, J.: 1984, Proc. IEE, 131F, 646.

Khokhlova, V.L.: 1975, Sov. Astr. 19, 576.

Khokhlova, V.L., Ryabchikova, T.A.: 1975, Ap&SS, 34, 403.

Lean, J.: 1990, these proceedings.

Maltby, P.: 1990, these proceedings.

Narayan, R., Nityananda, R,: 1986, ARA&A, 24, 127.

Piskunov, N.E., Tuominen, I., Vilhu, O.: 1990, AA, 230, 363.

Rice, J.B., Wehlau, W.H., Khokhlova, V.L.: 1989, AA, 208, 179.

Shore, J.E., Johnson, R.W.: 1980, IEEE Trans, IT-26, 26.

Skilling, J., Bryan, R.K.: 1984, MNRAS, 211, 111.

Spruit, H.C.: 1982, AA, 108, 356.

Spruit, H.C.: 1990, these proceedings.

Tikhonov, A.N.: 1963, Sov. Math. Dokl. 4, 1624.

Toner, C.G., Gray, D.F.: 1988, ApJ. 334, 1008.

Toner, C.G.: 1990, these proceedings.

Vogt, S.S.: 1981, ApJ. 250, 327.

Vogt, S.S., Penrod, G.D.: 1983, PASP, 95, 565.

Vogt, S.S., Penrod, G.D., Hatzes, A.P.: 1987, ApJ. 321, 496.

Vogt, S.S.: 1988, in Proc. IAU Symp. No. 132: *The Impact of Very High S/N Spectroscopy on Stellar Physics*, eds. G. Cayrel de Strobel, M. Spite, p. 253, Kluwer Acad. Publ. Dordrecht.

Discussion: Cameron

Young For some spectral lines, cool spots make them deeper (darker), so does that cause any *different* problems from those discussed?

Cameron No, it shouldn't make any difference *provided* that the line formation problem is treated correctly as a function of limb angle at the temperatures of both spots and photosphere.

Young Have you considered imaging features, like bright plage, which are purely in emission? Can that be done in a more or less similar fashion?

Cameron In principle it is possible, but there are severe problems with choice of image parameter. The chromospheric line formation problem can't be described in terms of a single image parameter. Different factors affect line widths and line intensities. Neff's talk tomorrow will address this kind of problem in detail.

Foing How does your observing strategy change for binaries with orbital smearing of the profiles that introduce the same top-hat convolved effect on all the spectral features, compared to the rotational smearing, which is latitude dependent?

Cameron The orbital smearing rate will change sinusoidally as a function of orbital phase, so it considerably complicates the planning of observations. The smearing of a given feature now depends on its distance from the centre of mass of the binary, so exposure times have to be shorter than in the single-star case. The smearing also affects the shape of the overall profile, not just that of the bump. This should be built into the image-data transformation, as it could lead to artifacts in the reconstructed image.

Landstreet As you have pointed out, errors in *line profile physics* can produce apparently reasonable fits to line profiles that are not correct. One powerful test for this effect that is not much used yet is to take a map derived from one spectral line and use it to predict the profile of another line, preferably of different strength, excitation profile, *etc.*

Cameron Yes, I agree. This test is particularly important if one is to stack the profiles of different lines in order to improve signal-to-noise. We have to make sure that all the lines are telling the same story when treated individually, before stacking their observed and computed profiles. This is also important for other types of data such as photometry. An image derived from spectroscopy alone cannot be trusted if it doesn't reproduce simultaneous photometric or other types of data correctly.

Landstreet The role of the regularising function used in Doppler Imaging is really to deal with the problem created by the desire to define an image with many pixels from line profiles which, when you consider the limited number of profiles and their limited resolution, have too few data to define a unique solution. A reasonable alternative would be to simply degrade the image resolution until a unique solution is found. This would result in big, unattractive pixels on the image, but would immediately give the viewer an accurate impression of the actual image resolution obtainable from the data.

Cameron That's a nice idea. I suspect that the image pixel sizes would have to vary with position on the star in a complex manner that depended on the amount of information in the particular dataset concerned. This would emphasise the fact that some points of the image are considerably less trustworthy than others!

COMPARISONS BETWEEN DOPPLER IMAGING AND STARSPOT MODELLING

E. Budding[1] and M. Zeilik[2]
[1]Carter Observatory, Wellington, NZ
[2]Department of Physics and Astronomy, University of New Mexico, Albuquerque, USA

1. Doppler Imaging with the program PFIT

It has been shown (Budding and Zeilik, 1990) that there is an essential similarity between "Doppler Imaging" and maculation wave modelling, in that both procedures can be based on the representation of features having a circular outline, which have a different surface intensity to the background light level of the spherical body on which they are located. These authors have previously developed programs (*FIT* and *SPOT*), by means of which such representations have been successfully applied to recover information to characterize the essential maculation parameters from photometric data on RS CVn stars.

Of course, such representations are bound to be some kind of ideal case — and in detail we can expect that neither will features be of circular outline, nor will the body be perfectly spherical, nor will various other assumptions about the distribution of surface intensities in the representation be likely to hold exactly true in reality.

The points about these procedures which we wish to emphasize is their *adequacy* in the context of available signal-to-noise in the kind of data to be investigated, and the *information limit* of such data, which relates to a certain (small) number of determinable parameters; suggesting that the extension of analysis to more sophisticated models in such data contexts is overextravagant. (Banks and Budding, 1990).

Budding and Zeilik's (1990) analysis adopted Doppler broadening, as indicated by Vogt and Penfold (1983), to be the predominating widener of the line. This produces the well-known 'dish-shaped' profiles, first discussed by Shajn and Struve (1929). In effect, Budding and Zeilik (1990) wrapped up all other minor contributions to line broadening by convolving the Doppler spread surface intensity with a delta function. In terms of the adopted hypothesis, this is quite a legitimate step in formulating the fitting function for lines of the expected type, since it is *only from the dish-shaped component to the generally broadened line that the Doppler imaging will retrieve the sought-after information.* Actually, however, real lines will be broadened by other agencies, and the fitting function has been developed to allow a gaussian inherent profile convolved with the Doppler spreading. This will introduce some extra parameters, though they are of no direct relevance to the quantities sought.

On the basis of the formulae spelled out in Budding and Zeilik (1990) an algorithm has been devised for derivation of parameters characterising 'spots' on rotationally broadened stellar spectral lines. The formulae appear in the constructed fitting function $PF(X)$, which also takes into account the 'partial' case, $d > 1 - k^2$, not included in the original paper. This fitting function is called by the subroutine which calculates the latest value of χ^2 as the fitting proceeds. This, in turn, is called by a χ^2 minimising subroutine, which is then built into the familiar program environment of the *FIT*-type programs. The whole program has been called *PFIT*. It requires an input file of the usual structure as for *FIT* and *SPOT*. One new point is that it is required to put in the mid-exposure orbital phase just before the parameter list.

In the parameter list a new entrant is the 'line central depth' parameter r_0. From the point of view of the user this is just another arbitrary quantity intended to match the central depth of the line, in units of the local continuum. Its value, for an absorption line, would be constrained by the inherent central depth of the line, as reduced by Doppler spreading, where

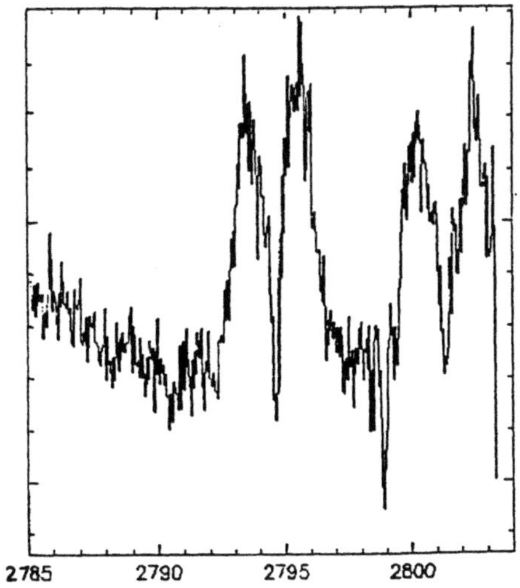

Figure 1. IUE image of the MgII h & k features for the RS CVn binary ER Vul.

the equivalent width of the line is required to remain constant. In the case of an emission line, the same parameter is given a negative number, which can be related to the flux ordinates of the spectrum. It is regarded as a superposed feature, produced essentially elsewhere than the absorption line spectrum over which it arises.

The other parameters have familiar counterparts in the photometric fitting function.

2. Application and results

Preliminary tests were made on data obtained from the Boulder IUE Data Center relating to some LWR observations of ER Vul in 1981. Keeping in mind the criteria of Vogt and Penfold (1983), we selected the central emission features of the MgII h & k lines as suitable material in which Doppler spreading is likely to have a relatively large influence. Our picture was of a more or less uniform, and relatively strong (compared with the Sun), chromospheric emission, onto which some enhanced region may be localised, giving rise to a 'blip' on the dish-shaped profile. At this stage, we should emphasize that our procedure is heuristic — we are interested to see what information about such a localized region may be recoverable. Our first rough trial indicated that a dish-shaped profile was an acceptable hypothesis, at the 95% level, for $S/N \sim 10$ (Fig 2(i)).

The essential similarities of line profile and photometric maculation wave fitting became clear after such initial experience. A key point is clearly what observational error should be attached to each datum. The lower this can be set, the more parameters we ought to be able to derive from the fit. Our estimation of this came from the level of *apparent* noise in other parts of the spectrum, *e.g.* the sloping wings of the absorption profile.

We next turned to some better reproduced MgII h & k profiles of the same star obtained in 1987 (Fig. 1). We adopted a value of 1 unit on the arbitrary ordinate scale for the standard deviation of measurement on this axis. The peak intensity is 17.2 units. It is inherently in the peak region of the profile where the efficiency of valid information retrieval will be greatest, because of the foreshortening effect, *i.e.* apart from S/N considerations. With 31 data points

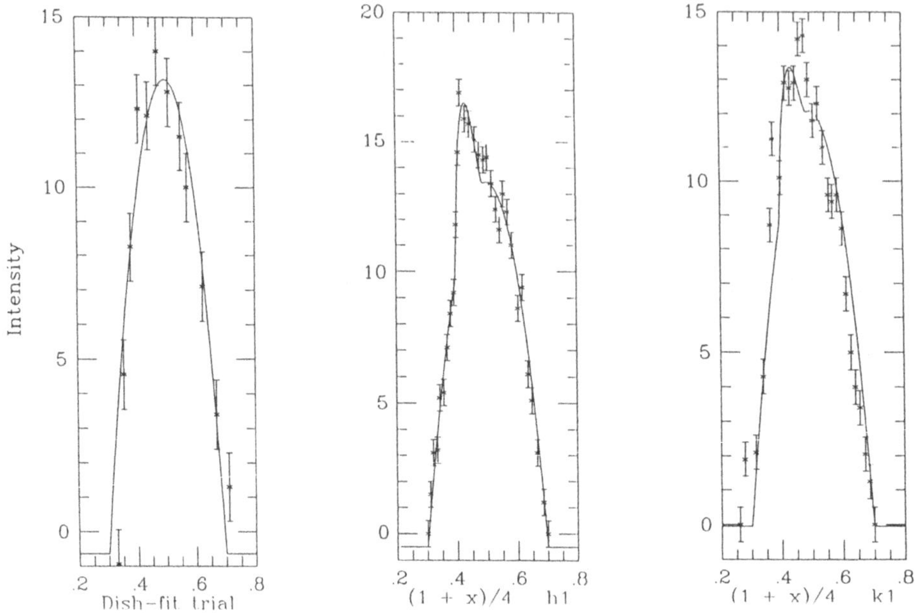

Figure 2. *PFIT* **fittings with** ***(a)*** **a 'dish-shaped' profile,** ***(b)*** **a better resolution of a similar feature, allows a 'spot' to be derived,** ***(c)*** **the corresponding h profile will not allow the same parameter resolution.**

a 'featureless' dish-shaped (k) profile could, at best, only return a χ^2 value of about 77, *i.e.* ≈ double what it should have been for an acceptable hypothesis.

It seemed apparent that a local feature was disturbing the upper left region of the profile, so it was decided to spot-model this. A drastic improvement in χ^2, down to 13.3, was thereby obtained. This strongly suggests that there really is a 'plage' feature on this k profile corresponding to the derived parameters, and the observation accuracy is rather better than that provisionally assigned, perhaps with peak S/N values of ~ 40 (Fig. 2(ii)).

We found, however, relatively large uncertainties in the parameter values. The smallness of the lowest eigenvalue of the curvature Hessian suggests that this solution is virtually on the brink of determinacy. Indeed the error of the spot intensity, which strongly correlates with those of longitude and background intensity compares with its already quite large value ($\sim$ 30 × background).

There is some indication that a feature of the same kind appears also on the more noisy seeming h-profile for this star. With a four parameter fit to the profile we found that the lowest eigenvalue, oriented closely in the direction of spot longitude, had dipped below zero, *i.e.* the longitude of the feature cannot be separated out when spot and backgound intensities, as well as spot size, are also included as unknowns (Fig. 2(iii)).

With these experiments we find some rather stringent limits on the information content of data such as the foregoing, *i.e.* S/N and $\nu \leq 100$, as with comparable photometric data sets, auguring the advisability of caution in the general determination of information by Doppler Imaging of such data.

References

Banks, T., Budding, E.: 1990, Ap&SS, in press

Budding, E., Zeilik, M.: 1990, Ap&SS, in press

Shajn, G., Struve, O.: 1929, MNRAS, 89, 222

Vogt, S.S., Penrod, G.D.: 1983, PASP, 95, 565

Discussion: Budding

Zeilik I have a comment and a sort of question/comment. To take the latter: I believe that the latitude determination (for an assumed circular spot) tends to have a sign ambiguity. This is generally so for the eclipsing systems. Presumably this is not so for II Peg with its lower inclination (I am thinking of Lazaro's diagram). Secondly, I think the position for SV Cam may not be as gloomy as you suggest. There was one sequence of spots which showed a longitude drift trend in a way which very nicely corroborated with a quite independent approach developed by an Italian group, I think Busso *et al* (1987), utilizing a differential rotation model and covering data for a similar timespan.

Linnell In the case of binary systems, with two possibly overlapping profiles, is there a change in analysis technique? When one component eclipses the companion, a part of the doppler profile is eliminated, at orbital phases other than quadrature. How does this affect the discussion?

Budding The underlying code can be applied to the eclipsed situation in a comparable way to the photometric problem. The cases we have investigated hitherto, however, refer to the situation at elongation. There are interesting situations, as Dr Eaton mentioned in his talk, where the circumstance of eclipse can add to the amount of derivable information. However, as to whether the spot itself is eclipsed, even though the eclipse profile be distorted, this is not too obvious to establlish; at least that was our experience with the photometric problem.

Foing The narrow absorption features observed in the Mg II h and k lines seem to arise from the interstellar medium as measured in other systems at similar distances (*cf* AR Lac (Neff *et al*)) or even closer. How do you correct for this before modelling the intrinsic stellar Mg II profiles?

Budding This is certainly a point worth making. In the particular profiles we have studied I believe it may not have been too serious an issue, because first, the star is not too distant and so the line should not be inherently that strong. But more significantly, the components are observed at elongation when they are both appreciably displaced from the rest wavelength. The contribution would appear as some disturbance to the right limb of the profiles under study, although, of course, it should be taken into account for the most accurate processing, I do not believe it seriously distrubs the present findings (where it was, in fact, disregarded).

THE POLAR-SPOT STRUCTURE ON HD26337

K.G. Strassmeier
Dyer Observatory, Department of Physics and Astronomy
Vanderbilt University, Nashville, TN 37235, U.S.A.

1. Introduction

Within the past several years, much effort has been spent on the development and refinement of indirect stellar imaging techniques (see, *e.g.* the review by Cameron in these proceedings). Their basic problem is to derive two-dimensional information from the analysis of light integrated over the stellar disk. This technique, commonly called Doppler Imaging, quickly became the most important tool for studying stellar surface structure throughout the Hertzsprung-Russell diagram.

2. Observations and analysis techniques

The spectroscopic observations were obtained at Kitt Peak National Solar Observatory (NSO) with the McMath solar telescope as part of the synoptic night time program during the observing season 1988/89. The instrumental resolution achieved was 43 000. The corresponding spectral resolution was estimated from the 0.15Å full-width at half-maximum of lines of the comparison lamp. The spectra have typical S/N ratios of 300:1. Simultaneous and near-simultaneous Johnson BV photometry was obtained with the 0.4 m Vanderbilt University Automatic Photoelectric Telescope on Mt. Hopkins, Arizona.

Three different computer codes to solve the inverse problem were used independently by three different groups, *viz.* S.S. Vogt and A.P. Hatzes (VH), J.B. Rice and W.H. Wehlau (RW), and N.E. Piskunov, I. Tuominen, T. Hackman, and M. Poutanen (PTHP). Detailed descriptions of their techniques are given in Vogt, Penrod, and Hatzes (1987), Rice, Wehlau, and Khokhlova (1989), and Piskunov, Tuominen, and Vilhu (1990). A fourth version based on a trial-and-error algorithm has been used by Strassmeier (1990a).

3. Results and Conclusions

1. The maps produced by all three groups show high-latitude spots of slightly different size, generally surrounding or covering the pole. RW and VH maps show a very cool polar cap for all three datasets. PTHP's map for 1988.00 also shows a polar feature but of much less contrast while their 1988.85 and 1988.94 maps have two, high-contrast spots situated at latitude $\approx 60°$ but no polar "caps". In comparing the maps in Figures 1, 2, and 3 (only one phase is shown), note that they have been produced with different graphical routiues representing brightness in RW's images, "threshholded" temperature in VH's images, and temperature in PTHP's images.

2. There are no bright spots in VH's and RW's solution (excluding the artifact at the pole for RW), while PTHP's calcium maps show several bright, plage-like features.

3. PTHP also attempted to use the somewhat weaker iron lines at 6430Å ($W_\lambda = 150$ mÅ) and 6411Å ($W_\lambda = 130$ mÅ). Generally, their Fe solutions are not satisfactorily consistent with the Ca images. This poses a serious problem since there is no obvious reason for such a discrepancy. Part of the problem might be severe blending and systematic continuum displacements of the observations (especially for FeI λ6430Å). Clearly, this point deserves future attention.

4. The high-latitude/polar spot on HD26337 seems to be long-lived (more than one year), its shape changes quite rapidly, comparable to the lifetime of low-latitude and equatorial features

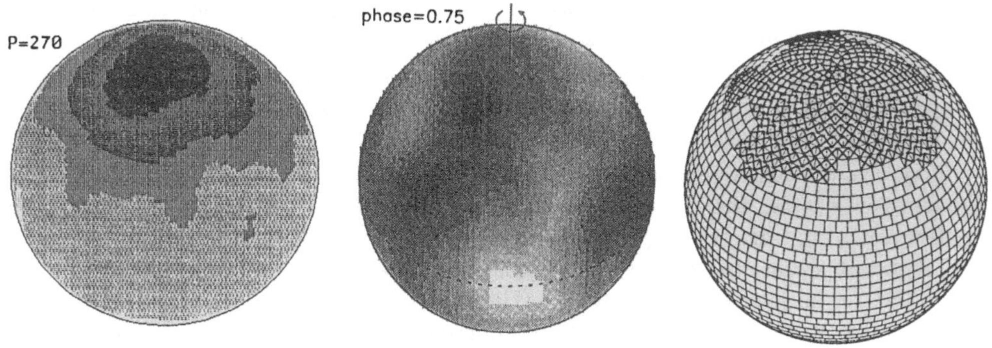

Figure 1: Calcium maps for 1988.00. From left to right: RW, PTHP, VH.

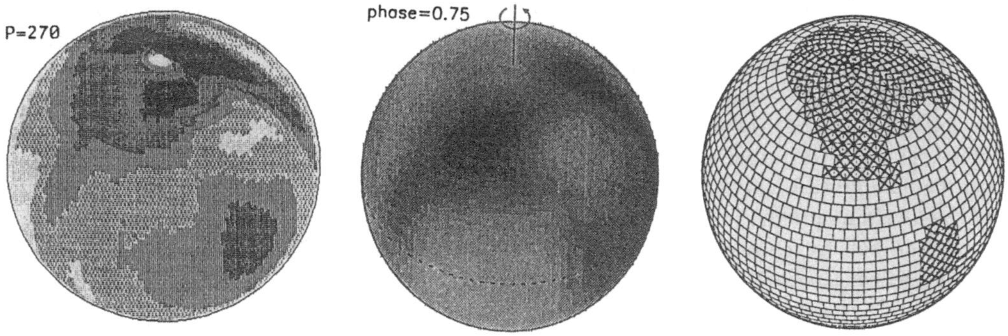

Figure 2: Maps for 1988.85.

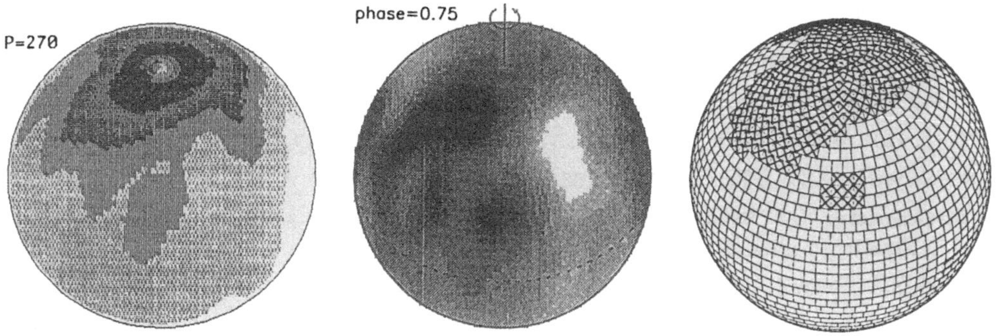

Figure 3: Maps for 1988.94.

of no more than a few weeks. These timescales are in agreement with findings from ten years of continuous V-band photometry (Strassmeier 1990a).

5. Simultaneous and/or contemporaneous broad-band photometry can be crucial to select "false" Doppler solutions in cases were the phase coverage of the spectroscopy is poor. Moreover, it helps to further constrain the final spot map.

In view of the controversy over the reality of high-latitude spots, we, of course, considered the question whether there is any reasonable way in which the dark polar spot could be an artifact, resulting from the mapping procedure or from the choice of parameters. Certainly, one argument against the polar-spot phenomenon is that the line used for Doppler imaging (CaI 6439Å) is fairly strong ($W_\lambda = 190\,\mathrm{m\AA}$). It is located on the damping part of the curve of growth and is formed in high layers of the atmosphere. In addition there is an obvious lack of knowledge of the damping constant and the degree to which an active chromosphere may contribute to the line core. On the other hand, if we form a mean profile for HD26337 by averaging over different phases and compare it with one from a spun-up inactive standard-star profile (Strassmeier 1990b) we find that the standard-star profile is much deeper in the line core but fits the wings above the half-maximum. Such a "null-star" technique (using μ Her as the standard) excludes all uncertainties concerning the choice of the adopted model atmosphere and line parameters but assumes identical photospheres for both G5 subgiants. Clearly, the average profile of HD26337 is much more flat-bottomed than either theoretical profiles or rotationally broadened standard-star profiles, and this argues strongly in favor of the reality of large, cool polar spots. A definitive answer to the controversy over the existence of cool polar starspots can be only given when we have better observations of more lines, and a better understanding of the local line profiles of strong lines.

In addition to the results stated above, we conclude that we found a fast-changing, highly inhomogeneous surface structure on HD26337 with cool spots on the equator and in excess of 60° latitude and even at the pole.

Acknowledgements. It is a pleasure to thank the mapping teams for their joint efforts. KGS wishes to acknowledge the financial support of the Austrian Fonds zur Förderung der wissenschaftlichen Forschung under project number P7993-PHY and the continuous support of the NSO McMath night-time program.

References

Cameron, A.C.: 1991, these proceedings

Piskunov, N.E., Tuominen, I., Vilhu, O.: 1990, A&A, 230, 363

Rice, J.B., Wehlau, W.H., Khokhlova, V.L.: 1989, A&A, 208, 179

Strassmeier, K.G.: 1990a, ApJ, 348, 682

Strassmeier, K.G.: 1990b, in NATO ASI, *Active Close Binaries*, ed. C. Ibanoglu and I. Yavuz, Kluwer, Dordrecht, in press.

Vogt, S.S., Penrod, G.D., Hatzes, A.P.: 1987, ApJ, 321, 496

Discussion: Strassmeier

Giampapa In Doppler Imaging techniques, what is the effect of the number of spots, as a parameter, on other parameters of the analysis such as latitude of starspots?

Strassmeier The resolution is not restricted to a certain number of spots, this is a parameter we derive from the profiles. Generally, the latitude of features at the equator is less well determined than at intermediate latitudes.

Hatzes It is our opinion that any temperature derived without using the photometry in the modelling is inaccurate. We feel that the spectral line profiles drive the location, size, and shapes of spots and that the photometry drives the contrast (i.e. temperature) between the spots and photosphere.

Strassmeier I agree, that's why we use simultaneous photometry as a further constraint for the Doppler image.

Hatzes Even though a spot feature has no Doppler shift, it still has Doppler information. Features lying at line centre are mapped into a strip at the centre of the line and thus have as much Doppler information as features near the limb with large Doppler shifts. Also features exactly at the pole have zero Doppler shift as a function of time, the only localized region for which this is true.

Strassmeier Yes, and this should make us especially careful when we talk about cap-like polar spots. At least we should keep it in mind.

Hatzes The O-C (observed-computed) of the line profiles from the Piskunov image show systematic trends indicating that his group is not adequately fitting the line profiles. In fact, when Piskunov relaxes his regularization in order to fit the profiles he produces an image more consistent with the Rice-Wehlau and Vogt-Hatzes image. This image, however, also produces spurious hot spots which do not fit the photometry, indicating that they are artifacts similar to the ones A.C. Cameron says occurs when you use temperature-dependent line profiles.

Strassmeier The Piskunov *et al* fits are good to 0.5% of the continuum and they indeed do show a systematic trend such that the theoretical profile is always too deep in the core and slightly too shallow in the wings. When they relax their regularization, their fits are good to 0.15% and the systematic O-C trend disappears (as it should). It seems to me that their W_λ and R_λ values for the local line profile changes too much when going from 4000 K to 5500 K,i.e., a factor 4 in W_λ and about 0.2 in residual intensity.

ULTRAVIOLET IMAGING OF PLAGE REGIONS ON LATE-TYPE STARS

James E. Neff
Laboratory for Astronomy and Solar Physics,
Code 681, NASA Goddard Space Flight Center,
Greenbelt, MD 20771, USA

1. Introduction

Because this is relatively young subject and because the quantity of published results is still quite small, I wish to concentrate on basic principles. The photospheric imaging problem now appears to be well understood (see the paper by Collier Cameron in this volume), and several groups are producing intriguing results (see Strassmeier *et al.* in this volume). There is a strong temptation to apply the same procedures to ultraviolet spectra in order to produce chromospheric images. Unfortunately, that won't work (at least not yet). Not only are the existing ultraviolet spectra of much lower quality than the visible spectra used as input to the photospheric imaging routines, the basic physics of chromospheres is yet to be developed. This is a natural result of our atmosphere blocking the ultraviolet radiation from stars. Mankind has thousands of years experience observing stellar photospheres visually and a century and a half of spectroscopic observations. By contrast, it has only been in the past decade that we have come to appreciate the preponderance of chromospheres and coronae in late-type stars and to observe these phenomena with space-based instruments. While the division of stellar atmospheres into three distinct layers (i.e. photosphere, chromosphere, corona) is a natural selection, not an artificial construct, our goal should be to understand the interrelationship of these layers and to produce realistic three-dimensional models. We must be very careful at this early stage to pose the correct questions.

Spots produce a *perturbation* on the photospheric spectrum, because they contribute only a small fraction of the overall visible light (this is the current paradigm, but it might not be consistent with the enormous spot areas seen in some of the photospheric images or with V-band amplitudes ranging up to 0.6 magnitudes for II Peg). Chromospheric active regions associated with spots, on the other hand, likely produce most or all of the ultraviolet light. To first order, we know the emitting volume of the photosphere, so we can use the observed spectrum to infer its physical properties. We know neither the surface area coverage nor the radial extent of the chromospheric active regions. These are the first properties we must attempt to measure. The actual distribution and shape of the active regions is of secondary interest. Nevertheless, the most straightforward way to measure the size of active regions is to *image* the chromosphere at several heights, thus synthesizing a three-dimensional picture.

This is not an easy task, but there are several advantages unique to chromospheres or to ultraviolet observations that tend to simplify the problem. The most obvious advantage is contrast: photospheric spots are dark features viewed against a bright background, whereas active regions ("plages") are bright regions viewed against a dark background. The contrast advantage implies that the same spatial detail can be inferred from spectra of far lower signal-to-noise than required for photospheric studies. The larger size expected for chromospheric active regions permits us to use lower resolution spectra. This is fortunate because currently available space-based, ultraviolet spectrographs produce lower-resolution, lower-S/N spectra than ground-based, visible spectrographs. Space-based observations are difficult to coordinate and telescope time is very expensive, but there are no sky constraints or day/night cycles in high-earth orbit. Therefore, we can monitor chromospheres continuously. Increased activity tends to produce a brighter and broader emission line profile from chromospheres, while it tends to "wash out" photospheric profiles.

As I will show, real stars conspire to defeat some of these advantages. Chromospheres are far more variable than photospheres, so continuous observations are a must. Systematic flows produce line shifts, introducing positional uncertainty. The high-temperature plasma tends to increase the intrinsic line widths, reducing spatial resolution.

In this paper, I will attempt to justify the following statements:

1. We do not know the atmospheric physics of chromospheres well enough to invert observed high-resolution profiles to images, as is being done for photospheres.

2. Our goal should be to determine the atmospheric physics of chromospheres first, and then use this as input to study the shape and motion of discrete active regions.

3. To attain this goal, we must measure active region spectra and sizes. To do so requires spatial resolution.

4. Outer atmospheres vary on many temporal and spatial scales, from flares that last several hours to stellar cycles lasting tens of years.

5. Chromospheric plasma dynamics affect the line profiles in ways that mimic spatial structure, so we must unravel both effects simultaneously.

6. We can do this only by obtaining high-resolution spectra of carefully chosen targets continuously for at least several rotational cycles and using these observations to map the surface flux distribution. These maps can be used in turn to separate the active and inactive as well as the spatial and temporal contributions to the observed spectra.

I will summarize the results from observations of AR Lac obtained at four epochs spanning six years, and I will present new results for two other stars: EI Eridani and HD 199178. I will end with some suggestions for future studies, both space-based and ground-based. Before reviewing these results, however, I will outline the basic principles of ultraviolet spectral imaging.

2. Basic principles of spectral imaging

2.1 Direct imaging

Stars are much too far away to be imaged directly using filled-aperture telescopes. At some future time, interferometric imaging might become possible. A few close-binary systems have been resolved using radio VLBI, but the resolution limit typically is on the order of a stellar radius. Mutel *et al.* (1984) used VLBI to probe the spatial structure of the radio emitting plasma, which likely corresponds with the coronal structure, of several RS CVn stars. Visible interferometry should eventually permit us to image directly *some* stars. Narrow-band observations, say in the CaII K line, will be required to image the chromospheres. For reference, the active regions on AR Lac described in the next section are less than 0.1 milli-arcsecond in diameter.

2.2 Photometric methods

If active regions are distributed non-uniformly in projected longitude on a star, then rotation will produce a modulation in the total received emission. Visible photometric modulation is, in fact, the chief observation explained by photospheric spots. To study chromospheres in this way, it is necessary to use spectro-photometry of emission line fluxes (because there is essentially no chromospheric continuum). Rodonò *et al.* (1987) compared the variation of ultraviolet emission line fluxes to the visible photometry for three RS CVn stars. They saw in at least one case a clear anti-correlation between the photospheric and chromospheric variations, as would be expected if the variation is due to large spot/plage complexes analogous to those seen on the sun. Walter, Gibson, and Basri (1983) and White *et al.* (1987) used the rotational and eclipse modulation of the total x-ray flux to probe the spatial structure of the corona of AR Lac. These techniques also can be applied to radio flux variability.

Rotational modulation thus provides our first clue to the spatial distribution of stellar activity. It must be understood, however, that a lack of modulation does not imply a lack of active regions, only a longitudinally symmetric distribution. Also, many of the interesting target stars are in close-binary systems, so we can't tell from ultraviolet spectro-photometry alone which star is varying (they probably both do).

The constraints imposed by photometric observations (intensity *vs* time) are not sufficient to *image* stars. Rather, they provide a measure of the global, hemispheric distribution of activity; essentially spot or plage ***area vs phase.*** The only advantage is that even low-resolution spectra can show rotational modulation. The rotational modulation of line fluxes can be used to determine the spectral contributions arising from active regions and flaring regions alone and to separate the global contributions of each star in eclipsing binaries.

2.3 Ultraviolet spectral imaging

I have mentioned several differences between photospheric imaging techniques (e.g., Vogt, Penrod, Hatzes 1987; Piskunov, Tuominen, and Vilhu 1990) and ultraviolet techniques (Neff 1987; Walter *et al.* 1987; Neff *et al.* 1989). For chromospheres, there is no single parameter that we can map from each grid point on a star to position on a model profile. Because of these differences, chromospheric spectral imaging must be based on a direct ***decomposition*** of emission line profiles. For a more complete comparison of chromospheric and photospheric imaging methods, see Neff (1990a).

A chromospheric emission line from a very rapidly rotating star provides a one-dimensional map of the star's chromospheric brightness. For example, consider a discrete region of high contrast on the surface of a rapidly rotating star. This "plage" region produces an effect on the line profile that moves from blue to red across the line profile as the star rotates. When the region is on the central meridian, there is no velocity shift. By measuring the relative velocity of the feature with respect to the center of the line profile as a function of rotational phase, its longitude and latitude can be determined. The amplitude of this velocity shift is $v_{rot} \sin i \cos \phi$ (where v_{rot} is the rotational velocity of the star, i is the inclination, and ϕ is the latitude of the region), if the region is on the stellar surface. If the region is spatially extended, the width of the spectral feature as a function of rotational phase can be used to determine the shape of the region on the stellar surface. Of course, real life is never so simple. Except for the case of a single, small region of very high contrast on an otherwise uniformly bright star, images can not be derived by simply identifying the wavelength of "bumps and wiggles" on the line profiles.

2.4 Technique

The basic premise of ultraviolet spectral imaging is that emission arising from a **uniform** distribution on a stellar surface will produce a **symmetric** profile *centered* at the stellar velocity. *Non-uniformities* of the surface brightness distribution will produce emission components that vary in wavelength due to the Doppler shift produced by the star's rotation. The net profile will be asymmetric, and a single-component fit to this profile will not lie at the stellar velocity. Therefore, fitting the observed profiles with a symmetric component centered at the stellar velocity accounts for the uniform part of the spatial distribution. If such a fit matches the observed profiles, then we can say nothing more about the spatial distribution. If, on the other hand, additional components are needed to account for the residual emission or to adjust the velocity centroid of the fit, then there must be a non-uniform component of the spatial distribution.

By applying a very basic concept, we have already separated the spatial distribution into uniform and non-uniform parts. The next step is to repeat this decomposition for spectra obtained at many (at least 10 or 12) different rotational phases. Each phase must be observed at least twice, not for redundancy but simply to produce a spatial map. With repeated phase-coverage, each spectrum component (i.e. uniform and non-uniform) can be further subdivided into time-dependent (i.e. variable) and time-independent (i.e. constant). The time-dependent parts subdivide one more time into phase-dependent (i.e. periodic) and phase-independent parts. Any feature that has not been seen to repeat at the same phase on at least two orbits cannot be interpreted as phase-dependent.

The principle itself is quite straightforward, but one must first select an appropriate target and then obtain observations more or less continuously throughout several rotational cycles. Having accomplished this, the first phase of the data analysis is to decompose the series of observed line profiles into 6 components. Each of these components represent different physical phenomena, not all of them spatial:

1. Uniform (i.e. symmetric about line center) and constant – This component tells us very little, because we cannot say whether emission arises *homogenously* from the entire surface or from a uniform distribution of active regions.

2. Uniform and phase-dependent – Applying only the constraint of symmetry, it possible for the uniform component to vary in intensity and width. If this variation is phase-dependent, then there could be a nonuniformity to the spatial brightness distribution that is too large-scale to produce a measurable influence on the line profile (i.e. a hemispheric difference). Binary effects, such as tidal distortion and unforseen eclipses, must also be considered.

3. Uniform and phase-independent variable – This could be due to intrinsic variability with an undetermined period or simply stochastic variability taking place on a global scale.

4. Non-uniform and constant – Asymmetries in the line profile that don't vary with time or phase are the most difficult feature to explain. If their position corresponds to the system center of mass, then they could reasonably be interpreted as emission from that region. If they are matched by a similar component on the opposite side of the line profile, then the presumed intrinsic profile should be called into question (e.g. it might not be gaussian).

5. Non-uniform and phase-dependent – Asymmetries that transit the line profile from blue to red as the star rotates map into discrete active regions on the stellar surface. Their radial velocity curve, relative to the mean stellar radial velocity, provides their position. Their line width, if they are resolved, can be used to measure their size. With the size of the region, we can directly measure its surface flux.

6. Non-uniform and phase-independent variable – These features must be due to short-lived phenomena taking place in localized regions. If several spectra are obtained during the lifetime of the event, then the size and location of the region can be measured as if it were phase-dependent. Several flares have been observed in this way (see Linsky *et al.* 1989; see also Neff 1990c for a review of ultraviolet flares).

The mapping of the non-uniform components into discrete regions can be complicated by the dynamics within the various regions. For example, during flares we see net red-shifts that presumably are caused by a net flow of the flaring plasma. The measured wavelength on the line profile maps directly to a radial velocity, but this radial velocity can be partly rotational and partly dynamical. The spatial resolution obtainable is proportional to the rotational velocity divided by the intrinsic width. If there are large randomly-directed flows, the intrinsic line width increases, thus decreasing the spatial resolution.

If you are not confused yet, bear in mind that frequently there are two stars in these rapidly-rotating systems. There can also be an interstellar component and eclipses to make life difficult (but interesting). Nevertheless, through careful bookkeeping it is possible to investigate all of these spatial, temporal, and dynamical phenomena simultaneously.

3. Results for AR Lacertae

AR Lac (=HD 219334) is an eclipsing RS CVn binary system (K0 IV + G2 IV). The two stars rotate synchronously with the orbital period (1.98 days), yielding an equatorial velocity of 72 km sec^{-1} for the K star and 39 km sec^{-1} for the G star. Because the system is bright (m_v=6.1) and because both stars are active, the system is an ideal candidate for ultraviolet spectral imaging.

3.1 October 1983

Walter *et al.* (1987) first developed these ultraviolet spectral imaging techniques to analyze a series of 8 high-dispersion spectra of AR Lac obtained with the IUE satellite in October 1983. The spectra were concentrated near the eclipse phases. This poor phase coverage and resolution and the lack of repeated phase coverage leads to ambiguous results. Their work nevertheless demonstrated the potential of the technique and led to a few surprising results. The primary conclusions of Walter *et al.* (1987) were that (1) both stars are active and globally variable, (2) discrete plage regions are resolvable on the K star surface, and (3) the far-ultraviolet surface flux spectrum of each component can be obtained from the observed low-resolution spectra. Figure 1 profoundly illustrates the need for additional asymmetric components to the MgII k fits.

3.2 September 1985

Armed with confidence in these techniques but requiring better phase coverage, we observed AR Lac continuously with IUE for 80% of a single orbital cycle is September 1985 (Neff 1987;

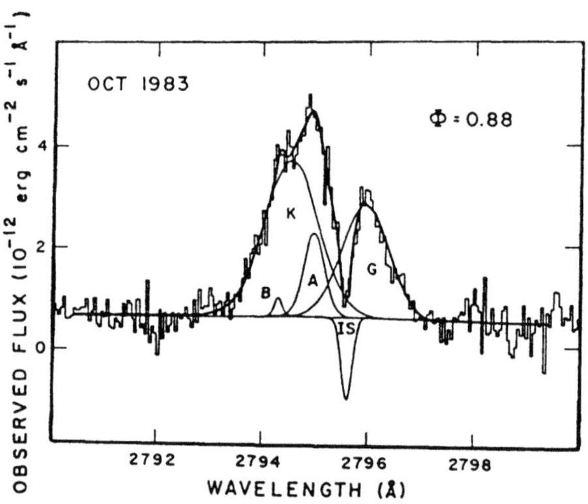

Figure 1: The fit to a MgII k profile from AR Lac consists of symmetric components, representing the uniform "global" emission from each star (labelled K and G), an interstellar absorption component (labelled IS), and several additional components (labelled A and B) to account for the asymmetry.

Neff *et al.* 1989). Eighteen high-resolution ($\Delta\lambda \sim 0.15$Å) LWP spectra of the MgII k (2795.5Å) line and 18 low-resolution ($\Delta\lambda \sim 6$Å) SWP spectra covering the range 1150 to 1950Å were obtained.

The velocities relative to K star of the non-uniform, phase-dependent emission components are shown in Figure 2. The components have been separated into three distinct groups, represented by different symbols. The corresponding plage longitudes were determined by the ***phase*** of the best fit solid-body (sinusoidal) rotation curves. The ***amplitude*** of the solid-body rotation curves are the product of the height above the photosphere, $v \sin i$, and the cosine of the latitude. The $v \sin i$ of AR Lac is well known, but there is no way to distinguish ***a priori*** between height and latitude. However, an amplitude less than $v \sin i$ indicates that the plage B (Figure 2 lies at high latitude. The minimum latitude can be determined by assuming that the plage lies on the photosphere. Radial velocity amplitudes greater than the stellar $v \sin i$ imply that the plages A and C (Figure 2) must lie well above the photosphere, with their minimum heights determined by assuming that the plages are equatorial. There is no way to distinguish between north and south latitudes, because the inclination of AR Lac is nearly 90°.

These radial velocity curves provide the *locations* of the surface features. To produce a crude image of the stellar surfaces, we must also measure the plage *sizes*. We used the line widths of the plage emission components, after deconvolution of the instrumental profile, to determine their extension in longitude. We assumed that all of the deconvolved component width is due to velocity smearing from a resolved region on the star. These measured sizes are upper limits, because there could be a substantial intrinsic width (i.e., turbulence and flows) and because the regions might not be homogeneous. There is no way to determine the latitudinal extent of the plages, because the inclination of AR Lac is ~90°, so we assumed that the plages are circular.

The final system model for AR Lac in September 1985 included five discrete chromospheric

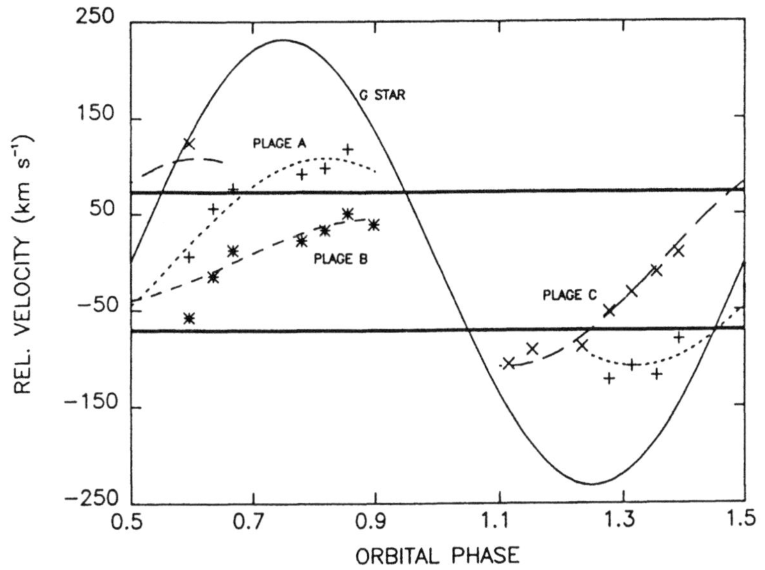

Figure 2: The velocities of the G star (thin solid line) and the additional emission components (plusses, asterisks, and X's) in September 1985 are shown here in the K star rest frame. Thick horizontal lines are drawn at ± 72 km sec^{-1}, the equatorial rotational velocity of the K star. The emission components fall along three distinct solid-body rotation curves, which are plotted only where the corresponding plage would be on the visible hemisphere.

regions: (1) Plage A located at longitude 205°, latitude 0° on the K star (longitude 0° corresponds to orbital phase 0.0, when the G star is totally eclipsed). It likely extended well above the stellar photosphere. The region covered 6% of the visible stellar surface. (2) Plage B located at longitude 240°, latitude±50° on the K star. It coverd 2% of the stellar surface. (3) Plage C located at longitude 130°, latitude 0° on the K star. It likely extended well above the photosphere, and it covered 9% of the stellar surface. (4) The G star was chromospherically *inactive* between longitudes 335° and 75°. This was derived by comparing the spectra obtained before and after the total eclipse of the G star. (5) A flare was observed on the G star. The flaring region covered at most 2% of the stellar surface. There was a systematic *redshift* and turbulent broadening of the flare emission.

The primary result from the 1985 observations was not the surface geometry but the *surface flux* within each region. As shown in Figure 3, the global (uniform) flux from both stars varied, as did the plage flux. All of this information can be presented most coherently as a series of images, which can be found in Neff *et al.* (1989).

3.3 September 1987

The goal of the September 1987 observing campaign was to observe AR Lac continuously with IUE for at least two full orbital cycles (≥ 4 days). Only in this way can the phase-dependence and time-dependence be sorted out. We obtained with IUE a series of 34 high-resolution LWP spectra and 33 low-resolution SWP spectra. This series provided uniform and complete phase coverage. More importantly, observations of the same phase were obtained during subsequent orbits. Preliminary results are given by Neff (1990b) and by Pagano *et al.* (this volume).

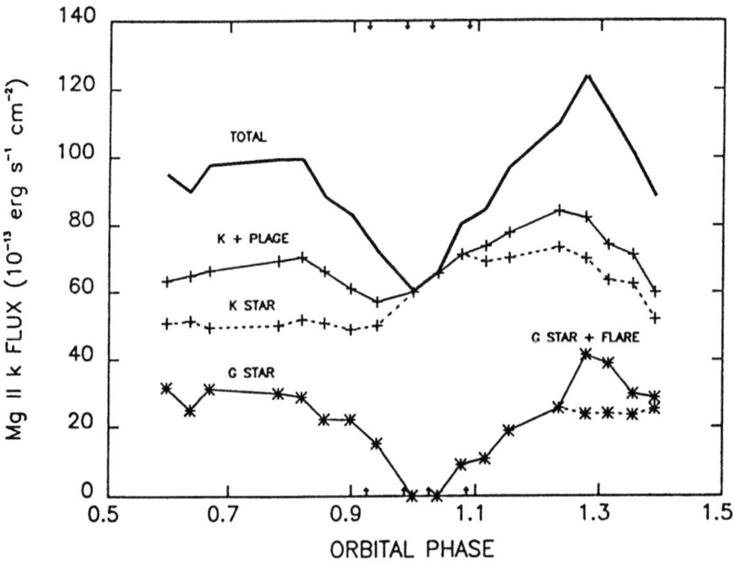

Figure 3: The integrated MgII k flux in each emission component in September 1985 is plotted as a function of time, with $\phi > 1$ following total eclipse to indicate the time sequence of the observations. The top curve is the total MgII k emission from the system (observed flux corrected for interstellar absorption). The plusses indicate emission from the K star, and the asterisks indicate emission from the G star. The solid lines show the total emission from each star (including plages and flare), while the dashed lines show only the emission from the global (uniform) components. The phases of 1st through 4th contacts of primary eclipse are indicated by arrows.

In Figure 4 I show the total emission from the system and the uniform emission from each star. Both stars are roughly 50% brighter in their trailing hemisphere than in their leading hemisphere. The overall line widths (Figure 5) show a hemispheric difference in the same sense as the flux. The structure that produces this variation must be very large-scale, or it would produce an asymmetric profile at some phases. This result is very firmly established, because spectra obtained at opposite quadratures, when the emission components are unblended, show the flux and width difference.

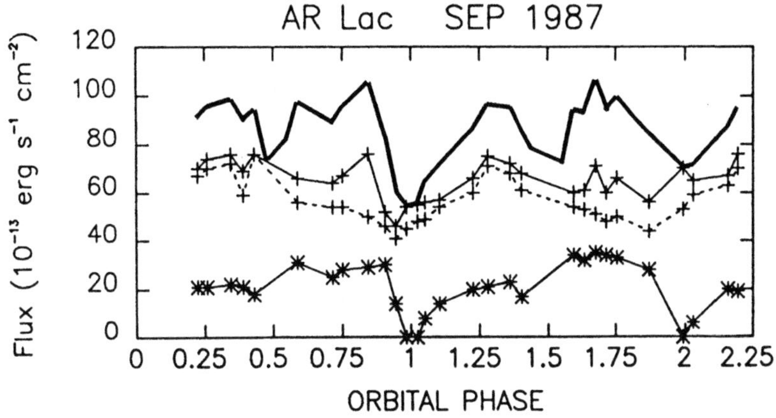

Figure 4: The integrated MgII k flux in each component in September 1987 is plotted as a function of time, or equivalent orbital phase (see Figure 3 for the key). The observations span two complete orbital/rotational cycles, or about 4 days.

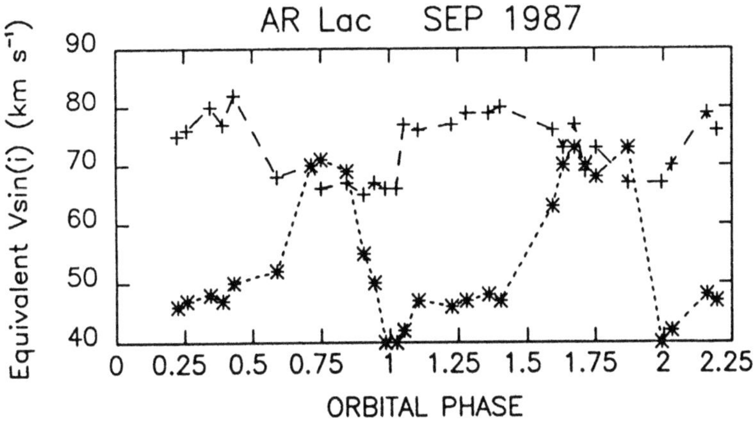

Figure 5: The measured line widths (with the instrumental profile deconvolved) of the uniform global emission components from the K star (plusses) and the G star (asterisks) of AR Lac in September 1987.

The features in Figures 4 and 5 that *repeat* from one cycle to the next are clearly spatially-induced. There were some noteworthy differences between the two orbits, showing that some of the features were transient. No major flares, however, were observed during this 4 day interval.

3.4 August/September 1989

We attempted to arrange full coverage of AR Lac in 1989 in order to study the evolution of

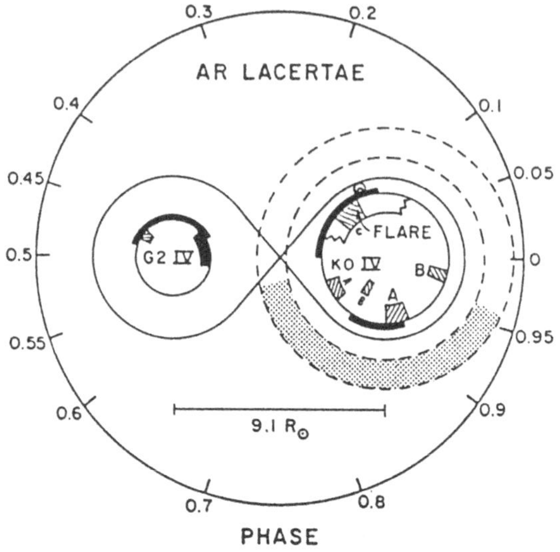

Figure 6: This scaled drawing of the AR Lac system shows the location and size of the discrete regions seen in 1983 (A′, B′, and FLARE) and 1985 (A, B, C, flare and inactive regions on G star). The 1980 coronal structure was determined by Walter *et al.* (1983).

the spatial structures. We settled for a series of observations obtained over a single orbital cycle, but these were nevertheless superior to the 1983 and 1985 observations. Pagano *et al.* (these proceedings), using the same algorithm and multiple-gaussian fitting routine, independently measured and analyzed the 1987 and 1989 data sets. The same basic features were observed in 1989 that were observed at other epochs, but there are some differences in the detail of the resulting map.

3.5 Summary of 1983–1989 Results for AR Lac

The 1983 data set included only 8 phases, nearly all clustered around the eclipses. The 1985 data set covered only 80% of a single cycle, completely omitting the secondary eclipse. The 1987 data set was the first obtained with the phase coverage necessary to conclusively demonstrate that the line profile variations are spatially-induced. The 1989 data again covered only a single orbital cycle, but completely and uniformly.

The global flux levels in 1987 and 1989 were approximately the same as in 1985. The change in width was more pronounced in the 1987 spectra. It is clear that the structure changed slightly between each epoch.

In Figure 6 I schematically overplot the chromospheric structure in 1985 and 1983 along with the coronal geometry in 1980, inferred by Walter, Gibson, and Basri (1983) from x-ray photometry. Pagano *et al.* (these proceedings) show a similar comparison of the 1987 and 1989 results and discuss the long-term trends. There appears to be a general drift toward lower longitude (i.e. opposite the sense of rotation) of the major, stable active regions, covering perhaps 200° in 6 years.

4. Results for other stars

4.1 EI Eridani

EI Eri (=HD 26337), a single-lined spectroscopic binary, is a favorite target for photospheric

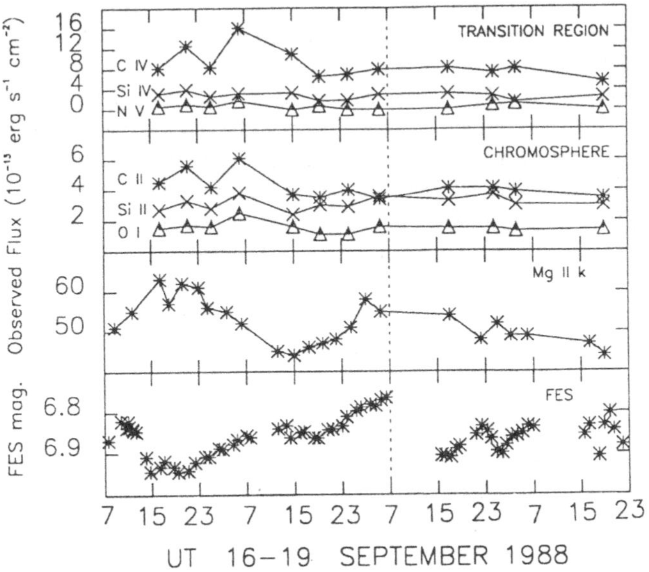

Figure 7: Emission fluxes for the transition region, chromosphere, lower chromosphere (MgII k), and photosphere (FES) are plotted against time during the September 1988 observing run of EI Eri. The dashed vertical line is drawn at the end of the first two days of observations, signifying the end of one rotational cycle and the beginning of the next. None of these diagnostics show the same behavior on both cycles. This is probably due in part to flaring during the first cycle, but the MgII k emission and the FES magnitudes show real variability between the cycles independent of flares.

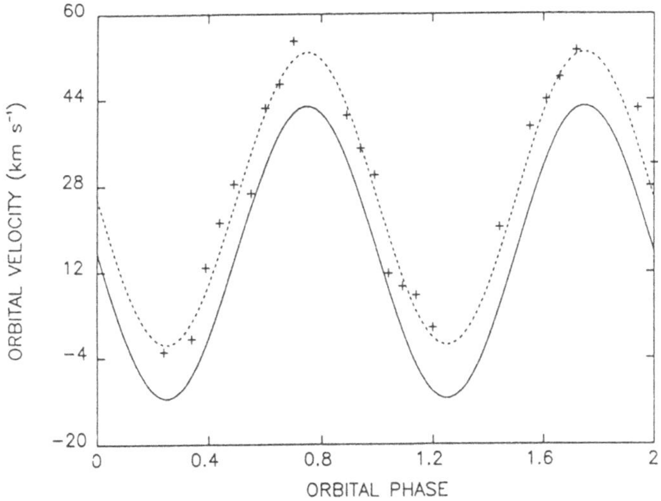

Figure 8: The measured (dashed) and predicted (solid) radial velocity curves of EI Eri (assuming a single, symmetric emission component) do not match. Such a systematic offset is likely to be instrumental. If so, a simple offset must be applied to the wavelength scale. If not, then there must be flows taking place on a global scale. The spatial mapping of discrete, phase-dependent asymmetries is unaffected by this offset.

imaging programs (see Strassmeier's paper in this volume). We observed EI Eri with IUE in September 1988 throughout nearly two complete rotational cycles (Neff *et al.* 1990), obtaining 23 high-resolution LWP spectra.

The measured ultraviolet line fluxes and visible magnitudes are shown in Figure 7. The first rotational cycle was startlingly different from the second cycle. Closer inspection reveals that much of this difference was likely due to flares during the first cycle, but much of it remains as simple time-dependence. The MgII k lines were highly variable in both flux and profile. When the MgII k lines were fitted with a single gaussian emission component, the measured positions did not coincide with the expected positions (see Figure 8). This can be due either to a systematic error in the IUE wavelength scale or to net flows on the star. We have not been able either to locate an instrumental cause or to rule one out, so we are forced to consider both possibilities (Neff *et al.* 1990).

4.2 HD 199178

Neff, Vilhu, and Walter (1988) presented all of the ultraviolet data for HD 199178. Final results are described by Neff *et al.* (1990). This star is another ideal imaging candidate, because it is single (and therefore has no radial velocity variations) and has a $v \sin i$ of 80 km sec^{-1}. We observed this star with IUE in September 1987 at 7 phases thoughout it's 3.337 day rotational cycle and in September 1986 at 6 phases.

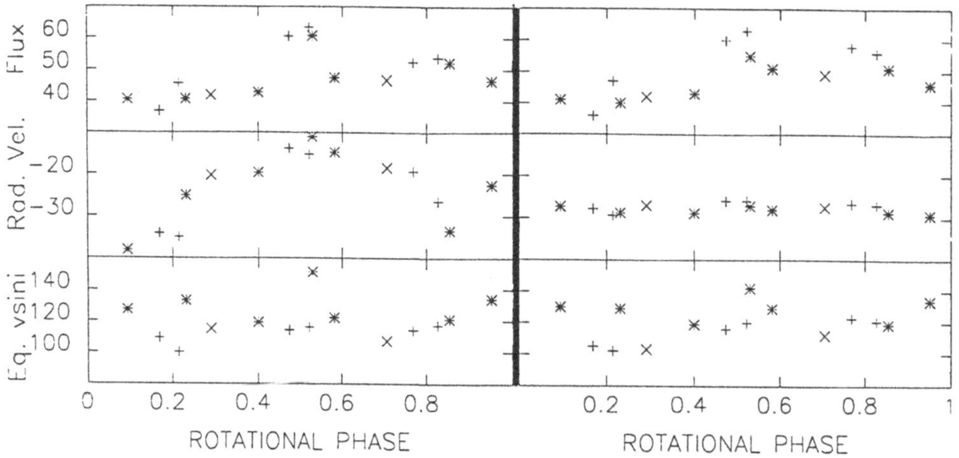

Figure 9: The left half of this figure shows the MgII k flux, position, and width of HD 199178 measured assuming a single symmetric profile. The apparent amplitude of the radial velocity curve is well beyond instrumental uncertainty, but is only 15% of the $v \sin i$ of HD 199178. The right half shows the same parameters when the position is constrained to be -28 km sec^{-1} (as measured from the visible lines) and an additional component added to the fits to account for the excess emission.

Figure 9 shows the results from the MgII k multi-components fits. When we represented the emission with a single gaussian component, we were forced to allow its position to vary with an amplitude of ~10 km sec^{-1} in order to match the observations. This is well outside the range permitted by the ground-based observations of Huenemoerder (1986) and well outside the range of instrumental uncertainties. There must therefore be some non-uniformity to the brightness distribution. The first step in mapping the non-uniformity is to constrain the position of the symmetric component to be the expected velocity of the star. When we did this, we found that all of the excess emission was always on the red side of the line profile. This remained a mystery until we considered the EI Eri results. If there is also a systematic $+10\,\mathrm{km\,sec^{-1}}$ redshift of the HD 199178 lines, then the excess emission would move from blue to red in the sinusoidal fashion expected for surface features.

Unfortunately, there is no way to know which of these two interpretations is correct. The measured position of the interstellar medium line provides a fiducial on the wavelength scale, but we need an independent measure of the flow velocity in this line of sight in order to apply this constraint. To my knowledge, there are no independent measures available.

5. Future directions

With the Goddard High Resolution Spectrograph on the Hubble Space Telescope, we will be able to derive images in other lines (e.g. C IV). This will permit us to probe the height-dependence of the chromosphere and transition region structure. Future spacecraft missions (e.g., LYMAN-FUSE and AXAF) will provide high-resolution spectra in the extreme ultraviolet and x-ray bands. Such spectra might permit spectral imaging of stellar coronae. I am analyzing ground-based spectra of the CaII K lines to see if they can be used to map chromospheric structure and dynamics. The CaII K lines should be similar to the MgII k lines, but they are viewed against a much brighter photospheric background. The chief advantage of these lines is that they can be observed from the ground.

6. Summary

AR Lac has proven to be an ideal target for ultraviolet spectral imaging. We have used this target to develop and refine the procedures, and it has reminded us that chromospheres are neither static nor homogeneous. Both stars in the AR Lac system are brighter and their profiles are broader from the trailing hemisphere than from the leading hemisphere. Several dominant active regions have been resolved on the surface of the K star at several epochs spanning a total of six years. The measured sizes of these regions can be used to calibrate the observed spectra into surface fluxes, which allows us to construct semi-empirical atmospheric models. If the regions seen at each epoch are long-lived, then they must be migrating in longitude by about -30° per year, opposite the sense of rotation.

EI Eri exhibited pronounced changes from one cycle to the next in September 1988, partially due to flares. Its MgII k profiles showed phase-dependent asymmetries, but work is still in progress to map these into surface structure. The chief hurdle to date has been the systematic redshift of the observed MgII k lines from EI Eri, which could be either instrumental in origin or intrinsic to the star. The fluxes from the transition region, chromosphere, and photosphere do not seem to correlate well. The 1988 observations of EI Eri clearly demonstrate the need to observe continuously for two cycles.

HD 199178 is an excellent photospheric imaging candiate, and it shows stable photospheric spotted regions. The MgII k data can be interpreted as either a time-dependent asymmetry of the red wing or as a large-scale non-uniformity of the chromospheric brightness. The HD 199178 database also includes evidence for temporal and dynamical phenomena.

While the progress in the past few years is very encouraging (partly because so little was known beforehand), much work remains to be done. We should continue to monitor these three systems throughout the remaining lifetime of IUE. This will permit us to study, among other things, the evolution of the spatial structure. Imaging of the transition region should become possible using the GHRS on the Hubble Space Telescope and moderate or high-resolution spectra of the C IV lines. Images of the photosphere, chromosphere, and transition region can then be combined to develop a three-dimensional model of magnetic activity.

Acknowledgements. This work was completed while the author was a National Research Council Research Associate at NASA/GSFC. Many of the observations described in this paper were obtained through an international collaboration of astronomers from Armagh, Catania, JILA/Colorado and several other institutions. I acknowledge support from NASA IUE research grants at GSFC and at the University of Colorado. Finally, I would like to thank the Scientific Organizing Committee for inviting me to present this review and for putting together such a stimulating program and the Local Organizing Committee for conducting a smooth and memorable meeting.

References

Huenemoerder, D.P., 1986, AJ, 92, 673.

Linsky, J.L., Neff, J.E., Brown, A., Gross, B.D., Simon, T., Andrews, A.D., Rodonò, M., Feldman, P.A. 1989, AA, 211, 173.

Mutel, R.M., Doiron, D.J., Lestrade, J.F., Phillips, R.B. 1984, ApJ, 278, 220.

Neff, J.E. 1987, Ph.D. thesis, University of Colorado, Boulder.

Neff, J.E. 1990a&b, in *Active Close Binaries*, eds. C. Ibanoğlu and I. Yavuz, (Kluwer: Dordrecht), in press.

Neff, J.E. 1990c, *Mem. Astron. Soc. Italy*, special edition, ed. B.R. Petterson, in press.

Neff, J.E., Strassmeier, K., Vilhu, O., Walter, F.M., Rodonò, M., 1990, in *The Sun and Cool Stars: Activity, Magnetism, and Dynamos*, ed. I. Tuominen, (Springer-Verlag, Lecture Notes in Physics), in press.

Neff, J.E., Walter, F.M., Rodonò, M., Linsky, J.L. 1989, AA, 215, 79.

Neff, J.E., Vilhu, O., Walter, F.M. 1988, in *A Decade of UV Astronomy With the IUE Satellite*, ESA SP-281, p. 291.

Piskunov, N.E., Tuominen, I., Vilhu, O. 1990, AA, 230, 363.

Rodonò, M., Byrne, P.B., Neff, J.E., Linsky, J.L, Simon, T., Butler, C.J., Catalano, S., Doyle, J.G., Andrews, A.D., Gibson, D.M. 1987, AA, 176, 267.

Vogt, S.S., Penrod, G.D., Hatzes, A.P. 1987, ApJ, **321**, 496.

Walter, F.M., Gibson, D.M., Basri, G.S. 1983, ApJ, 267, 665.

Walter, F.M., Neff, J.E., Gibson, D.M., Linsky, J.L., Rodonò, M., Gary, D.E., Butler, C.J. 1987, AA, 186, 241.

White, N.E., Shaefer, R., Parmer, A.N., Culhane, J.L. 1987, in *Proc. Fifth Cambridge Workshop on Cool Stars, Stellar Systems, and the Sun*, eds. J.L. Linsky and R.E. Stencel, (Springer-Verlag:Berlin), p. 521.

Discussion: Neff

Mullan What causes velocity to exceed $v \mathrm{Sin}\, i$? Is it extended material? If so why has one hemisphere larger $v \mathrm{Sin}\, i$ than the other hemisphere? If extended structure is present at one phase should it not be visible also 180 deg later in phase and therefore make the line broader there also (one should see a loop over the opposite edge also)?

Neff That's one of the most interesting questions raised by these results. I don't have an answer yet, but I should note that the hemisphere with the higher degree of line broadening is also 50% brighter and that the hemisphere difference is more profound for the G star. I have interpreted the plage radial velocity curves, which also exceed $v \sin i$, as extended material. That is an entirely separate problem from the global broadening.

Young Otto Struve noted and published the curious fact that, in double-lined SB such as AR Lac, ER Vul, and others, the *approaching* component always had the highest H and K emission core regardless of which star it was. Therefore the leading hemispheres are brightest. Why? Could it be running into circumstellar plasma (ie circumbinary) and heating?

Neff That is indeed what we see from AR Lac. The synchronized binary geometry produces two obvious differences between the leading and trailing hemispheres. The leading hemisphere is always facing the direction of the orbital motion and thus is sweeping up circumstellar material, and accretion streams terminate on the leading hemisphere. I can imagine a leading/trailing asymmetry in the magnetospheric properties as well. I don't want to re-discover the wheel, so I'll look into the references you suggest.

Giampapa What is the extent in stellar radii that you infer for these active regions?

Neff The amplitude of the "plage" radial velocity curves give 1.5 to 2.0 stellar radii, assuming solid body rotation, for two of the regions in 1985. The height would be greater if the regions are not equatorial. The other region seen in 1985 had an amplitude less than $v \sin i$. It would be on the surface at latitude 50 deg, but could be slightly extended if it is at a higher latitude.

Budding AR Lac is an eclipsing binary. Taking into account the role of eclipses on the net line profile structure is a non-trivial computational problem - I was wondering how you addressed this matter.

Neff For the secondary eclipse phases ($\phi \approx 0.5$), the method completely breaks down and we can derive very little information. The eclipse phases are far less of a problem, because it is a total primary eclipse. Most of the information can still be derived by removing the centroid velocity constraint on the global G star profile.

Budding You seem to have concentrated the analysis on the k-line profile. Do you find any corroboration with the h-line profiles?

Neff For the 1983 and 1985 data sets, the analysis was performed independently for both lines. The h-line results were always consistent with the k-line results, but they provided no new information. Because they are exposed at lower S/N, I have not used them for the analysis of 1987 and 1989 data sets.

Budding How about simultaneous photometric surveillance with AR Lac?

Neff AR Lac is a 2 day period system, so ground-based observations are difficult. I will ask Marcello Rodono to comment on our success at coordinating ground-based observations with our IUE observations.

Rodono Spot-plage spatial correlation has been studied (see poster by Pagano *et al.*, these proceedings). There are some positive and negative results. However, the general picture indicates both photospheric spots and chromospheric plages that migrate in longitude and latitude.

Linsky Your observations that the line width are larger on one hemisphere of each component of AR Lac is consistent with enhanced flows (upflows and downflows) on one hemisphere and thus many small plages with upflows and downflows. This situation will lead to misplacement of plages on the stellar surface.

Neff Systematic flows would change the derived location, and increased broadening caused by both upflows and downflows would decrease the spatial resolution.

Linsky Would you speculate on whether Doppler images that will be obtained in the transition region and coronal lines from HST, Lyman/FUSE, XMM and AXAF will be qualitatively different from those now being offered for the chromosphere using the Mg II lines?

Neff Of course they will be different, just to make our lives difficult. I appreciate your comment, because I didn't have time during the talk to describe ideas for the future. The next step will be to map the transition region using HST/GHRS to obtain a series of C IV (1550 Å) spectra.

Linsky Ayres and collaborators have shown that ultraviolet emission lines like C IV are red-shifted by 10 km/sec for active stars like λ And and Capella. These are global average values, so that red-shifts (presumably down-flows) could be much larger for plages. Not including such red-shifts could lead to large errors in the placement of plages on the stellar surface and perhaps explain why spots and plages often do not coalign on the stellar surface.

Neff The big problem is proving that the systematic shifts observed from EI Eri and HD199178 come from the star and not from the telescope or instrument. The double-lined velocity curve of AR Lac can be used to rule out systematic, global flows, but there could still be flows within the dominant plage regions.

As for the coaligament of spots and plages, I don't think there is much evidence for or against. Most methods only tell you where spots are, not where they aren't. It would be unwise, I feel, to use spot models, for example, to tell you where the plages are and then assume flows account for the observed difference.

(l to r) A. Young listens intently while B. Byrne explains how large polar spots might be!

A NEW APPROACH TO DOPPLER IMAGING OF LATE-TYPE STARS

M. Kürster and J.H.M.M. Schmitt
Max-Planck-Institut für Extraterrestrische Physik
D-8046 Garching bei München (FRG)

1. Introduction

While the first attempts to use line profiles for the mapping of surface structures on Ap stars date back to 1958 (Deutsch 1958, see Rice et al. 1989 for a review), the first Doppler image of a cool star was presented 7 years ago (Vogt and Penrod 1983). Various techniques for solving the ill-posed and ill-conditioned inversion or minimization problem of imaging the stellar surface from a series of distorted stellar line profiles have been employed. Regularization methods that use some additional constraint on the solution to select a particular image from the infinite number of possible images have been the most successful techniques so far. The "Maximum Entropy Method" (MEM) used by Vogt et al. (1987) and Hatzes et al. (1989) selects the image with the least amount of configural information. The "Tikhonov regularization" first used by Goncharsky et al. (1977) produces the image with the smallest gradients and consistent with the data (see also Khokhlova et al. 1986; Piskunov et al. 1990).

Both MEM and Tikhonov's method favour smooth brightness transitions from spot to photosphere. This, however, is not what sunspots look like. If the Sun is a good guide for the stellar case, a reasonable approximation to the photospheric brightness or temperature distribution is given by a 2-temperature model (umbra/photosphere). In this paper we present a new technique that uses this *a priori* constraint. One might wish to include penumbrae in a 3-temperature model, but usually there is no hope that the quality of the data allows to recover so much detail.

2. Properties of the Doppler Imaging problem

The assumption of two discrete brightness values for the stellar photosphere imposes a strong constraint on the image. However, discrete minimization problems in general are difficult to solve. Fortunately, stellar line profiles contain so much information about the spatial distribution of surface features that a solution can be found in a straightforward way using the algorithm described in Section 3. It exploits the following properties of the Doppler imaging problem.

Every individual rotationally broadened absorption line profile is essentially a 1-D projected image of the 2-D image of the visible hemisphere of the star, blurred by convolution with the specific stellar line profile and the instrumental profile. A single spot on the stellar surface causes a profile distortion at its radial velocity. If the product of line equivalent width, EW, and continuum flux, F, is smaller in the spot than in the photosphere, $EW_{sp} \times F_{sp} < EW_{phot} \times F_{phot}$, the spot will show as a pseudo-emission; it will show as excess absorption otherwise (Vogt and Penrod 1983). If the line equivalent width is the same in spot and photosphere, the pseudo-emission caused by a cool spot must be compensated by apparent excess absorption at other radial velocities.

As long as one selects an absorption line in which a spot will (for the range of temperatures in question) always appear as pseudo-emission or always as excess absorption, any profile distortion can be projected back onto the star in a unique way (accounting for blurring with the specific line profile): the less the radial velocity of a surface element deviates from that of the distortion, the more likely will that pixel be part of a spot. A combination of several line profiles taken at various rotation phases will narrow down the possible spot regions and disentangle the contributions

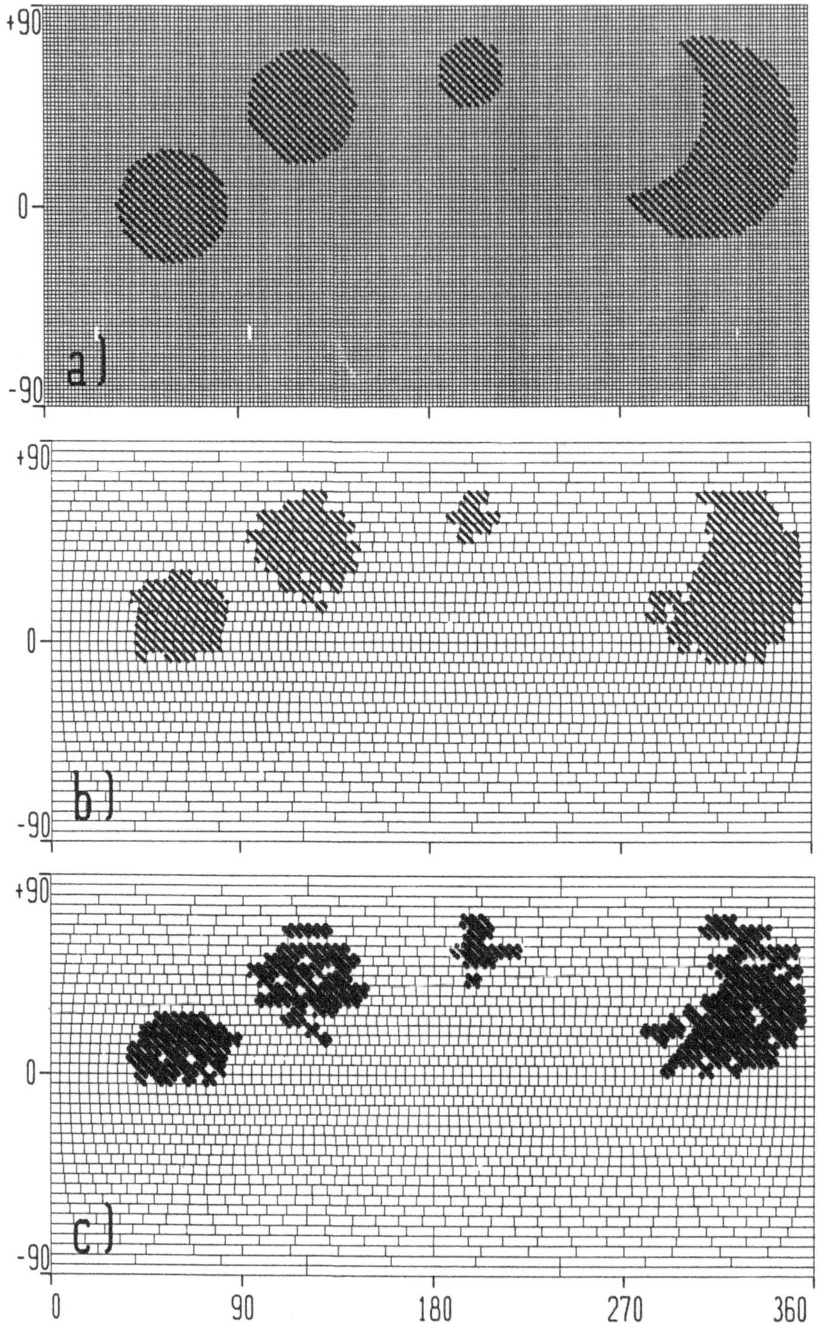

Figure 1. *a) The 4-spot-test:* A latitude *vs.* longitude plot of the input test image calculated on a 20,000 pixel grid with equal steps in latitude and longitude. The spot-to-photosphere brightness ratio is 0.3. Inclination, $i = 60°$.

b) The 4-spot-test: CLEAN reconstruction of *a)* calculated from 26 line profiles on a 2038 pixel grid with constant pixel areas. The correct value of the brightness or temperature ratio has been chosen.

c) The 4-spot-test: Same as *b)*, but for an assumed spot-to-photosphere brightness ratio of 0.0.

from different spots. With this back-projection a rough, but usable estimate of the probability for every pixel to be part of a spot can be obtained. We call this a "probability map."

3. The image reconstruction algorithm: a CLEAN procedure

The inversion technique that we propose here has several properties of the CLEAN procedure (Högbom 1974, 1979, 1984) extensively used in radio astronomy. Our version of CLEAN as used for the Doppler Imaging problem consists of the following steps:

1. Construct the "probability map" by comparing the distorted line profiles with theoretical profiles of a spotless star.
2. Identify the most probable pixel in the "probability map". Assign the lower temperature value to only this pixel ("minimum choice").
3. Update the "probability map" by comparing the data with theoretical line profiles of the star that contains just this one dark pixel.
4. Identify the most probable (still bright) pixel in the new "probability map" *etc.*
5. Stop, if the χ^2 agrees with the data errors or if the algorithm fails to converge.

Note that in the "probability map", only the most probable pixel needs to be identified correctly. The algorithm can also work with a larger number of discrete temperatures through the introduction of a "gain" factor. However, we use the 2-temperature approach to maximally constrain the solution. The correct value of the spot temperature has to be derived from photometry.

4. Simulated data: a test of the algorithm

It is impossible to demonstrate that the images produced with a particular reconstruction algorithm are always similar to the true image. Common practice among all groups doing Doppler Imaging is to test the algorithm for some artificial image from which simulated data are calculated, and to carry out a reconstruction. We show one such test example in Figures 1a - c.

Our "4-spot-test" is a somewhat idealized case with "infinite" signal-to-noise ratio and exactly known stellar parameters. The line profile is assumed independent of temperature. We used the same spectral resolution, geometrical parameters and number of line profiles as for the imaging of the K star AB Dor presented in Section 5. Similar to the results of Vogt et al. (1987) for their tests of MEM, the CLEAN method works well with errors in the inclination of $\approx 10°$ and requires only a few line profiles, well spaced over the rotational period, to produce a useful image. As indicated in Figure 1 simulation and reconstruction have been calculated with different grids.

We show a reconstruction for the correct choice of the spot temperature along with one which assumes spots that are too cool. The locations and approximate shapes of the spots are recovered well. The strongest deviation is a shift away from the equator (where latitude resolution is low) towards higher latitudes (*cf.* the "shadowing effect" described by Vogt et al. 1987).

In the reconstruction attempt with too cool spots the spot area is reduced in order to produce profile distortions of the correct amplitude. Defects close to the equator are more pronounced. Another property of the method can be seen in the spot at longitude 120° that is unaffected by the equator effect: since the algorithm does not use any smoothing or connectedness criterion, we are free to try and reconstruct the correct size of the spot by putting bright pixels inside of it.

5. Application to real data: AB Dor

We have applied our technique to a series of 26 profiles of the CaI λ6439Å line in the young, fast rotating, active K0 dwarf AB Dor (= HD 36705). The observations were carried out with the 1.4 m CAT at ESO, La Silla, in early February 1989. Spectral resolution was 50 000.

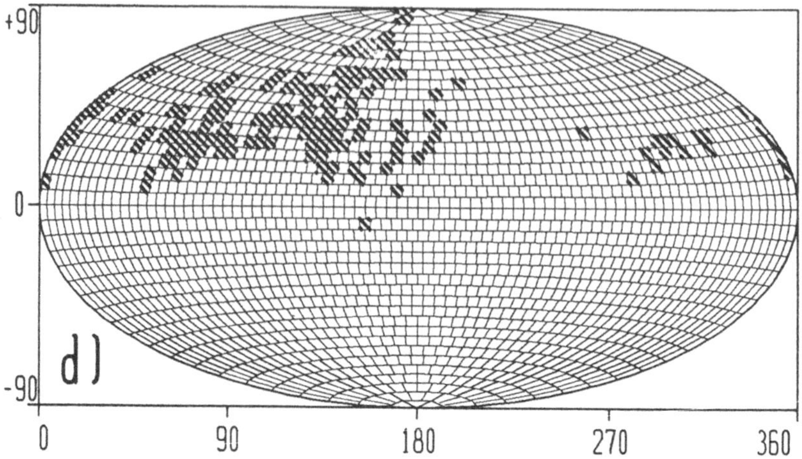

Figure 1. *d) Doppler image of AB Dor:* CLEAN reconstruction shown in Aitoff projection. Latitude and longitude scales apply only to the equator and to the central meridian, $i = 60°$. The area below latitude -60° is always invisible.

In Figure 1d we show a CLEAN reconstruction of the photosphere of AB Dor. Assumed rotational velocity and inclination were $v \sin i = 90\,\mathrm{km\,sec^{-1}}$ and $i = 60°$. We chose a spot temperature of 4300 K and a photospheric temperature of 5250 K. Allowing for a strengthening of the line equivalent width of $EW_{sp}/EW_{phot} = 1.5$, these values reproduce the V band light curve variations simultaneously observed by Cutispoto (1989, priv. comm.). Images with different spot temperatures are permitted by the line profiles, but in conflict with the photometry. Due to the small amplitude of the profile distortions, equivalent width variations which could possibly serve as a check for the spot temperature cannot be reliably detected.

We also calculated a MEM image of AB Dor with the software described by Skilling and Bryan (1984). Both the MEM and the CLEAN images agree very well on the positions of dark active regions. MEM is free to create bright regions as well as dark ones and produces smooth transitions between the various structures. The major difference between the images is the brightness constrast: while the ratio of the minimum and maximum brightness values is similar for both reconstructions, in the MEM image most pixel values are close to an intermediate value.

Both Doppler images show four dark active regions positioned at longitudes 0°, 60°, 130°, and 300°. While the always flat-bottomed line profiles of the evolved stars HR1099, HD199178, UX Ari, and EI Eri (imaged by Vogt and Penrod 1983, Vogt 1988, Vogt and Hatzes 1990, and Strassmeier 1990) all require a large cap-like polar spot, the profiles of the young main-sequence star AB Dor do not produce such a feature. Several pseudo-emission bumps enter the profile at its blue wing and disappear in the red wing. The corresponding spots in the Doppler image are concentrated around latitude +20°. Only the feature at longitude 130° extends to high latitudes.

6. Conclusions and future prospects

The presented CLEAN-like alogrithm is particularly suited for Doppler imaging of late-type stars. The 2-temperature model secures maximum use of the *a priori* knowledge of quasi-discrete surface temperatures and helps to considerably speed up the imaging process. The required computation time depends on the amount of structure to be recovered, but can be reduced by as much as two orders of magnitude when compared to MEM. This makes it possible to apply the algorithm to larger data sets such as several line profiles simultaneously.

While photometry is required to find the spot temperatures, the technique yields the locations and rough shapes of star spots. In this respect CLEAN and MEM reconstructions agree well.

We interpret the clustering of spots around latitude $+20^o$ in the AB Dor image as the detection of an active latitude. From the solar analogy one might expect a symmetric active latitude at -20°. However, the ability to detect features on the "southern" hemisphere with Doppler imaging is limited. The "shadowing effect" can make possibly existing features invisible.

The CLEAN image of AB Dor shows "spot groups" rather than single spots. This may be due to an imprecise choice of spot temperature, some instability of the method, and/or to the fact that we actually look at spot groups.

In an attempt to improve the stability of the method we are exploring the possibility to use the MEM image as an improved "probability map" (without further constraints it is the most probable image by definition). Computational expense prohibits the extensive "cleaning" that we use on the back-projection, but via a gradual change of the entropy definition (*cf.* Horne, 1985 for some applications), the reconstruction can be forced to approach a 2-temperature model.

References

Deutsch, A.J.: 1958, in *Electromagnetic Phenomena in Cosmical Physics*, IAU Symp. No. 6, B. Lehnert, (ed.), p. 209

Goncharsky, A.V.. Stepanov. V.V., Khokhlova, V.L. Yagola. A.G.: 1977, Pis'ma Astron. Zh. 3, 278 (English translation in Sov. Astron. Lett. 3, 147)

Hatzes. A.P., Penrod, G.D., Vogt, S.S.: 1989, ApJ, 341, 456

Högbom, J.A.: 1974, ApJS, 15, 47

Högbom, J.A.: 1979, in *Im. Form. from Coh. Funct. in Astron.* IAU Coll. 49, (ed.) C. Van Schooneveld, p. 237

Högbom, J.A.: 1984, in *Indirect Imaging,* (ed.) J.A. Roberts, Cambridge Univ. Press, p. 247

Horne, K.: 1985, MNRAS, 213, 129

Khokhlova, V.L., Rice, J.B., Wehlau, W.H.: 1986, ApJ, 307, 768

Piskunov, N.E., Tuominen, H., Vilhu, O.: 1990, AA, 230, 363

Rice. J.B.. Wehlau, W.H., Khokhlova. V.L.: 1989, AA, 208, 179

Skilling, J., Bryan, R.K.: 1984, MNRAS, 211, 111

Strassmeier, K.G.: 1990, ApJ, 348, 682.

Vogt, S.S., Penrod, G.D.: 1983, PASP, 95, 565

Vogt, S.S., Penrod, G.D., Hatzes. A.P.: 1987, ApJ, 321, 496

Vogt, S.S.: 1988, in *The Impact of Very High S/N Spectroscopy on Stellar Physics,* IAU Symp. No. 132. (eds.) G. Cayrel de Strobel, M. Spite, p. 253.

Vogt. S.S., Hatzes, A.P.: 1990, in *The Sun and Cool Stars,* IAU Coll. No. 130, Helsinki.

Discussion: Kurster

Zeilik If anyone wants a fast algorithm, I can recommend the program PFIT, which will provide an optimal Doppler image, together with a determinacy check, on a PC in 2 minutes or so.

The equivalent width test is an important methodological check. If we assume that the line limb-darkening coefficient is the same as in the photosphere, then the equivalent width for a pure absorption feature should stay constant.

Cameron What is your stopping criterion *i.e.* how do you decide when you have switched enough pixels to the spot temperature?

Kurster I tell the algorithm to stop when the estimated S/N level agrees with the χ^2 of the fit. This works for realistic S/N ratios of about 400 - 500 or so. In test cases where you have 'infinite' S/N ratios, one cannot fit the data much better than to a S/N level of 500 or so. This is due to the unstable nature of the discrete minimisation problem.

Cameron Are you confident that the scheme you propose for altering the Maximum Entropy default (to enforce a two-temperature model) will be stable?

Kurster I have played around with it using the 4-spot test star I showed to you, but trying the reconstruction with only a very coarse grid of pixels to save computation time. It works. Instabilities seem to occur (or maybe local minima), but Skilling's algorithm that I use is robust enough to cope with them.

Linsky Throughout astrophysics, systematic errors dominate over random errors. Could you identify the major systematic errors in your technique?

Kurster The continuum calibration. For Doppler imaging, S/N ratios of about 400 or more are required. It is difficult to achieve the same accuracy for the continuum calibration. Due to rotational broadening of the lines or blending (which can be severe for binaries) line-free continuum regions are hard to find.

(l to r) V. Tsikoudi, M. Mathioudakis, P. Panagi and B. Kellett discuss the quality of the tea.

FLUORESCENCE-INDUCED SURFACE INHOMOGENEITIES AND EJECTION AND TRAPPING OF PLASMA BY THE K DWARF COMPONENT OF V471 Tauri

Arthur Young and Lee Rottler
Astronomy Department, San Diego State University
San Diego, California, U.S.A.

Andrew Skumanich
High Altitude Observatory, N.C.A.R.
Boulder, Colorado, U.S.A.

The Hyades binary V471 Tau consists of a K2 dwarf star and a white dwarf in orbit with a period of about 12.5 hours. Due to its nearly synchronous rotation, the K dwarf exhibits significant chromospheric and coronal activity as judged by the usual radiative diagnostics. The white dwarf companion is sufficiently hot (35 000 K) and nearby to affect the ionization equilibrium of the chromosphere of the K dwarf on the hemisphere which faces the white dwarf. The input ultraviolet flux from the white dwarf induces a significant inhomogeneity on the surface of the K dwarf, such that the profile of the Hα line, which is normally observed to be in absorption on the outer hemisphere, is seen in emission on the inner (irradiated) hemisphere. Since both the white dwarf star and the irradiated hemisphere of the K dwarf are totally eclipsed during each orbital cycle, it is possible to observe them independently and thus to study the effects of the input radiation upon the otherwise active chromosphere.

In a previously published study by Young et al. (1988), a time-resolved series of spectra of the Hα region was presented and discussed. The spectra extended over one run of 6 consecutive nights in January, 1983, and one other night, eight months later, in September, 1983. While the powerful effect of the ultraviolet irradiation from the white dwarf was apparent in those observations, the profile of the Hα feature was complex, and frequently it failed to repeat its own structure from one cycle to the next. Knowing that flare activity was common and intense on V471 Tau (see Young, et al. 1983), the authors attributed both the complexities, and their tendency not to reproduce at common phases, to such activity.

The present paper reports a follow-up study conducted during two observing runs, each consisting of two nights, and separated from each other by nearly three months. Using a larger telescope and an improved detector, the new data have twice the resolution in wavelength of their predecessors, improved time-resolution by a factor of four, and better signal-to-noise characteristics by a factor of two in most cases. The surprising discovery revealed by these improved observations is that the Hα profile consists of multiple (3 or more) components. Resolution of the components permitted measurement of their radial velocities which revealed that at least two of them arise from plasma which cannot be on the surface of the K dwarf star. Re-examination of the original spectra from 1983 (published in 1988) showed that much of the complexity that was attributed to flares was actually due to external plasma.

Equipped with new insight, both sets of spectra were analyzed anew, revealing that flaring activity on a timescale of one day does occur on the K dwarf, and that it is probably associated with ejecting plasma. The external plasma is transient, also with a timescale of about one day. In one clear case, a flare was detected on one night (Nov. 3, 1985) by the 15% excess in the equivalent width of Hα which it induced. Meanwhile, external plasma was only weakly detected on that same night. On the next night we detected a considerably enhanced signature of external plasma with no indication of any flaring. Stochastic bursts, not unlike solar flares, probably erupt from active regions on the K dwarf, enhancing the ambient circumstellar (wind) plasma density in a highly non-homogeneous way. Our observations currently extend over three years, and they show that the maximum accumulations of plasma in 1983 were measurably larger than those that were observed in 1986. We infer from that difference that the general level of magnetic activity on the K dwarf was greater in 1983 than it was in 1986, and there may be some corroborating evidence for that inference when all of the optical photometry done during that interval is finally available in published form.

When they are detected, the external plasma components are nearly, but not precisely, symmetrically placed about the rest-wavelength of Hα in the rest-frame of the K dwarf star. Our first impression was that they were coming from concentrations of plasma at the inner and outer Lagrangian points, since their measured velocities were compatible with the expected values from those points. Under that assumption it is possible to derive the location of the putative L1 point, and thence the mass ratio of the system. The result, using values from the quadratures (to avoid flow velocity effects) was a value of 0.6, with the K dwarf being the more massive component. In a private communication, Dermott Mullan pointed out that this value is not compatible with the mass-function of the system which is well determined.

Further analysis of the external plasma components, at phases well removed from the quadratures, showed that there is little or no sinusoidal variation as would be expected of plasma that is nearly stationary at the L1 and L3 points. It now appears that a more appropriate interpretation is a ring or disk of material that is orbiting the K dwarf star itself. Irradiation of the gas by the ultraviolet flux from the white dwarf causes it to be more ionized and hence brighter in the region between the stars, and less ionized (by coronal X-rays) in the shadow of the K dwarf. The velocity near the inner region, in the rest-frame of the K dwarf, is about 190 km/sec. The outer region is more difficult to measure since it is always fainter, but it appears to be about 165 km/sec. Not surprisingly, the orbit is apparently not circular. Furthermore, the Keplerian velocity for material at the frame of the K dwarf.

The dynamics of the external plasma is puzzling. It is transient, but whenever it appears it seems to have about the same velocity characteristics, suggesting that it is ordered in some fashion. It decays (vanishes) on a time-scale of days, either falling back on the K dwarf, or perhaps leaking through the Lagrangian point regions. Magnetic control must be considered, but any simple rigid loop model places the plasma well beyond the center of mass of the system, and is not compatible with the lack of sinusoidal variation that is observed.

The velocities rule out any association with the white dwarf since they are much too slow in the rest frame of the K dwarf. More observations are planned for November 1990, and much more analysis is clearly necessary.

The collective evidence of optical photometry and spectroscopy, ultraviolet spectroscopy, and X-ray spectroscopy points to a picture of a K dwarf star whose surface is inhomogeneous on many different geometrical scales, from hemispheric to highly localized, and on many different time scales from minutes, to days, to years (*i.e.* activity cycling).

Acknowledgement. This research was supported by a grant from Research Corporation.

References

Young, A., Klimke, A., Africano, J.L., Quigley, R., Radick, R.R., Van Buren, D.: 1983, ApJ, 267, 655

Young, A., Skumanich, A., Paylor, V.: 1988, ApJ, 334, 397

Discussion: Young

van't Veer How do you explain the presence of material exactly at L1? It is not a stable situation and we should expect a mass flow everywhere around L1.

Young Indeed L1, L2 and L3 are unstable as opposed to the triangular points L4 and L5. X-ray results from EXOSAT, reported by Jensen and Swank, show absorptions of the white dwarf which are at the correct phase to be caused by plasma trapped at the stable L4 and L5 points. It is easier for ejected plasma to reach L1 and L3 than L4 and L5, and the instability of L1 and L3 is not an issue because the material flows through those points. It is NOT trapped at those points. The L1 and L3 points are locations where equipotential surfaces converge, and it is that property which probably accounts for concentration of plasma there.

Cameron The pattern of moving Hα absorption seen around phase zero is qualitatively very similar to the pattern produced by circumstellar clouds crossing the disk of AB Dor. Is the drift rate of the Hα absorption consistent with a stellar surface feature, or with scattering in a clump of neutral material above the stellar surface?

Young This is a very good point, particularly in view of your recent demonstration of such co-rotating opaque clouds above AB Dor. The answer is that we have not examined the absorption line phases carefully since our model becomes useful only when the bright fluorescent zone dominates the disk (phases 0.2 to 0.8). I should point out, however, that the cone of incident radiation from the white dwarf has a circle of tangency which is about 80° away from the line of centers or substellar point of the white dwarf. Thus, a cusp of that region comes onto the visible disk at a phase of 0.03 and is well on by 0.05. Unless the emissive region is severely limb-darkened, the cusp of emission should dilute the observed absorption core in an asymmetrical way due to the limb Doppler bias. The observed change in the asymmetry of the absorption line phase is entirely consistent with that expectation, and so we did not look for an alternate explanation. However, you can be certain that I will re-examine those spectra and do some calculations as soon as I get home!

Spruit Why do your "L1, L3" features have to be at L1, L3? Especially if these are temporary features that don't repeat well, there surely must be many other places in the system that would give equally good fits.

Young The features (in Hα) are temporary only in the sense that they appear and vanish on a day-to-day basis. However, it is not true that they do not repeat in any other way. When they appear, they have a relative radial velocity which is consistent at that phase with a unique location in the binary system. Our best argument for why those locations are actually L1 and L3 is that if we assume that those are the locations we can calculate the mass ratio of the system, independently of the points. Both points give a mass ratio of 0.6 to within their uncertainty, and that would be a remarkable accident if the two points are arbitrary in the system.

Mullan Our mass loss study of V471 indicated the strongest evidence of outflows on the "back"side of the K2 dwarf, and much less evidence around phase 0.5.

Young I think the key phrase here is "less evident" for flow near phase 0.5, and not necessarily less flow. I think this is a selection effect. The outside hemisphere, seen at phase 0.0, has chromospheric properties similar to "active" K dwarfs; but the inner hemisphere seen at phase 0.5 is greatly enhanced in all of the chromospheric lines, such as Mg II h and k. Detection of flow is more difficult since it depends on those lines in absorption. Our detection of plasma in the region of L1 implies that mass flow occurs from the inner hemisphere, and our detection of plasma near L3 corroborates your finding that mass loss does indeed occur from the backside as well. Most likely mass loss is not confined to any one hemisphere or region.

INFLUENCE OF STARSPOTS ON INTERNAL STELLAR STRUCTURE

H.C.Spruit
Max Planck- Institut für Astrophysik
Karl- Schwarzschildstr. 1, D-8046 Garching

1. Introduction

In order to interpret the photometric variations of spotted stars, one would like to know how large the effect of spots (cool areas) or faculae (bright areas) is on the structure of the star. How much of the luminosity change, for example, is due to a change of the temperature in the unspotted parts of the surface? How much does a possible radius change contribute? Are the color changes due only to the spot areas or does one have to correct for the color changes of the unspotted surface?

I will argue here that our knowledge of structure of, and energy transport in, stars is sufficient to give satisfactory theoretical answers to these questions. The reason is that the answers depend only minimally on uncertainties like the precise value of the convective efficiency (l/H ratio). They are even independent of the structure of the spots, as long as the thermal blocking effect of the spots extends to a depth that exceeds a certain (rather minimal) value. This contrasts sharply with our theoretical understanding of the spots themselves. For example, we have no way to predict the temperature of a starspot. For a more detailed account of the effect of spots on internal structure, see Spruit (1991).

An issue of central importance is that of thermal time scales (plural). When a dark spot area is suddenly 'switched on', the star is no longer in equilibrium because it emits less energy than it generates in the core. It starts a thermal adjustment process. Changes take place on a number of time scales. Some of these are on the order of the Kelvin contraction time of the star and therefore affect the position of the star in the HRD, but do not lead to changes that can be observed on historical time scales. Others are much shorter (days to years). The reason for the existence of these widely differing time scales lies in the very high thermal conductivity of a convective envelope (more precisely: conductivity for changes in entropy, since convective heat transport is nonlinear), combined with the very high heat capacity of a the convective envelope. A useful analogy for a stellar convection zone that I will exploit is that of an electric heater plate, imagined to be made of copper (high heat conductivity), and being thick (high heat capacity).

But before discussing these time dependent phonomena, consider the changes in structure on a very *long* time scale; long compared with the thermal time scale of the star as a whole (Kelvin-Helmholtz time scale).

2. Long term changes

On time scales long compared with the Kelvin- Helmholtz time scale, the star is thermally in equilibrium, emitting as much energy as is produced in the core (ignoring the slower change in structure due to nuclear conversion). The questions one is interested are:

- Do spots affect the luminosity generated in the core?
- How large are the radius changes?
- How large are the color changes?

2.1 Long term luminosity and radius changes

To answer these questions one can construct stellar models in the usual way, with changes in the description of the energy transport process in the convection zone to take into account the blocking effect of the spots, and a corresponding change in the outer boundary condition. Before proceeding, we note that in the presence of spots the effective temperature (defined in the usual way) is no longer equal to the surface temperature, and not a good measure of the spectral type. Instead we will use T_s, the *surface temperature of the unspotted surface.* If f is the effective fraction of the surface covered by spots (assuming them to be completely dark), we have

$$L = 4\pi R^2(1-f)\sigma T_s^4, \tag{1}$$

Thermally relaxed models of spotted main sequence stars have been made by Spruit and Weiss (1986). The results show that in stars with thin convection zones the luminosity is not affected by spots, while there is a significant luminosity reduction in fully and nearly fully convective stars. This can be understood qualitatively by appealing to some well known properties of radiative cores of stars (e.g. Kippenhahn and Weigert, 1990). Integrating inward into the core, solutions with different starting conditions rapidly converge to a single track. The characteristic distance for this convergence is of the order of the pressure scale height. If the radiative core spans a large number of pressure scale heights (as in F and early G stars), the dependence of the central conditions on changes taking place in the convection zone is small, so that the presence of spots does not influence the luminosity:

$$\Delta L/L \ll \Delta T_s/T_s. \qquad (F\ and\ early\ G\ stars,\ t \gg \tau_t)$$

The heat flux blocked by spots, in such stars, is all radiated by the non- spot area of the star, which therefore has a higher temperature. The higher temperature in the convective envelope leads, by hydrostatic equilibrium, to a larger radius. Since the temperature change is limited to the convective envelope (and maybe one pressure scale height below it), the radius change is of the order

$$\Delta R/D \sim \Delta \ln T_s|_p,$$

where D is the depth of the convection zone and the index $_p$ indicates a change at constant pressure. The radius change is therefore small in the F stars, which have thin convection zones.

The situation is different if the core is small (in the sense defined). A change in the depth of the convection zone due to spots now has a direct effect on the conditions in the core. This is illustrated best by the extreme case of fully convective stars (Spruit and Ritter, 1983). For the main sequence these are stars later than about M5 according to present models. In such stars the luminosity is rather sensitive to the central temperature (index around 5). The central temperature therefore changes little compared with the luminosity. But all fully convective stars are similar: they are (nearly) polytropes of index 1.5. Their central temperature is a known function of stellar mass and radius ($T_c \propto M/R$). Thus the radius change is also small compared with the luminosity change. The entire internal structure is therefore unchanged, except for a different heat flux. In particular the entropy of the convection zone is unchanged. Since the structure of the surface layers is uniquely determined by the surface temperature T_s (normally equated to the effective temperature, but not in the presence of spots!) and surface gravity g, this means that the change in temperature of the unspotted surface is small compared with the luminosity change. With (1) we have therefore

$$\frac{\Delta L}{L} \approx -f \qquad (fully\ convective\ stars,\ t \gg \tau_t),$$

$$\Delta R/R \ll \Delta L/L; \qquad \Delta T_s/T_s \ll \Delta L/L.$$

That is, the'missing flux' of spots in these stars is truly missing even on the Kelvin-Helmholtz time scale of the star. In reality the energy production is not an infinitely steep function of temperature, so that the central 'thermostat' is not perfect, and small changes in radius and surface temperature are present.

The behavior of stars of intermediate mass is between these extremes. Their surface temperature, radius, and luminosity are all affected by spots (see Spruit and Weiss, fig. 1,2).

2.2 Color changes

The absolute color changes, that is the differences with respect to the same, thermally relaxed star without spots, can only be observed in cases where very accurate mass determinations are available. This may perhaps be possible in some active binaries. More realistically, one can determine how far spotted stars are from the main sequence in a color- color diagram (cf. Eaton, this volume). Since all inhomogeneity in surface temperature makes the spectral energy distribution flatter, spotted stars are always on one side of the main sequence, even if there are bright spots as well as dark ones. We define a *color anomaly* as the distance to the main sequence measured parallel to the B-V axis, for example

$$\delta_{VK}^{BV} = \Delta(B-V) - \frac{\partial(B-V)}{\partial(V-K)}|_{MS}\Delta(V-K). \tag{2}$$

This anomaly is always negative (as in the case of unrecognized binaries). We can calculate how it depends on the assumed spot temperatures and spectral type. Not surprisingly, the magnitude of δ_{VK}^{BV} depends on details such as whether the spots have penumbrae or not. In the G and K star range, it is of the order −0.05 for spots covering 50% of the surface. This is rather small compared with the anomalies presently discussed (c.f. Radick et al. 1988; Campbell, 1984). Soderblom (1989) however shows that binarity has to be carefully excluded before interpreting the anomalies.

2.3 Bright rings

In the above we have tacitly assumed that the temperature of the unspotted stellar surface can be represented by a single value T_s. If part of the blocked heat flux reappears at the surface in the vicinity of the spot, as one might perhaps expect, this would not be justified, and there would be a 'bright ring' with its own color effects. The sun does not show any evidence of such rings around spots. The explanation for this has been found in the very high thermal conductivity of the convection zone due to turbulent convection (Spruit, 1977). In the mixing length model of convection, a change in heat flux $\Delta\mathbf{F}$ is related to the temperature by

$$\Delta\mathbf{F} = \kappa\rho c_p T\nabla(\Delta T/T), \tag{3}$$

where the turbulent diffusivity $\kappa \approx lv$ is found from the mixing length estimates of the eddy size l and convective velocity v. Due to the rapid increase of ρ with depth, only a small temperature difference is needed to carry a large heat flux. For example to carry a horizontal heat flux of the order $F_\odot = 6\,10^{10}$ over a length of 500 000 km (the size of a giant starspot) at a depth of 100 000 km (middle of the solar convection zone) a temperature difference $\Delta T/T = 0.001$ is needed. This is the expected temperature enhancement below such a spot. Since it is the quantity $\Delta T/T$ (proportional to the entropy fluctuation) that is carried by convection, the surface temperature fluctuation $\Delta T_s/T_s$ cannot be larger than this. Thus we would expect a giant sunspot of 500 000 km to show a bright ring with an intensity contrast of maybe 0.5% with respect to the average non- spot surface.

In making this estimate, however, we have assumed that *all* the blocked heat flux is reradiated elsewhere at the surface. This assumption is justified only if the spot lasts as long as it takes the star to reach thermal balance, which is hardly likely (since spots appear and disappear on much shorter time scales). The more realistic question therefore is how strong a bright ring would result around a spot that lasts for, say, ten years. This turns out to be smaller, as discussed in the next section.

The absence of bright rings can be made more understandable with a kitchen analogy. Consider an electric heater plate made of a nice thick piece of copper (rather than iron). I put a small piece of thermally insulating material on the plate, for example a piece of ceramic (see figure), switch on the electricity and wait until the heat lost from the plate equals the power dumped into the heater coil. Because of the poor thermal conductivity of the ceramic, its radiating top surface is cooler than the surrounding surface of the plate. Below the ceramic, the copper surface is a bit hotter than elsewhere because of the heat blocked by the ceramic. How hot is the resulting 'bright ring' around the ceramic? There is not much of a bright ring, because copper is a very good heat conductor. The heat flux blocked by the ceramic distributes quickly through the copper keeping the temperature inhomogeneity small. The surface is therefore also at a nearly uniform temperature, somewhat higher than it would have been without the piece of ceramic.

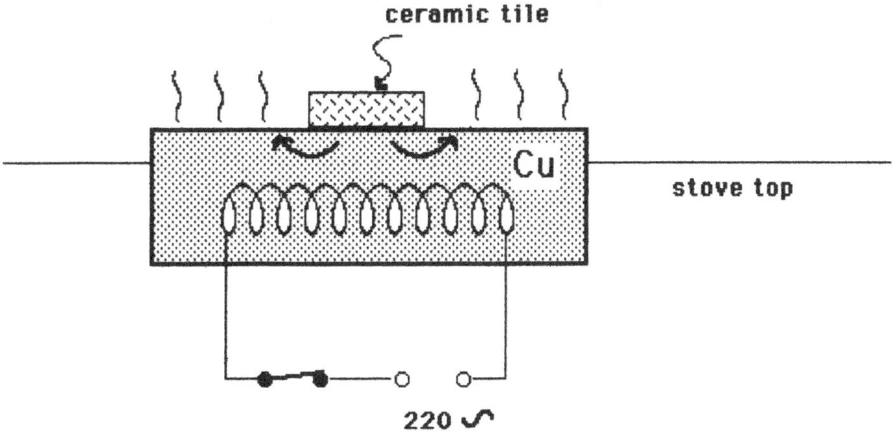

3. Time dependence

Suppose at $t = 0$ a starspot of radius r whose blocking effect on the heat flux extends to a depth d below the surface is suddenly 'switched on'. How does the temperature in the star respond to this, as a function of time? A first time scale of importance for a convective envelope is the ***diffusive time scale*** t_d:

$$t_d = D^2/\kappa_t, \tag{4}$$

where D is the depth of the convection zone and κ_t the turbulent diffusivity (see eq. (3)). This is the time it takes a disturbance in the entropy to 'smear out', through convective turbulence, over the entire depth of the convection zone. Since this process takes place by macroscopic motions, it is much faster than the time it takes a disturbance to diffuse through a *radiative* envelope. For the Sun, it is on the order of 1 yr. Because this time scale is so short, the temperature distribution in a convective envelope is kept close to a state of marginal instability, that is, it is nearly isentropically stratified ('adiabatically' stratified in stellar evolution nomenclature). An exception to this is the thin superadiabatic surface layer, where convection is not as efficient at smearing out entropy disturbances.

Because turbulence dilutes entropy disturbances through the entire convection zone, small temperature changes require large amounts of energy. Associated with this is a long time scale, the *thermal time* scale of the envelope:

$$t_t = U/L, \tag{5}$$

where U is the total thermal energy content of the convective envelope. This is the time it would take the temperature in the convection zone to change when the heat supply at its base were switched off. It is about $2\,10^5$ yr for the sun, more generally it is of the order $M_e/M\ T_e/T_c t_{KH}$ where M_e is the mass of the envelope, T_e the temperature at the base of the envelope, T_c the central temperature, and t_{KH} the Kelvin- Helmholtz time scale. In terms of the heater plate analogy of Fig.1, t_d is the time it takes the available thermal energy in the plate to spread out by conduction (on the order of maybe 10 s), t_t the time it takes this thermal energy content to change when the electricity is switched on or off (on the order of minutes). The ratio of these time scales is only more extreme in the case of a convective envelope.

According to these time scales, we have to divide the spottedness of the stellar surface into its long term ($\gg t_t$) average, and the short term fluctuations around this mean. In this sense, all observationally accessible time scales are 'short term'; including the 'activity cycle' time scales. On these time scales, the entropy of the convective envelope is practically *unchangeable.* When a spot appears on the surface, a part of the heat flux it blocks may appear around it as a 'bright ring', but the energy emitted in this ring will usually be only a small fraction of the blocked heat flux, the remainder being lost in the envelope, slowly (on the time scale t_t) heating it up. That is, there is a very efficient *storage* of heat in the envelope, and the divergence of the heat flux is nonzero. Though this storage is temporary, the time scale involved is infinite from an observational point of view. In terms of the heater plate analogy, the appearance of a spot is analogous to putting the ceramic tile on the surface, and removing it a few seconds later. A faint 'hot ring' does form around the tile, but during its presence it emits only a small fraction of the blocked heat flux, because we have not give the plate the time to reach equilibrium with the new boundary condition.

How much of the blocked heat flux returns to the surface in the form of a bright ring, and how large its maximum brightness is depends on the depth to which the blocking effect of the spot extends into the envelope. This is because the convective heat conductivity (unlike the case of the heater plate) is a steeply increasing function of depth.

The storage of heat that occurs on time scales $\ll t_t$ takes place throughout the convection zone. But since the heat capacity is largest in the deepest parts of the convection zone, an accurate approximate answer is obtained by ignoring the storage everywhere except in the deepest layers, and setting the entropy constant (in time) there. Since storage takes place in other places as well, this approximation overestimates the bright ring effects a bit. Such calculations were done in Spruit (1977) and Spruit (1982a,b). Foukal et al. (1983) and Chiang and Foukal (1984) did fully time dependent calculations, but with a convection zone of greatly reduced depth.

For spots whose radius r is equal to or less than their depth d, one find that the fraction α of the blocked heat flux that reappears at the surface in the form of a bright ring is of the order (Spruit, 1982a,b):

$$\alpha \approx 0.5[1 + d/(3H_0)]^{-2}, \qquad (t \ll t_t) \tag{6}$$

where H_0 is the pressure scale height at the surface. The fraction $1 - \alpha$ of the blocked heat flux is stored in the convection zone. For spots with a radius larger than their depth, the value of α is expected to be less than given by (6) by a factor of the order d/r, so that (6) can be regarded as an upper limit. As an example, for the Sun ($H_0 = 150$ km), spots with a depth of only 1000 km yield $\alpha = 0.05$: 95% of the blocked flux would be stored even for such shallow spots. Note that the bright ring effect decreases with the assumed depth of the spot. This is because a diversion of the heat flux is associated with a temperature change that decreases with this depth (c.f. the discussion near eq. (3)).

Though we do not know enough about sunspots to tell at which depth d their blocking effect on the heat flux disappears, these estimates show that it is highly unlikely that a significant amount of heat is radiated in bright rings for any kind of sunspot- like starspot.

3.1 Variations in color and stellar radius

The absence of bright rings implies that on the short term $t \ll t_t$ the temperature of the unspotted surface is unchanged:

$$\Delta T_s/T_s \ll \Delta L/L \qquad (t \ll t_t). \tag{7}$$

Since the temperature structure of the envelope does not change either, hydrostatic equilibrium implies that the stellar radius is unchanged:

$$\Delta R/R \ll \Delta L/L \qquad (t \ll t_t). \tag{8}$$

The photometric variability in all channels is therefore due entirely to the spots themselves.

4. Faculae

Faculae on the Sun are areas of enhanced emission. Their effect on the internal structure is in principle identical to that of spots, but with the opposite sign. The net effect on internal structure is just the sum of the two, which is smaller than that of spots alone.

The main difference is that the facular excess emission probably derives from a depth that is not as large as the blocking depth of spots, so that they may be surrounded by 'dark rings' (the equivalent of the spot's bright ring). Calculations (Spruit, 1976; Grossmann- Doerth et al. 1989) indicate that these bright rings are narrow, and compensate only a fraction of the excess emission in the facular structure. This agrees with the finding (Lean, this volume) that faculae and the active network account for an important part of the solar luminosity variation. They even dominate the variation on time scales of years, so that the Sun is brighter at maximum activity. One may wonder if this is also the case for very active stars, because it would affect the interpretation of photometric variations on longer time scales (longer than the rotation period, for which the identification with dark spots is supported by spectroscopic information).

Facular areas on the Sun are bright only as long as the filling factor of the magnetic field is less than about 30% (Frazier, 1971). In areas where more magnetic field erupts the faculae are replaced by dark pores. If this is also the case in active stars, one would expect faculae to be less important than they are for the Sun.

5. Conclusions

Except in the early F stars the convective envelope is deep enough that its thermal time scale is much longer than any observable time scale. As a result variations in spot coverage on the surface have no effect on the internal structure of the star. Only the average on very long time scales has an influence on the internal temperature, the temperature on the unspotted part of the surface and the radius of the star. Under minimal assumptions about the depth of heat blocking effect of starspots (which appear to be satisfied by the Sun) there are no bright rings around starspots. On observable time scales all photometric variations can be interpreted directly in terms of the fraction of the visible hemisphere covered by spots at each instant. Faculae have the opposite effects of spots, but they may not be as important on very active stars as they are on the Sun.

Changes in radius, luminosity and color on long time scales ($\gg 10^5$ yr for the Sun) due to large spots change the star's position in HRD and color-color diagrams. These changes are predicted to be measurable but rather modest even for very large spots.

References

Chiang, W.H., Foukal, P.: 1984, Solar Phys. 97, 9

Foukal, P.V., Fowler, P., Livshits, M.: 1983, ApJ, 267, 863

Frazier, E.N.: 1971, Solar Phys. 21, 42

Grossmann- Doerth, U., Knölker, M., Schüssler, M., Weisshaar, E.: 1989, in *Solar and Stellar Granulation*, R. Rutten, G. Severino (eds.). Kluwer, Dordrecht, p. 481

Kippenhahn, R., Weigert A.: 1990, *Stellar Structure and Evolution*, Springer, Berlin, Ch. 10.3

Spruit, H.C.: 1976, Solar Phys. 50, 269

Spruit, H.C.: 1977, Solar Phys. 55, 3

Spruit, H.C.: 1982a, AA, 108, 348

Spruit, H.C.: 1982b, AA, 108, 356

Spruit, H.C.: 1991, in C. Sonnett and M. Matthews (eds.) *The Sun in Time*, University of Arizona, Tucson, in press

Spruit, H.C., Ritter, H.: 1983, AA, 124, 267

Spruit, H.C., Weiss, A.: 1986, AA, 166, 167

Soderblom, D.R.: 1989, ApJ, 342, 823.

(l to r) J. Landstreet and J. Staude.

THE SYNCHRONIZATION OF BINARY COMPONENTS FROM THE MOTION OF STARSPOTS

Frans van't Veer[1] and Carla Maceroni[2]
[1]Institut d'Astrophysique - CNRS
98bis Boulevard Arago, F-75014 Paris, France

[2]Osservatorio Astronomico di Roma
Viale del Parco Mellini 84, I-00136 Roma, Italy

It is generally thought that for close late type binary components, spin-orbit synchronization is achieved for orbital periods of some 5 or 10 days and less (see Giuricin et al., 1984). The torque acting on the tidal bulge of the gravitationally deformed components is indeed the reason that the spin-down of the initially fast rotating components, due to angular momentum loss (AML), will be stopped by the action of transfer of angular momentum (AMT) from the orbit to the components. Therefore, we are confronted with the situation that at every instant of an angular momentum losing spin-orbit coupled (AMLOSC) binary, the dynamical evolution takes place according to the three equations expressing, (1) the AML of the components, (2) the AMT from orbit to spin, and (3) the decrease of the orbital period P_K as a consequence of the AMT.

These equations are given and commented on in three preceding papers on the subject (van't Veer and Maceroni, 1988, 1989, 1990) and will not be repeated here. They can be combined to give two differential equations describing entirely the variation with time of the period of both the components P and the orbit, and hence also the degree of synchronization $f = P_K/P$. This so called period evolution function (PEF) clearly depends on the parameters governing the AML and AMT (see Figure 1 for an example). On the other hand this PEF gives us the possibility to compute the present period distribution (PPD) from the initial period distribution (IPD) of AMLOSC binaries.

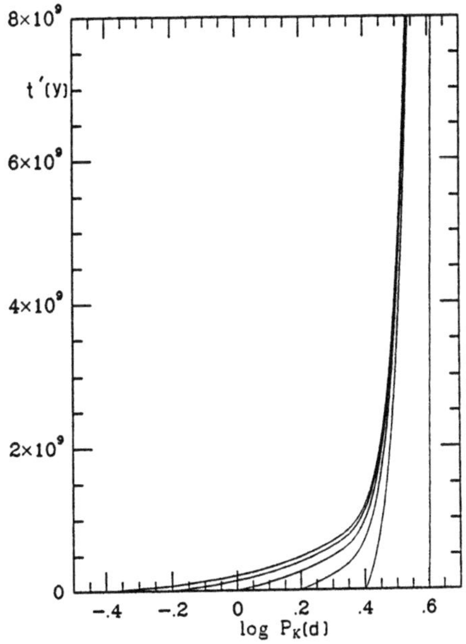

Figure 1. Variation of the orbital period with time (PEF) for G5V binaries calculated for AML law *a* shown in Figure 2.

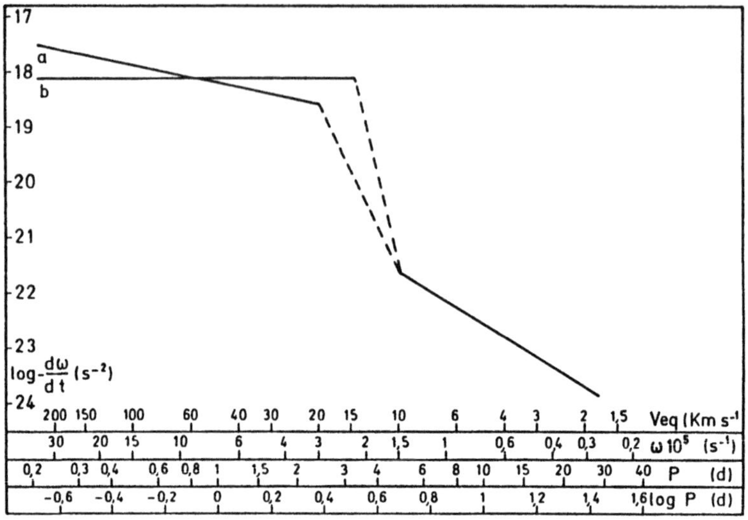

Figure 2. Two solutions for the spinning down of the outer layers of solar type stars (*see text*).

The PPD corrected for detectability and selection effects (van't Veer and Maceroni, 1988) shows a remarkable and significant bimodal character. In order to reproduce this bimodal distribution we absolutely need an AML for the fast rotating components of very close binaries, much higher than what is obtained after extrapolation towards rapid rotators of the canonical Skumanich relation, derived for slow rotators, and which suggests a power law of 3 ($\dot{\omega} \propto \omega^3$).

So we were obliged to introduce an AML function of the type as given in Figure 2, with an increase of AML from $P_K = 5^d$ to $P_K = 3^d$ by a factor greater than 100 with respect to the power 3 relation. This rapid increase is not without consequences on the synchronization. When there is a sudden increase of the AML of a binary component this component will immediately spin down more rapidly and, hence, the difference $\omega - \omega_K$ will become greater. From the

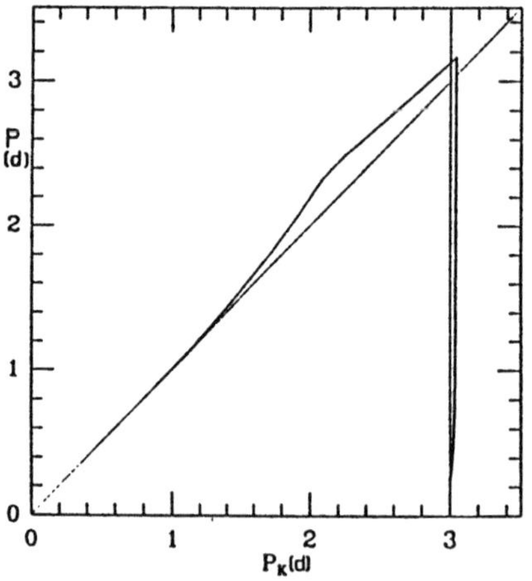

Figure 3. Variation of stellar spin and orbital revolution of an AMLOSC solar-type binary computed with AML law *a* (Figure 2). The timescale can be read from Figure 1.

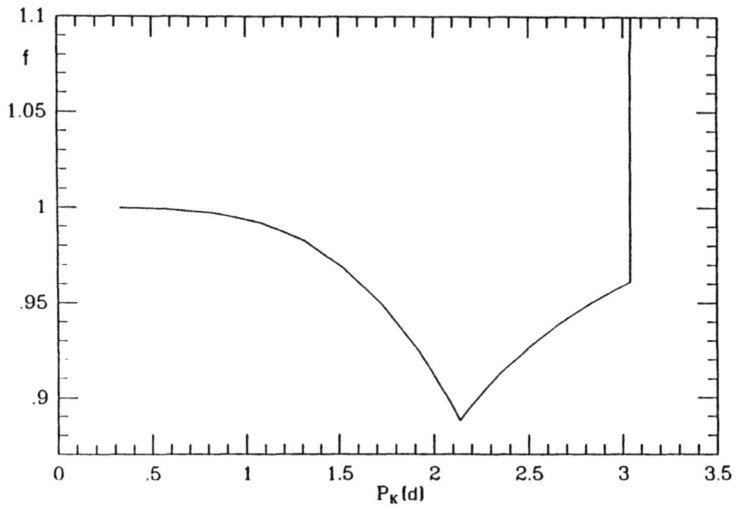

Figure 4. Variation of the degree of synchronization computed for the binary studied in Figure 3.

expressions for tidal synchronization (Zahn, 1977; Savonije and Papaloizou, 1985) we know that AMT increases at the same rate as $\omega - \omega_K$ and hence a new equilibrium will be reached with more rapid period evolution and a smaller f (Figures 3 and 4).

The sudden increase of AML we have found for the fast rotating components of very close solar type binaries is confirmed by studies of the AML of fast rotating T Tauri stars (Stauffer and Hartmann, 1987). Possibly in both cases the same type of mechanism switches on when the rotation exceeds a certain angular velocity. This mechanism could be related to the development of the stellar dynamo, the mass loss, the geometry of the magnetic field, or all three at the same time.

As we remarked already, the AML jump of a close binary component is accompanied by a sudden decrease of f and it is this change of f that needs to be measured in order to better understand the mechanism of AML. Measurements of the synchronization can be made by comparing the rotational velocities of the components with the orbital period of the system. In this case the radius of the component has to be estimated and the precision is certainly less than 10%.

A more precise method is given by the measurements of f with the aid of starspots or other surface inhomogeneities that can be detected and localized by spectral imaging methods or by the determination of the stellar rotation from a study of the photometric wave. Long series of observations have already permitted to detect the differential rotation with latitude. We predict that from observations of solar type binaries with periods around 2^d it will be possible to obtain information concerning f and hence AML from the motion of spot groups compared with the orbital motion. The measurements available at the present time (Strassmeier et al. 1988, 1989) are still too incomplete to permit definite conclusions.

What we can deduce from the still rather scarce data available at the present time is that synchronization for binaries with circularized orbits is good for the great majority of cases. The deviations rarely exceed some percent and when this image is confirmed, as we expect it, by future observations we will be forced to admit that the value of the parameter used to compute the AMT (Zahn 1977) is too small.

A reasonable explanation for this failure to confirm the theory by observations could be the interaction of strong magnetic fields of the components when they become very close to each other.

References

Giuricin, G., Mardirossian, F., Mezzetti, M.: 1984, A&A, 141, 227

Savonije, G.J., Papaloizou, J.C.B.: 1985, *Interacting Binaries,* eds. P.P. Eggleton and J.E. Pringle, (D.Reidel Publishing Company) p. 83

Stauffer, J.R., Hartmann, L.W.: 1987, PASP, 98, 1233

Strassmeier, K.G., Hall, D.S., Zeilik, M., Nelson, E., Eker, Z., Fekel, F.C.: 1988, A&AS, 72, 291

Strassmeier, K.G., Hall, D.S., Boyd, L.J., Genet, M.: 1989, ApJS, 69, 141

van't Veer, F., Maceroni, C.: 1988, A&A, 199, 183

van't Veer, F., Maceroni, C.: 1989, A&A, 220, 128

van't Veer, F., Maceroni, C.: 1990, *Proceedings NATO-ASI, Kusadasi, Turkey* eds. C. Ibanoglu and I. Yavuz (Kluwer-Dordrecht), in press

Zahn, J.-P.: 1977, A&A, 57, 383

Discussion: van't Veer

Young In the Strassmeier *et al* catalog of RS CVn binaries, the vast majority are synchronous, but of those that are not synchronous by significant amounts, about 1/2 of them are super-synchronous and 1/2 are sub-synchronous. Why does such a bifurcation exist? Can we generalize anything about these two categories, such as age?

van't Veer Super-synchronous rotation may be expected in certain cases from differential rotation effects, but not more than a few precent. It may also be expected in a very limited number of cases for young systems with components that are still slowing down. What is expected from theory is completely different: a non-negligable number of systems with $f < 90\%$.

Linnell There is not universal agreement that contact binaries are the evolved descendants of detached components. We lack good data on angular momentum loss rates for contact systems. It is uncertain whether the Skumanich law applies to these short period systems.

Cameron This concept of a switch to a more efficient but period-independent braking mechanism at high rotation rates has also been invoked to explain the distribution of rotation rates among the G dwarfs in the α Per cluster and the Pleiades K dwarfs. Can the same empirical braking laws be used to explain both the bimodal binary period distribution and the cluster rotation distributions?

van't Veer This is exactly what we are thinking, but the mechanisms of nature are always richer than we can guess.

Spruit Cannot selection effects distort the apparent number distribution of binary periods, and make a derived angular momentum loss function unreliable?

van't Veer The period distribution we used was corrected for selection and detectability effects by us (van't Veer and Maceroni 1988 AA, 199, 183).

Budding You did not refer to it by name, but I think your talk covered the phenomenon of the so-called 'Svechnikov-gap'. Of course this should only apply for the cooler, convective envelope type stars, and I think its explanation has been linked to the role of magnetic braking. Where I find the scenario somewhat troubling is with regard to the apparent over-abundance of contact binaries. According to most of the originally used formulae AML should now really run away with itself. Is this a problem for you?

van't Veer The overabundance of contact binaries with respect to detached binaries is only apparent. They are much easier to detect for the light is continuously varying and the probability to show eclipses is much greater. When we properly correct for these two effects they have normal abundance in agreement with their predicted duration of life.

(l to r) M. Kürster asks A. Collier Cameron not to bite his finger. A. Hatzes is amused but S. Brown decides to walk away!

SPECTROSCOPIC SIGNATURES OF ACTIVE REGIONS ON MAIN SEQUENCE STARS

Mark S. Giampapa
National Solar Observatory
National Optical Astronomy Observatories
P.O. Box 26732, Tucson, AZ 85726-6732, USA

I. Introduction

I will briefly comment on the implications of some recent datasets for the morphology of the chromospheres of late-type, main sequence stars. I will assume that chromospheres themselves are spatially associated with sites of emergent magnetic flux on stellar surfaces, as is the case for the Sun. These sites correspond to thermal inhomogeneities as perceived in white light (cool spots) or in a chromospheric line (*e.g.* CaII H and K). An understanding of the line formation processes that give rise to key chromospheric diagnostics can yield clues to both the level of non-radiative heating which is present, and the surface distribution and physical properies of magnetic "active regions". In the following, I will offer interpretations of recent Hα observations of M dwarfs in the context of the nature of the thermal inhomogeneities associated with the magnetic field configurations that may characterize the atmospheres of these low mass stars.

2. Discussion

2.1 Lateral inhomogeneities vs. radial segregation in M dwarf stars

A comprehensive observational study of M dwarfs in the solar neighborhood by Stauffer and Hartmann (1986) disclosed systematic trends in chromospheric activity as a function of fundamental stellar properties such as mass and chemical composition. Among these trends is a decline in chromospheric Hα absorption line widths toward later colors (see Stauffer and Hartmann 1986, their Figure 11). This property of the Hα line in dM stars is reminiscent of the Wilson-Bappu effect which, in the case of the dwarf M stars, has been observed in the CaII K line (Giampapa et al. 1981; Fleming and Giampapa 1989) and in transition region diagnostics (Ayres et al. 1983).

There is, however, another striking aspect to their diagram, namely, the substantial scatter in Hα absorption line widths at a given (R - I) color. The origin of the scatter merits consideration in the context of stellar chromospheric structures. In particular, if the Hα absorption line originated in the same kind of active region on M dwarfs of a given color then the profile shape should remain constant. The *equivalent width* would vary in proportion to the fractional area coverage. or "filling factor", of these regions on the stellar surface (Giampapa 1985). Conversely, if the surface is composed of a range of active regions that may include both Hα absorption and emission "plages", analogous to solar plage, then both the line *shape* and the equivalent width can vary. More specifically, the filling in of an Hα absorption profile, similar to that seen in a normal dM star, by Hα emission from another active region may yield, as seen in the integrated stellar spectrum, a net weak absorption profile of greater width than the absorption line alone. Hence, we would predict that line widths should *increase* with decreasing absorption equivalent widths. Alternatively, globally uniform chromospheric heating rates and the resulting *radial structure,* rather than lateral inhomogeneities, may be the principal factor that determines chromospheric Hα line profile shapes and strengths in the field dM stars.

In order to test this hypothesis, I construct in Figure 1 a diagram of Hα line widths *versus* line strengths (equivalent widths) utilizing the data tabulated by Stauffer and Hartmann (1986,

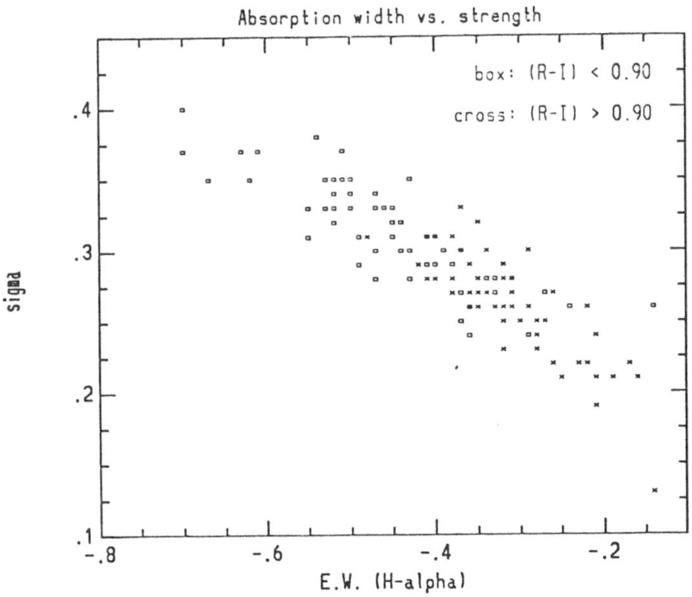

Figure 1. Gaussian width of the Hα absorpion line *versus* equivalent width. Data from Stauffer and Hartmann (1986).

their Table 1). I differentiate between stars with colors earlier or later than (R - I) = 0.90, or about dM2, in order to examine any color-dependent effects. Such effects may be reflected in Figure 1 given (1) the general decline in Hα absorption equivalent widths toward later spectral types and (2) the aforementioned decline in line widths toward lower luminosities (see Stauffer and Hartmann 1986, their Figures 8 and 11, respectively).

Inspection of Figure 1 reveals a clear trend of *decreasing* line widths toward weaker chromospheric Hα absorption line strengths. While there is still considerable scatter, the trend is opposite to that expected from an inhomogeneous atmosphere composed of both Hα emission and Hα absorption regions. The Hα line profile, which essentially represents a mapping of the source function, is determined in part by Doppler broadening and the total optical thickness of the chromosphere (Cram and Mullan 1985). The trend in Figure 1 is, in fact, consistent with line broadening on microturbulent scales. From the perspective of stellar chromospheric line formation, the results imply that it is principally the ***radial segregation*** of the atmosphere that determines the properties of the net chromospheric spectrum seen in low mass stars. Any lateral inhomogenities on the surfaces of the dM stars must therefore be characterized by relatively small spatial scales.

2.1 The "marginal" BY Draconis stars

I will now discuss aspects of the outer atmospheres of an apparently rare class of dwarf M stars that Bopp (1987) has referred to as the "marginal" BY Draconis stars. These intriguing objects exhibit a weak Hα feature yet have the kinematics and metallicities representative of the young disk galactic population. Because of the nature of Hα line formation in cool dwarfs, a weak feature can indicate a relatively low level of chromospheric heating or it can represent a level of activity which is intermediate between a dMe star and a dM star with strong Hα absorption. Subsequent observations of the CaII resonance lines by Giampapa et al. (1989) confirmed the conjecture that these stars are, in fact, characterized by an intermediate degree

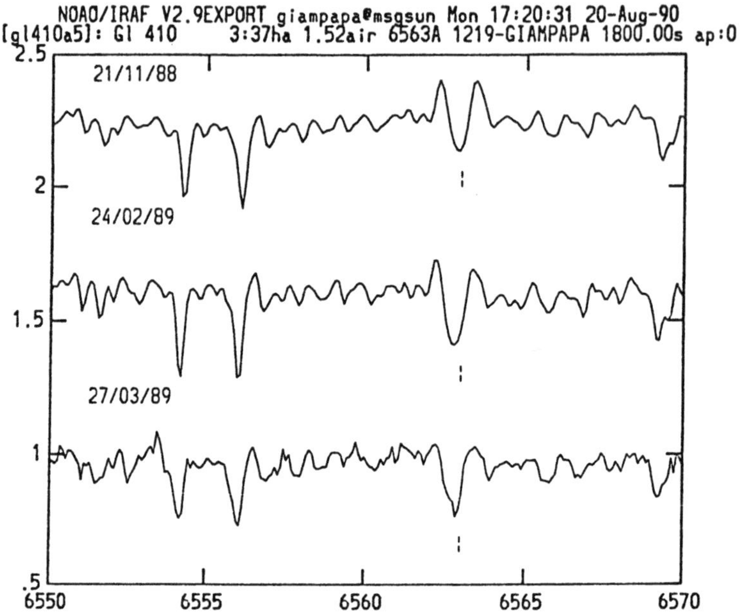

Figure 2. The Hα absorption in Gl 410 as seen on three different UT dates. The line is marked. The spectra were obtained with the NSO McMath/CCD and solar-stellar spectrograph system.

of chromospheric activity.

In order to gain further insight on the nature of the outer atmospheres of this class of dwarf M stars, I implemented a program of synoptic spectroscopic observations of the Hα line in selected objects (Gl 410, Gl 900 and Gl 907.1) using the NSO McMath 1.5 m telescope and the solar-stellar spectrograph in conjunction with a CCD detector. The spectral resolution of these observations was $\lambda/\Delta\lambda \simeq 42\,000$ at Hα. Over 70 spectra were acquired during two observing seasons. An example of a spectrum in the Hα region for the prototype, Gl 410 (see Young et al. 1984), is displayed in Figure 2. At these high spectral resolutions we see that the line is composed of weak dual-peaked emission and a central absorption feature that extends below the local continuum level.

The shape of the profile suggests three different schematic models for the structure of the chromospheres in these stars. In particular, such a feature might arise from (1) a stellar surface consisting of both Hα emission and Hα absorption active regions; (2) extensive Hα prominences that are seen in absorption against the disk but emit blue-shifted and red-shifted emission off the limb; or (3) a single-component, compact chromosphere. While the synoptic observations revealed variability in both the total strength of the Hα line and the position of prominent features in the profile, no periodic variability has as yet been detected. This would appear to argue against model (1). In addition, simulations of representative Hα absorption and emission profiles combined to yield a net "observed" profile fail to satisfactorily fit the Hα profiles observed in Gl 410.

The velocity positions and shifts of the peaks of the emission are consistent with prominences filled with Hα-emitting material extending $\sim 1\,R_*$ above the limb (assuming an "intermediate" rotational velocity of about $5\,\mathrm{km\,sec^{-1}}$). Actual calculations of Hα profiles in prominences will have to be performed to explore this model further. In the case of (3), single-component, semi-empirical model chromospheres can likely be constructed which will yield profiles similar to the observed profile. Following Cram and Mullan (1979), such profiles would arise only in a narrow range of possible chromospheric physical conditions. I note an additional observation that a static model chromosphere would have difficulty explaining, namely, the occasional appearance of asymmetric emission.

Further observations combined with realistic calculations will be required to distinguish between models (1) - (3). Regardless of which model is correct, the fundamenal question that still remains is the following: what are the special conditions that produce the chromospheric structures in these stars and why are these enigmatic objects rare?

3. Summary

The observed characteristics of Hα in M dwarf stars combined with a consideration of chromospheric line formation in cool dwarfs suggests that it is radial structure, rather than lateral inhomogeneities, which is the predominant factor in controlling "chromospheric activity" in low mass stars. This conclusion implies, in turn, that chromospheres determined by a single heating rate are relatively widespread on the stellar surface. The implications for magnetic field structures depend upon whether there is a unique association between chromospheric heating rate and specific kinds of magnetic field configurations.

Synoptic observations of the "marginal" BY Draconis stars revealed irregular variability in the total strength and the positions of features in the their unique chromospheric Hα line. Further work must be done in order to ascertain the kind of chromospheric structures that yield the range of observed profiles. The special, and apparently rare, conditions that give rise to this subset of the low mass dwarfs remaim elusive.

Acknowledgements I acknowledge the ongoing, dedicated efforts of our nighttime synoptic observer at the NSO/McMath telescope, Mr. Paul Avellar, who obtained the Hα observations of the intermediate M dwarfs discussed herein. I am grateful to Mr. E. DeFonso for his assistance with the reduction of the McMath data. I thank the staff of the Armagh Observatory, especially Brendan Byrne and Gerry Doyle, and the Royal School for their kind hospitality and excellent meeting organization. Finally, I must tell the reader that the NOAO is operated by the Association of Universities for Research in Astronomy, Inc., under cooperative agreement with the National Science Foundation.

References

Ayres, T.R. et al.: 1983, ApJ, 270, L17

Bopp, B.W.: 1987, ApJ, 317, 781

Cram, L.E., Mullan, D.J.: 1979, ApJ, 234, 579

Cram, L.E., Mullan, D.J.: 1985, ApJ, 294, 626

Fleming, T.A., Giampapa, M.S.: 1989, ApJ, 346, 299

Giampapa, M.S., et al.: 1981, ApJ, 246, 502

Giampapa, M.S.: 1985, ApJ, 299, 781

Giampapa, M.S., Cram, L.E., Wild, W.J.: 1989, ApJ, 345, 536

Stauffer, J.R., Hartmann, L.W.: 1986, ApJS, 61, 531

Young, A., Skumanich, A., Harlan, E.: 1984, ApJ, 282, 683

PHYSICAL MODELS OF SOLAR AND STELLAR SPOTS

Andrew Skumanich
High Altitude Observatory
National Center for Atmospheric Research
Boulder, Colorado 80307 U.S.A.

1. Introduction

The basic observational parameters that one has in the case of star spots are, in the mean, their *total area, temperature*, and, in some cases, *magnetic field* strength. Thus, any theoretical analysis of spot structure should bear in mind the issue of how such parameters might scale from star to star. Most, if not all, theoretical studies address the nature of spots under the solar condition because this is the circumstance which is known in resolvable detail. In this review I consider such studies, but from the point of view of how their results might scale to other than solar conditions. This may be somewhat presumptuous given the continuing situation that "sunspots remain an absurdity after so many years in that the structure remains unexplained by anything more than unproven conjectures" (Parker 1978).

The present observational picture of the magnetic structure of sunspots is reviewed first. It is argued that the continued use of the Beckers-Schröter (1969) magnetic profile in spot models is untenable. The dipole representation of the recent magnetic profile data, as well as the magnitude of the field aligned currents for 'twisted' spots, is also reviewed. Next, the theoretical picture of both the subsurface and above-surface magnetic structure is reviewed for a variety of theoretical "conjectures" including both current sheet and distributed current models. It is found that lateral force balance leads to a fairly simple relation in which the product of the square-root of the pressure excess and the area of the flux tube is essentially constant with depth and equal to the emerged magnetic flux. Thus, the *mean* field strength (flux per unit area) should scale as the square-root of the pressure *not only* for thin flux tubes *but also* for spots. Observational evidence for such a scaling is reviewed. The issue of the magnitude (on the average) of the erupted flux, and hence of spot area, must be addressed by dynamo models and is outside of the scope of the present review.

2. Observational Picture of Spots

The observation by Beckers and Schröter (1969, hereafter BS) of a spot magnetic profile in which the surface field magnitude falls off with the lateral radius as $(1+(R/R_p)^2)^{-1}$, where R_p is the observed penumbral radius, continues to have currency, in spite of many other observations to the contrary, due, perhaps, to the model dependent nature of the inversion of the Stokes line polarization profiles. However, recent *direct*, i.e., model-free Zeeman *splitting*, measurements by Deming *et al.* (1988, hereafter DBJW), are found to confirm the Stokes results.

In Figure 1, I compare the field magnitude profile derived by Lites and Skumanich (1990, hereafter LS) from Stokes II profiles for the penumbra of their 1980 March 09 spot with the direct measurements of a penumbral field by DBJW. The LS profile, given by the solid curve, has been scaled by a constant factor 1.134 to allow a comparison with the DBJW data (indicated by open circles). It is apparent that within the scatter due to departures from azimuthal asymmetry, the field profiles are quite consistent implying that the LS shape is not significantly influenced by model dependencies. Recent observations by Adam (1990) supports the LS results. Further independent confirming observations are cited by LS and are not repeated here.

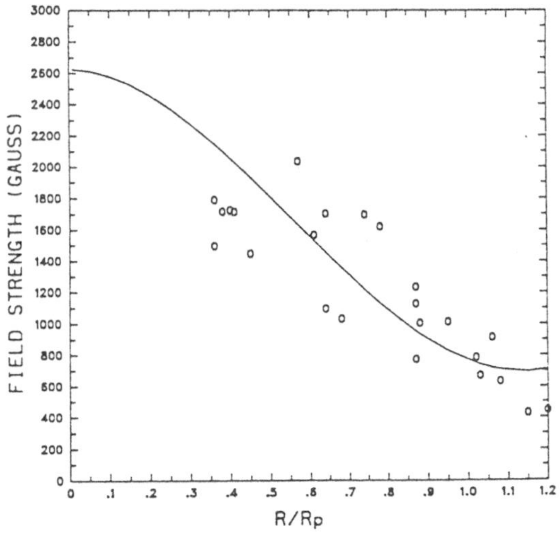

Figure 1. Comparison of Stokes inversion penumbral field with direct measurement.

The BS profile cited above could be interpreted as due to a magnetic monopole buried at a depth $Z = -R_p$ below the photosphere, $Z = 0$. However, the LS (and other) profile data cannot be so represented. In Figure 2 we show the LS radial profile for the vertical (to the solar surface) field $B_Z(R, Z = 0)$, given by the inverted triangle points, as well as the radial field, $B_R(R, Z = 0)$, given by the open circles, for the LS 1980 Sept. 15 spot. These are based on inclinations whose radial behavior is well-confirmed by others, see LS and Adam (1990). The solid curve represents a *dipole* fit to the B_Z profile for $R \leq R_p$.

Taking the flux potential $u(R, Z) = \tilde{p}R^2/((Z + |Z_d|)^2 + R^2)^{3/2}$ where $\tilde{p}$ is the dipole strength and $-|Z_d|$ the depth of the virtual dipole-source (i.e., of a sufficiently small ring current), LS

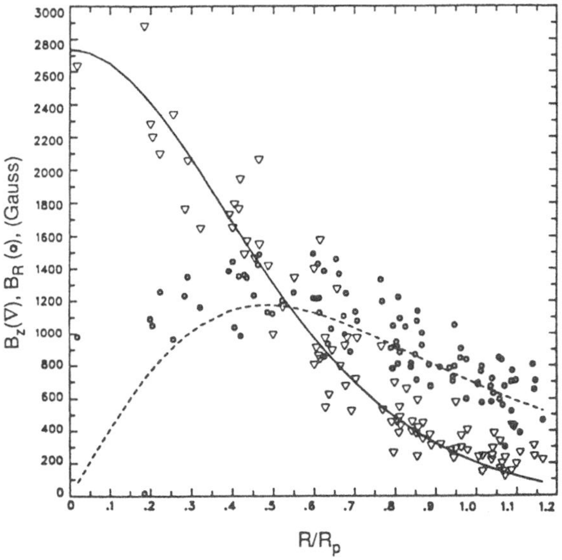

Figure 2. Solid line — Dipole fit to $B_Z(0, R)$ field for 1980 Sept. 15 spot. Dashed line — Predicted dipole radial field component.

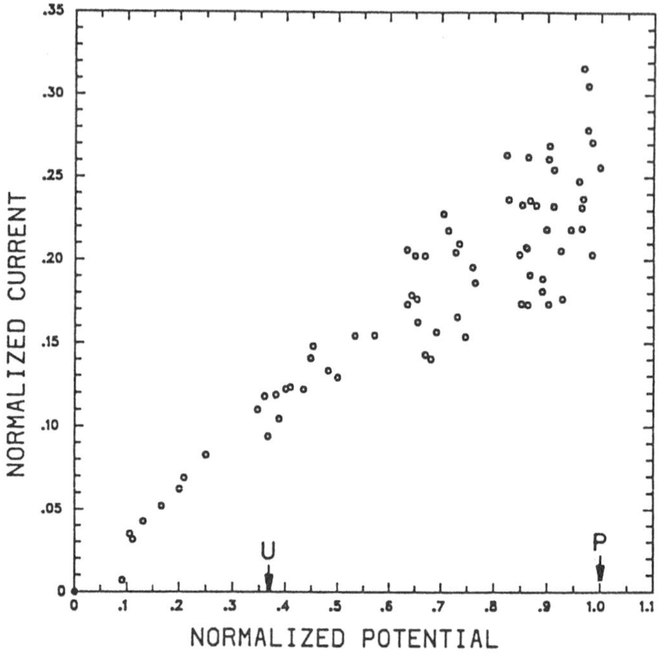

Figure 3. Observed field aligned current for 1980 Sept. 15 spot.

find that $\tilde{p} = -\frac{1}{2}B_{00}|Z_d|^3$ with $B_{00} = 2740$ G and $Z_d = -0.96R_p$. Note that the definition of u is such that $B_Z(R,Z) = -1/R\ \partial u/\partial R$ and $B_R(R,Z) = 1/R\ \partial u/\partial Z$. The emerged flux $\Phi = \int_0^{R_p} B_Z 2\pi R dR$ turns out to be 10^6 GMm^2 (or 10^{22} Maxwells) with $B(0,0) \equiv B_Z(0,0) \equiv B_{00}$. The observed penumbral radius was 19 Mm.

The dashed curve in Figure 2 gives the dipole radial field distribution and appears to represent the radial data fairly well. Note that different azimuths are plotted at a given R and that such azimuthal scatter due to the imperfect azimuthal symmetry precludes an accurate measure of the departures from the dipole. The new HAO Stokes polarimeter with its $\approx 0.''8$ resolution should allow a more accurate study of such departures. Thus, to a first approximation, the surface distribution of a spot magnetic field appears to be more dipolar than monopolar.

Some single isolated spots show an azimuthal field $B_\phi(R,0)$, i.e., they have a twist. If one interprets the azimuthal field in terms of field aligned Ampere current, i, one can write $B_\phi = i(u)/R$ to allow for variations from field line to field line identified by constant values of u. In Figure 3, we present the Ampere current, normalized by u_p/R_p ($= 0.163B_{00}R_p = 8.48$ TeraAmps), for one of the LS spots, in terms of the derived normalized dipole potential u/u_p where u_p represents the field line at the penumbral radius, $u_p = u(R_p, Z)$. Note that the dependence is approximately linear and thus similar to that used by Yun (1971) in his theoretical study of twisted spots. Note that the observed linear slope is $A = R_p di/du = 0.3$.

In this review we will not consider the more general case of spot pairs or other interacting three-dimensional configurations. A recent review by Low (1985) covers these cases.

3. Theoretical Picture of Spots

To the theoreticians, spots become idealized objects that are described by the MHD equations for force and energy balance. Thus, equations such as

$$\frac{1}{4\pi}(\nabla \times \mathbf{B}) \times \mathbf{B} = \nabla P + \rho g \hat{Z} \ , \tag{1}$$

for force balance and

$$0 = -\nabla \cdot \mathbf{F}_{\mathrm{cond}} - \nabla \cdot \mathbf{F}_{\mathrm{rad}} - \nabla \cdot \mathbf{F}_{\mathrm{conv}} + G(P, \rho, \mathbf{B}) - L(P, \rho, \mathbf{B}) \ , \tag{2}$$

for energy balance, rather than sugar plums, dance in their heads. In the case of the radiative flux-density they know that $\mathbf{F}_{\mathrm{rad}} = \int_{4\pi} I(\mathbf{x}, \boldsymbol{\Omega})\boldsymbol{\Omega} d^2\Omega(\theta, \phi)$ with I given by the equation of transfer

$$\boldsymbol{\Omega} \cdot \nabla I = j - \kappa I \ , \tag{3}$$

while they write for the heat flux-density that $\mathbf{F}_{\mathrm{cond}} = -\mathrm{K}(P, \rho, \mathbf{B})\nabla T$, with a tensor conductivity $\mathrm{K}(P, \rho, \mathbf{B})$. However, less is understood about the remaining thermodynamic effects and these are parameterized, e.g., a reduced mixing length is used to model convective heat transfer in a magnetic field or Alfvén waves are used to model the field generated gain and loss terms G,L. Suitably idealized boundary conditions close this picture. It is not surprising that Parker refers to theoretical pictures as mere conjectures.

Given the difficulties in a suitable understanding and treatment of the energy equation, most theoretical attention has been directed at the implications of force balance. Not surprisingly, the cylindrically symmetric picture has received the most attention. In this case, the radial force balance, which effectively relates the electric current density to the presence of pressure gradients across magnetic field lines, becomes, refer Low (1975),

$$\frac{1}{4\pi}\left(\frac{\partial B_R}{\partial Z} - \frac{\partial B_Z}{\partial R}\right) B_Z = \left(\frac{\partial P}{\partial R}\right)_Z \tag{4}$$

while vertical force balance, which gives the hydrostatic relation along a field line, reads

$$\rho g = -\left(\frac{\partial P}{\partial Z}\right)_{\textit{along a field line}} \tag{5}$$

In terms of the flux potential $u(R, Z)$, now more general than the dipole form used above, (4) and (5) become, with $P(u, Z) = P(R(u, Z), Z)$,

$$\frac{\partial^2 u}{\partial Z^2} + R\frac{\partial}{\partial R}\left(\frac{1}{R}\frac{\partial u}{\partial R}\right) = -4\pi R^2 \left(\frac{\partial P}{\partial u}\right)_Z \tag{4'}$$

$$P(u, Z) = P(u, Z_l) \exp\left\{ -\int_{Z_l}^{Z} dZ'/h(u, Z') \right\} \tag{5'}$$

with the thermodynamics specified by a given function

$$h(u, Z) = \frac{k}{g}\frac{T(u, Z)}{\mu(u, Z)} \ .$$

The value of $P(u, Z_l)$ represents the boundary condition at some depth Z_l. The boundary conditions on u must also be specified. One usually requires u to become force-free, if not potential, above the photosphere, since the pressure drops below the magnetic pressure very rapidly. A constant field with depth is the lower condition.

The simplest force-balance models of spots are the current sheet (CS) or "solenoid" models. Here the right-hand side (rhs) of equation (4′) is zero everywhere except for being singular on the boundary of the spot, i.e., at the current sheet. It is easy to show that the force-balance condition requires that the (tangential) magnetic field $B_t(R_s(Z), Z) = \sqrt{8\pi\Delta P(Z)}$ where $\Delta P(Z)$ is the jump, $P_{ex}(Z) - P_{in}(Z)$, in pressure across the current sheet at radius $R_s(Z)$.

A particularly simple example (Low 1990) is a dipole field which is truncated at some field line. We consider the LS spot and cut the field at the field line, u_p. Using the LS dipole parameters one can then derive the height variation of the pressure jump that is in equilibrium with a flux of 10^6 GMm^2 having the properties of the LS spot at the surface. This $\Delta P(Z)$ is illustrated in Figure 4 by the curve marked by *a*. The shape of the flux tube, i.e., $R_s(Z)$, is indicated in Figure 5, while the shape of the surface magnetic profile, $B(R,0)/B_0$ (the LS profile), is indicated in Figure 6, in both cases, by *a*.

Schmidt and Wegmann (1983, hereafter SW) considered a more general variation of $\Delta P(Z)$, refer to curve marked by *b* in Figure 4. In this case, the field is non-dipolar and one must solve equation (4′) for $u(R, Z)$ with the specified surface value $B_t(R_s(Z), Z) = \sqrt{8\pi\Delta P(Z)}$ except that the location of the surface, i.e, $R_s(Z)$, is unknown and must also be solved for. For $\Delta P(Z)$, SW assume that $P_{in}(Z) = 0$ and that $P_{ex}(Z)$ is the normal quiet sun pressure stratification. Their solution for the flux tube radius $R_s(Z)$ and the resulting surface field profile is indicated in Figures 5 and 6 by the curve label *b*. We note that the consequence of an excess pressure larger than that for the cut-dipole case is a significant constriction of the flux tube. The two cases are not quite comparable as SW consider a flux of $\Phi = 2 \times 10^6$ GMm^2 so that the one would expect even more constriction for $\Phi = 10^6$ GMm^2, than pictured.

Jahn (1989) extended the work of SW and introduced a finite internal pressure calculated from an energy equation along the same prescription that Deinzer (1965) and Yun (1968) used in their study of a distributed current model of sunspots. One introduces a mixing length parameter as an eigenvalue and solves for a convective envelope model for the spot atmosphere

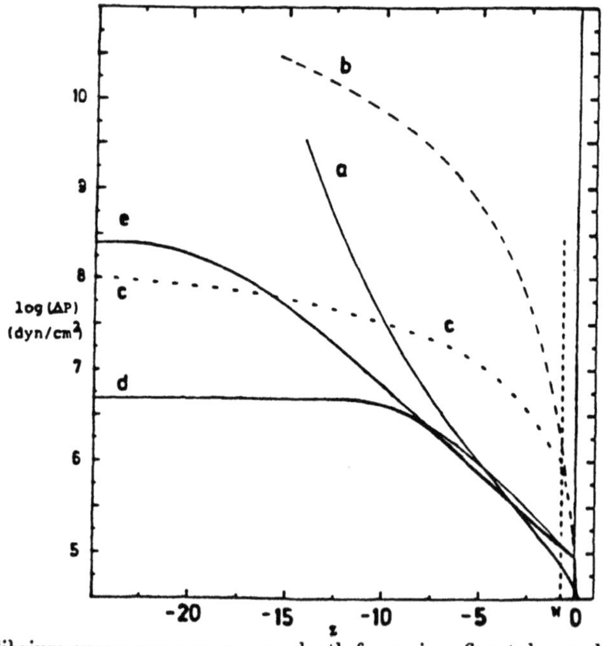

Figure 4. Equilibrium excess gas pressure vs. depth for various flux tube models: a) cut dipole fit to LS profile, b) SW model, c) Jahn convective model, d) and e) Jahn fit to BS profile. Quiet photosphere at $Z = 0$; W = Wilson depression.

with the appropriate boundary conditions. The curves marked by c indicate the resulting pressure distribution, flux tube radius and surface field profile in Figures 4, 5, and 6, respectively. Note that the pressure excess is similar to SW in the surface layers, i.e., equal to the quiet sun pressure, and differs significantly in the subsurface layers. Jahn studies the case $\Phi = 10^6$ GMm^2 and hence his solution is directly comparable to the cut-dipole case. Thus the deviations in the surface field in Figure 6 from the dipole distribution map into significant subsurface $\Delta P(Z)$ and $R_s(Z)$ variations. Finally, Jahn considers other pressure distributions as indicated by cases d, e. These were chosen to fit the BS surface field profile again with $\Phi = 10^6$ GMm^2 and to explore the role of the lower boundary condition.

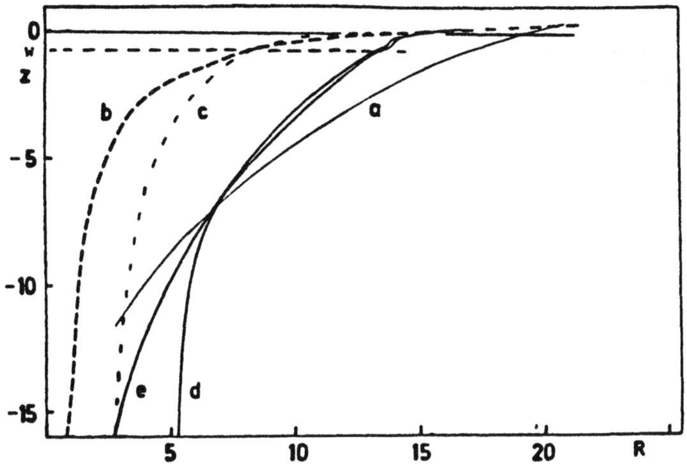

Figure 5. Equilibrium flux tube radius vs. depth for various models. Labels as in Figure 4.

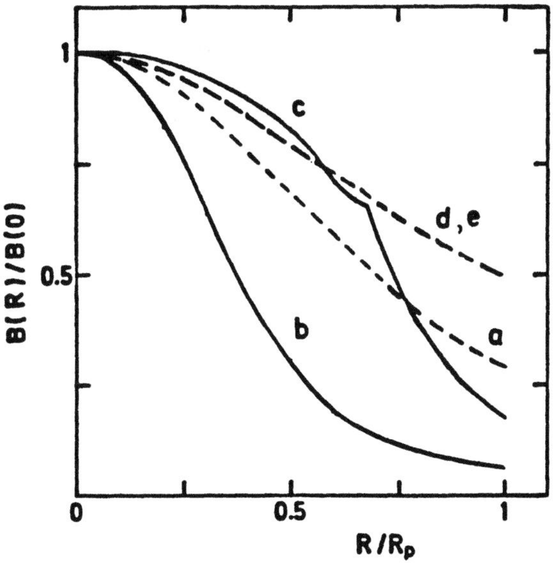

Figure 6. Radial profile of surface magnetic field for various models. Labels as in Figure 4.

Table 1.
Ratio of pressure-excess field to mean field ($\sqrt{\Delta\beta}$)

Case\Z(Mm)	0.3	0	$-W^1$	−2.5	−5	−10	−15	−25
LS Dipole (a)		0.80		0.85	0.89	0.95	1.00	1.00
SW (b)	0.50	0.80	0.70	0.95	1.20	1.10	1.00	
Jahn (c)	0.90	0.70	1.00	1.00	–	–	–	1.00
Jahn (e)	0.90	0.98	0.96	1.04	0.95	1.00	–	1.00
Yun (Distrib.)			4.35	3.64	3.28	2.85		

[1]W is the Wilson depression

A remarkable property of all these solutions emerges if we consider the quantity $\sqrt{\Delta\beta(Z)} \equiv \sqrt{8\pi\Delta P(Z)}\pi R_s^2(Z)/\Phi$ which may be interpreted as the ratio of the 'jump' field B_t to the mean flux tube field at Z. Note that $\Delta\beta(Z)$ can be interpreted as a *mean excess* plasma beta associated with the pressure excess. Table 1 presents this ratio for different depths as determined from the curves in Figures 4 and 5. One finds that this ratio is almost a constant! If one considers the inverse ratio, it would appear that lateral force balance yields a mean magnetic field with height that scales as the square-root of the confining pressure excess. In the surface layers, as we have seen, this is essentially the quiet sun photospheric pressure. Thus, we find that the *mean spot magnetic field scales with the square-root of the photospheric pressure*, i.e., $\bar{B} = \Phi/\pi R_s^2 = \sqrt{8\pi P(\tau \simeq 1)}$. This implies that *spot areas* are determined by the amount of *emerged flux*, the latter quantity presumably being controlled by the nature of the dynamo mechanism.

Note that the large 'contrast' in $B(R)/B(0)$ between $R = 0$ and $R = R_s(\equiv R_p)$ for the SW magnetic profile in Figure 6 seems inconsistent with the mean field given by the entry in Table 1 at $Z = 0$ (which may be in error), given that $B_t(R_s, 0) = B(R_s)$. Since the magnetic data was not published, I am unable to determine the reason for this inconsistency except to note that a curved $\tau = 1$ surface mixes different Z levels.

I believe that the above scaling relation is a universal *approximate* property of spot force balance. It should hold for distributed currents and hence distributed pressure models if one replaces the sheet radius by that for some suitably chosen flux surface to represent the umbral boundary. The simplest distributed models are ones with a single degree of freedom (hereafter 1D) based on Schlüter-Temesvary (1958, hereafter ST) similarity assumption which reduces equation (4′) to an equation solely in Z. Deinzer (1965) and later Yun (1968) applied such an approach to study the *subsurface* structure of spots. It was applied to a study of the above-surface or boundary-layer structure by Jakiemiec (1965), Yun (1972), Landman and Finn (1979), and Flaa, Osherovich, and Skumanich (1982, hereafter FOS). Only the last authors impose the appropriate asymptotic potential condition. More recently, distributed current studies of the boundary-layer structure with two degrees of freedom (hereafter 2D) have been carried out by Pizzo (1986), while 2D subsurface models with both a current sheet and internally distributed currents have been studied by Jahn (1989).

We consider Yun's 1968 1D study as it has an improved upper boundary condition compared to Deinzer. Yun uses the ST assumption with a radial shape for the field which is Gaussian in R but with an unknown Z-dependent width and defines the radius, R_p, of the flux tube to be given by the field line $u(R_p, Z) = 0.934u_0$, i.e., $\Phi = 0.93(2\pi u_0) \equiv 0.35B(0,Z)\pi R_p^2(Z)$ where the constant $2\pi u_0$ fixes the total flux. The factor 0.35 comes from the *observed* radial shape at the surface (refer to Yun 1968) and yields a radius at the surface which represents the penumbral radius in the photometric signature of such model spots, Pizzo (1986). The *magnetic flux* Φ(or u_0), normalized *surface gradient* $D = ((R_p/B_0)(dB_0/dZ))_{-W}$ (at the Wilson depression) where $B_0(Z) = B_Z(0,Z) = B(0,Z)$ is the axial field, and *effective temperature* of the spot T_{sp} are fixed parameters.

I have calculated $\sqrt{\Delta\beta(Z)} = \sqrt{8\pi\Delta P(Z)}\pi R_p^2(Z)/\Phi$ from Yun's tables (Yun 1968) of $\Delta P(Z)$ and $B_0(Z)$. Here the confining pressure excess is now given by $\Delta P(Z) = P(R \rightarrow \infty, Z) - P(0,Z)$ where $P(R \rightarrow \infty, Z)$ is the quiet sun pressure and $P(0,Z)$ is the spot axial pressure which is obtained from the simultaneous solution of the 1D energy equation. The case $T_{sp} = 4000$ K, $D = 4.75$ and $\Phi = 0.93 \times 10^6$ GMm^2 is listed in Table 1. A comparison with the CS results indicate that $R_p(Z)/\sqrt{2.85} = 0.59R_p(Z)$ is the equivalent to $R_s(Z)$. Note that $R_{\text{umbra}} = 0.4R_p$.

We see that $\sqrt{\Delta\beta(Z)}$ is less slowly varying than in the current sheet case but none-the-less, an approximate square-root scaling holds. If I consider the other parametric values of D, Φ, and T_{sp}, studied by Yun, I find that the *surface value* scales as

$$\sqrt{\Delta\beta(-W)} = \sqrt{8\pi\Delta P(-W)}/\bar{B}(-W) = 2\sqrt{(4000/T_{sp})D} \tag{6}$$

to within an accuracy of 2% or so. Here, as before, the mean field $\bar{B} = \Phi/\pi R_p^2$. Note that the variation of the rhs is a weak one, amounting to 10% at $T_{sp} = 5000$ K. This then is the scaling law in the distributed current case except that the scaling of D with quiet photospheric parameters must be known for the correct upper boundary condition.

We can study this scaling by considering the boundary-layer structure studies of Flaa, Osherovich, and Skumanich (1982, hereafter FOS), and Pizzo (1986, hereafter VP). In these studies, equation (4′) is solved, but from the spot surface upward. In the 1D FOS study one specifies at the lower boundary, $Z = -W$, an axial field strength, B_{00}, and the radial profile of the field with the scale factor given in terms of R_p viz. $u(R,-W)/u_0 \sim 1 - exp(-R^2(1.63)^2/R_p^2)$ for the ST-Gaussian case. In the ST return-flux case u has an extra term due to the return flux. For $Z \rightarrow \infty$ the field is required to become potential. The confining pressure is still as in Yun's case but now $P(0,Z)$ is the *observed* umbral (axial) pressure with photosphere at $Z = -W$. Thus, $\Delta P(Z)$ depends functionally on the assumed Wilson depression, W. Note that assigning an *arbitrary* gradient D at $Z = -W$, as Yun does, is not the correct boundary condition as imposed in either the current sheet or in the FOS boundary-layer studies. We can use the boundary-layer studies to determine the correct value of D to be used in Eq. (6).

One finds that the FOS solutions for the flux potential depart from the potential solution only in a boundary layer of thickness $\Delta Z \sim W$. One also finds that the vertical gradient of the axial field $(dB_0/dZ)_{-W}$ becomes an eigenvalue of the force balance equation with the scaling (refer FOS)

$$D = -\left(\frac{R_p}{B_0}\frac{dB_0}{dZ}\right)_{-W} = D_0 - C\beta_{eff} \tag{7}$$

where the potential solution gradient is given by D_0 and $\beta_{eff} \equiv 8\pi\Delta P(-W)/B_{00}^2 \equiv (0.35)^2\Delta\beta(-W)$. One finds $(D_0, C) = (3.260, 0.357)$ in the Gaussian profile case (STG) and $(2.823, 0.288)$ in the return-flux case (STRF). Equation (7) can be derived by a Z-integration of the (radially integrated) force-balance equation (a second-order equation), refer to FOS, where one finds that D_0 is fixed by matching the constant of integration to the asymptotic potential solution while $C = 8h_{ex}/R_p$ where h_{ex} is the scale height of $\Delta P(Z)$, basically the external pressure scale height. For $R_p = 10$ Mm, the value used by FOS, and $h_{ex} = 0.36$ Mm, (VP), one finds $C = 0.288$, the same as that determined from the numerical analysis.

Note that in this distributed current model, both the mean field $\bar{B} = \Phi/\pi R_p^2 \equiv 0.35B_{00}$ *and* the spot radius R_p $(= 10$ Mm$)$ are being held fixed for a particular solution, so that it is not surprising that $(dB_0/dZ)_{-W}$, rather than $\bar{B}$ (i.e., R_p for fixed Φ) represents the eigenvalue of the force-balance equation.

Pizzo drops the ST assumption and returns to the 2D solution of (4′). He uses essentially the same upper and lower boundary conditions on u as FOS but must now specify $P(u,Z)$ everywhere. Pizzo assumes as given the quantities

$$h(u,Z) = h(u_0,Z) - \Delta h(Z)\left(1-\frac{u}{u_0}\right)^2 \tag{8}$$

$$P(u,Z_l) = P(u_0,Z_l) - \Delta P(Z_l)\left(1-\frac{u}{u_0}\right)^2 \tag{9}$$

with $u_0 = u(R\to\infty,Z)$ and $u(0,Z)=0$, while $\Delta P(Z_l)$ is the confining base pressure excess, as in FOS, while $\Delta h(Z) = (k/g)\{(T/\mu)_{\infty,Z} - (T/\mu)_{0,Z}\}$. Thus, the temperature difference profile between spot umbra (axis) and quiet sun must now be explicitly specified. These are *observationally* specified and are essentially the same as used by FOS. Note that $P(u=0,Z)$ and $P(u_0,Z)$ are the umbrae and quiet sun pressure profiles as in FOS while $Z_l = -0.12$ Mm $-W \approx -W$.

Pizzo determines $(dB_0/dZ)_{Zl}$ for a variety of $\Delta P(Z_l)$ and R_p with B_{00} kept fixed. Fitting his resulting gradients by Eq. (7) for $R_p = 4.9$ Mm and 16.3 Mm I find that $D_0 = 2.88$ and $C(R_p = 4.9) = 0.53$ while $C(R_p = 16.3) = 0.14$. These values yield $C \simeq 6.3h_{ex}/R_p$. These results are consistent with the FOS values. I note that by assuming $D = 1$ Pizzo derives an algebraic expression for β_{eff} ($\equiv (D_0-1)/C$ in this case) from integrals over an approximate expression for force-balance. If we express $C = D_1 h_{ex}/R_p$ then we can determine the residual dependence of D_1 on atmospheric parameters by means of Pizzo's expression. My numerical fit values for D_1 agree with those calculated from his expression. Whether his expression for D_1 is necessary in practice is moot, given the insensitivity of our final scaling law to this coefficient.

With the equation for D given by (7), we can now derive the final expression for the scaling of the surface value of the mean field from Eq. (6). One finds that

$$\frac{\bar{B}}{\sqrt{8\pi\Delta P(-W)}} = \frac{\Phi/\pi R_p^2}{\sqrt{8\pi\Delta P(-W)}} = \left(\frac{1+(D_1/2)(h_{ex}/R_p)(4000/T_{sp})}{4D_0(4000/T_{sp})}\right)^{1/2}. \tag{10}$$

To introduce the quiet photospheric pressure one can take $\Delta P(-W) \simeq P_{ex}(-W) = exp(W/h_{ex})P_{ex}(\tau\approx 1)$. Yun finds that W depends strongly on T_{sp} but is insensitive to Φ and D. I find that $W\,(\mathrm{Mm}) \approx 0.860\;(4000\;/\;T_{sp})^3$ from his tabular values. Further study is needed to see if the numerical value of 0.860 Mm can be expressed in more basic terms.

For the Yun case, $T_{sp} = 4000$ K, $R_p = 14$ Mm so that with $D_0 \approx 3, h_{ex} = 0.36$ Mm and $D_1 = 8$ one finds that $\sqrt{\Delta\beta(-W)} \simeq 3.3$. I believe this value, which is now consistent with a potential boundary condition, should replace Yun's 4.35 in Table 1 since the correct $D(\approx 2)$ is lower than the value of 4.75 used by Yun. Thus, $\sqrt{\Delta\beta(Z)}$ should not vary as much as Yun's results indicate. The lower BC fixes $\sqrt{\Delta\beta(Z_l)}$ to 2.85 at sufficient depth. This conclusion yields a near constancy of $\sqrt{\Delta\beta}$ as in the CS results. I note that to correct for the fact that the surface field zenith angles at the penumbra γ_p ($\equiv$ arctan $(D/2)$) are not large enough in the STG case, Osherovich (1982) introduced a return flux component into u to flatten the field. This, as we have seen, has a minor effect on our scaling parameters (D_0, D_1) yet yields a flatter emergent field, perhaps too flat if one considers the LS observations.

Note that the rhs of Eq. (10) is a slowly varying function of both R_p and T_{sp} and that, if stellar spots have $T_{sp} \simeq T_{sp}(\odot) \simeq 3800$ K, one has that exp $(0.5W/h_{ex}) \approx 4$ so that, one has that

$$\bar{B} = \Phi/\pi R_p^2 \approx \sqrt{8\pi P_{ex}(\tau\simeq 1)} \tag{11}$$

as the *scaling law for star spots*. Since such a scaling also holds for thin-flux tubes (network fields) one expects a universal square-root scaling to apply whether one is *dealing with spots or network fields.*

There is another aspect of Pizzo's study that is of interest to us. By working in two-dimensions he has a degree of freedom not available in the ST approach. A bright umbral-penumbral ring is found in the photometric signature of Pizzo's model, as well as in the STG case used by Jakimiec and Zabża (1966) and Landman and Finn (1979, hereafter LF), as well as in the STRF generalization, Murphy (1990). LF use an ad hoc Bernoulli effect due to a field-aligned flow to modify the internal pressure field to eliminate the brightness ring. The magnetic field, i.e. $u(R, Z)$, is not updated since $\Delta P(Z)$ is not affected, however, the required flow has an Alfvén Mach number of order unity so that its effect must be included. This is not possible in the ST approach.

Pizzo eliminates the brightness ring by adjusting the internal pressure field, i.e., $h(u, Z)$ without changing $\Delta h(Z)$ or $\Delta P(Z_l)$, i.e., the boundary values. He then solves for the new field configuration. It is found that changes in $T(R, Z)$ only near the $\tau = 1$ surface eliminates the ring without *appreciably changing the field topology*. This is because near $\tau = 1$ lateral force balance is dominated by the magnetic field so that minor changes in the field imply significant changes in $P(u, Z)$.

Figure 7 illustrates the changes in the topology of the field due to the adjustment of $T(R, Z)$. Note that the $\beta = 1$ surface lies very near the $\tau = 1$ surface. Note also that the shape of the modified field is quite like that of the neighboring initial field even though neighboring field lines refer to different flux, i.e. u, values. This has important implications for radiative transfer inversions since the topology for hydrostatic equilibrium is insensitive to the temperatures and pressures being solved for. Thus, one can use any of the extant field models, e.g., the LS dipole, to specify the topology, i.e., connectivity, for hydrostatic equilibrium and, though a three-dimensional formal transfer integral is needed, the *geometry*, as in the plane-parallel approximation, *is known*. Further understanding of the thermodynamics of sunspots will certainly occur when $T(R, Z)$ and $P(R, Z)$ is known at other places in addition to the umbral axis and the quiet sun. This is an important study that is now quite feasible.

Finally, I review the effect of field aligned currents, $i(u)$, on sunspot structure and the scaling law. The forces introduced by such currents add the (presumed known) term $i(u)di/du$ to the lhs of Eq. (4′). The effects of this term was studied by Yun (1971) for the subsurface structure in the STG case and by Osherovich and Flaa (1983) for the boundary-layer structure for both the STG and STRF cases.

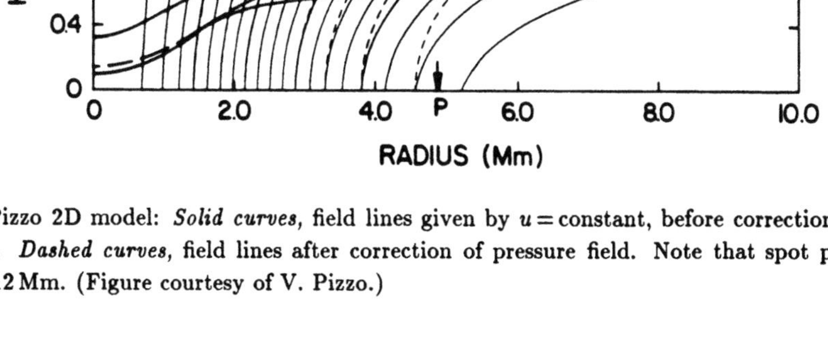

Figure 7. Pizzo 2D model: *Solid curves*, field lines given by u = constant, before correction for bright umbral ring. *Dashed curves*, field lines after correction of pressure field. Note that spot photosphere occurs at 0.12 Mm. (Figure courtesy of V. Pizzo.)

Yun's assumed current distribution is nearly linear in u, reaches a maximum at $R \simeq 0.6R_p$ and then drops rapidly to zero by R_p. Yun uses a mean slope $A = \langle R_p di/du \rangle = 1.20$, which is four times larger than the LS value. The exact shape and the degree of drop-off turns out to be unimportant.

If one considers Yun's solution for the (only) case $T_{sp} = 4000$ K, D= 7.73 and $\Phi = 0.93 \times 1.6 \times 10^6$ GMm^2, one finds that the *thermodynamic* structure is significantly affected, contrary to statements appearing in the literature, but the field structure is not. Thus, the axial field, $B_0(Z)$, and hence the mean field, is increased by $\lesssim 6\%$ throughout the spot so that the flux-tube 'radius' $R_p(Z)$ is decreased by $\lesssim 3\%$. However, I find that the umbral (axial) pressure is increased by $\delta P^{(A)}(Z)$ such that

$$P^{(A)}(Z) \equiv P(Z) + \delta P^{(A)}(Z) \simeq P(Z) + 5\left(\frac{A}{1.2}\right)^2 \left(\frac{B_0(Z)}{B_0(-W)}\right)^2 \left(\frac{B_0(Z)^2}{8\pi}\right) \tag{12}$$

where $P(Z)$ is the solution with $A = 0$. This increase presumably compensates for the increased tensile constriction due to B_ϕ^2. The numerical coefficient varies slowly about the value 5 by $\pm$ 20% throughout the flux-tube. I presume it is insensitive to D, T_{sp}, and Φ and scales as A^2.

The consequence of the increased pressure is to raise the $\tau_{Rosseland} = 2/3$ surface, i.e., decrease the Wilson depression, by 75 km or a scale height. In the immediate subsurface layers $P^{(A)}(Z)$ is a factor 2 larger than $P(Z)$ but rapidly drops to be 2% larger by $Z \sim -R_p/2$, enough larger however to cause $\Delta P^{(A)}(Z)$ to become negative! The associated temperatures are higher by 60% and drop to 4%.

As a consequence of (12), the associated mean excess plasma beta becomes

$$\Delta\beta^{(A)}(Z) \simeq \Delta\beta(Z|D, T_{sp}) - 5\left(\frac{A}{1.2}\right)^2 \left(\frac{B_0(Z)}{B_0(-W)}\right)^2 \tag{13}$$

where $\Delta\beta(Z|D, T_{sp})$ is the A $= 0$ solution. Note that the effect of B_ϕ^2 can also be considered as a simple augmentation of $P_{ex}(Z)$, i.e., $\Delta\beta^{(A)}(P_{ex} + \delta P^{(A)}) \simeq \Delta\beta(D, T_{sp})$. In the surface layers one finds for Yun's case that $\sqrt{\Delta\beta^{(A)}(-W)}$ is $\sim$ 10% less than $\sqrt{\Delta\beta(-W)}$. Thus, the scaling constant is changed by only $\sim$ 10%. Since the value of A^2 used by Yun is an order of magnitude larger than LS find, the correction for B_ϕ^2 appears to be negligible. In the interest of completeness however, one can show that

$$\frac{\bar{B}}{\sqrt{8\pi\Delta P(-W)}} \simeq \left(\frac{1 + (D_1/2)(h_{ex}/R_p)(4000/T_{sp})}{4D_0(4000/T_{sp}) - (3.5 + 2.48(4000/T_{sp}))A^2}\right)^{1/2} \tag{14}$$

where the scaling of $D^{(A)}(D_0, D_1)$ with A has been taken from the boundary-layer studies of Skumanich, Flaa, and Osherovich (1991).

I do not, for lack of space, consider the mixed CS and distributed current models studied by Jahn (1989).

4. Evidence for square-root scaling

One might ask what evidence exists that starspots are similar to sunspots. Other papers at this Bicentenary Colloquium are directed to the nature of starspots so that I'll restrict my attention to only a few comments on this issue.

In the case of G2V stars consider HD 1835 (9 Cet) which has the same effective temperature as the Sun but a more active chromosphere, viz. $R'_{HK} = 3.3R'_{HK}(\odot)$, and a more rapid rotation rate $\omega = 3.3\omega(\odot)$. Campbell and Cayrel (1984) have detected starspot *umbrae* via TiO and CaH spectra and find that they cover "about 3% of the visible stellar hemisphere" if they have similar physical conditions to those on the Sun. Such spots must be responsible for the luminosity variations seen by Chugainov (1980) who measures an amplitude variation of $0^m.032$. In the case of the Sun on DOY 99, 1980, one finds an *observed* fractional (total) *umbrae* area ($A'_u = \Sigma(A_u/2\pi R_{\odot}^2)$ to be bounded by $(\bar{\mu})^{-1}A_{sp}^{\rm proj} \geq A'_u \geq (R_u/R_p)^2\bar{\mu}^{-1}A_{sp}^{\rm proj}$ where the projection factor $\bar{\mu} = 0.73$, and the total projected spot area $A_{sp}^{\rm proj} = 0.32\%$ (Hoyt *et al.* 1983). With $R_u/R_p = 0.4$ (Allen *et al.* 1963), one has that an umbral area of *e.g.* $4\% \geq A'_u \geq 0.07\%$ darkened the sun by $0^m.002$ (Willson, refer Hoyt). The inequality is due to the fact that spot groups have less penumbra than isolated spots. Thus, the ratio of darkening to fractional umbral area between the two stars agree to better than a factor of 2, hence, spot temperatures agree to better than 20%. Saar (1987) has measured the mean magnetic field on this star and found $\bar{B}/\sqrt{8\pi P(\tau \simeq 1)} = 0.95$. Whether this is due to network (thin flux tubes), as Saar argues, or spots, is immaterial by our arguments in section II. In a most interesting paper Saar and Linsky (1986) have shown that the square-root scaling holds for dwarfs of later spectral types.

In the case of the RSCVn stars the situation is confused. Gondoin, Giampapa, and Bookbinder (1985) report $\bar{B}/\sqrt{8\pi P} = 1$ for HD222107 (λ And, G8 III-IV), while Giampapa *et al.* (1983) report 2. In the latter case, Giampapa (1984) argues that an admixture of spot umbrae with their higher fields corrupted their value. My arguments in section II above show that this suggestion is untenable since one is measuring mean fields, i.e., integrated over the disk. For HD 17433 (K1-2 IV), Bopp *et al.* (1989) find a ratio of 1.5. Using an entirely different technique, Donati *et al.* (1990) measure a mean spot field of 1 kG for HR 1099 (K1 IV). This yields a ratio of 0.6 if one uses $P(\tau \simeq 1) = 10^{4.82}$ dynes/cm^2 as estimated by Bopp *et al.* However, this value may be too large. I note that Allen (1963) lists P (K0V)= $10^{4.9}$ dynes/cm^2 while P (K1 III)= $10^{3.7}$ dynes/cm^2. Strangely enough, the Donati *et al.* spot seems to be *bright* and not dark so that a correction for T_{sp} may need to be made! It is obvious more observations are needed of the RSCVn stars.

In conclusion, it would appear that star spots, at least in the dwarfs, are homologous to those on the Sun and satisfy lateral force balance.

Acknowledgments

The author is indebted to H.S. Yun for a personal copy of his thesis, without which much of the analyses would not have been possible. He thanks B.C. Low for his illuminating discussions, for suggesting the truncated dipole model, and for a careful reading of the manuscript. Thanks to Linda Croom who made a magnificent translation of my handwritten manuscript into TEX. The author's trip to Armagh was partially supported by the Armagh Observatory and he expresses his sincerest thanks for the hospitality and generosity of its Director, Dr. Mart de Groot and of Dr. P.B. Byrne and the entire Armagh staff.

References

Adam, M.G. 1990, Solar Phys., 125, 37.

Allen, C.W. 1963, Astrophysical Quantities (London: The University of London, Athone Press).

Beckers, J.M., Schröter, E.H. 1969, Solar Phys., 10, 384.

Bopp, B.W., Saar, S.H., Ambruster, C., Feldman, P., Dempsey, R., Allen, M., Barden, S.P. 1989, ApJ, 339, 1059.

Campbell, B., Cayrel, R. 1984, ApJ, 283, L17.

Chugainov, P.F. 1980, Izv. Krymsk. Astrofys. Obs., 61, 124.

Deinzer, W. 1965, ApJ, 141, 548.

Deming, D., Boyle, R.J., Jennings, D.E., Wiedemann, G. 1988, ApJ, 333, 978.

Donati, J.-F., Semel, M., Rees, D.E., Taylor, K., Robinson, R.D. 1990, AA, 232, L1.

Flaa, T., Osherovich, V.A., Skumanich, A. 1982, ApJ, 261, 700.

Giampapa, M.S., Golub, L., Warden, S.P. 1983, ApJ, 268, L121.

Giampapa, M.S. 1984, in Space Research Prospects in Stellar Activity and Variability, eds. A. Mangeney and F. Praderie (Meudon: Obs. de Paris), p 309.

Gondoin, Ph., Giampapa, M.S., Bookbinder, J.A. 1985, ApJ, 297, 710.

Hoyt, D.V., Eddy, J.A., Hudson, H.S. 1983, ApJ, 275, 878.

Jahn, K. 1989, AA, 222, 264.

Jakimiec, J. 1965, Acta Astr., 15, 145.

Jakimiec, J., Zabża, M. 1966, Acta Astr., 16, 73.

Landman, D.A., Finn, G.D. 1979, Solar Phys., 63, 221.

Lites, B.W., Skumanich, A. 1990, ApJ, 348, 747.

Low, B.C. 1975, ApJ, 197, 251.

Low, B.C. 1985, Solar Phys., 100, 309.

Low, B.C. 1990, private communication.

Murphy, G.A. 1990, The Synthesis and Inversion of Stokes Spectral Profiles, Cooperative Thesis No. 124, University of Sydney and National Center for Atmospheric Research.

Osherovich, V.A. 1982, Solar Phys., 77, 63.

Osherovich, V.A., Flaa, T. 1983, Solar Phys., 88, 109.

Parker, E.N. 1978, in The New Solar Physics, ed. J.A. Eddy (Boulder, Colo.: Westview Press), p 1.

Pizzo, V.J. 1986, ApJ, 302, 785.

Saar, S.H. 1987, Observations and Analysis of Photospheric Magnetic Fields on G, K, and M Dwarf Stars, PhD thesis, University of Colorado.

Saar, S.H., Linsky, J.L. 1986, Adv. Sp Phys, 6, No. 8, 235.

Schlüter, A., Temesvary, S. 1958, in IAU Symp. 6, Electromagnetic Phenomena in Cosmical Physics, ed. B. Lehnert (Cambridge: Cambridge University Press), p 263.

Schmidt, H.U., Wegmann, R. 1983, in Dynamical Problems in Mathematical Physics, eds. B. Brosowski and E. Martensen (Frankfurt: P. Lang Verlag), p 137.

Skumanich, A., Flaa, T., Osherovich, V.A. 1991, to be published.

Yun, H.S. 1968, Theoretical Sunspot Models, PhD thesis, Indiana University (Ann Arbor, Mich.: University Microfilms).

Yun, H.S. 1971, Solar Phys., 16, 398.

Yun, H.S. 1972, Solar Phys., 22, 137.

Discussion: Skumanich

Giampapa Now that you have outlined the stability conditions that describe spots can you make any comments on the factors that may determine spot evolution and decay?

Skumanich My discussion assumed lateral force balance and gives no insight into stability conditions. One needs to perturb such states in order to study their stability along the same lines as Meyer *et al* have done.

Gurman How would the magnetostatic model have to be modified to account for the Evershed circulation in the penumbra?

Skumanich Such fluxes add a dynamic pressure term that would modify the pressure term in the lateral force balance equation. Indeed Landman and Finn did just that to correct for an umbral bright ring they found in the Schluter-Temesvary model. Unfortunately the flow was close to Alvenic, hence not a small perturbation. Thus, to first order, spots can adjust by dropping their thermal pressure to compensate or by expanding laterally (smaller ΔP, hence larger R^2).

Maltby Could you explain/comment on why the "true" model is more true than other models?

Skumanich I used the word "true" when I implied "truer", *i.e.* make complete - or to be more accurate, more complex!

Mullan What is the effective temperature of a sunspot? Does your Stokes instrument measure T_{eff} as a function of total magnetic field strength?

Skumanich We have not evaluated the monochromatic brightness temperatures (rather than the effective temperature *i.e.* bolometric brightness) of our observed spots although we could do so. However we would need to model the atmosphere to go from monochromatic temperature to effective temperature.

Foing Current magnetic field measurement by Zeeman (Saar), or Zeeman Doppler relate to the facular component which dominates the signal. Do you expect a change in the $\phi/(8\pi\Delta P)^{1/2}\pi R^2$ ratio when going from network elements facular elements, pores to spots, due to increase of the magnetic tension.

Skumanich No, the ratio is of order 1 in all three elements. It is trivially correct, i.e. exactly 1, in small-scale features without curvature. The quantity $\phi/(8\pi\Delta P)^{1/2}\pi R^2$ represents the mean field and not the axial field and thus includes the effects of curvature.

LITHIUM IN RS CVn BINARIES AND RELATED CHROMOSPHERICALLY ACTIVE STARS

R. Pallavicini[1], S. Randich[2], M.S. Giampapa[3]
[1]Osservatorio Astrofisico di Arcetri, Florence, Italy
[2]Dipartimento di Astronomia, University of Florence, Italy
[3]National Solar Observatory, Tucson, Arizona. USA

1. Introduction

It is well known that Lithium is progressively depleted in Population I late-type stars through the action of convective motions that transport surface Lithium to the hotter interior where it is rapidly destroyed by nuclear reactions (*e.g.* Soderblom 1984). Late-type evolved stars, such as the cooler components of RS CVn binaries, are therefore not expected to have an appreciable abundance of Lithium. However, there have been occasional reports in the literature on the presence of the LiIλ6708Å line in some active stars of very late spectral types, including stars classified as RS CVn binaries (see Pallavicini et al. 1987 and references therein). In this paper we report the results of an extensive search of the Li line in a sample of southern RS CVn binaries and related chromospherically active stars. To our knowledge, this is the first time that the question of the abundance of Lithium in RS CVn stars has been addressed systematically using a large data sample.

The results of our survey confirm the presence of Lithium in a large fraction of the programme stars, although a strong Li feature does not appear to be a general property of all RS CVn stars as a class. We find a broad range of different Li strengths for active stars of similar effective temperature and luminosity class, which suggests either different physical natures for the stars in our sample or different evolutionary histories. We consider several possibilities to explain the presence of the Li line in chromospherically active stars and we show that studies of Lithium in these objects bear important implications for understanding their evolutionary status. We also show that modulation of the Li line by surface activity is not a viable explanation for the presence of a strong Li line in RS CVn binaries and other "spotted" stars.

2. Observations and data sample

Our sample stars are primarily from the list of southern active stars of Bidelman and MacConnell (1973). This list is based on the appearance of the CaII H and K lines in low-resolution objective prism spectra. Since it uses only one indicator of chromosspheric activity, it is likely to include objects that may be of different physical nature and in different evolutionary phases. A second group of sources have been taken from the catalogue of Strassmeier et al. (1988) by selecting objects that were visible from the southern hemisphere. In this case, the RS CVn nature of the program stars is better established. Finally, other sources were taken from the lists of southern RS CVn candidates of Weiler and Stencel (1979) and Hearnshaw (1979). These lists too are mainly based on the appearance of the CaII lines in low-resolution spectra. In total, the sample comprises more than 60 southern stars of spectral types G and K and luminosity classes V, IV and III. Several inactive stars of various spectral types and luminosity classes were also observed for comparison.

The observations were carried out at the European Southern Observatory (ESO), La Silla, Chile, in several observing runs (Nov. '86, Dec. '87, Jan '89, June '89 and April '90; the run of June '89 through the courtesy of Dr. L. Pasquini). The instrument used was the Coudé Echelle Spectrometer (CES) fed by the 1.4 m CAT telescope. We used the short-camera and a cooled

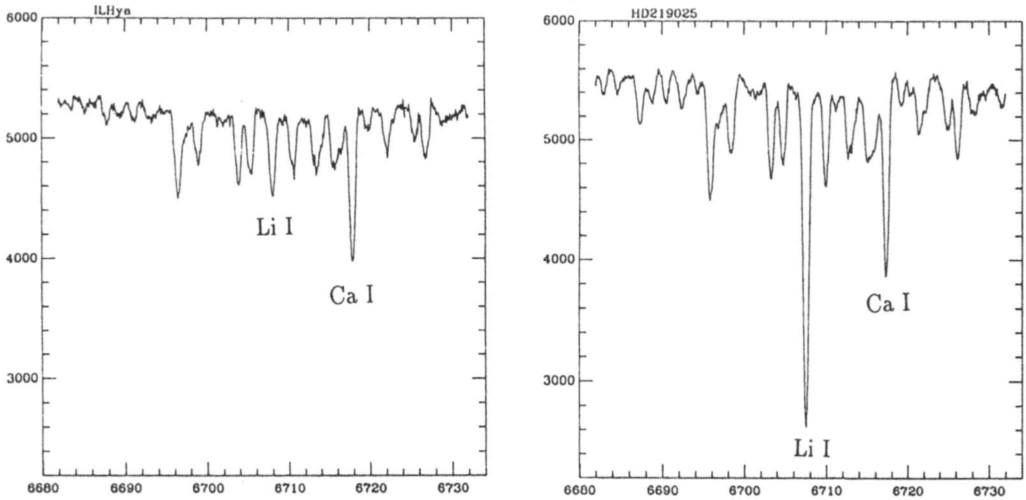

Figure 1. Spectra of IL Hya and HD219025 in the Lithium region. The measured equivalent width of the Li blend is 113 mÅ for IL Hya and 430 mÅ for HD219025. For both stars, the contribution of the FeI line at 6707.4Å can be estimated to be 37 mÅ at most.

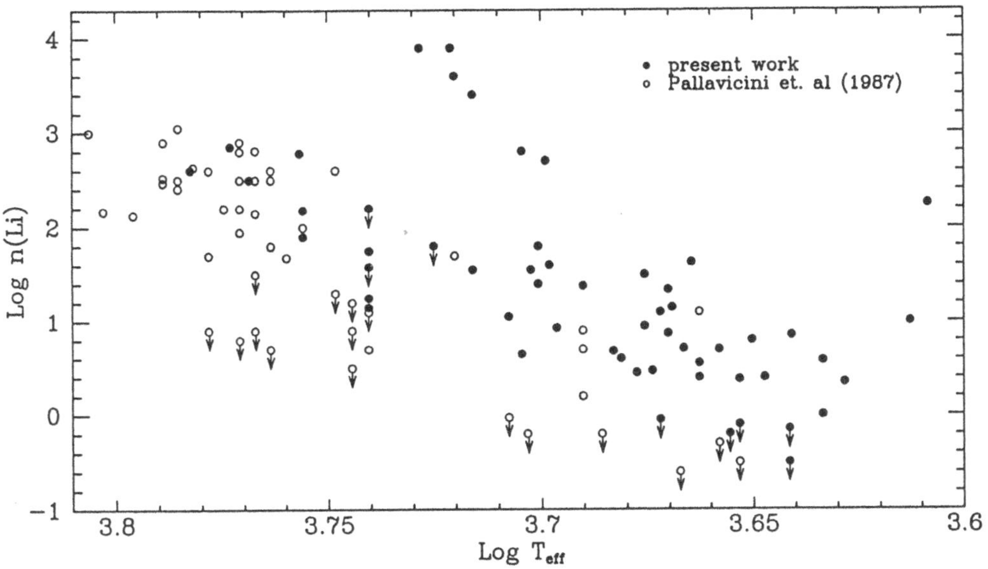

Figure 2. Lithium abundance *vs* effective temperature for the stars in our sample and in Pallavicini et al.(1987). Note the excess Li abundance for most stars in our sample.

CCD detector. The resolving power was R = 50 000 and the S/N ratio was in all cases > 100. Examples of the spectra are shown in Figure 1. The spectral range available (≈ 50 Å) allows simultaneous observation of the LiI line at 6707.8Å, of the CaI line at 6717.7Å and of a number of FeI lines. Since many of our stars are fast rotators (with $v \sin i \geq 10\,\mathrm{km/sec}$), the LiI line at 6707.8Å is usually blended with a nearby FeI line at 6707.4Å. In case of blends, we have estimated the contribution of the Fe line to the Li blend by comparison with a number of narrow-line standard stars of different spectral types and luminosity classes.

Measured equivalent widths were converted to Li abundances using the curves of growth of Pallavicini et al. (1987), which were based on Bell and Gustafsson's atmospheric models. The major uncertainty on the derived Li abundances comes from the effective temperatures which were estimated from the color index (B-V). Although effective temperatures derived from (B-V) colours may be quite uncertain (especially for late-type binaries), our main conclusions are not significantly affected by the uncertainty in the stellar parameters. Also the correction for the contribution of the FeI line at 6707.4Å is not critical in most cases, since the Li line. when present, is often much stronger than the estimated contribution of the nearby FeI line.

3. Results

Our results are summarised in Figure 2 where we plot the derived Li abundances (on a scale where $\log n(H) = 12.00$) *vs.* effective temperature (*filled symbols*). We also plot for comparison (*open symbols*) the results of a survey of Li in a sample of field stars carried out with the same instrument at ESO (Pallavicini et al. 1987). The latter sample shows the typical behaviour for field stars. Li abundance rapidly decreases towards cooler temperatures (there are only a few rare detections of Lithium at spectral types later than G8), while there is a large spread of Li abundances for stars of the earlier spectral types. The latter spread is usually interpreted as due to the different ages of field stars. The comparison with the results of the present survey clearly shows excess Li abundance in a large number of chromospherically active stars in our sample. Also evident are the extremely large Li abundances derived from a small group of stars in the sample.

Roughly, nearly two-thirds of the programme stars appear to have an anomalously high Li abundance, including five stars for which the Liλ6708Å line is stronger that the CaI line at 6718Å. Four of these stars have Li abundances comparable to, or even larger than, the canonical cosmic value for Population I stars (*i.e.* $\log n(Li) \geq 3.1$). In contrast to what typically observed for field stars, we find that only a small fraction of the cooler stars in our sample show a very weak or absent Li line. Finally, the small group of stars of higher effective temperatures that were present in our sample show the Li line, but the derived Li abundances are consistent with the typical spread in Li strengths observed for late-F and early-G stars.

4. Discussion

We have considered various possibilities to explain the strong Li line observed in many cool stars in our sample:

a) the Li line may be strengthened as a consequence of the presence of large cool spots (Giampapa 1984);

b) some chromospherically active stars may have evolved from late-A or early-F progenitors with shallow convective zones (Fekel et al. 1987);

c) the current lists of southern active stars may contain young objects and even pre-main sequence stars, in addition to evolved post-main sequence stars.

The first possibility is ruled out by simultaneous photometric and spectroscopic observations of a number of "spotted" stars that we have carried out at different phases (see Pallavicini, Cutispoto and Randich elsewhere in this volume). These observations have shown no variations of the LiI equivalent width in excess of what expected from measurement errors (which are typically $\leq 5-10\%$), in spite of the fact that strong photometric variations with phase were observed at the same time.

The other two possibilities are more likely and both are viable explanations for different objects in our sample. The stars with the strongest Li line (particularly those with a Li line comparable t,o or stronger than, the CaIλ6718Å line) are most likely members of young stellar associations and/or may be post-T Tauri stars. This is probably the case for AB Dor, HD139084, HD174429, HD219025 and others. Comparison with spectra of post-T Tauri candidates recently obtained by us at ESO shows a strong similarity in the Li region with the stars in this group.

Stars with a weaker Li line but still an unsually high Li content for their colour have more likely evolved from progenitors that have suffered little Li depletion while on the main-sequence. This is probably the case for HR1099, YY Men, IL Hya, TW Lep and others. These stars are mostly giants and subgiants of spectral types later than G8 and many of them are classified as RS CVn binaries.

In any event, it is clear that current lists of chromospherically active stars in the southern hemisphere are highly heterogeneous. Observations of the Li line in these stars may be of great help for elucidating their true physical nature and evolutionary status.

References

Bidelman, W.P., MacConnell, D.J.: 1973, AJ, 78, 687

Fekel, F.C., Quigley, R., Gillies, K., Africano, J.L.: 1987, AJ, 94. 726

Giampapa, M.S.: 1984, ApJ, 277, 235

Hearnshaw, J.B.: 1979, IAU Coll. No. 46, 371

Pallavicini, R., Cerruti-Sola, M., Duncan, D.K.: 1987, A&A, 174, 116

Soderblom, D.R.; 1984, *Lecture Notes in Phys.* 193, 205

Strassmeier, K.G., Hall, D.S., Zeilik, M., Nelson, E., Eker, Z., Fekel, F.C.: 1988. A&AS, 72, 291

Weiler, E.J., Stencel, R.E.: 1979, AJ, 84, 1372

Discussion: Pallavivini

Innis Two comments on this very interesting survey. First, one way to distinguish the young stars may be kinematically, as seems demonstrated with AB Dor and PZ Tel (both Li strong), which are members of the Local Association (age ≈ 70 million years). Second, for the evolved stars, earlier observations (unpublished) by the Monash group showed similar results, *i.e.* higher Li than in slowly rotating giants. Some of these stars were single. I think Andrew Cameron showed that they share the same kinematic properties as main sequence A and F stars. I think these Li-rich stars are being observed during a short-lived evolutionary phase as the stars change from fast rotating A-F main sequence stars to slowly rotating K giants depleting their Li on the way.

Budding (*a propos* of binarity discussion) Isn't it the case that it is very difficult to prove that a star is not a binary? Even the Sun has been considered to have a possible binary companion.

Pallavicini I agree. In fact I prefer to say that there is no evidence in our spectra that these stars are binaries, (*i.e.* the group with the strongest Li line which we think is made of post T Tauri stars), rather than saying that they are single.

Byrne From the spectra you showed it is not obvious that the neutral metal lines are invarient in EW between slow and rapid rotators. If this is not so, it would have implications for correcting the Li EW for blending with the FeI 6707 line in rapid rotators using the value from slow rotators.

Pallavicini I do not see how rotation can change the equivalent width of lines. the impression you had is probably due to the fact that the spectra I showed were not normalised to 1 and are in arbitrary units. In any case, the correction for FeI is not critical in most cases, since the equivalent widths of the Li blend in these stars are on the order of 100 mÅor more, while the contribution of the FeI line is $\approx 30 - 40$ mÅat most. So the presence of Li in many of these stars is out of question.

Pettersen I would like to remark that even further down the main sequence there is no rotational modulation in the Li line detected so far. The latest spectral type where Li 6707A is seen is M0-M1, namely in the star V1005 Ori = Gl 182. At a time when there was notable rotational modulation detected in UBVR filters, it was not possible to detect changes in the Li profile in coude spectra with $S/N \approx 50$.

(l to r) J. Gurman walks away in disgust while G. Tagliaferri, L. Pasquini, G. Cutispoto and R. Pallavivini are amused by the idea that Li I is enhanced in spots.

RELATIONSHIP OF STARSPOTS TO OTHER INDICATORS OF STELLAR ACTIVITY

Jeffrey L. Linsky[1]
Joint Institute for Laboratory Astrophysics
National Institute of Standards and Technology
and the University of Colorado
Boulder, Colorado 80309-0440, USA

1. Introduction.

Most of the papers presented at this meeting discuss various aspects of sunspots and their stellar analogues – methods for detecting starspots, physical models for starspots, magnetic fields, lifetimes, differential rotation, and spectroscopic indicators of the plasma within and above starspots. Here I will discuss the evidence that starspots (the generic term that includes sunspots) are not isolated phenomena, but are rather centers of activity on stars. While it is true that coronal x–ray emission and chromospheric ultraviolet emission are present on the Sun far away from sunspots, the strongest manifestations of activity – flares, high velocity flows, and very bright x–ray and ultraviolet emission – generally occur near sunspots or in the active regions that surround or are located near sunspots. The simple, and probably correct, explanation for this identification of sunspots with the most violent aspects of activity is that the magnetic fields are strongest, most complex, and most rapidly changing near sunspots.

If the Sun were an accurate paradigm of stellar activity, then we would expect starspots to be spatially correlated with other indicators of stellar activity, and stars with large spot coverage would be spectroscopically distinct from stars without large spot coverage. This paper will explore the evidence for and against both hypotheses. After we have discussed both possibilities, we will return to the question of whether the Sun truly provides a useful paradigm for stellar activity. We begin by discussing the objectives and difficulties of multiwavelength observing campaigns that form the basis for testing the relationships of starspots (best observed in the optical) to other indicators of activity (best observed outside of the optical).

2. The importance and objectives of multiwavelength studies.

Many participants at this conference have played major roles in planning multiwavelength (also called *panchromatic*) observing campaigns to study the surface structures and phenomena that occur in the atmospheres of active late-type stars. Indeed, astronomers at Armagh Observatory have been pioneers in applying this research technique to understanding stellar activity. Since such campaigns provide the most useful data for understanding starspots and their relationship to other indicators of activity, it is useful to remind ourselves of the major questions that these campaigns should address.

- **What is the global energy balance for each component (structure) of the stellar atmosphere?** The objective here is to identify the nonradiative heating mechanism and heating rate as a function of height (or position) by determing the total radiative power and other energy loss rates that must be balanced by the local heating rate. Since the radiation emerges over a broad spectral range from x-rays to the infrared, multiwavelength observations are required.

[1]Staff Member, Quantum Physics Division, National Institute of Standards and Technology

- **What is the momentum balance for each component (structure) of the stellar atmosphere?** Instead of assuming hydrostatic equilibrium, it is important to measure the energy flux in winds, downflows, flows within flux tubes, and other dynamic events that can contribute to the energy balance and change the interpretation of some important spectral diagnostics (cf. Linsky 1990).

- **What is the geometry of the stellar atmosphere?** Even though most stars other than the Sun cannot be resolved by direct imaging techniques, there are indirect techniques like Doppler imaging, rotational modulation, and time resolved observations during eclipses that can be used to estimate the area of bright and dark features on the stellar surface. Information on the surface area of atmospheric components is essential to infer surface fluxes and thus the total radiative power per unit area and the local heating rate in the structure. We anticipate that $\beta = P_{gas}/(B^2/8\pi) \ll 1$ for identifiable structures in the outer atmospheres of active stars, and that the structure shapes and gas flows are controlled by the magnetic field.

- **What are the magnetic field strengths and how do they evolve with time?** There is now considerable evidence that outside starspots, the magnetic field strength in the photosphere can be estimated from the relation $\beta \approx 1$ and the total magnetic flux in the photosphere may be determined using line broadening techniques (Linsky 1989, Saar 1990). The magnetic fluxes within active regions may be inferred from Zeeman Doppler imaging (e.g. Semel 1989), linear polarization (e.g. Huovelin, Saar and Tuominen 1988), or circular polarization (e.g. Mürset, Solanki and Stenflo 1988) techniques, but as yet we have no reliable method for measuring the magnetic field strengths or fluxes in starspots. It has been assumed that the magnetic properties of a star depend only on T_{eff}, g, Ω (the angular rotation rate), and perhaps tidal forces for close binary systems. The time evolution of magnetic fields in active stars may be important and should be studied.

- **Does the Sun provide a useful paradigm (role model) for stellar activity?** Until now the Sun has been widely assumed to be the appropriate role model, but this assumption must be tested. I will return to this question at the end of the paper.

Multiwavelength observing programs are difficult to organize, but can be highly rewarding scientifically. Some examples of the difficulties and rewards for research programs across the broad range of astrophysics are reviewed in the symposium volume ***Multiwavelength Astrophysics*** (Córdova 1988). Fortunately, some of the long standing technical problems are now being resolved with the launch of a sensitive x-ray satellite (ROSAT), the commissioning of a major radio telescope array in the Southern hemisphere (the Australia Telescope), and worldwide networks of photometric and spectroscopic telescopes for continuous coverage of variable stars during observing campaigns. The remaining problems are typically human; in particular, how to acquire and schedule simultaneous observing time on many major telescopes given the vagaries of the peer review system and scheduling complexities. These difficulties are real, and a better system is needed for supporting worthy multiwavelength observations. My focus in this paper, however, is on the interpretation of such data sets. Here it is important to avoid circular reasoning by not presuming the solar paradigm or simplistic geometries.

3. Spatial relations between spots, active regions, flows, and flares using rotational modulation and Doppler imaging techniques.

In this section I summarize the available Doppler imaging and rotational modulation data that correlate starspots spatially to other manifestations of stellar activity. Although many observing campaigns have aimed at answering this question, the two campaigns that provide the clearest results remain the October 1981 observations of II Peg and V711 Tau.

3.1 Coordinated observations of the October 1981 plage on II Peg.

Rodonò *et al.* (1986, 1987) observed the single-lined RS CVn system II Peg for nearly a complete 6.7 day orbit with IUE and obtained coordinated optical photometry. Since the inclination angle (i) between the rotation axis and the line of sight is uncertain for this system, the location and sizes of the inferred spots depend on the assumed angle. For low angles, such as $i = 35°$, the photometry are best fit by two high latitude spots visible at all phases, whereas for large angles, such as $i = 90°$, the photometry are best fit by three spots, one of which is at low latitude. This equatorial spot (Spot No. 2) had a radius of 17° and covered 8.5% of the apparent stellar surface at central meridian crossing.

What was noteworthy about II Peg at this epoch was the striking enhancement in the flux in all emission lines during the phase interval $\Delta\phi \approx 0.50$ with very sudden onset and decline in the flux at either end. The flux enhancement was a factor of 5 for the transition region lines and a factor of 2.4 for the chromospheric lines like Mg II. Rodonò *et al.* (1987) argued that the brightening was not a flare but rather a long-lived plage, because it was observed over nearly the full 6.7 day rotation period of the star. The rapid onset and disappearance of the plage and its visibility for $\Delta\phi \approx 0.50$ is simply interpreted as the rotation of a small, very bright region located near the equator for $i \sim 90°$ or at a latitude $l \approx -i$. If we adopt $i \approx 90°$, then the plage lies over the trailing edge of Spot No. 2. That is, the central longitude of the plage is about 22° in longitude behind the center of the spot, which has a radius of 17°. Alternatively, the two-spot solutions for $i = 60°$ or 35° both result in two high latitude spots that are always visible and thus uncorrelated in phase with the plage. Solar analogy points to the $i \approx 90°$ solution which we adopt, but the case is not conclusive.

What is the area of the plage? We know that the plage appeared and disappeared within a phase interval $\Delta\phi \leq 0.073$, corresponding to a longitude interval of $\leq 26°$. Byrne *et al.* (1987) computed models for the plage assuming that its conversion factor (from observed flux to surface flux) was 1.75×10^{19}, which corresponds to 1.3% area coverage or a circular area near disk center with a radius of 10°. With this assumed area, the surface fluxes for the transition region lines in the plage are 150-300 times larger than for the mean atmosphere outside the plage. Byrne *et al.* (1987) estimated the volume of a loop completely filled with gas emitting the C IV and Si IV lines consistent with the observed fluxes in these lines and the electron density was estimated from density-sensitive lines ratios. The derived loop dimensions imply that the plage emission comes from an even smaller area with an extent of only 2° in longitude. The completely filled loop model may not be realistic, however, if the Sun is a useful guide, since solar active region loops are filled with coronal gas with the transition region gas located typically near the footpoints.

Table 1 compares the sizes and relative locations of the plage and nearby spot assuming that the inclination angle $i \approx 90°$. The plage appeared to be displaced by 22° in longitude from the spot, but the plage overlies a portion of the spot unless the plage radius is smaller than 5°.

3.2 Coordinated observations of the 3 October 1981 flare on V711 Tau.

The RS CVn system V711 Tau=HR1099 was observed intensively for 5 days in October 1981. Using optical photometry, Rodonò *et al.* (1986) derived the location of two spots on the K1 IV star of this system. At that time one large spot (radius 22°) was at low latitude (+10°) and

Table 1. Comparison of spots with nearby plage and flare for two well-studied systems.

Parameter	II Peg Plage 1981 October 1-7	V711 Tau Flare 1981 October 3
Spot angular radius	17°	23°
Spot latitude	0°	60°
Spot longitude	230°	60°
Spot phase (ϕ)	0.64	0.17
Spot area (% of disk)	8.5	15
Feature radius	$\leq$ 13° in UV	27° in radio
Feature latitude	0° in UV	$\sim$ 60° in radio
Feature longitude	252° in UV	$\sim$ 60° in radio
Feature phase (on central meridian)	0.70 in UV	$\sim$ 0.18 in radio
Feature area (% of disk)	$\leq$ 5.0 in UV	21 in radio

rotated onto and away from the visible hemisphere of the K1 IV star, which has an inclination angle $i = 33°$. The other large spot (radius 23°) was at high latitude (+60°) and thus always in view. An independent Doppler imaging study by Vogt and Penrod (1983), using contemporaneous data, also found two spots with nearly the same locations and sizes.

The IUE spectra (Rodonò *et al.* 1987) show marginal evidence for a maximum flux in the ultraviolet emission lines at the time of optical minimum, which occurred at the time of central meridian passage of the high latitude spot. The most interesting aspect of the IUE data set is the detection of a large flare near phase 0.18 when the high latitude spot was located near the center of the visible disk, covering 15% of the visible surface. Linsky *et al.* (1989) analyzed all of the flare data, which include a total of eight IUE spectra obtained over a period of 16.5 hours. At the beginning of the flare the chromospheric lines (e.g. Mg II k) were 1.5 times brighter than before the flare, the transition region lines were 2.4 times brighter, and the ultraviolet continuum was also enhanced. The surface brightness of the flare in these emission lines and thus the total radiative power and heating rates depend on the surface area of the flare.

A critical piece of information for understanding the flare is the detection of broad Mg II h and k line profiles with enhanced red wings in the earliest spectrum. A fit to the initial Mg II k line profile using four Gaussians with the constraints that the centroid velocities for three of the Gaussians must be at the known radial velocities of the G star, K star, and interstellar Mg II absorption feature, yields parameters for the fourth Gaussian representing k line emission only from the flare. The centroid velocity for the flare was 90 ± 30 km s^{-1} relative to the K star. A weakly exposed high resolution spectrum obtained at least 7 hours after the onset of the flare indicates that the C IV 1548 Å is red-shifted relative to the K1 IV star by a similiar amount. Red-shifted emission in the Mg II lines, probably indicating downflowing material, was previously detected during flares on two other RS CVn systems – the 1 January 1979 flare on UX Ari (Simon *et al.* 1980), the 19 September 1985 flare on AR Lac (Neff *et al.* 1989). On the other hand, Houdebine *et al.* (1990) detected high velocity mass outflows with speeds up to 5800 km s^{-1} in the Balmer lines during a flare on the dMe star AD Leo.

What makes the V711 Tau flare unique is that we know the location of the spots at the time of the flare. If the flare occurred near disk center, the material would flow down toward the surface with a velocity of 90 ± 30 km s^{-1}. If it occurred near the receding limb, the flow would have been tangential to the surface at 50 km s^{-1}, and if it were located near the advancing limb, the flow would also have been tangential to the surface at 130 km s^{-1}. At the time of the flare, the high latitude spot was located at disk center and the low latitude spot was at the advancing limb. While these data do not place hard constraints on the location of the flare, the location of strong downflows over sunspots (Kjeldseth-Moe *et al.* 1988), and the presence of a large spot at disk center strongly suggest that the flare was in the vicinity of the spot at disk center if the solar activity paradigm is valid. In particular, the solar data indicate long-lived downflows of 40–80 km s^{-1} for spectral lines formed between $4.40 \leq \log(T) \leq 5.40$ with no apparent dependence on temperature.

The peak 6.4 GHz (4.7 cm) flux of 170 mJy observed near the maximum in the ultraviolet emission lines provides another interesting clue about the location of the flare. Table 2 summarizes the brightness temperature (T_b) that results from different assumptions about the emitting area. If we assume that the flare was typical of the flares resolved by VLBI techniques (cf. Mutel *et al.* 1985), then $T_b \approx 2 \times 10^{10}$ K and the emitting area was 1.4 times the area of the spot at disk center. This is consistent with incoherent gyrosynchrotron emission from mildly relativistic electrons (mean energy E_o=2 MeV) trapped in many magnetic loops emerging from a spot. This emission mechanism has been identified as the likely microwave flare mechanism by many authors (e.g. Brown and Crane 1978, Dulk 1985). Variations in the radio flux on time scales of several hours indicates radiative decay of the relativistic electrons in magnetic fields ~ 100 Gauss, which is consistent with field strengths in large loops above sunspots. The similiarity in size between the radio emitting loops and the starspot at disk center strongly supports (but does not prove) the identification of the radio flare with this spot.

Table 2. Radio emission during the 3 October 1981 flare on V711 Tau

Assumption	T_b (K)	Conclusion
Fills one loop	3.5×10^{12}	Unlikely because $T_b \gg$ VLBI results
Covers full disk	5×10^9	Unlikely because $T_b \ll$ VLBI results
T_b has a typical value for flares determined from VLBI images	2×10^{10}	$A_{radio} = 6 \times 10^{22}$ cm^2 $= 1.4 A_{spot}$ $E_o = 2$ MeV V(10^5 K plasma)=0.0001 V(loops)

What is the size of the flaring plasma seen in the ultraviolet emission lines? Density-sensitive line ratios; in particular, the Si IV (1396+1403 Å)/C III 1909 Å, Si III 1892 Å/C III 1909 Å, and C III 1909 Å/C III 1175 Å ratios all indicate a factor of 20 enhancement of the electron density in the transition region to $n_e = 1 \times 10^{11}$ cm^{-3}, corresponding to a flare gas pressure of 1.6 dynes cm^{-2} at T=60,000 K. This indicates that the flare volume in such collisionally excited lines as C IV 1550 Å was 400 times smaller than the nonflare transition region or 10^{-4} times the volume of the radio emitting loops. Without additional data it is not possible to distinguish between a model with the C IV emitting plasma located at the footpoints of the radio emitting loops or in separate 10^5 K loops.

While the evidence is not conclusive, the downflowing plasma and microwave emission fit consistently with a model in which the ultraviolet and radio emission occur above the large spot at disk center.

3.3 Doppler imaging studies of plages on AR Lac.

The Doppler imaging technique has been used to determine the longitudes, latitudes and areas of two plages on AR Lac for the October 1983 observing campaign (Walter *et al.* 1987) and three plages for the October 1985 campaign (Neff *et al.* 1989). These plages are all located on the K0 IV star in the system. A comparison between these plages and starspots on the surfaces of the stars in the AR Lac system is difficult, however, because of the 1.98 day period of the system, the small amplitude of the photometric wave, ellipticity of the stars, and the possibility of spots on both stars. Nevertheless, Kang and Wilson (1989) were able to infer the location of two spots at six different epochs from 1972 to 1981, which showed steady migrations toward decreasing longitude as a function of time. The extrapolated positions of these two spots are given in Table 3. Also listed are the spot locations derived by Rodonò *et al.* (1986) for the 1981.7 epoch.

Table 3. Comparison of Plages and Spots on AR Lac

Year	P	Plage Long.	Plage Lat.	A_{plage}	Spot1 Long.	Spot1 Lat.	A_{spot}	Spot2 Long.	Spot2 Lat.	A_{spot}
1981.7	–	–	–	–	190°	±50°	0.043	83°	±20°	0.063+
1981.7	–	–	–	–	249°	±20°	0.13	92°	±20°	0.053*
1983.8	–	–	–	–	178°	±20°	0.13	8°	±20°	0.053*
1983.8	A	280°	0°	0.01	–	–	–	–	–	–
1983.8	B	345°	0°	0.01	–	–	–	–	–	–
1985.8	–	–	–	–	110°	±20°	0.13	288°	±20°	0.053*
1985.8	A	205°	0°	0.06	–	–	–	–	–	–
1985.8	B	240°	±50°	0.02	–	–	–	–	–	–
1985.8	C	130°	0°	0.09	–	–	–	–	–	–

Notes: P = Plage designation. +From Rodonò *et al.* (1986). *From Kang and Wilson (1989) assuming no change in spot latitude or radius and their migration law.

In some cases there is reasonable agreement between the plages and spots in Table 3. For the 1983.8 data set, for example, Plage B lies only 23° in longitude from Spot No.2, although the sizes of the plage and spot differ considerably. For the 1985.8 data set, Plage C and Spot No. 1 are only 20° apart in longitude and the areas are similiar. Nevertheless, three of the plages do not have close spot counterparts and two of the spots have no plage counterparts. The disagreement could be due in part to the absence of contemporary photometry, and the assumption that spots do not drift in latitude or change size, which is clearly in error (Eaton 1990). Also, the determination of spot positions by different authors and methods even at the same epoch can differ significantly. For example, the spots on AR Lac in 1981.7 as derived by Rodonò *et al.* (1986) and by Kang and Wilson (1989) are not in agreement (see Table 3), and the latitudes of spots on EI Eri determined by photometry and Doppler imaging differ significantly (Strassmeier 1990). Thus the case for spatial connection between plages and spots on RS CVn systems has not yet been made conclusively.

4. An excess flux technique for studying the relation between spots and active regions.

In the previous section, I discussed the best examples of stars for which a spatial comparison can be made between spots and active regions using rotational modulation and Doppler imaging techniques. These techniques are successful *only* when no more than three active regions dominate the integrated stellar spectrum, the star is observed throughout at least one rotational period and preferably longer, transient flares are easily distinguished from long-lived active regions, and the star is rapidly rotating (required for Doppler imaging). Here I propose a new spectroscopic technique for estimating the fractional coverage by active regions, which may then be compared with the fractional coverage of starspots. Since this technique does not require that only a few active regions dominate or that the star be rapidly rotating, it should be applicable to many more stars than the Doppler imaging technique.

The technique, which I call the *excess flux* technique utilizes Vilhu's (1987) discovery that the surface fluxes of the Mg II 2800 Å doublet, the C IV 1550 Å doublet, and soft x-rays each reach a maximum value (called *saturated* surface fluxes) for the most active stars. Vilhu interpreted the *saturated* surface fluxes as those corresponding to the maximum rate of steady-state heating possible given the available mechanical energy in the convective motions in a rapidly-rotating star. I interpret this to refer to stars with equipartition magnetic fields completely covering their surfaces. Only the T Tauri stars and flares exceed these *saturated* values, presumably because of the extra emission from a turbulent boundary layer between the circumstellar disk and the chromosphere for T Tauri stars or the rapid release of magnetic energy accumulated over a long time for flares. The empirical *saturated* surface fluxes for early K subgiants are listed in Table 4. The observed surface fluxes in solar plages lie about a factor of 10–30 below the *saturated* limits. This is consistent with typical solar plage magnetic flux densities of 100 ± 20 Gauss, which corresponds to the filling factor of equipartition photospheric magnetic fields (about 1500 Gauss) in plages in the range of 5–10% (Schrijver 1987).

Table 4. Saturated and plage surface fluxes

Line	Region on a star	Surface flux*	Reference
MgII h+k	Saturated value	100	Vilhu (1987)
MgII k	AR Lac K0IV Plages A+B (Sep 85)	109	Neff *et al.* (1989)
MgII k	AR Lac K0IV Plages A (Oct 83)	92	Walter *et al.* (1987)
MgII h+k	AR Lac K0IV Plages A+B (Sep 85)	≈ 200	Neff *et al.* (1989)
MgII k	F^{plage}_{MgIIk}	109	Adopted value
CIV 1550 Å	Saturated value	10	Vilhu (1987)
CIV 1550 Å	Solar plages	0.3–1	Vilhu (1987)
CIV 1550 Å	AR Lac K0IV Plages A+B (Sep 85)	13.7	Neff *et al.* (1989)
CIV 1550 Å	AR Lac K0IV Plages A (Oct 83)	4.8	Walter *et al.* (1987)
CIV 1550 Å	F^{plage}_{CIV}	13.7	Adopted value

*Units 10^5 ergs cm^{-2} s^{-1}.

Neff *et al.* (1989) identified three distinct plages on the surface of the K0 IV star in the AR Lac system observed in September 1985 by Doppler imaging the MgII k line. They found that plages A and B cover 8% of one hemisphere of the star with the plage surface flux listed in Table 4. Also listed is the CIV doublet surface fluxes assuming that the plage area is the same as for MgII. A previous study of AR Lac observed in October 1983 (Walter *et al.* 1987) derived somewhat smaller surface fluxes, but the data are sparse and the plage areas less well determined. I adopt, as the maximum nonflare surface fluxes, the plage values for AR Lac in September 1985. These values are 40-100% larger than the saturated values proposed by Vilhu (1987) for the most active spatially-averaged stars. This difference can be explained by the most active stars not being completely covered by bright plages.

I assume that active stars consist of plage regions covering a fraction A_{plage} of their surface and quiet regions covering a fraction $(1 - A_{plage})$. The plage covering fraction is given by

$$A_{plage} = c\frac{f^{max}_{obs} - f^{min}_{obs}}{F_{plage} - F_{quiet}},$$

where c is the conversion factor from observed to surface flux for the entire stellar disk (Rodonò *et al.* 1987). The quiet value of the surface flux, $F_{quiet} = cf_{obs}^{min}$, is an estimate of the line surface flux in the absence of plages. Like the visual magnitude of an unspotted (*immaculate*) star, the value of F_{quiet} derived from the smallest observed value of f_{obs}^{min} will likely be an upper limit, since some plage regions are probably present on the disk at all times.

Many observations of a star over a long period of time are needed to provide a sensible estimate for F_{quiet}. For example, Dorren and Guinan (1990) find a minimum value of $f_{CIV}^{min} = 2.5 \times 10^{-12}$ from 60 observations of V711 Tau obtained over a 6 year period. For the maximum nonflare CIV fluxes observed from V711 Tau in October 1981 (Rodonò *et al.* 1987) and February 1983 (Andrews *et al.* 1988), the values of A_{plage} are 0.17 and 0.23 (see Table 5). For the K0 IV star in the AR Lac system, the lowest observed MgII k line flux (corrected for interstellar absorption) is $f_{MgII\ k}^{min} = 4.3 \times 10^{-12}$ (Walter *et al.* 1987). The derived values of A_{plage} are 0.032 for October 1983 and 0.12 in September 1985. For II Peg the lowest values for the MgII k line and CIV flux are those observed in September 1986 by Doyle *et al.* (1989), although the July 1985 data set provides fluxes nearly this low (Doyle 1988).

The values of A_{plage} derived from the MgII fluxes for AR Lac, 0.032 and 0.12, are consistent with the plage areas of 0.02 and 0.08 derived from the Doppler imaging studies using the October 1983 and September 1985 data sets, respectively. This gives us confidence that the *excess flux* technique provides sensible estimates for plages areas, at least for the MgII lines.

Table 5 lists the fractional area of the stellar disk covered by spots, A_{spot}, at the phase of maximum line emission as derived from contemporary optical photometry. The cited values of A_{spot} were obtained from the spot radii and locations given in the same papers as the emission line fluxes with the following exceptions. For the October 1981 observations of II Peg, only Spot No. 2 was included since this spot is correlated in phase with the strong increase in emission line fluxes during this observing campaign. For the September 1986 observations of II Peg, only the area of the smaller spot, Spot No. 2 (Byrne 1987), is listed in the table, since the peak MgII and CIV flux occurred when the giant spot was on the invisible hemisphere of the star! For AR Lac Spot No.2 was near disk center at the time of maximum MgII flux in October 1983, and Spot No.1 was near disk center at the time of maximum MgII flux in September 1985. I have estimated the spot areas for V711 Tau from the Doppler-imaged spot maps produced by Vogt (1988).

Table 5. Estimates of the fractional stellar disk coverage of active regions

Star	When	c	Line	f_{obs}^{min}	f_{obs}^{max}	F_{quiet}	A_{plage}	A_{spot}
AR Lac (K0)	Oct 83	3.03(17)	MgIIk	4.3(-12)	5.3(-12)	1.3(6)	0.032	0.053
AR Lac (K0)	Sep 85	3.03(17)	MgIIk	4.3(-12)	8.0(-12)	1.3(6)	0.12	0.13
II Peg	Oct 81	3.58(17)	MgIIk	4.75(-12)	6.8(-12)	1.7(6)	0.080	0.085+
II Peg	Feb 83	3.58(17)	MgIIk	4.75(-12)	7.1(-12)	1.7(6)	0.091	≈ 0.11
II Peg	Jul 85	3.58(17)	MgIIk	4.75(-12)	6.9(-12)	1.7(6)	0.084	?
II Peg	Sep 86	3.58(17)	MgIIk	4.75(-12)	6.0(-12)	1.7(6)	0.049	≈ 0.12†
II Peg	Oct 81	3.58(17)	CIV	5.2(-13)	4.4(-12)	1.9(5)	1.2	0.085+
II Peg	Feb 83	3.58(17)	CIV	5.2(-13)	2.5(-12)	1.9(5)	0.60	≈ 0.11
II Peg	Jul 85	3.58(17)	CIV	5.2(-13)	1.17(-12)	1.9(5)	0.20	?
II Peg	Sep 86	3.58(17)	CIV	5.2(-13)	6.5(-13)	1.9(5)	0.04	≈ 0.12†
V711 Tau	Oct 81	1.04(17)	CIV	2.5(-12)	4.4(-12)	2.6(5)	0.17	≈ 0.13
V711 Tau	Feb 83	1.04(17)	CIV	2.5(-12)	5.0(-12)	2.6(5)	0.23	≈ 0.17

Notes: +Assuming $i = 90°$ and only Spot No. 2. †Assuming only Spot No. 2.

What can we conclude from the available data? The October 1981 and February 1983 II Peg MgII data sets and the AR Lac data sets indicate that $A_{plage} \approx A_{spot}$. However, in September 1986, the II Peg plage and spot areas differ by a factor of 2.5 and the phase of maximum MgII emission agrees with the small spot, not the very large spot on the opposite hemisphere. The CIV data for II Peg and V711 Tau imply plage areas very much larger than the spot areas, and in one case larger than the stellar disk itself. This could be due to a large underestimate in the saturated value of the CIV surface flux or, more likely, to flaring. On the other hand, $A_{plage} \approx A_{spot}$ for both of the V711 Tau CIV data sets. The data in hand lead me to conclude that the *excess flux* technique appears to be a sensible way to infer plage areas at least from the MgII fluxes, but the spatial correlation of plages and spots is unclear and requires further work.

5. Is the Sun a good paradigm for active stars?

I compare in Table 6 some of the magnetic and phenomenological properties of active stars with those of the Sun. My purpose here is not to come to a conclusion, but rather to stimulate more thought on this fundamental question. In the table the subscripts N and S refer to the chromospheric network and spots, respectively, and f is the filling factor for equipartion magnetic fields in the photosphere.

Acknowledgements.

This work is supported by NASA grants NAG6–82 to the University of Colorado and S–56460–D to the National Institute of Standards and Technology.

References.

Andrews, A.D. *et al.* : 1988, AA, 204, 177

Brown, R.L., Crane, P.C.: 1978, AJ, 83, 1504

Byrne, P.B.: 1987, in *Cool Stars, Stellar Systems, and the Sun,* Lecture Notes in Physics 291, ed. J.L. Linsky, R.E. Stencel, Springer–Verlag, Berlin, p.491.

Byrne, P.B., Doyle, J.G., Brown, A., Linsky, J.L., Rodonò, M.: 1987, AA, 180, 172

Córdova , F.: 1988, *Multiwavelength Astrophysics*, Cambridge Univ. Press, Cambridge.

Dorren, J.D., Guinan, E.F. 1990: ApJ, 348, 703

Doyle, J.D.: 1988, AA, 192, 281

Doyle, J.D., Butler, C.J., Byrne, P.B., Rodonò, M., Swank, J., Fowles, W.: 1989, AA, 223, 219

Dulk, G.A.: 1985, ARA&A, 23, 169

Eaton, J.A.: 1990, in this volume.

Houdebine, E.R., Foing, B.H., Rodonò, M.: 1990, AA, in press.

Huovelin, J., Saar, S.H., Tuominen, I.: 1988, ApJ, 329, 882

Kang, Y.W. and Wilson, R.E.: 1989, AJ, 97, 848

Kjeldseth–Moe, O. *et al.* : 1988, ApJ, 334, 1066

Linsky, J.L.: 1989, Solar Phys., 121, 187

Linsky, J.L.: 1990, in *Mechanisms of Chromospheric and Coronal Heating,* ed. E. Priest, R. Rosner, P. Ulmschneider, to appear.

Linsky, J.L., Neff, J.E., Brown, A., Gross, B.D., Simon, T., Andrews, A.D., Rodonò, M., Feldman, P.A.: 1989, AA, 211, 173

Table 6. A comparison between the Sun and active stars

How Active Stars Differ from Sun	Consequences of the Difference
(1) A factor of 10-100 greater magnetic flux/surface area $\mid f_N B_N \mid_{\odot} \approx 0.02 \times 1500G = 30G$ $\mid f_N B_N \mid_{dM_e} \approx 0.5 \times 4000G \approx 2000G$ $\mid f_N B_N \mid_{RS} \sim 0.3 \times 1000G \approx 300G$	• "isolated" flux tubes outside of spots $f_N \approx 0.02$ (quiet Sun) $f_N \approx 0.1$ (plage) • "dense packed" flux tubes on active stars
(2) Dense packing of magnetic flux tubes on active stars	• Slower divergence of B with $R/R_\star$ $\Rightarrow B_{AS}(R/R_\star) \gg B_{\odot}(R/R_\star)$ –larger magnetic fields in coronae –extended x-ray emitting coronae –bright microwave emission • $A_{fluxtube}$ increases slowly with $R/R_\star$ –MHD waves are damped lower in flux tubes –higher temperatures occur where n_e is large $\Rightarrow$ bright chromospheres, TR, and coronae –The Parker "nanoflare" mechanism efficient • Does this explain the onset of strong microwave emission at M 0V?
(3) Large high latitude starspots occur on active stars	• dynamo processes occur differently –differential rotation pattern must be qualitatively different
(4) Granulation patterns may differ • For RS CVn systems lower gravity increases convective velocities • For M dwarfs H_2 dissociation increases superadiabadic gradient and convective velocities	• Could enhance acoustic wave heating
(5) The acoustic energy frequency spectrum changes along the Hertzsprung-Russell diagram	• Electrodynamic resonances may lead to enhanced coronal heating near spectral type M0 V (Mullan 1984)
(6) Strong tidal forces are present in close binary systems like RS CVn, W UMa and Algol systems	• May alter the differential rotation distributions which result in qualitatively different dynamo activity
(7) The magnetic flux in starspots is a larger contribution to the total flux Assume $\frac{B_S}{B_N} \approx \left(\frac{B_S}{B_N}\right)_{\odot} \approx \frac{3500G}{1500G} \approx 2.3$ • For active stars $\frac{f_S B_S}{f_N B_N} \approx \frac{0.25 \times 2.3 \times B_N}{0.5 \times B_N} \approx 1$ • For the Sun $\frac{f_S B_S}{f_N B_N} \leq \frac{0.001 \times 2.3 \times B_N}{0.02 \times B_N} \leq 0.1$	• The microwave, x-ray and UV emission will exhibit large rotational modulation when a large low-latitude starspot (or spot group) rotates onto or off from the visible hemisphere

Mullan, D.J.: 1984, ApJ, 282, 603

Mürset, U., Solanki, S.K., Stenflo, J.O.: 1988, AA, 204, 279

Mutel, R.L., Lestrade, J.-F., Preston, R.A., Phillips, R.B.: 1985, ApJ, 289, 262

Neff, J.E., Walter, F.M., Rodonò, M., Linsky, J.L.: 1989. AA, 215, 79

Rodonò, M. *et al.* : 1986, AA, 165, 135

Rodonò, M. *et al.* : 1987, AA, 176, 267

Saar, S.H.: 1990, in *Solar Photosphere: Structure, Convection, and Magnetic Fields*, ed. J.O. Stenflo, Kluwer Academic, Dordrecht, p.123.

Semel, M.: 1989, AA, 225, 456

Schrijver, C.J.: 1987, AA, 180, 241

Simon, T., Linsky, J.L., Schiffer, F.H. III: 1980, ApJ, 239, 911

Strassmeier, K.G.: 1990, ApJ, 348, 682

Vilhu, O.: 1987, in *Cool Stars, Stellar Systems, and the Sun,* Lecture Notes in Physics 291, ed. J.L. Linsky, R.E. Stencel, Springer–Verlag, Berlin, p. 110.

Vogt, S.S.: 1988, in *The Impact of Very High S/N Spectroscopy on Stellar Physics*, ed. G. Cayrel de Strobel, M. Spite, Kluwer Academic, Dordrecht, p. 253.

Vogt, S.S., Penrod, G.D.: 1983, PASP, 95, 565

Walter, F.M. *et al.* : 1987, AA, 186, 241

Discussion: Linsky

Giampapa Can you identify any physical process among those that you discussed that has no counterpart in the Sun?

Linsky No.

Gurman You've made the point that we should ignore the solar paradigm for a variety of reasons, yet you use solar flare observations as a basis of comparison. Along that line, it should be quoted that while solar hard X-ray and microwave bursts often have very similar light curves, there is usually a time delay between the two that is thought to represent a propagation time for the exciting population of electrons. If this is so, the delay would be larger in the larger loops you have discussed.

Linsky Obviously I have not been fully consistent with admonitions to avoid assuming *a priori* the solar paradigm. Yes, the delay time should be larger for the larger loops on the active stars. This should be searched for when stellar observations of hard X-rays become more feasible.

Mullan The existence of 10^3 times more energy in a stellar flare than in the largest solar flares is not a convincing argument for a physical difference between solar and stellar flares. Even in the Sun, flare energies span 9 orders of magnitude (nanoflares to flares). Therefore the extra 3 orders can be accommodated without necessarily abandoning the solar paradigm.

Linsky I agree that the physics may be the same in solar and the more energetic flares, although the case is far from proven at this time. However, the very different boundary conditions, for example the much larger magnetic fluxes on the active stars, produce phenomena with much larger scales (in this case energy) than for the Sun.

Skumanich OSO-8 observations of flare kernels have shown that the Mg II lines are blue-shifted during the rise phase, $\approx 15\,\mathrm{km/sec}$, and become red shifted in the decay phase, $\approx 5\,\mathrm{km/sec}$. This was the first determination of flare induced shifts in the Mg II lines.

SUNSPOTS: AN OBSERVATIONAL OVERVIEW

P. Maltby
Institute of Theoretical Astrophysics
University of Oslo

1. Introduction

Our empirical knowledge of sunspots is considerable (*e.g.* Bray and Loughhead 1964; Cram and Thomas 1981; Obridko 1985; Schmidt 1985; Schröter et al. 1987; Stix 1989; Cox, Livingston and Matthews 1990). Even so, Parker's (1979) conclusion that sunspots are a phenomenon lacking scientific explanation is still valid.

A selection of topics has been made, with the intention of discussing some observational aspects of sunspots that may be relevant to other stars. In the space and time available it is unfortunately not possible to include radio observations and the interesting high energy part of the sunspot phenomenon. The focus will be on large sunspots. Although this means that an average size sunspot will not be discussed, the interest for stellar work is presumably on the large sunspots.

2. Sunspot groups and sunspot nests

The well-known Maunder "Butterfly diagram" displays the distribution of sunspots/spot groups in time and in heliocentric latitude throughout the solar cycle. Less attention has been given to the study of the pattern in solar longitude, although the motion of a large-scale pattern in the revivals of sunspots and complexes of activity has been discussed (*e.g.* Becker 1955). More recently Gaizauskas et al. (1983) studied complexes of activity from magnetograms, whereas Castenmiller, Zwaan and Zalm (1986) have used sunspot groups as tracers for activity complexes.

The followimg comments are mainly based on a paper in press by Brouwer and Zwaan (1990). The study is carried out by representing each sunspot group with three coordinates: longitude (L), latitude (ϕ) and time (t). They find that the sunspot groups appear in "nests". In many cases more than one sunspot group has been recorded in the same nest during one disk crossing. However, in a still active nest there may be no sunspots visible for one full disk crossing. In order to be member of a nest the sunspot group must satisfy a clustering criterion that combines deviations in L and ϕ and corresponds to r.m.s. deviations of 1.85 and 1.03 degrees, respectively. This small deviation in latitude strongly suggest that the notation "sunspot nests" may give a better representation than "active longitudes". In this sense the phrase "active longitudes" is directly misleading.

An interesting part of the study by Brouwer and Zwaan (1990) is their investigation of what fraction of the data represent real sunspot nests. To answer this question the authors estimate the fraction of chance coincidences by using concocted samples with randomized longitudes. Comparing the results from the concocted sample with 5.5 years of real data their conclusion is as follows. About half of the 128 nests with 2 members represent real nests; at least 0.70 of all nests with 3 members are genuine sunspot nests. For 4 or more members the corresponding fraction is 0.91. One third of all sunspots are members of sunspot nests.

For sunspot nests with 4 or more members the amplitude of any large-scale latitude dependent flow like meridional circulation is ≤ 0.02 deg per day, corresponding to ≤ 3 m/sec. The maximum deviation from a smooth differential rotation is 0.05 deg per day or 7 m/sec.

3. Continuum observations

Numerous papers have been published showing one measured parameter as a function of sunspot size. Several of these studies are valuable but investigations dealing with the determination of physical parameters like the sunspot brightness are usually very uncertain since stray light tends to reduce the quality of the measurements (*e.g.* Zwaan 1965). Not all observers agree with this statement and claim to have detected a size dependence or, for instance, a difference between growing and stable sunspots (*e.g.* Chou 1987). We note that extremely good observing conditions and space observations have not been used freqently enough to determine with accuracy the intensity and other physical parameters of small and medium size sunspots. The discussion below of sunspot continuum intensities will therefore be based mainly on our knowledge of large sunspots.

3.1 Methods of observation and reduction

Our present knowledge rests on the introduction of photoelectric photometry (Makita and Morimoto 1960; Jensen and Ofstad 1963) and the development of a proper correction methodology for stray light (Stumpff 1961; David and Elste 1962; Birkle and Mattig 1965; Zwaan 1965). One may illustrate the development in observing technique over time by plotting published values of the umbra/photosphere intensity ratio versus wavelength. It should be emphasized that such a plot may be explained by improvements in the observations and in the correction method. The origin of the stray light is small amouts of light scattered from each position in the solar image which contributes to the observed intensity of the sunspot. In passing we note that a test of the correction method is actually possible by using the transit of Mercury as an artificial spot on the solar disk (Maltby 1971; Mattig 1971).

The observed fine structures in the form of umbral dots (*e.g.* Adjabshirzadeh and Koutchmy, 1983) and penumbral filaments (*e.g.* Muller 1973; Collados, del Toro Iniesta, and Vázques 1988) show dimensions comparable to the scale height and are accordingly important for understanding the physics of sunspot atmospheres. Extremely good observing conditions or space observations are required to improve the estimates of penumbral filament temperatures (*e.g.* Kjeldseth-Moe and Maltby 1974) and the temperature of umbral dots (*e.g.* Adjabshirzadeh and Koutchmy, 1983; Obridko and Staude, 1988). In the following our attention will be on the average values of the penumbra and of the darkest part of the umbra, as the aim of this section is to present values of parameters which may be compared with other stars.

Two methods in photoelectric photometry have been used to measure sunspot continuum intensities: 1) observations in "continuum windows" between spectral lines, and 2) observations in broad spectral bands, corrected for spectral lines. One problem with the first approach is that continuum windows are scarce, especially in the visible umbral spectrum where the high spectral line density depresses the continuum. Furthermore, since high spectral resolution and relatively long exposure times are required, extremely good and stable observing conditions are necessary for the method to work. With broad spectral band observations good signal-to-noise ratios are obtained with short integration times. The observations are corrected for spectral lines by using the best available knowledge of the sunspot spectrum within each spectral band. The problem with this approach is that possible differences in spectral characteristics between different sunspots cannot be taken into account without supplementary observations at high spectral resolution.

During the period 1967 - 1986 observations of sunspot intensities were carried out through 10 months of each year at the Oslo Solar Observatory. The broad band pinhole photometers recorded intensities in a total of 11 wavelength regions in the spectral range 0.387 - 3.8 μm.

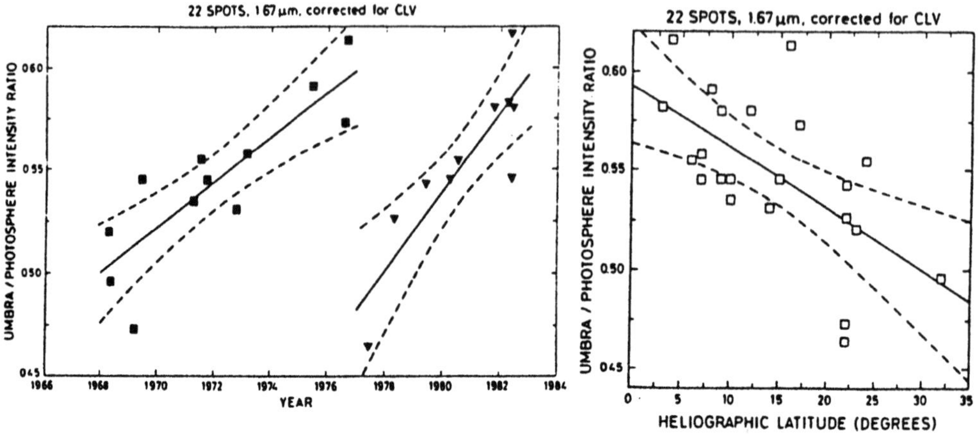

Figure 1. Umbra/photosphere intensity ratio at 1.67 μm, corrected for the center-limb variation, is give as a function of time for cycle 20 (*squares*) and cycle 21 (*triangles*). The ratio is also shown as a function of spot heliographic latitude. All large sunspots observed during good seeing conditions are included. Least squares regression lines as well as 95% confidence limits are given.

Only measurements with an accuracy of 0.015 in the umbra/photosphere intensity ratio were retained. The observations allowed studies of spectral energy distribution, center-limb variation, solar cycle variation, as well as comparison with other sunspot parameters, both for the umbrae and the penumbrae.

3.2 Results

Most of the results from these observations have been published earlier; in this presentation only a summary of the findings will be given.

(i) The observations confirm the finding of Rossbach and Schröter (1970) that both umbra and penumbra intensities are independent of sunspot size for umbral radii larger than 4 arc sec. This contrasts with the earlier view (Bray and Loughhead 1964; Stenflo 1976) of a systematic increase in umbral brightness with decreasing size. Note, however, that this does not imply that all large sunspots show the same intensity. Real differences in the radiation temperature between different sunspots have been detected for large sunspots, and may also exist for smaller sunspots (*i.e.* umbral radii ≤ 5 arc sec).

(ii) The observed penumbral intensities (Maltby 1972) may, to a first approximation, be described by a single parameter value of $\Delta\theta = constant$, where $\Delta\theta = 5040(1/T_p - 1/T_{ph})$ and T_p and T_{ph} are respectively the temperature of the penumbra and the photosphere, as averaged over penumbral filaments and granules. The average value for penumbrae is $\Delta\theta = 0.055$.

(iii) Differences in penumbral intensities between different sunspots are observed and found to be real. Sunspots with a darker than average umbra usually show a relatively dark penumbra. The correlation coefficient between penumbra and umbral intensities is 0.9 at a wavelength of 1.67 μm.

(iv) Within the accuracy of the observations it has not been possible to detect any center-limb variation in the penumbra/photosphere intesity ratio.

(v) By observing the very same sunspot on several days during its passage over the solar disk it is found that all sunspots studied for more than one day show a real and significant decrease in the umbra/photosphere itensity ratio towards the limb (Albregtsen, Jorås, and Maltby 1984). This observed limb darkening of the umbra/photosphere intensity ratio contradicts the findings

of earlier observers; one reason is evidently that the relative umbral limb darkening is simplest to detect in the infrared, a wavelength region not easily available to earlier observers.

(vi) The hypothesis (*e.g.* Adjabshirzadeh and Koutchmy 1983) that a variation in the relative number of umbral dots from one sunspot to another may account for the observed difference in umbra/photosphere intensity ratio has been studied. A change in the number of bright umbral dots corresponds to a change in umbral intensity from one sunspot to another that is wavelength dependent. Whereas the hypothesis would give the largest change in the visible, the observations show the largest change in the infrared.

(vii) A solar cycle variation in the umbra/photosphere intensity ratio was detected (Albregtsen and Maltby 1978). The umbra/photosphere intensity ratio appears to be a linear function of the phase in the solar cycle (Figure 1). The phase is measured as the time t elapsed since the last minimum in the cycle, the duration of which is t_0. At 1.67 μm the correlation coefficient between the umbra/photosphere intensity ratio and t/t_0 is 0.86 (+0.07, -0.12). The variation is caused by differences in *umbral* intensity; the observed change in the intensity ratio is too large to be explained by a variation in the photospheric intensity.

(viii) It is well known that the average heliographic latitude of sunspots varies throughout the solar cycle. Figure 1 also shows the connection between the umbra/photosphere intensity ratio and the heliographic latitude, with a correlation coefficient of -0.63 (-0.12, +0.16) at 1.67 μm. Figure 1 and the corresponding correlation coefficients show that the umbral intensity is better correlated with t/t_0 than with the sunspot heliographic latitude.

The tendency for high latitude sunspots to be darker than low latitude sunspots may possibly have a counterpart in other stars. In light curve starspot modelling, one neglected result is the high latitude of huge spots (Vogt 1983; Rodonò et al. 1985). For this reason it may be of value to mention possible interpretations of the observed connection between epoch/latitude and sunspot brightness. One argument goes as follows. Since the sunspot temperature appears to be a function of the phase in the solar cycle, the responsible subphotospheric flux tubes probably have a lifetime comparable to the solar cycle. The umbral brightness may be determined by the age of the subphotospheric flux tube (Schüssler 1980). An alternative interpretation, where the temperature depends on the height of the flux tube within the convection zone at the time of formation, has been presented (Yoshimura 1983). Nordlund and Stein (1990) have recently performed a preliminary simulation of the central part of a sunspot umbra. Considering the total mass in the flux tube as approximately conserved, they argue that the relative pressure difference might control the "looseness" of a sunspot, and thus determines the umbral brightness dependence on cycle phase.

3.3 Sunspot flux deficit and irradiance modelling

The modulation of the solar irradiance has been discussed by several authors (*e.g.* Smith and Gottlieb 1974; Willson et al. 1981; Foukal and Lean 1986; Hudson 1988). In these papers one aim is to calculate the sunspot flux deficit. For starspot modelling, similar questions are relevant. From the intensity measurements presented above, some points may be of interest.

(i) It would have been nice if we could find one parameter that did describe the radiative output from sunspots. The observed center-limb variation of the umbra/photosphere intensity ratio shows that the corresponding $\Delta\theta$ value is not constant. In fact, it is difficult to characterize the average umbral intensity with a single parameter, for instance by scaling the solar temperature T as function of optical depth, τ with a constant factor.

(ii) The observations show that sunspots differ in intensity. For sunspots with umbral radii > 5 arc sec the variation with area is not apparent, but this does not imply that the sunspots show the same radiation temperature. A careful study would have to take into account both the center-limb and latitude variations of the intensity ratios.

(iii) Although the average penumbral temperature may be characterized by a single parameter, $\Delta\theta = 0.055$, significant changes in the penumbral intensity from one sunspot to another have been measured.

(iv) In some flux deficit studies the time variation in sunspot area is taken into account, assuming that the ratio between penumbral and umbral area is constant, independent of time. The variation in the ratio between total area and umbral area as a function of age for 30-day sunspots has been investigated by Ringnes (1964). His Figure 23 shows that the ratio between the total area and the umbral area varies between 3.5 and 6.2 as function of age for a 30-day sunspot.

Recently measuremments of the umbra/penumbra ratio, sunspot and plage photometry have been carried out by Brandt, Schmidt, and Steinegger (1990), Steinegger et al. (1990) and by Steienegger, Brandt, and Schmidt (1990) in order to clarify the effects on solar irradiance variations. Further discussion of this topic is given by Lean during this colloquium.

3.4 Effect on helioseismology measurements

Consider observations of the Sun as a star. The presence of a sunspot on the solar disk means that, compared with the "unspotted" disk, the contribution to the intensity from the sunspot area is (in most cases) less. Consider velocity measurements using, say, the potassium $\lambda 769.9\,\mathrm{nm}$ line. Before the sunspots central meridian passage (CMP) this corresponds to less contribution from an area that has a line-of-sight velocity directed towards the observer. In the integrated line profile this corresponds to a slight displacement in the opposite direction, that is to the red. After CMP the line-of-sight velocity of the sunspot is directed away from the observer, whereas the corresponding velocity signal is negative. This line of argument was used by Andersen and Maltby (1983) to explain the 13.1 ± 0.2 day velocity signal detected by Claverie et al. (1982). Additional influence on the velocity signal may come from other parts of the active region (Durrant and Schröter 1983).

4. The magnetic field

The magnetic field observed on the Sun emerges from the solar interior and has already been broken up into separate flux tubes when reaching the visible surface. It was pointed out by Parker (1955) and independently by Jensen (1955) that the confining force of gravity unavoidably introduces magnetic buoyancy. The magnetic field outside sunspots also separates into individual flux tubes with field-free regions between them (*e.g.* Zayer, Solanski, and Stenflo 1989).

4.1 Line Formation Theory

The magnetic field in sunspots are most commonly inferred from measurements of the polarization in spectral lines split by the Zeeman effect (see Zeeman 1913). It may be of some interest to know that the splitting of the D_1 and D_2 lines of Na by a magnetic field was first found by Fievez (1885), 11 years before the effect was discovered by Zeeman in 1896.

The interpretation of the observations rests on an understanding of the theory of spectral line formation in the presence of a magnetic field. The line formation theory in terms of Stokes parameters was formulated originally by Unno (1956) and extended by Rachkovsky (1962) to include magneto-optical effects like Faraday rotation. Considerable effort has been devoted to the solution of the Unno-Rachkovsky equations both in LTE and non-LTE, for a review see Rees (1990). Although more general non-LTE formulations, including quantum interference and partial redistribution exist (Landi degl'Innocenti 1983), the practical applications are usually limited to electric dipole transition, LS-coupling, no quantum interference between the Zeeman states, complete redistribution on scattering, stimulated emission treated as negative absorption, and steady state atmosphere.

In this review attention will be given to another limiting factor, *i.e.* the need for a better treatment of molecular and atomic blends as well as depressions in the continuum caused by line haze. Consider the FeIλ5250.22Å line that has been extensively used for magnetic field measurements, partly because of its simple Zeeman triplet pattern and large Landé factor, $g = 3$. The presence of a strong π-component in the umbra, nearly independent of the aspect angle was pointed out already by Evershed (1944). The blend was identified by Kjeldseth Moe (1973) as the λ5250.24Å P_1 (71) line of TiO belonging to the (0,0) band of the α-system. His calculations showed that estimates of the inclination of the magnetic field vector - using the line depths of the π- and σ-components - will easily yield values 20 to 30 degrees in error if the molecular blend is neglected. In this example a molecular blend disturbs only one of the line components, whereas all three components are influenced by the line haze. More generally, in a comparison of observed and calculated line profiles it is necessary to:

- include molecular and atomic lines that blend with the spectral line in question,
- include additional absorption responsible for a depression of the continuum.

This suggests that the observed line haze in the spectra of umbrae and late type stars may seriously influence the results of magnetic field measurements. These spectra are so crowded with lines that care should be taken in all three cases of polarimetry measurements; *i.e.* broad-band, magnetograph and high spectral resolution observations. The broad-band polarization observed in sunspots (Leroy 1962; Illing et al. 1974) has, so far, had limited diagnostic value. A better theoretical calculation of the spectrum, taking into account the blends present may improve the situation. This approach could possibly be of value in the discussion of a differentially twisted magnetic field in sunspots (*e.g.* Makita 1986).

4.2 Measurements

The magnitude of the magnetic field near the centre of the sunspot is usually between 2000 and 3500 G. Several authors including Beckers and Schröter (1969), Wittmann (1974), Kawakami (1983), Lites et al. (1987), Lites and Skumanich (1990), and Lites, Scharmer, and Skumanich (1990) have presented measurements of the magnetic induction $B(r)/B(0)$ as function of the distance r from the centre of the sunspot, measured in units of the penumbral radius. The dependence $B(0)/(1+r^2)$ of $B(r)$ from Beckers and Schröter has become accepted as a standard, although several of the more recent studies suggest a steeper gradient and accordingly a smaller value of B at the rim of the penumbra. In comparisons of this kind, care should be taken since:

- the umbra/penumbra ratio and most probably the related field configuration may differ from one sunspot to another,
- sunspots with circular symmetry are extremely difficult to find in nature.

The determination of the magnetic field configuration requires that the inclination of the field is derived from the ratio of the π- and σ- components. These measurements are sensitive to scattered light and to the adopted line formation theory. Inclination angle to the vertical of the magnetic field vector versus the distance r from the centre of the sunspot have been published by several authors including Beckers and Schröter (1969), Wittman (1974), Gurman and House (1981), Hagyard et al. (1983), Kawakami (1983), Mickey (1985), Lites aud Skumanich (1990), and Lites, Scharmer, and Skumanich (1990). The more recent papers suggest an inclination of 60 - 70 degrees close to the rim of the penumbra, whereas earlier papers like Beckers and Schröter (1969) found the field to be much closer to being horizontal. However, the uncertainty in these determinations is large (see Lites, Scharmer, and Skumanich 1990) and we may ask if this situation may be improved in the near future. The magnetically sensitive infrared line of MgI at 12.32 μm may be quite valuable in this respect (see below).

The field measurements at photospheric heights have also been studied with the intention of determining the vertical gradient of the magnetic field (*e.g.* Hofmann and Rendtel 1989) and the deviation from a potential field (*e.g.* Hagyard 1988). These measurements indicate that the umbral magnetic field can extend with strong field strength into the chromosphere and corona. A review of the extrapolation of photospheric magnetic field into the corona is given by Sakurai (1990). Direct measurements of the CIVλ1548Å line by the UVSP instrument on board the Solar Maximum Mission by Hagyard et al. (1983) shows that the sunspot field may be of the order of 10^3 G at, say, a temperature of 10^5 K.

4.3 The MgI 12.32 μm line

It appears from the list of vector magnetographs presented by Makita (1990) that considerable effort is already devoted to this branch of solar physics. Part of the difficulties stem from instrumental polarization (see Lites, Scharmer and Skumanich 1990). The THEMIS and LEST projects (Rayrole 1987; Engvold 1990) will hopefully bring new knowledge to this field.

Most of the difficulties in vector magnetic field measurements originate from the fact that the magnetic splitting is relatively small in the visible spectral region. The sensitivity of measurements to the magnetic field depends on the ratio of the line splitting to the thermal Doppler width, which scales as the product of the wavelength and the Landé g-factor. The MgI line at 12.32 μm (Murcray et al. 1981; Brault and Noyes 1983; Chang and Noyes 1983) is therefore of special interest. By using a grating postdisperser, incorporating a blocked impurity band detector, Deming et al. (1988) have succeeded in using this line for sunspot magnetic field measurements. Although the MgI line appears to be difficult to measure in the umbra, this technique appears very promising. The line may also be of interest in stellar spectra (Jennings et al. 1986).

Whereas Lemke and Holweger (1987) suggested that the infrared MgI line was formed in the photosphere, Zirin and Bopp (1989) claimed that the line was formed in the chromosphere above the temperature minimum. A more comprehensive study of the line formation is in preparation by Carlsson, Rutten, and Shchukina (1990), confirming the suggestion by Lemke and Holweger (1987).

The reader may have noticed that the term Zeeman splitting has not been used in the description of this line. The reason is that for sunspot magnetic fields the MgI level fine structure is much smaller than the magnetic splitting, which accordingly follows the Paschen-Back pattern, with a Landé factor of unity. This has been confirmmed by direct laboratory measurements of the 12.32 μm line by Lemoine, Demuynck, and Destombes (1988).

5. Flows

The wavelength displacement of spectral lines in the spectra of sunspots is known as the Evershed effect; occasionally it is called the Evershed-Abetti effect. The wavelength shifts are usually interpreted as material motion (Evershed 1909) and referred to as the Evershed flow; for a review see Sivaraman (1984). The observed line shifts could also be caused by other processes, for instance by compressible waves. At photospheric heights the flow is nearly horizontal, directed from the umbra towards the photosphere. For sunspots with a complicated structure the velocity field also becomes more complicated and may show tangential velocity components (Abetti 1932; Maltby 1964). Generally the flow field is not circularly symmetric; even so, this assumption is used frequently in deducing the flow field from the observed line-of-sight velocities. Contrary to the suggestion by Kinman (1952), the flow is restricted to the penumbra. The flow in the moat surrounding sunspots may be derived from studies of moving magnetic features and is of another character (Brickhouse and LaBonte, 1988).

Some high spatial resolution oberations suggest that the highest flow speeds occur in the dark penumbral filaments (*e.g.* Wiehr et al. 1986; Bida et al. 1990), whereas one recent study (Lites, Scharmer, amd Skumanich 1990) fails to confirm this finding. The magnitude of the flow at this level is usually a few km/sec, but speeds of up to 6 km/sec have been reported. Recently, a time variation in the flow has been reported (Shine et al. 1990).

The observational picture is not as simple as indicated above. White light observations show penumbral grains moving in the opposite direction, *i.e.* towards the umbra with a speed of the order 0.5 km/sec (Muller 1973; Shine et al. 1987; Zirin and Wang 1989). These motions are possibly not material. Spruit (1981) has suggested that the flux tubes, comtaining the Evershed flow, move up and down with respect to an optical depth of unity. Another approach may be to regard the penumbral filaments as individual flux tubes and then suggest that the apparent motions observed in the white light pictures reflect the propagation of distortions at the flux tube boundary (Maltby 1989).

5.1 Inverse Evershed flow

The reversal in flow direction in the chromosphere compared to that of the photosphere is well known (*e.g.* St. John 1913). Observations of chromospheric lines show that the flow is directed towards the umbra, along flow channels that may be identical to the superpenumbral filaments (*e.g.* Beckers 1962). The large horizontal extent of the flow in the chromosphere suggests motion along field lines that extend approximately two sunspot radii from the spot centre.

Dialetis, Mein and Alissandrakis (1985) emphasize that their observations show a subsonic flow in magnetic flux tubes, whereas Beckers (1962) and Maltby (1975) have presented observations that were interpreted as showing a siphon flow where the downflow becomes supersonic, at least for some flux tubes. This difference in opinion may not be important, since the downflow speed in different flux tubes will depend on conditions like the variation in pressure along the tube. It seems unlikely that individual flux tubes should have exactly the same distribution of physical parameters, such as the pressure. Hence, differences in the driving force for individual flux tubes may give both subsonic and supersonic solutions in the same sunspot region.

Very few observations of time variations have been presented; Maltby (1975) found that the flow channels alter with characteristic times of 8 and 70 min; most likely corresponding to a characteristic time of perturbations and to the lifetime of the flow channels themselves. It is interesting to note that the observed 70 min lifetime of the flow channels is comparable to the time it takes for the gas to pass from one end of the loop to the other. This suggests that the nature of the flow is closely linked to the time variation of the superpenumbral filaments.

5.2 Flows in the sunspot Transition Zone

Spectrograms obtained with the High Resolution Telescope and Spectrograph (HRTS), both on rockets and on Spacelab 2, show strong mass flows through the transition zone that are particularly prominent above sunspot regions. Most characteristic for the flow associated with sunspots are regions with supersonic downflows, but upward flowing gas is also observed with generally smaller velocities (*e.g.* Kjeldseth-Moe et al. 1988). The flow pattern is certainly not circularly symmetric; it changes from one day to the next and even on a time scale of minutes, but the general character is maintained and appears always to be present.

Another characteristic property of the flow is the presence of several distinctly different flow speeds within the one arc sec resolution element. Multiple flow speeds are also observed in areas with upflows, except at some locations inside the sunspot, where only a single velocity component is seen (Maltby et al. 1990). An apparent lack of balance between up- and downflowing mass fluxes is found.

Measurements of the flow velocities in the transition zone between the chromosphere and corona above sunspots have also been carried out by the Ultraviolet Spectrometer and Polarimeter (UVSP) on the Solar Maximum Mission (SMM) spacecraft. The results from UVSP and HRTS may at first sight appear to be partly in conflict. However, one should be aware of the fact that the supersonic downflows detected by HRTS are practically undetectable by the available modes of operation of the UVSP instrument (Woodgate et al. 1980; Alissandrakis et al. 1988).

6. Concluding remarks

The sunspot magnetic field influences the energy transport processes and accordingly the sunspot temperature distribution. The magnetic field also alters the gas pressure distribution; the gas is nearly in hydrostatic balance along the field lines. The sunspot brightness is primarily determined by the energy transport processes, but the energy equation is mathematically coupled to the momentum equation.

The largest contribution to the lateral gas pressure difference between the photosphere and the umbra comes from the magnetic field, that acts both through the magnetic pressure $B^2/2\mu$ and through the magnetic tension. The importance of this relation was first pointed out by Alfvén (1943), who found that static equilibrium in a vertical magnetic field is only possible if the temperature in the field is lower than in the environment. Using semi-empirical models one finds that the lateral pressure difference reaches a maximum that exceeds the magnetic pressure at heights relatively close to an optical depth of unity in the umbra. The exact value depends on the adopted value for the Wilson depression, *i.e.* the height difference between the levels with optical depth equal to unity in the photosphere and the umbra. Although analytical magetostatic models are of interest the focus has shifted to numerical solutions (*e.g.* Pizzo, 1987). An approach that involves numerical solutions of a free boundary value problem has been introduced (Schmidt and Wegmann 1983; Jahn 1989). Recently numerical simulations have been presented (Nordund and Stein 1990).

The observed flows show that the sunspot is not in perfect magnetostatic equilibrium. In order to illustrate the physics involved, consider flows along the magnetic field lines and assume that the collisions are frequent enough to permit a magneto-hydrodynamic approach. Let the magnetic flux tube be regarded as rigid and unperturbed by the flow. This low-β approach is based on studies of the siphon flow (Meyer and Schmidt 1968; Pikelner 1971; Cargill and Priest 1980, 1982; Noci 1981; Mariska and Boris 1983). The gas has stellar abundance, pressure P, mass density ρ and moves with velocity v along the magnetic field in a flux tube with cross section A, that changes with position, z. Assume that the temperature distribution is known and restrict the problem to the equations of mass and momentum conservation together with the equation of state. Eliminating P, we find after rearrangement that,

$$\left(\frac{v^2}{a^2}-1\right)\frac{1}{v}\frac{\partial v}{\partial z} = \frac{1}{A}\frac{\partial A}{\partial z} - \frac{1}{T}\frac{\partial T}{\partial z} - \frac{g_e}{a^2} + \frac{1}{v}\left(\frac{1}{A}\frac{\partial A}{\partial t} + \frac{1}{\rho}\frac{\partial \rho}{\partial t}\right) \tag{18}$$

where the effective gravity force, ρg_e is written as,

$$g_e = g_{\|} - D + \frac{\partial\, v}{\partial\, t} \tag{19}$$

and $a^2 = RT/\mu$ is the square of the isothermal sound speed. The phenomenological acceleration D (*e.g.* Leer and Holzer 1980) enters into the problem in the same way as the gravitational acceleration along the field, $g_{\|}$. A time change in A, represented by the term $1/(vA)\partial\, A/\partial\, t$, has a similar effect on the flow to that of the $1/A\,\partial\, A/\partial\, z$ term or the acceleration term D/a^2.

If the flux tube cross section A increases/decreases with time this corresponds formally to a momentum addition/subtraction to the flow.

For a steady flow it appears that the effect of each term on the right hand side of the equation (*i.e.* effects caused by a change in cross section with position, change in temperature with z and effective gravity) reverses in action as v changes from less/larger than to larger/less than the isothermal sound speed a. Consider the effect of a change in cross section along the flux tube. An increase in the cross section leads to a reduction in flow speed for a subsonic flow. However, a supersonic flow is analogous to the flow of traffic in a multilane roadway; if the roadway widens to more lanes, the traffic speed is increased. A supersonic downflow may be accelerated by gravity and/or a temperature increasing upwards, whereas an increase in the flux tube cross section with height may slow the supersonic downflow down.

We note that driving a flow requires a pressure difference, unless additional accelerations, for example a wave pressure, are present. Consider now solely the effect of a velocity shear in the flow. It can be shown that Alfvén waves are stable provided the maximum flow velocity is less than twice the Alfvén speed (Lau and Liu 1980). The reason for this stability is that the magnetic field prohibits free exchange of vorticity, which is the cause of the instability in classical shear flow hydrodynamics. This implies that neighbouring flows may be quite different without disturbing one another, provided no other effects like, for instance, wave-wave interaction occur.

In several aspects, the results of observations are compatible with the predictions of a field aligned flow. These characteristics are the tendency for upflows and downflows to occur in different regions, redshifts usually stronger than blueshifts, and the detection of shocks may be possible in some of the downflows. However, from recent studies of line profiles it is apparent that a few points require special attention in the transition zone (Kjeldseth-Moe et al. 1988).

References

Abetti, G. 1932, Mem. R. Oss. Astr. Arcetri, 50, 47.

Adjabshirzadeh, A., Koutchmy, S.: 1983, AA, 122, 1.

Albregtsen, F., Jorås, P.B., Maltby, P.: 1984, Solar Phys. 90, 17.

Albregtsen, F., Maltby, P.: 1978, Nature, 274, 41.

Alfvén, H.: 1943, Arkiv Mat. Astr. Fysik, 29, No.11, 1.

Alissandrakis, C.E., Dialetis, D., Mein, P., Schmieder, B., Simon, G.: 1988, AA, 201, 339.

Andersen, B.N., Maltby, P.: 1983, Nature, 302, 808.

Becker, U.: 1955, Z. Ap. 37, 47.

Beckers, J.M.: 1962, Austral. J. Phys. 15, 327.

Beckers, J.M., Schröter, E.H.: 1969, Solar Phys. 10, 384.

Bida, T., Lites, B.W., Johannesson, A., Scharmer, G.: 1990, BAAS, 22, 880.

Birkle, K., Mattig, W.: 1965, Z. Ap. 60, 264.

Brandt, P.N., Schmidt, W., Steinegger, M.: 1990, Solar Phys. submitted.

Bray, R.J., Loughhead, R.E.: 1964, *Sunspots* (London: Chapman Hall).

Brault, J., Noyes, R.: 1983, ApJ, 269, L61.

Brickhouse, N.S., LaBonte, B.J.: 1988, Solar Phys. 115, 43.

Brouwer, M.P., Zwaan, C.: 1990, Solar Phys. in press.

Cargill, P.J., Priest, E.R.: 1980, Solar Phys. 65, 251.

Cargill, P.J., Priest, E.R.: 1982, Geophys. Ap. Fluid Dynamics, 20, 227.

Carlsson, M., Rutten, R.J., Shchukina, N.G.: 1991, in *The Dynamic Sun*, 6th Europ. Meeting on Solar Phys. (Debrecen, Hungary), in press.

Castenmiller, M.J.M., Zwaan, C., van der Zalm, E.B.J.: 1986, Solar Phys. 105, 237.

Chang, E.S., Noyes, R.W.: 1983, ApJ, 275, L11.

Chou, D.-Y.: 1987, ApJ, 312, 955.

Claverie, A., Isaak, G.R., McLeod, C.P., van der Raay, H.B., Pallé, P.L., Roca Cortes, T. 1982, Nature, 299, 704.

Collados, M., del Toro Iniesta, J.C., Vázquez, M.: 1988, AA, 195, 315.

Cram, L.E., Thomas, J.H. (eds.): 1981, *The Physics of Sunspots*, Sacramento Peak Obs. Conf., Sunspot, NM, USA (Sunspot, N.M.: SPO).

Cox, A.N., Livingston, W.C., Matthews, M.: 1990, *The Solar Interior and Atmosphere*, Univ. Arizona Press, in press.

David, K.-H., Elste, G.: 1962, Z. Ap. 54, 12.

Deming, D., Boyle, R.J., Jennings, D.E., Wiedemann, G.: 1988, ApJ, 333, 978.

Dialetis, D., Mein, P., Alissandrakis, C.E.: 1985, AA, 147, 93.

Durrant, C.J., Schröter, E.H., Nature, 301, 589.

Engvold, O.: 1990, in *Solar Photosphere, Convection and Magnetic Fields*, ed. J.O. Stenflo, IAU Symp. No. 138, 469.

Evershed, J.: 1909, MNRAS. 69, 454.

Evershed, J.: 1944, Observatory, 65, 190.

Fievez, Ch. 1885, *Repertorium der Physik*, ed. F. Exner (München: Oldenbourg), 21, 766.

Foukal, P., Lean, J.: 1986, ApJ, 302, 826.

Gaizauskas, V., Harvey, K.L., Harvey, J.W., Zwaan, C.: 1983, ApJ, 265, 1056.

Gurman, J.B., House, L.L.: 1981, Solar Phys. 71, 5.

Hagyard, M.J.: 1988, Solar Phys., 115, 107.

Hagyard, M.J., Teuber, D., West, E.A., Tandberg-Hanssen, E., Henze, W., Beckers, J.M., Bruner, M., Hyder, C.L., Woodgate, B.E.: 1983, Solar Phys. 84, 13.

Hofmann, A., Rendtel, J.: 1989, Astron. Nachr. 310, 61.

Hudson, H.S.: 1988, ARA&A, 26, 473.

Illing, R.M.E., Landman, D.A., Mickey, D.L.: 1974, AA, 35, 327.

Jahn, K.: 1989, AA, 222, 264.

Jensen, E.: 1955, Ann. d'Ap. 18, 127.

Jensen, E., Ofstad, P.: 1963, Inst. Theor. Ap. Oslo, Rept. No. 15.

Jennings, D.E., Deming, D., Wiedemann, G.R., Keady, J.J.: 1986, ApJ. 310, L39.

Kawakami, H.: 1983, PASJ, 35, 459.

Kinman, T.D.: 1952, MNRAS, 112, 425.

Kjeldseth-Moe, O.: 1973, Solar Phys. 33, 393.

Kjeldseth-Moe, O.: 1989, in *MINI-Workshop on Flux Tubes in the Solar Atmosphere*, eds. E. Leer, P. Maltby (Oslo: Inst. of Theor. Ap.), p. 77.

Kjeldseth-Moe, O., Brynildsen, N., Brekke, P., Engvold, O., Maltby, P., Bartoe, J.-D.F., Brueckner, G.E., Cook, J.W., Dere, K.P., Socker, D.: 1988, ApJ, 334, 1066.

Kjeldseth-Moe, O., Maltby, P.: 1974, Solar Phys. 36, 101.

Landi degl'Innocenti, E.: 1983, Solar Phys. 85, 33.

Lau, Y.Y., Liu, C.S.: 1980, Phys. Fluids, 23, 939.

Leer, E., Holzer, T.E.: 1980, J. Geophys. Res. 85, 4681.

Lemke, M., Holweger, H.: 1987, AA, 173, 375.

Lemoine, B., Demuynck, C., Destombes, J.L.: 1988, AA, 191, L4.

Leroy, J.L.: 1962, Ann. d'Astr. 25, 127.

Lites, B.W., Skumanich, A.: 1990, ApJ, 348, 747.

Lites, B.W., Scharmer, G.B., Skumanich, A.: 1990, ApJ, 355, 329.

Lites, B.W., Skumanich, A., Rees, D.E., Murphy, G.A., Carlsson, M.: 1987, ApJ, 318, 930.

Makita, M.: 1986, Solar Phys. 106, 269.

Makita, M.: 1990, in *The Solar Iterior Atmosphere,* eds. A.N. Cox, W.C. Livingston, M. Matthews, University of Arizona Press, in press.

Makita, M., Morimoto, M.: 1960, PASJ, 12, 63.

Maltby, P.: 1964, Astr. Norvegica, 8, 205.

Maltby, P.: 1971, Solar Phys. 18, 3.

Maltby, P.: 1972, Solar Phys. 26, 76.

Maltby, P.: 1975, Solar Phys. 43, 91.

Maltby, P.: 1989, in *MINI-Workshop on Flux Tubes in the Solar Atmosphere,* eds. E. Leer, P. Maltby, Inst. of Theor. Ap., Oslo, p. 3.

Maltby, P., Brekke, P., Brynildsen, N., Kjeldseth-Moe, O., Bartoe, J.-D.F., Brueckner, G.E.: 1990, in *The Dynamic Sun,* 6th Europ. Solar Phys. Meeting, Derecen, Hungary, in press.

Mariska, J.T., Boris, J.P.: 1983, ApJ, 267, 409.

Mattig, W.: 1971, Solar Phys. 18, 434.

Meyer, F. Schmidt, H.U.: 1968, Zeit. Angew. Math. Mech. 48, 218.

Mickey, D.L.: 1985, in *Measurements of Solar Vector Magnetic Fields,* ed. M. Hagyard, NASA Marshall Space Flight Center, Huntsville, USA, p. 183.

Muller, R.: 1973, Solar Phys. 32, 409.

Murcray, F.J., Goldman, A., Murcray, F.H., Bradford, C.M., Murcray, D.G., Coffey, M.T., Mankin, W.G.: 1981, ApJ, 247, L97.

Noci, G.: 1981, Solar Phys. 69, 63.

Nordlund, Å, Stein, R.F.: 1990, in *Solar Photosphere, Convection and Magnetic Fields,* ed. J.O. Stenflo, IAU Symp. No. 138, 191.

Obridko, V.N.: 1985, *Sunspots and Activity Complexes,* (Russian) Nauka, Moscow.

Obridko, V.N., Staude, J.: 1988, AA, 189, 232.

Parker, E.N.: 1955, ApJ, 121, 491.

Parker, E.N.: 1979, ApJ, 230, 905.

Pikel'ner, S.B.: 1971, Solar Phys. 17, 44.

Pizzo, V.J.: 1987, in *Theoretical Problems in High Resolution Solar Physics II,* eds. G. Athay, D.S. Spicer, Proc. Workshop, Boulder, USA, (NASA Conf. Publ. 2483), p. 1.

Rachkovsky, D.N.: 1962, Izv. Krymsk. Astr. Obs. 28, 259.

Rayrole, J. in *The Role of Fine-Scale Magnetic Fields on the Structure of the Solar Atmosphere,* eds. E.-H. Schröter. M. Vázquez, A.A. Wyller, Proc. Inaugural Workshop DES Telescope, Canary Islands, Cambridge Univ. Press, P. 367.

Rees, D.E. in *The Solar Interior and Atmosphere,* eds. A.N. Cox, W.C. Livingston, M. Matthews. Univ. Arizona Press, in press.

Ringnes, T.S.: 1964, Astr. Norvegica, 8, 303.

Rodono, M.: 1986, in *Cool Stars, Stellar Systems, the Sun,* eds. M. Zeilik, D.M. Gibson, Lecture Notes in Physics (Berlin: Springer), p. 475.

Rossbach, N., Schröter, E.H.: 1970, Solar Phys. 12, 95.

Sakurai, T.: 1990 in *The Solar Interior and Atmosphere,* eds. A.N. Cox, W.C. Livingston, M. Matthews. Univ. of Arizona Press, in press.

Schmidt, H.U. (ed.): 1985, *Theoretical Problems in High Resolution Solar Physics,* Proc. MPA/LPARL Workshop, München, (MPA 212).

Schmidt, H.U. Wegmann, R.: 1983, *in Dymamical Problems in Math. Phys.* eds. B. Brosowski and E. Martensen (Frankfurt: Peter Lang), p. 137.

Schröter, E.-H., Vázquez, M., Wyller, A.A. (eds.): 1987, *The Role of Fine-Scale Magnetic Fields on the Structure of the Solar Atmosphere.* Proc. Inaugural Workshop D-E-S Telescope Install. Canary Islands, Cambridge Univ. Press.

Schüssler, M.: 1980, Nature, 288, 150.

Shine, R.A., Title, A., Tarbell, T., the SOUP team 1987, in *Theoretical Problems in High Resolution Solar Physics II,* ed. G. Athay and D.S. Spicer (NASA Conf. Publ. 2483), p. 133.

Shine, R., Smith, K., Tarbell, T., Title, A., Scharmer, G.: 1990, BAAS, 22, 878.

Sivaraman, K.R.: 1984, Kodaikanal Obs. Bull. 4, 11.

Smith, E.v.P., Gottlieb, D.M.: 1974, Space Sci. Rev., 16, 771.

Spruit, H.C.: 1981a, in *The Physics of Sunspots,* eds. L. Cram, J.H. Thomas, Sacramento Peak Obs. Conf. (Sunspot N.M.: SPO), p. 359.

Steinegger, M., Brandt, P.N., Schmidt, W.: 1990, *The Dynamic Sun,* 6th Europ. Meeting on Solar Phys., Debrecen, Hungary, in press.

Steinegger, M., Brandt, P.N., Pap, J., Schmidt, W.: 1990, Ap&SS, in press.

Stenflo, J.O.: 1976, in *The Energy Balance and Hydrodynamics of the Solar Chromosphere and Corona,* eds. R.M. Bonnet, Ph. DeLache, IAU Coll. No. 36, 143.

St. John, C.E.: 1913, ApJ, 37, 322.

Stix, M.: 1989, *The Sun,* (Berlin: Springer).

Stumpff, P.: 1961, Z. Ap. 52, 73.

Unno, W.: 1956, PASJ, 8, 108.

Vogt, S.S.: 1983, in *Activity in Red Dwarf Stars,* eds. P.B. Byrne, M. Rodonò, IAU Coll. No. 71 (Dordrecht: Reidel), p. 137.

Wiehr, E., Stellmacher, G., Knolker, M., Grosser, H.: 1986, AA, 155, 402.

Willson, R.C., Gulkis, S., Janssen, M., Hudson, H.S., Chapman, G.A.: 1981, Science, 211, 700.

Wittmann, A.D.: 1974, Solar Phys. 36, 29.

Woodgate, B.E., Tandberg-Hansen, E.A., Bruner, E.C., Beckers, J.M., Brandt, J.C., Henze, W., Hyder, C.L., Kalet, M.W., Kenny, P.J., Knox, E.D., Michalitsianos, A.G., Rehse, R., Shine, R.A., Tinsey, H.D.: 1980, Solar Phys. 65, 73.

Yoshimura, H.: 1983, Solar Phys. 87, 251.

Zayer, I., Solanki, S.K., Stenflo, J.O.: 1989, AA, 211, 463.

Zeeman, P.: 1913, *Researches in Magneto-optics,* Macmillan, London.

Zirin, H., Bopp, B.: 1989, ApJ, 340, 571.

Zirin, H., Wang, H.: 1989, Solar Phys. 119, 245.

Zwaan, C.: 1965, Rech. Astron. Obs. Utrecht, Vol. 17, Part 4, p. 1.

Discussion: Maltby

van't Veer What do we know about the gas pressure in sunspots? How is it related to the magnetic pressure and where do we find force free fields (if we find them)?

Maltby At the height of the photosphere the lateral difference in gas pressure is compensated by the magnetic pressure and magnetic tension; i.e., far from a force-free field. However, as one moves up the atmosphere the field soon, above say 2000 km, becomes force-free.

Byrne In view of the preponderance of high latitude spots in late-type stars would you like to comment on whether the latitude dependence of spot/photophere contrast is predominantly a latitude effect or a cycle effect?

Maltby The observations show a better correlation between intensity and epoch than between intensity and latitude. As the two correlations are not independent, one will effect the other. One may speculate about the cause and which is the most important parameter

Mullan What are the ranges of magnetic field strength in spots?

Maltby 2000 - 4000 Gauss.

Mullan What is the effective temperature of the umbra (you showed brightness temperature as a function of wavelength: presumably someone has created a model atmosphere for an umbra to give a realistic T_{eff})?

Maltby It may be an advantage to work with the brightness temperature as a function of wavelength. If you need a value for the effective temperature, you may use 4000 K.

Mullan Have you measured temperatures of umbral dots?

Maltby Not in the observations I presented.

Lean It's been suggested that there are 'active' and 'passive' sunspots, in the sense that active spots contribute to irradiance variations whereas passive spots don't. Therefore active spots are presumably darker than passive spots - but this appears to be inconsistant with observations that the intensity of these spots is approximately constant.

Maltby We have found differences in intensities of sunspots, i.e. we cannot regard the intensity of the spots as approximately constant.

Innis I have two comments on the 13.1 day velocity period (from the passage of sunspots over the solar disk) from helioseismology. First, historically it was known that spots would affect the velocity signal - a paper in *MNRAS* from 1974 or 1975 discusses this - actually it is Doppler Imaging!

Second, of more current interest, Bob van der Raay (Birmingham) has analysed around a decade of data on the 13.1 day effect, and found phase coherence over this time. I think my memory is right if I say the rotation rate derived is that of the sunspots, not the solar surface - implying phase coherence of the longitudes of the spots over this time.

Maltby Thank you for calling attention to your earlier results that, in fact, were well known to me.

Foing You have shown the typical curve for the temporal variation of the spot area during its development. Can you also obtain the development of the spot intensities during its lifetime?

Maltby The intensity as function of age has been searched for, without finding a systematic trend.

Landstreet For comparison with cool spotted stars could you tell me what the distribution of spot sizes on the Sun is? How large are the largest observed spots?

Maltby The sunspots for which I gave the results of intensity measurements are among the largest sunspots and have umbral radii larger than 5". The distribution of sunspots versus size is such that we only have data for a very small part of the whole distribution.

(l to r) L. Pasquini speaks with J. Linsky while A. Skumanich has to be elsewhere!

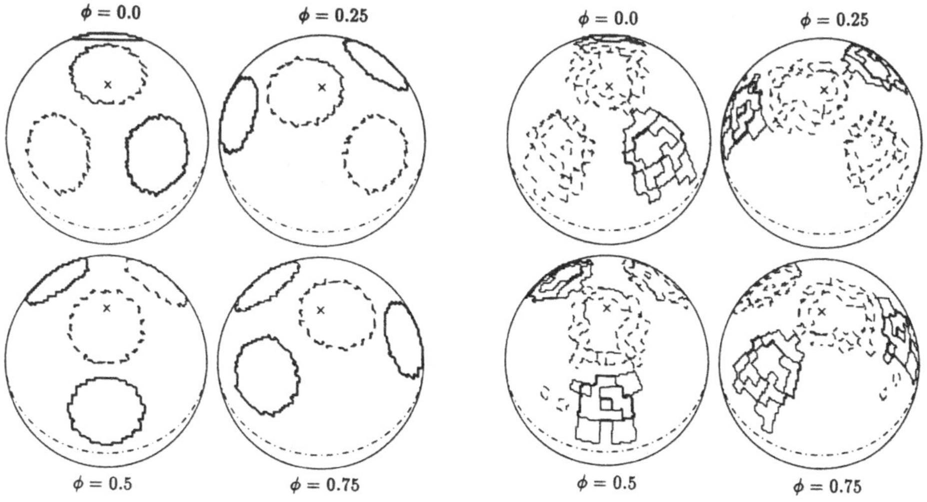

Figure 1. The *left panel* shows the test star used to generate the "observed" data set at four rotation phases, ϕ. The *right panel* shows the recovered image of the magnetic flux distribution on the test star. The *solid contours* enclose fields directed radially outwards. The *dashed contours* enclose fields directed radially inwards. The contour interval is 250 Gauss. The field strength in each of the magnetic patches on the test star is 1000 Gauss and the surface is uniformly bright.

four magnetic patches are accurately recovered. However, the value of the magnetic flux inside the patches is noisy and it is difficult to determine a unique value for the flux in each patch. These characteristics are typical of our simulations. The spatial characteristics of the field are always accurately recovered but the derived flux values in the magnetic patches vary considerably from one image pixel to the next.

3. Discussion

Our simulations have shown that ZDI can produce maps of the magnetic flux distribution on rapidly rotating stars. By using the unpolarised spectra to recover a conventional Doppler Image, we can begin to directly study the relationship between the observed magnetic fields and surface inhomogeneities. The ZDI technique works best for magnetic regions with large Stokes V signatures, *i.e.* bright, unipolar regions. Conversely, fields in dark star spots and mixed polarity regions will only produce small signatures in Stokes V and may go undetected.

References

Donati, J.-F., Semel, M. & Praderie, F.: 1989, A&A, 225, 467

Donati, J.-F., Semel, M. & del Toro Iniesta, J.C.: 1990, A&A, in press

Donati, J.-F., Semel, M., Rees, D.E., Taylor, K. & Robinson, R.D.: 1990, A&A, 232, L1

Landolfi, M. & Landi Degl'Innocenti, E.: 1982, Solar Phys. 78, 355

Piskunov, N.E., Tuominen, I. & Vilhu, O.: 1990, A&A, 230, 363

Rice, J.B., Wehlau, W.H. & Khokhlova, V.L.: 1989, A&A, 208, 179

Robinson, R.D.: 1990, ApJ, 239, 961

Saar, S.H.: 1989, in *Solar Photosphere: Structure, Convection and Magnetic Fields,* IAU Symp. No. 138. ed. J.O. Stenflo, Reidel:Dordrecht, p. 427

MAPPING MAGNETIC FIELDS ON RAPIDLY ROTATING STARS

S.F. Brown[1] and J.-F. Donati[2]
[1] Department of Applied Mathematics,
University of Sydney 2006, Australia
[2]Départment de Recherches Spatiales
Observatoire de Paris, Section de Meudon, France

1. Introduction

The most widely used method for measuring the magnetic fields that are present on late-type stars involves the study of Zeeman broadening of spectral lines observed in unpolarised light. The technique, pioneered by Robinson (1980), has been recently reviewed by Saar (1989). It has led to the detection of magnetic fields on many late-type stars. Its application is, however, limited to slowly rotating stars.

Recently, Semel (1989) and Donati, Semel and Praderie (1989) have developed a technique, called Zeeman Doppler Imaging or ZDI, capable of producing images of stellar magnetic fields on rapidly rotating stars. The technique is similar to conventional Doppler Imaging but uses circularly polarised light. Specifically, a high resolution measurement of the circular polarisation Stokes V parameter across a spectral line is a one-dimensional map of the longitudinal component of the surface magnetic field. Using several of these maps gathered at different rotation phases, Donati, Semel and del Toro Iniesta (1990) and Donati et al. (1990) have been able to detect magnetic fields on the Ap star ϵ UMa and the RS CVn star HR1099. However. the models of the field they derive from the one-dimensional images rely on *a priori* assumptions about the field geometry.

In this paper we describe numerical simulations from a code designed to produce two-dimensional images of stellar magnetic fields without making any assumptions about the field geometry. Our goal is to place ZDI on a similar footing to conventional Doppler Imaging, where two-dimensional reconstructions of surface brightness and abundances on rapidly rotating stars have been performed for some time (*e.g.* Vogt, Penrod and Hatzes 1987, Rice, Wehlau and Khokhlova 1989 and Piskunov, Tuominen and Vilhu 1990).

2. Calculation and Results

The inversion process is very similar to that used in Doppler Imaging but polarised radiative transfer calculations must be used to find the local line profile and Stokes V profile. We use the spectral line model of Landolfi and Landi Degl'Innocenti (1982). It assumes a plane-parallel Milne-Eddington model atmosphere with a constant ratio between the line and continuum absorption coefficients and a linear dependence of the source function on optical depth. The magnetic field is assumed to be constant throughout the line forming region.

We define the image vector to be values of the radial component of the magnetic field measured at approximately 500 grid cells distributed evenly over the stellar surface. The inversion routine iteratively adjusts the image vector until the fit between the model spectra and the observed spectral profiles is within the uncertainty level of the data. Simultaneously, an image entropy function is maximised. The effect of the entropy function is to smooth the resulting image. The algorithm used is adapted from that presented by Skilling and Bryan (1984).

As an example, the left panel of Figure 1 shows an image used to create an "observed" data set consisting of 16 Stokes V profiles, each profile consisting of 32 data points. The right panel of Figure 1 shows the image recovered by the inversion code. The location and polarity of the

Semel, M.: 1989, A&A, 225, 456

Skilling, J. & Bryan, R.K.: 1984, MNRAS, 211, 111

Vogt, S.S., Penrod, G.D. & Hatzes, A.P.: 1987, ApJ, 321. 496

Discussion: Brown

Foing You are mapping locally along 1D a quantity (your signal $\propto f \times B \times I_{mag}/I_{quiet}$). How can you derive independently the effect of the local filling factor, and intensity contrast of the magnetic features vs quiet atmosphere? The use of unpolarised light profile might be leading to an estimate of I_{mag}/I_{quiet} of spots rather than of the magnetic patch component giving rise to the Zeeman Doppler Imaging signal.

Brown For weakly split magnetic lines we cannot separate the field strength from the filling factor and only recover the magnetic flux. The surface brightness will be determined from the unpolarised spectrum and errors in the derived brightness will give errors in the derived magnetic flux. Like all Doppler Imaging techniques, the locations of inhomogeneities are found much more accurately than the physical conditions in the inhomogeneities.

Budding With regard to systematic errors, I guess part of this would relate to possible errors built into the model. We heard just now about thermodynamics, but I was more familiar, from my own experience, with assumptions about rigid-body rotation, limb darkening and so on. Another thing I would like to say is that I tend to be a bit sceptical when I see some discussions of the inverse problem which recovers very detailed information. There ought to be some formal relationship between the scatter which we have in data space, and the confidence with which we can specify values in parameter space. We did hear about the very high S/N requirement for your particular example, but do you have some feeling about whether this quality of data is in reasonable accord with the confidence you feel about your derived parameters?

Brown Our technique has the same assumptions as conventional Doppler Imaging with regard to solid body rotation, *i.e.* no significant change in surface structure over a few rotation periods, etc. If the data are more uncertain, then the final image will be determined more by the entropy function and will be smoother and contain less detail. However it is difficult to give error estimates for the image pixels.

Staude Results of Stokes polarimetry depend critically on the assumed atmospheric structure and the details of line formation. How did you determine the parameters of the Milne-Eddington model in the Unno solution used? Did you carry out some fitting to observed profiles?

Brown In the simulations shown here the use of a more detailed line formation model is unlikely to change the conclusions. Results from solar polarimetry show that the Milne-Eddington model gives good results when modelling magnetic fields. When inverting real data we intend to find the Milne-Eddington parameters by fitting to detailed model calculations at a variety of limb angles and temperatures. During the inversion we will use these values in the profile calculations.

Mullan To detect fields in spots (as opposed to the plages), it would be preferable to observe in the infra-red. Do you have plans to use the 12μ lines for Zeeman Doppler Imaging?

Brown Observing in the infra-red would give less contrast between the spots and unspotted regions, making spot fields easier to detect. The Zeeman splitting is also larger in the infra-red which may make it possible to derive both the field strength and the filling factor rather than just the surface flux. However, I am not sure that we could get the required S/N and resolution in the infra-red. I am not aware of any plans to observe in the infra-red.

CaII K LINE OBSERVATIONS OF SOLAR STARS

L.Pasquini
European Southern Observatory, La Silla
Casilla 19001, Santiago 19, Chile

1. Introduction

A knowledge of the physical properties of spatially resolved regions on the Sun is fundamental for our understanding of activity phenomena in late-type stars. The study of the solar cycle, as well as the detailed investigation of active regions and spots on the solar surface has provided the basic data for the interpretation of photometric and spectroscopic variability in late type stars.

In the past few years many advances have been made in high resolution stellar spectroscopy, mainly due to the use of echelle spectrographs and new high sensitive detectors. These improvements have allowed stellar astronomers to obtain data with a quality closer to that of solar observations. As a step towards a detailed quantitative comparison of solar and stellar data, it appears natural to compare the solar observations with those of stars of similar spectral type and luminosity class.

In this contribution I present high resolution (R = 60 000), high S/N (S/N ≈ 50 at the bottom of the line) CaII K line observations of G0 - G5 stars, classified as dwarfs. The K line of CaII was chosen because it is the best studied chromospheric line and has been well monitored for the Sun during the last solar cycle (White and Livingston 1981, Oranje 1983). The aim of this study is twofold: 1) To provide a set of high resolution CaII observations for an unbiased sample of nearby solar type stars and 2) To compare the behaviour of CaII K line features observed during the solar cycle with the corresponding features observed in field stars.

2. Observations

The observations were obtained in two runs during October and November 1988 at ESO, La Silla using the CES spectrograph with CCD detector at the Coudé focus of the CAT 1.4 m telescope. The resolving power, as measured in Th-Ar lamp comparison spectra, was R = 60 000. Although higher resolution would have been desirable, this choice was considered an optimal compromise between the wavelength range covered, the brightness of the program stars and the requirement of obtaining a S/N of ≈ 50 at the bottom of the K line.

The sample consists of all the Gliese (1969) stars in the range G0 - G5 comprised between $20^h < \alpha < 8^h$ and $-10° > \delta < -80°$, brighter than V = 7.5 and classified as dwarfs, giving a total of 44 objects.

Since no true continuum appears in the 30Å region covered by the CES spectra, the pseudo-continuum at 3950.5Å has been taken as reference point (Pasquini etàl. 1988); the absence of a true continuum implies that in order to get quantitative spectroscopy the recorded spectra must be carefully corrected for the telescope plus instrument response function. This response is the product of several causes, *i.e.* the grating blaze function, vignetting at the edge of the spectra introduced by the camera, a possible misalignement of the light path and the effects due to variable polarization caused by the third (flat) mirror of the telescope. All these factors produce, after flat field correction, a 'tilt' in the spectra which not only changes from night-to-night (see *e.g.* Rebolo etàl. 1989), but also depends on the apparent position of the object in the sky.

In order to reduce these effects I have developed a procedure, which consists of observing

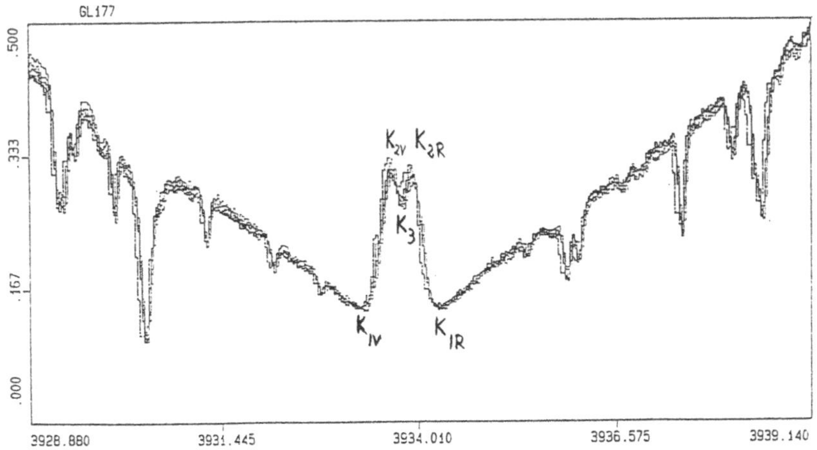

Figure 1

each night several B stars properly distributed in declination. Both the B star and the object observations were performed within 1 - 2 hour from the meridian.

Each star spectrum, bias subtracted, has been divided by a flat field obtained with an internal quartz lamp, in order to eliminate high frequency features. The same procedure has been applied to the B-star spectra; these have been further corrected for stellar or interstellar lines interpolating the adjacent continuum and, finally, heavily smoothed.

The spectrum of each program star has been then divided by the corrected spectrum of the closest B star, in order to eliminate the residual 'tilt'; note that with this procedure possible errors introduced by the internal flat field are cancelled out.

For 3 bright stars I have obtained spectra on different nights, with the aim of searching for possible spectral variations produced by surface inhomogeneities, which modulate the chromospheric core of the line as they transit across the stellar disk.

These repeated observations can also be used to check the level of reproducibility obtained with the adopted technique, giving an estimate of the uncertainties expected for the majority of the other stars, for which only one spectrum was obtained.

In Figure 1 six spectra of Gl 177 are shown: no further correction beside those previosly explained has been applied. The photospheric part of the spectra coincide at a residual level of $\approx 5\%$ in the region of the K line core. In Figure 1 are also marked the main K line features, according to the standard nomenclature.

3. Comparison with the Sun as a star

The CaII K line has been followed during the last solar cycle by several groups (White and Livingston 1981, Oranje 1983, Keil and Worden 1984, Sivarasan etál. 1987) which have produced a quite detailed analysis of the variations showed by the K line features in relation to the solar activity level. In analogy with White and Livingston 1981 (hereafter WL81) I will use the intensity of the central K3 as the main activity indicator.

WL81 reported that on the Sun the K1 minima tend to increase their separation with increasing activity, while the K2 show the opposite behaviour. These two trends are consistent with the observed K1 and K2 variations in plage spectra, supporting the hypothesis that chromospheric variations during the solar cycle are mainly due to the presence of active regions (WL81). In Figures 2a,b the K2 and K1 separations are shown for my targets as a function of the K3 intensity (*squares*) together with the values for the Sun at minimum and maximum activity as

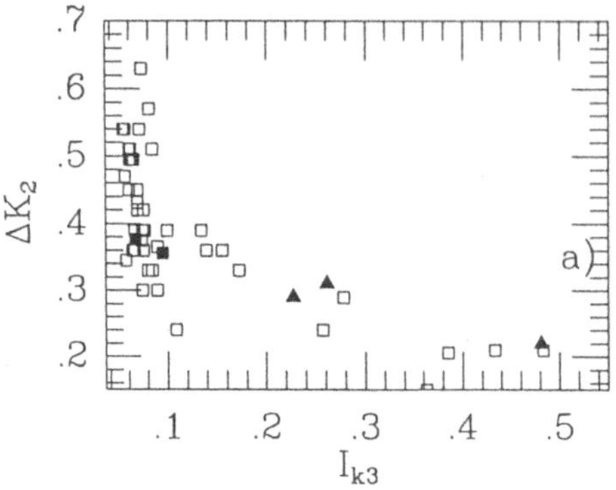

Figure 2a

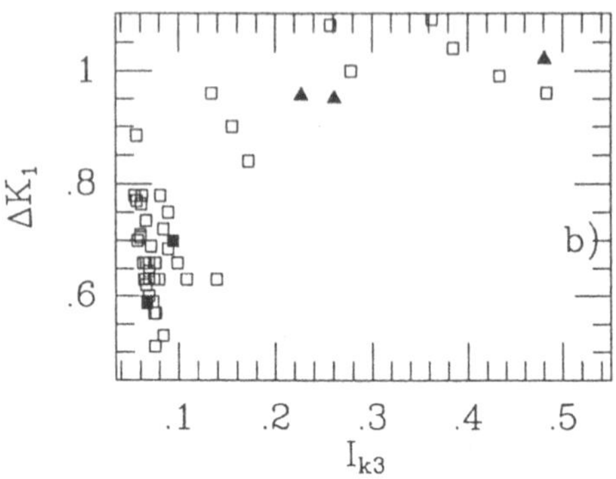

Figure 2b

given by WL81 (*open triangles*). The trends observed by WL81 are clearly visible, although there is a big scatter for non-active stars.

The reasons of this scatter are various, ranging from the difficulty in precisely identifying the K1 and K2 features in very low activity stars, to the effects due to changes in the line core induced by the Wilson-Bappu law (Wilson and Bappu 1956), whicch are already noticeable in this restricted G0 - G5 sample. In both Figures 2a and b there is also present a possible 'flattening' of the observed trends towards very active stars; this flattening is not too surprising if we consider that the K1 and K2 widths for these objects are very similar to those observed in spatially resolved plages on the Sun; in fact the filled triangles, which represent 3 solar plages as observed by Shine and Linsky (1972) show parameters very close to those of the most active stars.

A flattening (or saturation) effect similar to that observed in Figures 2a and b, is present in Figure 3, where the 1Å index (the residual area at the center of the line in 1Å bandpass) is plotted as a function of K3.

The line drawn is not the best fit to the data, but the WL81 relationship, scaled to the λ3950.5Å continuum and extrapolated to higher K3 values. While low activity stars are consistent with the solar relationship, the most active ones show a departure.

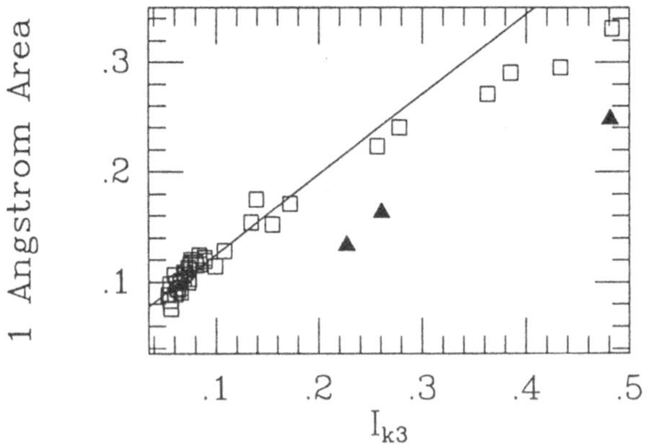

Figure 3

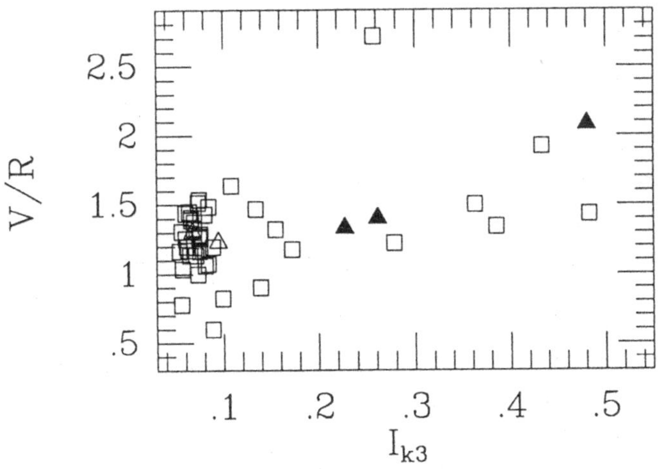

Figure 4

Finally, I analyse the asymmetry of the K core, through the V/R parameter, defined as $V/R = (I_{k2v} - I_{k3})/(I_{k2r} - I_{k3})$. WL81 found that this quantity in the Sun changes from a value of 1.30 at the solar minimum to a value of 1.23 at the maximum; *i.e.* the line core tends to be more symmetric with increasing activity. In Figure 4 V/R is shown as a function of K3; the violet K2 is, in general, higher than the red K2, but the trend observed by WL81 is not obviously present. Note, however, that the use of this index may be questionable for very active stars, in which I_{k3} is very close to I_{k3}. In this case the index V/R is the ratio of two small numbers obtained as differences of similar quantities, and it is affected therefore by relatively large errors.

4. Conclusions

In this paper I have presented high-resolution, high-S/N data concerning an umbiased sample of solar-type stars. It is possible to get high quality stellar observations which allow quantitative comparison between solar and stellar spectral features.

Similarities between the stellar results and solar plages suggest that activity in solar stars is most likely due to the presence of plages similar to those observed on the Sun, although a sort of "saturation" effect may be present and the CaII emission for the most active stars of the sample can probably be explained only by having strong solar-like plages covering a large fraction of the stellar surface.

Some of the CaII spectral features do not seem to follow the behaviour observed during the solar cycle. A detailed study of these deviations will be extremely important for understanding the possible differences between the Sun and other stars.

References

Gliese, W.: 1969, *Catalogue of Nearby Stars*

Keil, S.L., Worden, S.P.: 1984, ApJ, 276, 766

Oranje, B.J.: 1983, A&A, 122, 88

Pasquini, L., Pallavicini, R., Pakull, M.: 1988 A&A, 191, 266

Rebolo, R. et al.: 1989, A&AS, 80, 135

Shine, R.A., Linsky. J.L.: 1972, Solar Phys. 25, 357

Sivarasan, K.R., Singh, J., Bagare, S.P., Gupta. S.S.: 1987, ApJ, 313, 456

White O.R., Livingston, W.C.: 1981, ApJ, 249, 798

Discussion: Pasquini

Giampapa Do you observe an increase in K1 seperation with increasing K-line strength for stars of the same spectral type? This would imply that both heating rates and plage area coverage increase from quiet to active stars.

Pasquini Yes, as shown in one of the viewgraphs, K is increasing with activity. I believe (see conclusions) that activity for the most active stars in the sample can be explained only with the presence of both, strong plages and high area coverage.

Skumanich How did you correct for scattered light?

Pasquini Having 2-D CCD data, I have subtracted "real bias levels" directly from the observations frame. The instrument used (CES + short camera) is known to have very low levels of scatted light in the blue. In fact the agreement between daily sky spectra and published solar data is good. The fact that many stars show parameters (K3 intensity, 1Åindex) lower than the quiet Sun, also suggests that the contribution of scattered light must be small.

Mullan Shine and Linsky's work indicated strong macroturbulence in young plages. Do you have any information on macroturbulence in young stars?

Pasquini No, I have not yet done a profile analysis to determine it.

UV OBSERVATIONS OF SUNSPOTS

Joseph B. Gurman
NASA Goddard Space Flight Center
Laboratory for Astronomy and Solar Physics
Greenbelt, MD 20771, USA

1. Introduction

The Sun was observed with a variety of rocket- and satellite-borne ultraviolet spectrographs for nearly 25 years before the spatial resolution of the UV telescopes caught up sufficiently with spectrograph development to allow imaging of individual sunspots. I refer, of course, to the Harvard and Naval Research Laboratory spectrographs on Skylab. Since then, the yet finer resolution of the OSO-8, HRTS, and UVSP instruments has made it possible to investigate the structure and physical properties of the sunspot chromosphere and transition region. Even now, the picture of the sunspot transition region is unclear. Is it characterized by bright plumes or more nondescript emission? Are there supersonic downflows everywhere, or perhaps just in light bridges? Why are oscillations in the transition region over umbrae confined to a narrow frequency range starting just below the acoustic cutoff frequency at the temperature minimum? Since these rather basic, physical issues are in some doubt, I will concentrate on the observations as they pertain to these problems, rather than maintaining a strict chronological approach. Table 1 describes the primary sources of ultraviolet observations of sunspots to date.

Table 1.

UV Instruments that have produced Sunspot Observations

Instrument	Spacecraft	Year(s)	Spatial res. arc sec	Spectral Res. (Å)	Spectral Range (Å)
S055	SKYLAB	5/1973 - 2/1974	5 x 5	1.6 - 8	280 - 1340
S082-B			2 x 60	0.060	1175 - 1950
CU (HRUS)	OSO - 8	1975 - 1978	2 x 20*	0.015	1150 - 1850
LPSP			1 x 3*	0.025	Lyα, Lyβ MgII *k*, *h* CaII *K*, *H*
HRTS	rockets SPACELAB 2	1975, 1978, *etc* 1985	1	0.060	1175 - 1700
UVSP	SMM	1980 - 1989*	3*	0.010 0.020	1150 - 1800 1800 - 3200

*Adjustable; cited resolution is most common for sunspot observations.
*Wavelength drive operational in 1980 and 9/1984 - 4/1985.

2. The temperature minimum and chromosphere

The UVSP data set includes a substantial number of raster images of active regions at wavelengths around 3090Å. The continuum and the many weak absorption lines are formed within a scale height or so of the temperature minimum. Since these observations are free of seeing effects (the spacecraft jitter is considerably smaller in amplitude than the 3 arc sec pixel size), the only spatially scattered light arises from the wings of the instrumental modulation transfer function. A small, random sample of these rasters yields umbra-to-photosphere intensity ratios of as little as 0.035. Ignoring a number of critical considerations, such as whether the continuum and weak lines at 3090Å are formed in LTE, possible differences in opacity sources between umbra and quiet Sun, and so on, one can make a simple minded estimate of the minimum umbral temperature by simply assuming LTE and taking ratios of Planck functions. The umbral temperature obtained by this oversimplification is approximately 3200 K.

Using the OSO-8 spectrometers to examine the umbral chromosphere and lower transition region, Lites and Skumanich (1982) obtained line profile observations of Lyα, Lyβ, CaII *K* and *H*, MgII *k* and *h*, and the C IV resonance line at 1548Å, in a sunspot umbra, penumbra, and nearby plage. Their Mg II profiles (Figure 1) show quite clearly the distinctions between spot and plage: the spot profiles are much narrower and show no central reversal, while the peak plage profile is several times brighter than the umbral emission core. Observations of the same spot with the LPSP spectrometer on the following day by Kneer et al. (1981) show that the centrally unreversed umbral profile has about the same peak intensity as the centrally reversed profile in the quiet Sun. Gurman (1984), reporting Mg II h observations of umbrae in five regions, noted variations of over a factor of three in the emission core intensity (Figure 2). Since these measurements appear to be free of both spatially and spectrally scattered light, it appears that there is a fairly broad range of sunspot chromospheres.

This impression is borne out by the range of observed Lyα intensities relative to the quiet

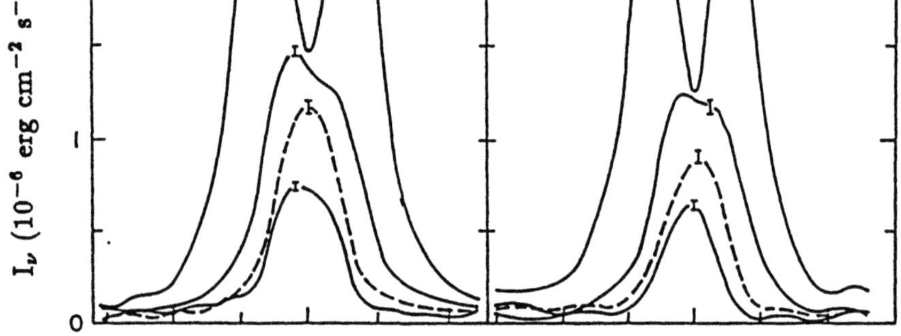

Figure 1. MgII *k* and *h* profiles obtained by Lites and Skumanich (1982) with OSO-8 in a large umbra in McMath region 13937, 1975 November 16. In order of increasing intensity, the profiles were obtained in the umbra (lowest *solid* line and *dashed* line), penumbra, and plage. Each tick mark on the abscissa represents 0.5Å.

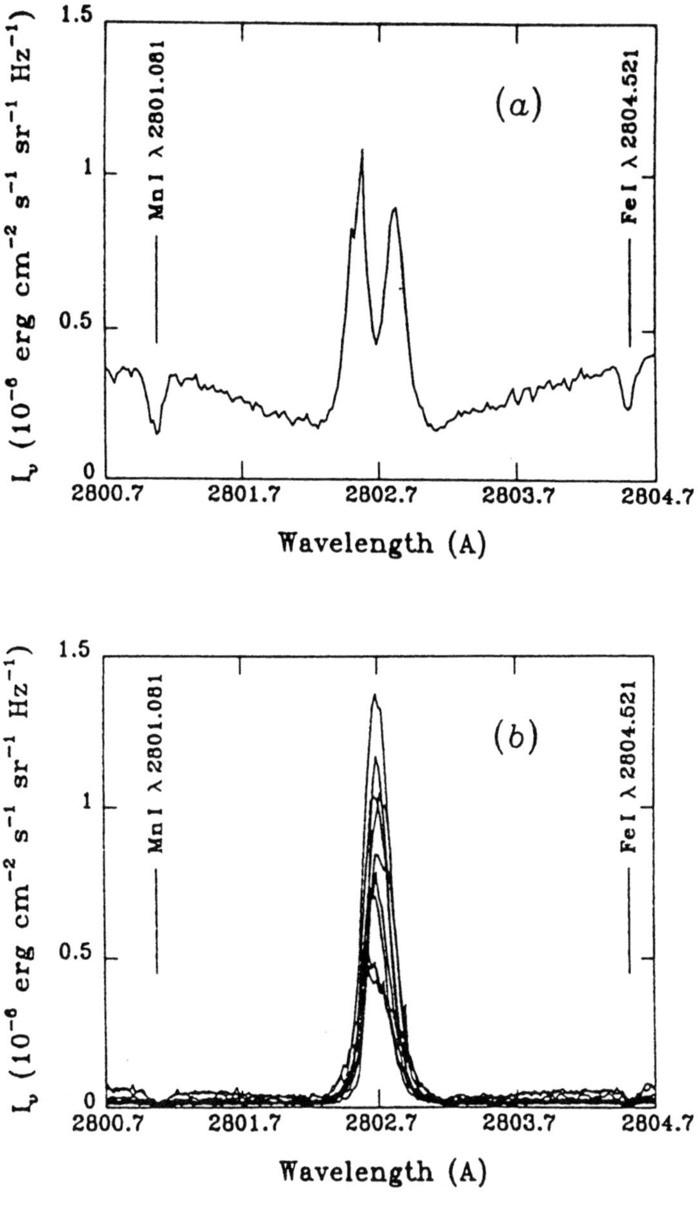

Figure 2. Mg II *h* profiles obtained with the Ultraviolet Spectrometer and Polarimeter (UVSP) on SMM in (*a*) the quiet Sun and (*b*) umbrae in five distinct active regions during 1980 (Gurman 1984).

Sun and plage. Lites and Skumanich found that the umbral Lyα intensity in their spot was "significantly less than observed in the quiet Sun," and a factor of 1.6 to 3 times less than the umbral intensity measured by HRTS in a dark umbra (Basri et al. 1979) and a light bridge (Nicolas et al. 1982). The OSO-8 (Kneer et al. 1981) and HRTS I (Basri et al. 1979) observations, and at least one UVSP measurement (Fontenla et al. 1989), however, shows Lyα intensities 3 - 4 times higher in spot umbrae than in the quiet Sun. As Figure 3 shows, it is not uncommon to have a

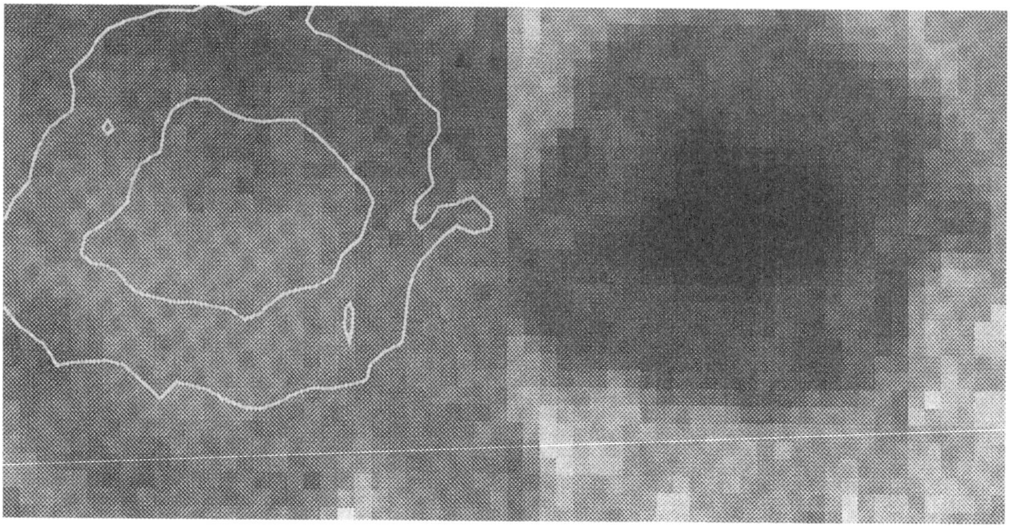

Figure 3. Simultaneous HI Lyα (*left*) and 2426Å (*right*) rasters obtained by UVSP in AR 2665, 1980 September 13, 07:31 UT. The images were obtained with a 3 arc sec × 3 arc sec entrance aperture and 30 mÅ wide exit apertures. The outer umbral and penumbral contours of the 2426Å image have been superposed on the Lyα image.

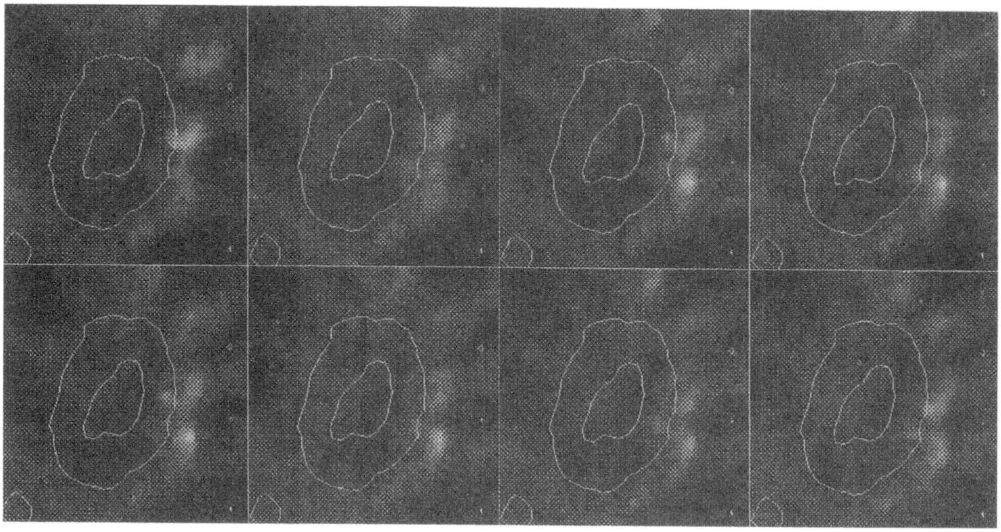

Figure 4. UVSP CIV 1548.19Å raster images of a sunspot, with umbral and penumbral contours from a simultaneous 3090Å raster superposed. Each raster consists of 30 × 30 pixels of size 3 arc sec × 3 arc sec, and the time step between rasters (running *left* to *right*, the *top* to *bottom*) is 127 sec.

substantial variation in Lyα emission over a single sunspot. In this case, one could not even recognize the location of the spot without the longer-wavelength contours. We expect that the

Lyα source function will be dominated by the Hα source function - that is, by the population of the $n = 2$ level - and the spatial distribution of Lyα emission will be dominated by the non-local, Hα radiation field. Inspection of an Hα image obtained on the same day as these UV data confirms this view: the Hα plage is located south and east of the spot, but little or no plage lies north or northwest of it; thus the brighter Lyα over the southern and eastern parts of the spot.

Using their line profile observations to construct a model chromosphere, Lites and Skumanich (1982) found that, in contrast to Lyα, the chromospheric region in which the Mg II lines are collisionally excited is *effectively thin;* that is, the line source function is determined by local conditions (τ, T) alone. Their model also required that the nonthermal broadening be less than 3 km sec^{-1} throughout the chromosphere to reproduce the narrow k and h emission cores. Finally, Lites and Skumanich note that the H Ba continuum from the photosphere *outside* the umbra controls the H ionization, and thus the electron density in the umbral chromosphere. They therefore expect that the Ca II and Mg II flux from extended starspots depends directly on the size of those spots.

3. The transition region

3.1 Penumbral activity

Bonnet et al. (1982) obtained broad-band images around 1600Å that show little or no sign of penumbrae. This may be an indication of the vertical extent of the penumbral fibril structure. In transition region lines, the *outer edge* of the penumbra is usually quite active, even in unipolar spots in quiescent active regions. Figure 1 shows a short time series of images in C IV 1548Å (with superposed umbral and penumbral contours from simultaneous 3030Å images), scaled to bring out the looplike structures around the penumbral circumference. Even on a time scale of minutes, these structures change dramatically in intensity. On closer inspection, there is even a suggestion of interaction among the various loops. It is also obvious that parts of the penumbra are much more intense, and active, than others. Dere, Schmieder and Alissandrakis (1990) point out that the bright knots surrounding the penumbra may be associated with emerging magnetic flux.

3.2 Nonthermal line broadening

Many of the observations I will mention in discussing the sunspot transition region pertain only to the umbra; others, to both umbra and penumbra. I will attempt to make clear in each instance just where in the spot the observational conclusions are valid. For instance, Cheng, Doschek and Feldman (1976) used the SKYLAB SO82-B spectrograph to measure a number of line profiles in a spot, but their 60 arc sec slit width meant that they integrated umbra and penumbra together. (They do note, however, that no plage was included in the area covered by the slit.) These observations are noteworthy because they represent the first measurements of line *widths* in a sunspot; these authors found that nonthermal broadening was significantly less than in the quiet Sun. Varying from 9 km sec^{-1} at SiII (formed around $1.6\,10^4$ K) to 25 km sec^{-1} at OV ($2.5\,10^4$ K). At CIV (10^5 K), the nonthermal broadening was 13 km sec^{-1}, quite comparable to the 14.6 km sec^{-1} Gurman and Athay (1983) found for an average of over 1000 umbral profiles. By comparison, the nonthermal broadening of CIV in the quiet Sun is 26.5 km sec^{-1} (Athay, et al. 1983). One possible explanation for the difference is that the strong, umbral magnetic field suppresses two of the three degrees of freedom accessible to the disordered magnetic features averaged into the 3 arc sec $\times$ 3 arc sec aperture size of the UVSP observations. The umbral nonthermal velocity would thus be a factor $\sqrt{3}$ smaller then the quiet Sun value, or about 15 km sec^{-1} (Gurman and Athay 1983).

3.3 Emission measure

One of the most persistent problems in the characterization of the sunspot transition region is simply whether sunspots are bright or dark in transition region emission lines. The SO55 observations of Foukal et al. (1974) found that the transition region above umbrae was the brightest feature in active regions in lines formed at 10^5 to 10^6 K. Examining a larger sample (some 22 large spots), Foukal (1976) found that all the spots showed strong EUV emission at some time, but there were several instances in which large umbrae were invisible in lines formed at temperatures of $3\,10^4$ K to $6\,10^5$ K for up to several days.

This picture is strikingly different from that given by observations of lines in the range λ1100-1800Å. Cheng, Doschek, and Feldman (1976) found that the OI, CII, SiII, SiIV, CIV,and NIV

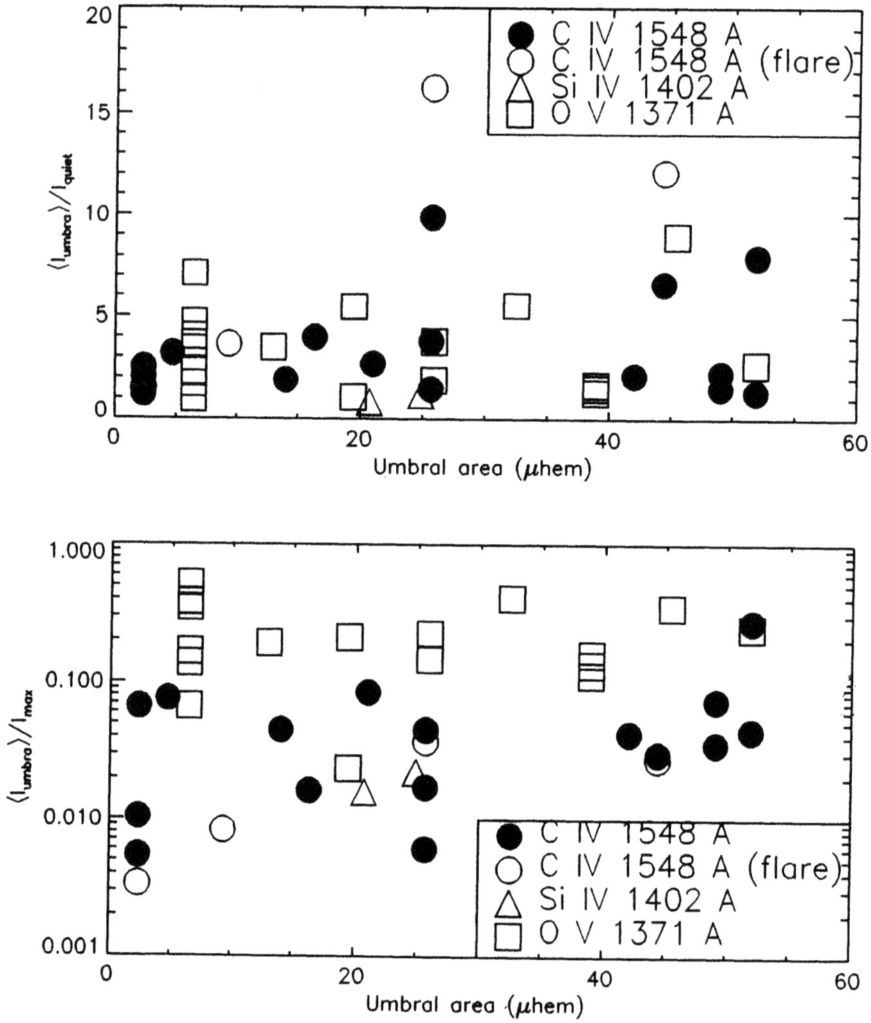

Figure 5. Mean umbra-to-quiet Sun (*top*) and umbra-to-plage bright point (*bottom*) contrast for 40 umbrae observed by UVSP in OV 1371Å, SiIV 1402.77Å, and CIV 1548.19Å. The quiet Sun values refer to an average of quiet network and cell interior regions, and the plage bright points are the maxima in the respective raster images.

lines they studied were roughly as bright over the sunspot as in the quiet network, despite the fact that their slit was so long as to include both umbra and penumbra. Similarly, the OSO-8 CIV observations of Lites (1980) show sunspot intensities 2 - 3 times those in the quiet Sun, and a factor of three lower than in a plage. Nicolas et al. (1982), using 1 arc sec HRTS II observations of an umbra with a prominent light bridge, noted that the ratio of umbral to quiet Sun intensity grew steadily with temperature of formation from near unity for Si II ($3\,10^4$ K) to 20 for OV ($2.5\,10^5$ K). Gurman and Athay (1983), studied umbral CIV emission in eight spots in 1980 with a spatial resolution of 3 arc sec × 3 arc sec (nearly 300 independent measurements); they found a mean intensity seven times that of the dark network cell interiors, but only about 50% greater than the mean quiet Sun (bright network plus cell interiors). In an attempt to see if the phase of the solar cycle could explain the differences between the SO55 and the longer-wavelength measurements, I have examined the 1984 - 1985 UVSP data set for rasters including sunspots. Figure 5 shows the umbra-to-quiet Sun and umbra-to-plage contrast for 10 umbrae in five active regions. Here, while there is considerable scatter, the contrast is often even lower than in Nicolas et al.'s spot. The systematically higher umbra-to-plage ratios in OV may be real, or it may be due to the 10 arc sec × 10 arc sec aperture used in the OV observations. The other lines were observed with 3 arc sec × 3 arc sec (CIV) and 4 arc sec × 4 arc sec (SiIV) apertures.

How do the longer and shorter wavelength observations fit together? The emission measure (EM), $\int N_e^2\, dh$, calculated by various authors are plotted in somewhat confusing fashion in Figure 6. The solid line represents Raymond and Doyle's (1981) quiet Sun EM, and the dot-dashed line, the sunspot EM of Noyes et al. (1985). Both of these curves represent the products of Skylab SO55 observations. The EM values calculated from longer-wavelength measurements are plotted as individual symbols. The open squares are from Cheng, Doschek, and Feldman's (1976) SO82-B data, and the various circles, from the UVSP observations of Kingston et al.(1982). The SO55 emission measure was derived by summing (to enhance signal-to-noise) six spectra from two 5 arc sec × 5 arc sec pixels, one over the spot umbra, the other, more intense, over a penumbral plume. The SO82-B spectrum, as already noted, averages umbra and quiet penumbra, and the UVSP observations are of separate, 3 arc sec × 3 arc sec pixels over the umbra (*filled circles,* 19 August; *open circles,* 1 September, 1980) and penumbra (*double circles, filled and open* as for the umbral pixels) 10 arc sec away. (The location of the UVSP observations is checked against 3090Å rasters obtained with the same optics as the shorter wavelength data.) In the, probably vain, hope of preserving some clarity, I have omitted Kingston et al.'s plage emission measures; they uniformly lie about an order of magnitude above the penumbral emission measures.

If we can forgive the mixing of ionization concentration calculations, radiative emissivity curves, instrumental calibrations, and even temperature ranges over which the emission measures are calculated that occurs in such a comparison, we may conclude that:

- At temperatures below about $2\,10^5$ K, both umbral and penumbral brightnesses vary by factors of 2 to 10 from spot to spot.
- At the same temperatures, the penumbral EM in the brighter Kingston et al. spot is consistent with the primarily penumbral plume EM of Noyes et al.
- For temperatures below 10^5 K, the fainter Kingston et al. umbra is indistinguishable from the quiet Sun emission measure, but exceeds it by a factor of ≈ 2 at temperatures around $2\,10^5$ K.
- None of the longer wavelength (SO82-B, HRTS, UVSP) data is incompatible with the enhanced emission at temperatures of 10^5 K to 10^6 K, but the indication is that there, too, there is a factor of 2 to 5 variation in both umbral and penumbral emission measures.

3.4 Electron pressure

The difficulty in distinguishing sunspots from the rest of the transition region is also evident in determinations of the electron pressure. Several diagnostic methods have been used to determine the electron number density n_e, and thus the electron pressure: line center optical depth, intensity ratios of lines from the same ions, and intensity ratios of lines from different ions formed at similar temperatures. Of these, the most reliable is probably the intersystem line ratio method, since both lines are formed at essentially the same temperature, and thus the same physical location. With the exceptions of Foukal's (1976) measurements, which are compromised by one weak line in each ratio, however, all the methods give much the same results: the electron pressure lies in the range 0.1 - 0.4 dyn cm^{-2}, with values ranging up to 1.0 dyn cm^{-2} in the light bridge observed by Nicolas et al. (1982). For comparison purposes, Nicolas et al.'s quiet Sun electron pressure of 0.2 dyn cm^{-2} is also included in the Table 2.

The primary sources of uncertainty in these diagnostics are instrumental sensitivity and the atomic data. Doyle et al. (1985) point out that the effects of departures from a Maxwellian distribution on the ionization concentration of the various species can be neglected in plumes, where the temperature gradients appear to be an order of magnitude smaller than in the quiet Sun. According to Cheng and Kjeldseth-Moe (1977) and Gurman et al. (1982), the relationship

Table 2.

Sunspot Transition Region Electron Pressure from Emission Line Diagnostics

Reference	Feature	Diagnostic	P_e (dyn cm^{-}2)
Cheng et al. (1976)	umbra, penumbra	SiIII	0.13
Foukal (1981)	Plume	CIII OVI NeVII	0.01 0.02 0.11
Foukal (1981)		OV	0.34
Nicolas et al. (1982)	umbra, quiet Sun	CIV/NiIV	0.20 - 1.00 0.20
Kingston et al. (1982)	umbra	CIV/NIV SiIV/OIV	0.13 - 0.27
Gurman et al.	umbra	OIV	0.12 - 0.24
Doyle et al. (1985)	penumbral plume	CIII	0.06 0.13 0.43

[1] n_e derived from line center optical depth
[2] Ratio of intersystem lines

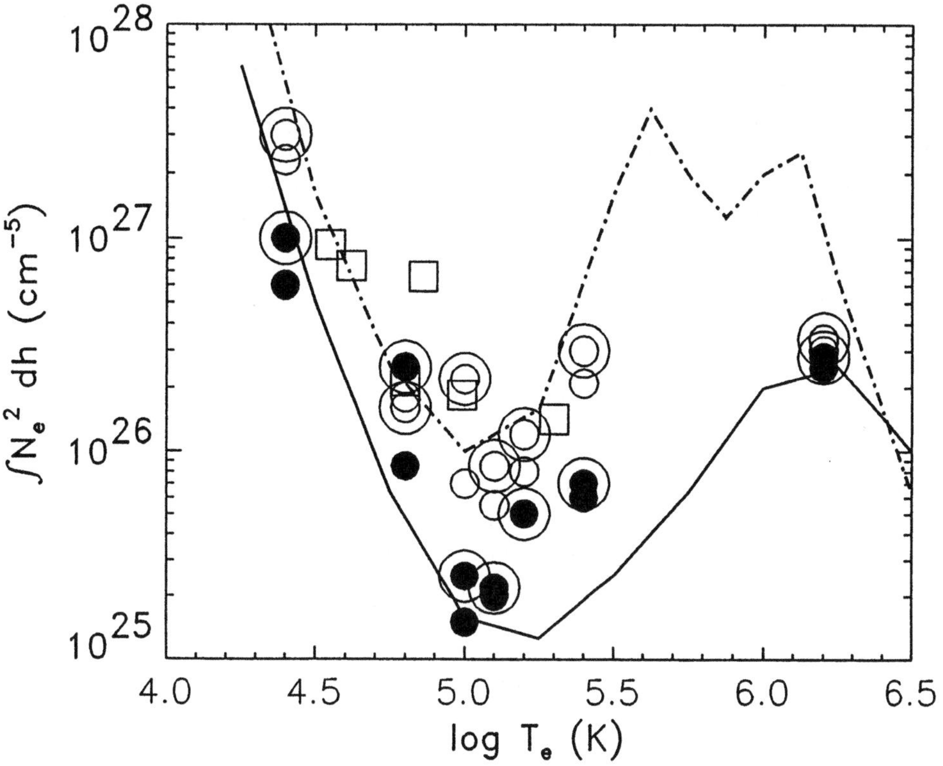

Figure 6. Empirical emission measures: *solid line:* quiet Sun (Raymond and Doyle 1981); *dot-dashed line:* sunspot penumbral plume (Noyes et al. 1985); *open squares:* umbra and penumbra (Cheng and Kjeldseth-Moe 1977); *circles:* umbra (*single circles*) and penumbra (*double circles*) of Kingston et al. (1982). *Filled circles* represent 1980 August 19 observations; *open circles,* 1980 September 1.

of temperature gradients is similar for apparently plumeless, lower-temperature observations as well ($\approx 10^3$ K km^{-1} *vs.* $\approx 10^4$ K km^{-1}, at 10^5 K).

3.5 Comparisons with starspots

The emission measure and electron pressure determinations lead us to a warning for those who wish to apply ultraviolet observations of sunspots to other stars. *It is not consistent with the longer wavelength observations to assume that bright plumes cover sunspot umbrae at the temperatures where lines observable by IUE and GHRS are formed.* Plumes appear to be transient phenomena, and the only observations of umbrae that appear to be noticeably brighter than the surrounding quiet Sun below $2\,10^5$ K were obtained in light bridges. *Ordinary, dark umbrae are usually indistinguishable from the quiet Sun in the emission lines observed by SO82-B, HRTS, UVSP, - and IUE.*

4. Mass flows

4.1 Horizontal flows

The Evershed flow in the photosphere (Evershed 1913) is directed outward from the umbra, and continues up to several thousand km beyond the outer edge of the penumbra. In the chromosphere, the direction of the flow is reversed, and this inward, "reverse Evershed" flow extends upward into at least the lower transition region. A large, round spot observed by HRTS II (Nicolas and Kjeldseth-Moe 1981) and several more examples in the UVSP dataset (Athay et al. 1982, 1983) show superpenumbral inflow velocities of the order of 4 km sec^{-1}. Using Spacelab 2 HRTS observations, Dere et al. (1990) have mapped *spiral* inflow patterns in both the Hα chromosphere and the CIV transition region, with velocities of up to 17 km sec^{-1} in the transition region. In all cases, as with the photospheric outflow, the transition region and chromospheric inflow starts thousands of km outside of the penumbra.

4.2 Vertical flows

One of the most remarkable features of the body of observations of the sunspot transition region is that it provides clear examples of both the unmistakable presence and complete absence of supersonic downflows over umbrae. HRTS I (Brueckner 1981), HRTS II (Nicolas et al. 1982, Brekke et al. 1987), and Spacelab 2 (Kjeldseth-Moe et al. 1988) observations all show substantial downflow velocities in the sunspot transition region. Redshifted components are seen in all lines formed at transition region temperatures, and even in chromospheric lines such as the OI multiplet at λ1301Å. The largest velocities were recorded by HRTS II; Figure 7 shows two velocity components measured in lines formed in a wide range of temperatures. Both of these components are supersonic at the temperatures of maximum equilibrium ionization concentration, but it is possible that in such flows, line formation takes place at significantly different temperatures. By contrast, the Spacelab 2 spectra typically show redshifted components at 15, 40, and 100 km sec^{-1} (Maltby et al. 1990).

Assuming a filling factor of 0.01 by area, Maltby et al. (1990) have estimated the net mass flux in a 50 arc sec $\times$ 60 arc sec region including the Spacelab 2 spot (but excluding regions of emerging magnetic flux outside the spot) to be $1.7\,10^{-9}$ and $2.3\,10^{-9}$ gm cm^{-2} sec^{-1} on two successive days. They further note that within this net downflow are extensive regions of upflow, primarily over the umbra. After about a day, however, a strong downflow appears in the umbra, at the location of a photospheric light bridge. Similarly, the supersonic downflows observed by HRTS II also occurred over a light bridge. One should note, however, that other, quite strong downflows in the Spacelab 2 spot occurred at the outer edge of the penumbra.

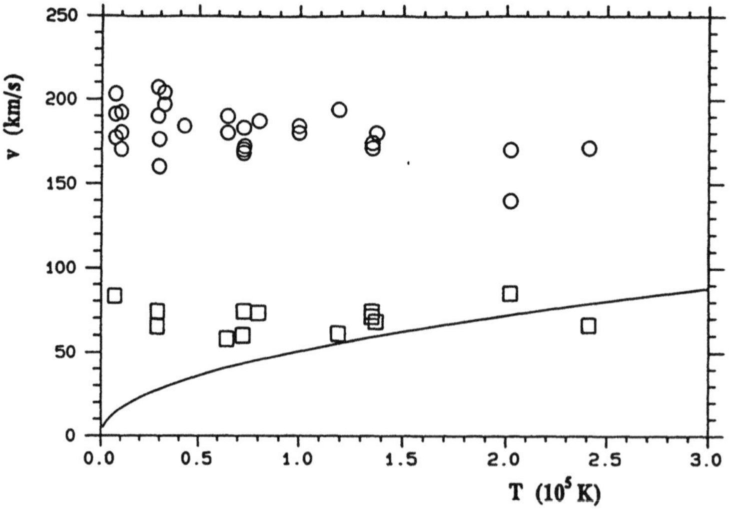

Figure 7. HRTS II measurements of downflow speeds in lines formed at a variety of temperatures. *Circles:* highest-velocity components; *squares:* second dominant components; *solid line:* adiabatic sound speed. From Brekke et al. (1987).

Surprisingly, the UVSP data set shows only much weaker signs of supersonic downflows. Originally, we thought that our Dopplergrams, with velocity bandpasses of only about $\pm$ 30 km sec^{-1}, simply excluded any sign of supersonic components in either direction. The broad, multicomponent HRTS II line profiles, however, indicated that there was a range of upflow velocities over spots, including substantial components within the UVSP Dopplergram bandpass. To clear up any confusion that might have been caused by the use of narrow-slit Dopplergrams (*e.g.* Athay et al. 1982), Gurman and Athay (1983) studied Dopplergams obtained with a 0.6Å wide exit slit bisected into two 0.3Å segments by beamsplitter prisms, with the output recorded by two separate photomultipliers. Thus we obtained all the photons, and at least some velocity information, from each pixel in a raster. In eight sunspots, we observed no net umbral velocities over 15 km sec^{-1} in either direction, and the mean vertical velocity of all the umbrae was found to be 1.2 ± 5.6 km sec^{-1} *upward*, relative to the mean of nearby quiet Sun areas. (Since, as Gebbie et al. (1981) and others have pointed out, the quiet transition region as a whole is characterized by a downflow of 4 km sec^{-1}, our mean umbral upflow measurement is really ≈ 5 km sec^{-1}, relative to the photosphere.)

Since the unshifted line profile is routinely the strongest component in the HRTS observations, these results could still be misleading: we were just measuring the average line shift, and had no knowledge of the real appearance of the line profiles. I therefore examined a set of 28 identical spectral scans, obtained in nine sunspots, of the λ1550Å region; these spectra were obtained in 3 arc sec $\times$ 3 arc sec pixels corresponding to the darkest umbral pixels of λ3090Å rasters. Part of a typical scan is shown in Figure 8(a). We can see at once that the broad, multicomponent red-shifts of the HRTS data are simply absent here. Indeed, the only systematically present feature to the red of the primary λ1548.19Å line lies at an average distance of 0.181 ± 0.037Å, and its average strength is 0.08 ± 0.05 that of the primary. The results for the λ1550.77Å line, which is only half as strong, are 0.122 ± 0.063Å and 0.08 ± 0.06. The differences in the separation indicate that these weak, redshifted features are blends that are undetectable in the normally much broader lines. There are some cases, however, such as Figure 8(b), that clearly show what could be supersonic components with $\Delta v \approx 50$ km sec^{-1}, but they are much weaker relative to the unshifted component than in the HRTS observations.

Is there any way out of this apparently contradictory situation? At one time, I wondered whether the sunspot transition regions observed by UVSP during the solar maximum in 1980 were systematically different from those observed by HRTS, which were all well away from solar maxima. The similarity of the results of the 1984 - 1985 UVSP observations to those from 1980

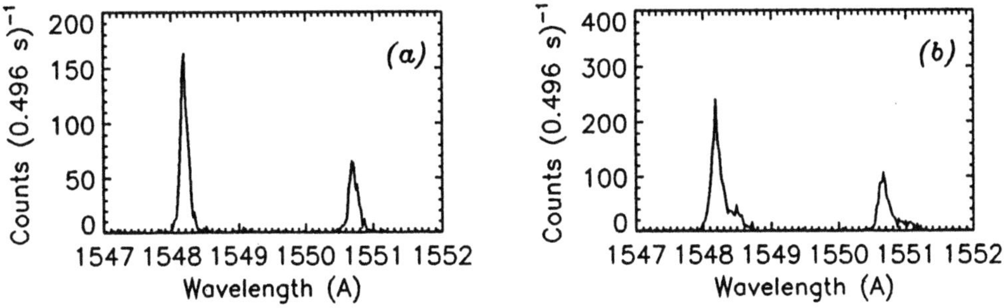

Figure 8. UVSP spectra of sunspot umbrae in (*a*) AR 4598, 1984 November 22 and (*b*) AR 4637, 1985 March 31. Both spectra were obtained with a 3 arc sec $\times$ 3 arc sec entrance aperture, 30 mÅ exit slit, and 25 mÅ sample size. Case (*b*) is one of the few obvious cases of supersonic umbral downflows obtained with UVSP; (*a*) is much more common.

would seem to exclude this possibility. I would like to suggest that the UVSP and HRTS umbral downflow data sets are, respectively, small and very small samples of a very variable phenomenon. The Spacelab 2 HRTS results show quite distinctly that strong downflows are not always present over a given umbra. Sometimes they are visible only at the outer edge of the penumbra (Dere et al. 1990), sometimes directly over the umbra, and sometimes both. The UVSP data concentrate on the umbra, which would certainly reduce the likelihood of observing supersonic downflows. Perhaps it is even possible to associate the EUV plumes with the downflows; the models of Foukal (1978) and Doyle et al. (1985) both associate substantial (albeit subsonic) downflows with coronal sunspot plumes. If both plumes and downflows are transient phenomena, this association is at least possible. I would thus further suggest that strong downflows occur in umbrae almost exclusively but sporadically in light bridges, which are far from ubiquitous, and in the bright, dynamic penumbral loops. In quiescent regions of the sunspot umbra, which UVSP observations were designed to study, supersonic downflows would be unlikely. One might even speculate that the light bridges only appear when transition region material is falling at supersonic velocities and producing shocks lower in the solar atmosphere. Perhaps light bridges are even replenished by such flows. Clearly, this is a matter for extensive, systematic future observations, as well as modelling of the effect of such downflows on the white-light appearance of the umbra.

5. Oscillations

Unmanned spacecraft have an inherent advantage over manned spacecraft in pointing stability. Likewise, orbiting spacecraft simply have much more time to observe time-dependent phenomena than rocket flights. It is not surprising, therefore, that the first observations of transition region oscillations in sunspots were reported by members of the UVSP team (Gurman et al. 1982; Henze et al. 1984) on SMM. We stumbled on the oscillations while making repeated Doppler and magnetic maps of active regions, and convinced ourselves of the reality of the oscillations by obtaining time series of line profiles in a single, umbral pixel. Figure 9 shows the time series of central intensity and Doppler shift obtained from least-squares Gaussian fits of profiles of the CIV resonance line at $\lambda 1548.19$Å in an umbra, as well as the power spectrum of each time series. It should be noted that virtually identical results for total intensity and lineshift can be obtained by simply using the zeroth and first moments of the emission line. The 30-minute time series show obvious power peaks above the 99.9% confidence level (corresponding to 3.05σ in a normal distribution) in both intensity and Doppler shift. We have detected oscillations in CIV in the frequency range 5.8 mHz to 7.8 mHz (129 sec - 173 sec periods) in all the spots (well over a dozen) in which we have attempted the observation. The periods and velocity amplitudes appear to be independent of spot size, over a range of a factor of nearly 20 in area. The oscillations are present in OIV 1401.16Å and SiIV 1402.77Å as well. Outside of sunspot umbrae, there is no signal in the 5.8 - 7.8 mHz range, so we are confident that the oscillation is solar in origin.

The coincidence of blueshifts (negative excursions in λ_0) and maxima in intensity, as well as the absence of any obvious correlation of oscillation parameters with spot dimensions, led us to consider whether the oscillations can be characterized solely by the vertical stratification of the umbral atmosphere; that is, as upward propagating magnetoacoustic waves. If we assume that the magnetic field is purely vertical, in fact, we can treat the oscillations simply as acoustic waves confined to the vertical axis. The fractional intensity fluctuation, $\delta I_0/< I_0 > = 0.085$, agrees reasonably well with the temperature and density fluctuations expected for an adiabatic sound wave with the measured mean velocity amplitude, 2.0 km sec^{-1} at $T = 10^5$ K. (We should note that in a thin transition region we really expect sound waves to be *isothermal*, rather than adiabatic, as is borne out by observations of more "saw-toothed" variations in velocity and intensity, but the data presented here are, oddly, more typical. In any case, the relative

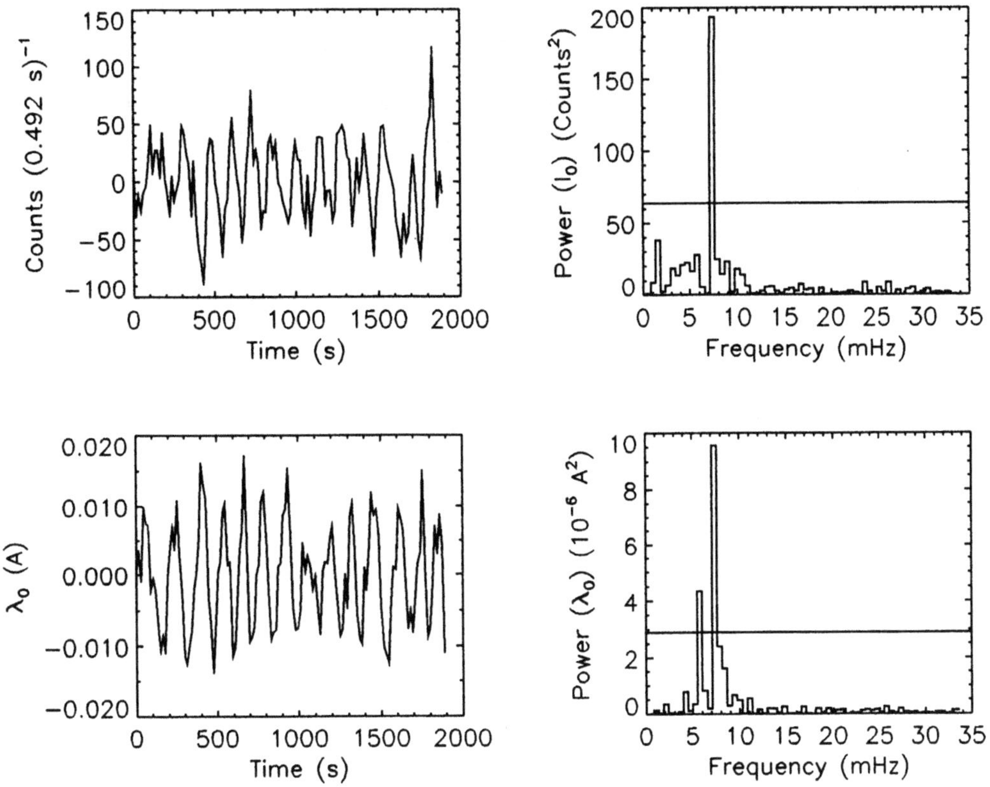

Figure 9. Time series of (*left*) central intensity (*top*) and line center position (*bottom*), and their power spectra (*right*), for line profiles of the CIV resonance line at λ1548.19Å in the umbra of a large spot in AR 2522. The UVSP observations were obtained on 1980 June 26 with a 3 arc sec × 3 arc sec aperture, 30 mÅ exit slit bandpass, and a cadence of 14.88 sec.

variations of intensity and velocity are consistent with isothermal oscillations as well.)

This rather unexpected discovery led us, at about the same time as Žugžda, Locans, and Staude (1983: ZLS), to examine whether the similarity of the oscillations in spots of different sizes could be explained by the existence of a resonant acoustic wave cavity in the umbral chromosphere. Rather than perfect trapping, this cavity was found by ZLS and Gurman and Leibacher (1984) to effect *resonant transmission* of acoustic waves that tunnel through the umbral temperature minimum at just below the acoustic cutoff frequency. (Higher-order transmission peaks represent a broader cavity, bounded below by the subphotospheric temperature gradient.) This process is analogous to the wave optics case of a Fabry-Pérot filter. Since the chromospheric cavity would be expected to contain both upward- and downward-propagating waves (reflected from both the temperature minimum and the steep temperature gradient in the upper chromosphere), we expected the phase difference between velocity and intensity in lines formed in the chromospheric cavity to be more complex than those in the transition region lines. When we obtained times series of line profiles of MgII *k* after the repair of the SMM spacecraft in 1984, the oscillation spectrum agreed well with the resonant transmission predictions (Gurman 1987). It should be noted, however, that only the lowest frequency Mg II power peak displays the predicted phase relationship between intensity and velocity. Although the interpretation of intensity phase as pressure phase is probably valid in the chromosphere, the disagreement with prediction may also be an indication that the photospheric, MHD wave cavity described by Thomas (1985) cannot be ignored in a complete description of umbral oscillations. In any case, the resonant transmission mechanism appears to allow us the exciting prospect of an independent validation of details of radiative transfer models of the umbral chromosphere.

6. The magnetic field

Determining the vertical gradient of the umbral magnetic field from visible wavelength observations has always been difficult: optically thick lines are formed over a range of heights, chromospheric lines are not easily interpreted by the Unno grey-atmosphere formalism, and so on. The first opportunity to estimate the field gradient more directly came with the UVSP measurement of the longitudinal magnetic field in the transition region over a sunspot. The CIV observations of Henze et al. (1982) were obtained with a 10 arc sec × 10 arc sec entrance aperture. These measurements yielded umbral longitudinal field strengths of 900 to 1200 Gauss, with uncertainties typically $\approx 10\%$.

Hagyard et al. (1983) found that either by using the transverse component of the photospheric field, $(\partial B/\partial z)_{z=0}$ as a boundary condition in solving $\nabla . B = 0$ or by a potential field extrapolation from the photospheric field measurements to find the height at which the longitudinal field matched the observed transition region values, the UVSP measurements put the umbral transition region at least 4000 km above the photosphere, roughly twice the height predicted by hydrostatic radiative transfer models. The mean field gradients found by these methods were in the range of 0.20 Gauss km^{-1}.

7. Conclusions

Among the conclusions we can draw from ultraviolet observations of sunspots are:

- Non-local radiation fields and excitation are critical in describing strong, chromospheric line emission over spots.
- Non-thermal broadening of transition region lines is significantly reduced over umbrae.
- The transition region over the penumbra is often quite dynamic, but
- the umbra is often indistinguishable from the quiet Sun in transition region emission lines formed below $\approx 2 \times 10^5$ K, and
- it is not yet clear whether the bright plumes observed at higher temperatures really overlie most umbrae. It cannot be emphasized too strongly that
- in the wavelengths accessible to IUE and GHRS, there is no evidence for particularly strong emission or high densities over umbrae.
- The HRTS observations of transient, supersonic downflows over spot umbrae, as well as penumbrae, can only be reconciled with the absence of strong umbral downflows by further observations. Such observations should be pursued vigorously with the SUMER and CDS instruments on SOHO, so that we can determine whether such flows might be expected in starspots.
- Intensity and velocity oscillations measured with ultraviolet emission lines provide a useful diagnostic tool for verifying umbral atmospheric structure predicted by hydrostatic, radiative transfer models, as do the field gradients inferred from comparing measurements of photospheric and transition region magnetic fields.

Acknowledgements I am indebted to Pål Brekke, Peter Foukal, Kenneth Dere, and Brigitte Schmieder for generously providing me with material for this review, and to Stephen Drake for many helpful discussions during its gestation. I should also like to thank the local organizing committee for helping with local expenses, and Daniel Spicer for arranging funding for my travel here. My deepest appreciation goes to the designers, fabricators, and operators of the Skylab, rocket, OSO-8, SMM, and Spacelab 2 instruments that have provided us with so many novel views of the sunspot atmosphere. I trust that the successors to these devices will continue to confound and excite us in the future.

References

Athay, R.G., Gurman, J.B., Shine, R.A., Henze, W.: 1982, ApJ, 261, 684

Athay, R.G., Gurman, J.B., Shine, R.A., Henze, W.: 1983, ApJ, 265, 519

Basri, G., Linsky, J., Bartoe, J.-D., Brueckner, G., Van Hoosier, M.: 1979, ApJ, 230, 924

Bonnet, R.M., Bruner, M., Acton, L.W., Brown, W.A. Decaudin, M., Foing, B. 1982, AA, 111, 125

Brekke, P., Kjeldseth-Moe, O., Bartoe, J.-D.F., Brueckner, G.E.: 1987, in Proc. 8th ESA Symp. on Europ. Rocket & Balloon Programmes & Related Research, ESA SP-270, p. 341

Brueckner, G.E.: 1981, in *Solar Active Regions,* ed. Orrall, F.Q. (Boulder: Colorado Associated University Press), p. 113

Cheng, C.-C., Doschek, G., Feldman, U.: 1976, ApJ, 210, 836

Cheng, C.-C. Kjeldseth-Moe, O.: 1977, Solar Phys., 52, 327

Dere, K.P., Schmieder, B., Alissandrakis, C.E.: 1990, AA, in press

Doyle, J.G., Raymond, J.C., Noyes, R.W., Kingston, A.E., ApJ, 297, 816

Evershed, J.: 1913, *Kodaikanal Obs. Bull.* 15, 63

Fontenla, J., Reichmann, E., Tandberg-Hanssen, E.: 1988, ApJ, 329, 464

Foukal, P.V.: 1976, ApJ, 210, 575

Foukal, P.V.: 1978, ApJ, 223, 1046

Foukal, P.V.: 1981, in *The Physics of Sunspots,* eds. Cram, L.E., Thomas, J.H. (Sacramento Peak Observatory: USA), p. 191

Foukal, P.V., Huber, M.C.E., Noyes, R.W., Reeves, E.M., Schmahl, E.J., Timothy, J.G., Vernazza, J.E., Withbroe, G.L.: 1974, ApJ, 193, L143

Gurman, J.B.: 1984, Solar Phys. 90, 13

Gurman, J.B.: 1987, Solar Phys. 108, 61

Gurman, J.B. Athay, R.G.: 1983, ApJ, 273, 374

Gurman, J.B. Leibacher, J.W.: 1984, ApJ, 283, 859

Gurman, J.B., Leibacher, J.W., Shine, R.A., Woodgate, B.E., Henze, W.: 1982, ApJ, 253, 939

Hagyard, M.J., Teuber, D., West, E.A., Tandberg-Hanssen, E., Henze, W., Beckers, J.M., Bruner, M., Hyder, C.L., Woodgate, B.E.: 1983, Solar Phys. 84, 13

Henze, W., Tandberg-Hanssen, E., Hagyard, M.J., Woodgate, B.E., Shine, R.A., Beckers, J.M., Bruner, M., Gurman, J.B., Hyder, C.L., West, E.A.: 1982, Solar Phys. 81, 231

Henze, W., Tandberg-Hanssen, E., Reichmann, E.J., Athay, R.G.: 1984, Solar Phys. 91, 33

Kneer, F., Scharmer, G., Mattig, W., Wyler, A., Artzner, G., Lemaire, P., Vial, J.-C.: 1981, Solar Phys. 69, 289

Kjeldseth-Moe, O., Brynildsen, N., Brekke, P., Engvold, O., Maltby, P., Bartoe, J.D.-F., Brueckner, G.E., Cook, J.W., Dere, K.P., Socker, D.G.: 1988, ApJ, 334, 1066

Lites, B.W.: 1980, Solar Phys. 68, 327

Lites, B.W. Skumanich, A.: 1982, ApJS. 49, 293

Maltby, P., Brekke, P., Brynildsen, N., Kjeldseth-Moe, O., Bartoe, J.-D.F., Brueckner, G.E.: 1990, in preparation

Nicolas, K.R. Kjeldseth-Moe, O.: 1981, in *The Physics of Sunspots*, eds. Cram, L.E. and Thomas, J.H. (Sacramento Peak Observatory, USA), p. 336

Nicolas, K.R., Kjeldseth-Moe, O., Bartoe, J.-D., Brueckner, G.: 1982, Solar Phys. 81, 231

Noyes, R.W., Raymond, J.C., Doyle, J.G., Kingston, A.E.: 1985, ApJ, 297, 805

Raymond, J.C. Doyle, J.G.: 1981, ApJ, 247, 686

Thomas, J.H.: 1985, Australian J. Phys. 38, 811

Žugžda, Y.D, Locans, V., Staude, J.: 1983, Solar Phys. 82, 369

Discussion: Gurman

Linsky As I recall solar plages are typically larger than spots, but for dMe and RS CVn systems the plages are typically smaller than the spots. Is my characterization of the solar data correct?

Gurman In C IV the most extensive spot groups seem to be associated with quite concentrated areas of plage nearby; smaller, looser spot groups are indeed smaller than the C IV plage. Obviously, though, the visible hemisphere of the Sun has much more plage that the spot coverage.

Maltby The difference in observability of (down)flow by the HRTS and the UVSP is probably the reason for the apparent disagreement. Another point is that for the HRTS data chromospheric lines have been used to determine the wavelength scale. Would you like to comment?

Gurman Since the UVSP line profiles were obtained in only single, 3 x 3 arcsec pixels, and the HRTS profiles along a slit with a stigmatic spectrograph, there is an obvious selection effect. The velocity zero relative to patches of quiet transition region is repeatable to ± 5 km/sec, and the unshifted profile occurs at the same grating step quite repeatably, once allowances are made for orbital Doppler corrections.

Mullan Your estimates of fields in umbrae show only a factor 2 weakening in field stength over a height range of 4000 km. Giovanelli suggested canopy in field at 700 km. Why is the field in the spots not spreading out more rapidly with height?

Gurman The UVSP results show the field strength decreasing by this factor in a spot whose umbra is ≈ 1500 km across; this is a considerable "fanning out". Over the penumbra, one expects from solutions like Pizzo's that the deflection would be much more pronounced.

Byrne Having heard many cautions about applying the solar model to stars it was very encouraging to listen to your talk. In stellar UV line emission we see rotational modulation of cooler, chromospheric lines, such as Mg II, in antiphase with visible light, but hotter lines, such as C IV fail to show modulation. C IV is very scattered which we have attributed to "microflaring". These results suggest that starspots behave similarly to sunspots.

STARSPOT LIFETIMES FROM LONG-TERM PHOTOMETRY

G.J. Anders, D.W. Coates, K. Thompson,
Department of Physics, Monash University,
Clayton, Victoria,
Australia, 3168
and
J.L. Innis,
School of Physics and Space Research, University of Birmingham,
Edgbaston, Birmingham,
England, B15 2TT

1. Introduction

Part of the observing programme at the Monash Observatory is dedicated photometry of a small number of southern active-chromosphere stars. Together with data from other workers, these observations form a fairly extensive history of the photometric behaviour of such objects. Our analysis of these data is leading to a general understanding of some of the properties of the large, long-lived starspots present on these stars. This paper will present some of the recently published findings of the Monash group on two of these stars, and also some further results from newly obtained data.

2. History

The two stars studied in detail are AB Dor (HD36705) and PZ Tel (HD174429). Both are single, late-type young stars showing highly variable lightcurves, and activity over a wide range of wavelengths. The rotation period of AB Dor is 0.514 day, while for PZ Tel it is 0.945 day (See Innis, Thompson and Coates, 1986 (ITC); Innis, Coates and Thompson, 1988 (Paper I); Innis, Thompson, Coates and Lloyd Evans, 1988 (Paper II); Innis, Coates, Thompson and Lloyd Evans, 1990 (Paper III), and references therein).

The light curves of AB Dor show, at times, very rapid changes even from night to night. This indicates that the starspots must also be changing (size and/or temperature) on these timescales. However at other times the changes are not so great, and aspects of the overall behaviour can be followed. Light curves for this star are presented in Figure 1 (from Thompson and Thompson, 1989; Anders, 1990; and new data) for 1988 December, 1989 October/November, and 1990 January/February. The range in V light of the star from late 1978 to early 1990 is shown in Figure 2 (Paper II and new data). A general decline in mean brightness is evident.

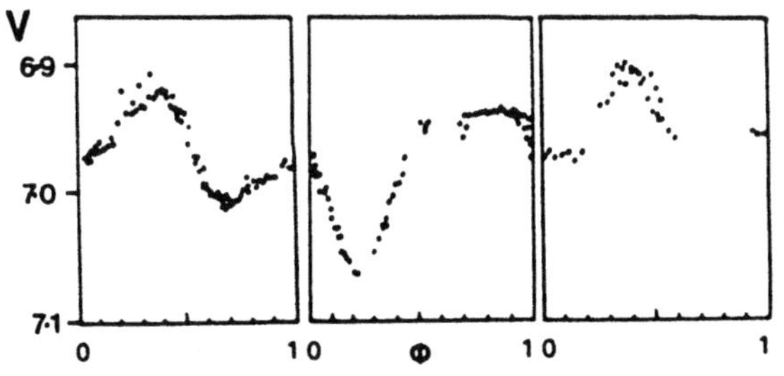

Figure 1. V versus phase (P = 0.51479 day) for AB Dor in (left) 1988 Dec, 1989 Oct/Nov and 1990 Jan/Feb.

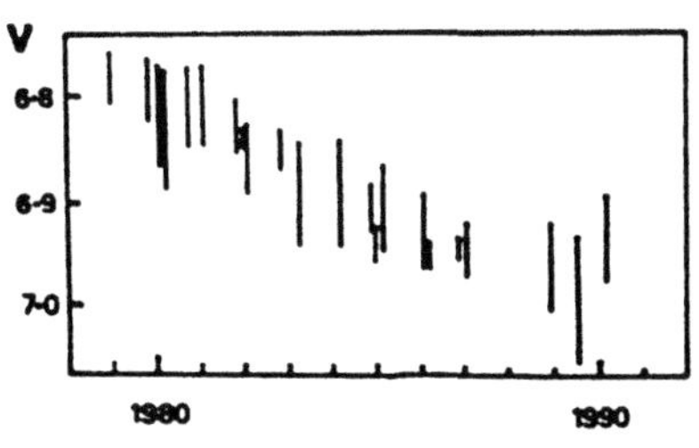

Figure 2. Range of V light versus year for AB Dor.

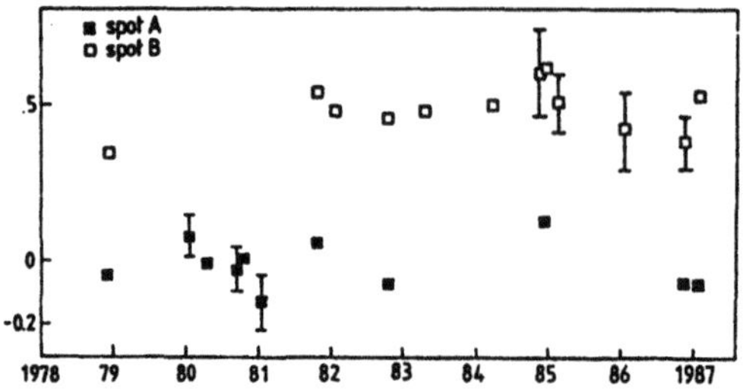

Figure 3. Observed phases of minimum light versus year for AB Dor.

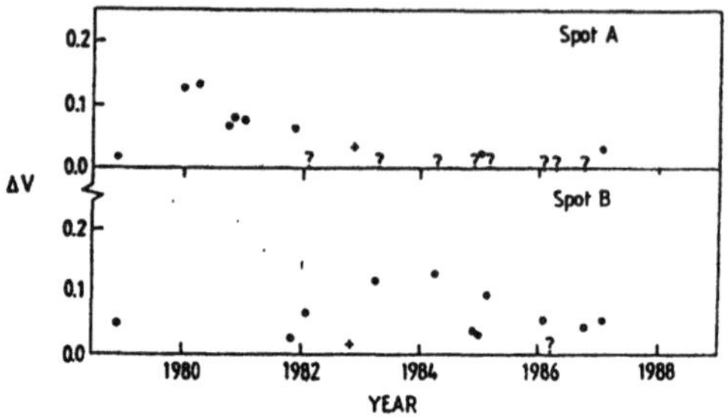

Figure 4. Contribution to the observed ΔV of AB Dor of spots A (upper) and B versus year.

We interpret the gross features of the photometric behaviour of AB Dor from 1978 to early 1988 in a qualitative model involving just two large spots (or spot groups) separated by 180° in longitude (phase) on the star. This is illustrated in Figures 3 and 4 (from Paper II), showing respectively the phase of minimum light (*i.e.* the approximate mid-longitude of the spot group), and the estimated contribution (in V magnitude) to the observed variation made by each spot (denoted as A and B). If the phase identification is correct this suggests the presence of "active longitudes" where spots or spot groups form, intensify, and decay on timescales of 2 to 4 or more years (see Figure 4). It appears though that the "active longitudes" have remained regions of spot formation throughout the 10 years of observations. The 180° separation of the active longitudes is reminiscent of an "oblique-dipole rotator" (see Paper II).

For PZ Tel, the data obtained indicate similar behaviour for the spots as seen for AB Dor. Figure 5 plots the V range against time for PZ Tel. In contrast to AB Dor, the mean V light level has stayed roughly constant for this star. The deep minimum of mid-1982 precedes a bright maximum of about 1 year later. We believe there is evidence to indicate that the dissolution of the spot or group that caused the minimum of mid-1982 left behind a bright remnant, which resulted in the observed bright maximum of mid-1983 (see Paper III). Our data indicate a spot lifetime of at least 4 years. There are unfortunately no data prior to late 1979.

Two interpretations of the photometric behaviour of PZ Tel were offered (Paper III). One involved the presence of three spots. Two of these were 180° apart in longitude (as for AB Dor), and the other showed a relative phase drift, interpreted as evidence for differential rotation. The

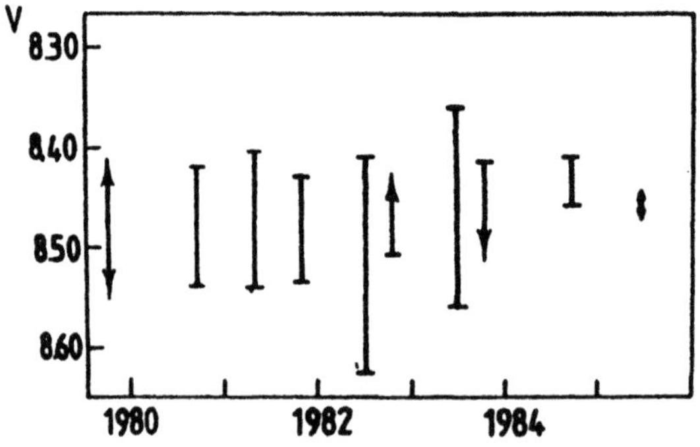

Figure 5. Range in V light versus year for PZ Tel.

estimated size of this shear was similar to the solar shear at low latitudes. The alternative interpretation was for two spots only, stationary in phase relative to each other, but approximately 120° apart in longitude. We were unable to distinguish between these two possibilities.

3. New Data

Observations of these stars are continuing. The data on PZ Tel are not as yet able to select between the two hypothesis for this star, but do suggest that the active longitude where the "deep minimum" spot of 1982 occurred is still active.

For AB Dor, the photometry continues to show rapid changes in the light curves (Figure 1). Of great interest is the flat-topped light curve of 1989 Oct/Nov shown in the middle panel, perhaps indicating an unspotted view of the star at these phases. However, maximum light at this time is substantially fainter than the star has been in previous seasons (Figure 2). Either the star was uniformly spotted at these phases in 1989 Oct/Nov, or, if unspotted, another mechanism must be operating to reduce the overall brightness of the star.

It is interesting to compare the phase of minimum light in 1988 December (about 0.7) and 1989 Oct/Nov (about 0.2). If we speculate that this is indeed due to the same spot (or group), we must allow for a drift in phase (forward or backward) of about 0.5 over the approximately 320 day interval. If this was a shift back in phase, then we would expect to see a minimum around phase 0.0 - 0.05 in 1990 Jan/Feb, which may be consistent with the observations. From the rate of drift over this time we can estimate a rotation period of about 0.51436 day, as opposed to the rotation periods derived from the 1978-1988 data of 0.51479 ± 0.00001 day. Interpreting the new data as showing the presence of a spot at a latitude different to that for the spots previously seen on AB Dor (and for an assumed spot separation of 30° - correct within a factor of 3 or so!), we obtain a value for the differential shear comparable to the solar value at low to mid latitudes. This result is of course very tentative, but indicates that further comprehensive photometry may be able to address some fundamental questions concerning spot behaviour on such stars.

References

Anders, G.J., 1990. IBVS No. 3437.
Innis, J.L., Coates, D.W., and Thompson, K., 1988. MNRAS, 233, 887 (Paper I).
Innis, J.L., Thompson, K., and Coates, D.W., 1986. MNRAS, 223, 183 (ITC).
Innis, J.L., Coates, D.W., Thompson, K. and Lloyd Evans, T., 1990. MNRAS, 242, 306 (Paper III).
Innis, J.L., Thompson, K., Coates, D.W. and Lloyd Evans, T., 1988. MNRAS, 235, 1411 (Paper II).
Thompson K., and Thompson, I., 1989. IBVS No. 3320.

Discussion: Innis

Mullan You say Ca II is maximum in AB Dor at spot visibility maximum, but Hα is strongest at spot visibility minimum (Cutispoto and Pallavicini). Is there a contradiction?

Innis The observation that Ca II emission was maximum at spot visibility maximum came from simultaneous photometry and spectroscopy over 8 days (I think) in Febuary 1985 (see paper 1 from ref. list). I'm not certain that such a correlation would always be seen, (from both geometric effects and intrinsic Ca II variability). The main problem with Hα is that it is complicated by the "co-rotating clouds" of magnetically confined material discovered with high resolution data by Andrew Cameron and Rich Robinson, which show as features moving through the Hα profile. Andrew could I'm sure give a better account of this work than I could.

through the $H\alpha$ profile. Andrew could I'm sure give a better account of this work than I could.

Cameron Our time-series spectra of AB Dor have a time resolution of 3 minutes, and show (a) bright $H\alpha$-emitting features drifting through the profile at the rate expected for surface features, and (b) absorption features with a drift rate consistent with a location 2 - 9 R_* from the rotation axis. The duration of each absorption transient is typically 1 hour. The absorption seen by Cutispoto and Pallavicini is just what you'd see if you integrated for $\frac{1}{2}$ hour during one of these events.

Byrne I would like to draw attention to a very important aspect of the V variations in AB Dor and that is the almost monotonic decrease in mean V magnitude. This would possibly imply an increase in global spottedness.

Innis This is quite possible, but perhaps I'd add that maybe there are other mechanisms for producing the variation. I will be interested to hear Judith Lean's talk (on the Bolometric Variation of the Sun) this afternoon.

Young Concerning AB Dor, you show evidence for two spots about 180 deg apart. Was there ever a time in its observed history when only one spot was in evidence, and might that account for the secular decrease in total output of the star (i.e. the development of the second spot)? On Gl 890 we have found a definite growth of a second spot, also 180 deg from the original single spot so I am interested in just how similar are the behaviours of these two rapidly rotating cool dwarf stars.

Innis From the interpretation we've presented for AB Dor, we think that the initial data were taken when there was only one spot on the star. The development of the second spot took place about the time the first spot was disappearing. Perhaps we've misinterpreted this, but I can't obviously relate the growth of the second spot with the overall mean decline. I'd echo the comment I made earlier on Brendan Byrne's question, *i.e.* perhaps there are other causes of variability too (faculae?).

THE EFFECT OF SURFACE INHOMOGENEITIES ON TOTAL SOLAR IRRADIANCE

Judith Lean
E.O. Hulburt Center for Space Research
Naval Research Laboratory, Washington DC 20375, USA

1. Introduction

Simultaneous observations from the Active Cavity Radiometer (ACRIM) on the Solar Maximum Mission (SMM) satellite (Willson and Hudson, 1988) and the Earth Radiation Budget (ERB) experiment on Nimbus 7 (Hickey et al. 1988) have demonstrated conclusively that the sun's spectrally integrated emission from its entire disc is indeed variable. That the solar "constant" is, in fact, not constant, is of great astrophysical and geophysical interest; implications for the astronomy and astrophysics community of the sun's irradiance variations on all time scales have been reviewed recently by Hudson (1988).

ACRIM and ERB observations are primarily of solar visible radiation. This is because, for the sun's effective temperature of 5770 K, 48% of the total photon output is emitted at wavelengths from λ400 - 800 nm. Emissions which originate in the sun's hotter and less dense outer atmosphere, such as the corona and upper chromosphere, have long been known to respond dramatically to solar magnetic activity (White, 1977; Lean, 1987). That even visible radiation, which is formed in the solar photosphere, the lowest layer of the sun's atmosphere, responds to solar magnetic activity, is an important key in understanding the sun as a variable star. By combining recent knowledge of the sun's irradiance variations with long term, ground-based observations of the radiative output from other solar-like stars, a broader context of stellar activity has begun to emerge (Radick et al. 1990). This allows, conversely, glimpses of possible historical solar irradiance variations by association with older and younger solar-type stars.

Solar visible radiation penetrates the earth's atmosphere, reaching the earth's surface where it heats the land and the oceans. Solar forcing of the earth's climate has long been speculated on the basis of correlations between a variety of climatic indicators and solar variability (e.g., Friedman, 1986). In particular, observational evidence for total irradiance variations in phase with solar activity indicates that times of solar quiescence, such as in the Maunder (1646-1716) and Sporer (1450-1550) minima, are therefore times of reduced total irradiance. The conjunction of these minima in solar activity with cold terrestrial temperatures in the Little Ice Age (1450-1850) supports the hypothesis of solar activity as a climate driver. Not just total radiative output, but also its spectral dependence is important for climate studies, since solar emissions at ultraviolet wavelengths, which are significantly more variable than visible emissions, are absorbed in the earth's atmosphere which may, via radiative, chemical or dynamical coupling, also modulate tropospheric climate (*Studies in Geophysics,* 1982).

Since ACRIM and ERB first demonstrated real variations in solar total irradiance, much has been learned about the origin of these variations by utilizing a wealth of observations of the sun, not as a star, but with spatial resolution of a few arc sec. In white light spectroheliograms, which have been made for more than a century, dark sunspots are the most prominent surface inhomogeneities. Faculae, which are only a few percent brighter than the background white-light sun, are more easily detected by observing their chromospheric couterparts, called plage, at λ393 nm in the core of the CaII K line. Observations of the sun's irradiance variations at UV wavelengths, which are dominated by facular emission alone, have also proved to be useful tools

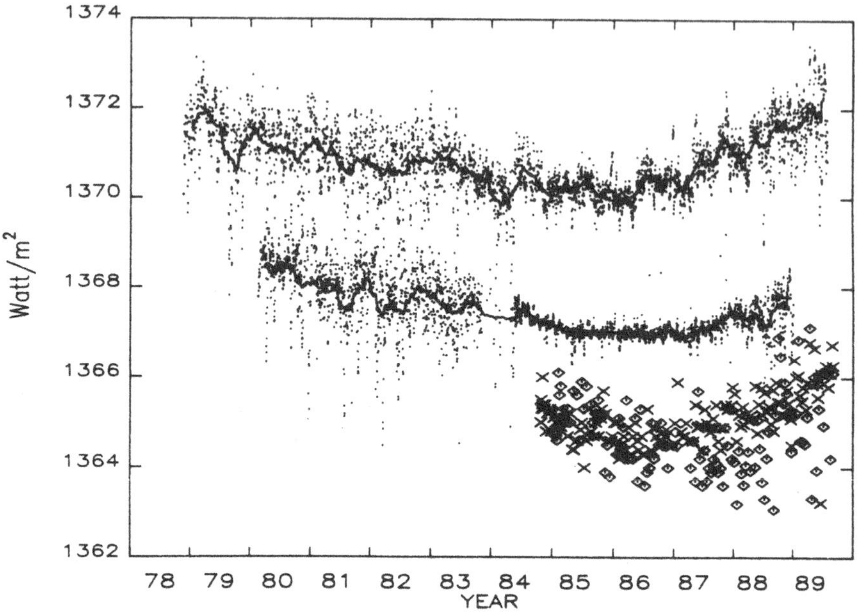

Figure 1. Measurements of the sun's total irradiance made by ERB (*upper curve*), ACRIM (*middle curve*) and, more recently, by ERBS (*crosses*) and NOAA-9 (*triangles*).

in interpreting the long term total irradiance variations.

2. Measurments of total solar irradiance

Space-borne cavity pyrheliometers, which have neither spatial nor spectral resolution, have been measuring the total solar irradiance since 1978 (Hickey et al. 1988; Willson and Hudson, 1988; Hudson, 1988). Data from four different satellite radiometers are shown in Figure 1. Since the precision of solar radiometers currently surpasses their accuracy, relative changes of $< 0.1\%$ throughout the 11-year solar cycle have been better determined than the absolute value of the total solar irradiance which remains uncertain by 0.25% ($\pm 3.5\,\mathrm{W\,m^{-2}}$). The decrease in irradiance from the maximum of solar cycle 21 (1979 - 1981) to minimum activity in 1986, and the increase with the onset of solar cycle 22, are evident.

Although the ACRIM and ERB data are in general agreement about the long term solar cycle decrease of $\approx 0.08\%$ during the descending phase of solar cycle 21, there are important differences between them in the ascending phase of solar cycle 22. These differences are thought to be associated with the ERB data, and are currently being resolved (Hoyt, 1990). ACRIM data ceased in 1989, with the demise of the SMM satellite. Although ERB is still collecting data, its lifetime remains uncertain. NOAA-9 and ERBS continue to monitor the total solar irradiance, but only a few times per month (Lee et al. 1987).

Details of day-to-day changes which are superimposed on the longer-term solar cycle total irradiance variations are illustrated in Figure 2. Both the ACRIM and ERB data, which are daily observations, provide unambiguous records of the variations in the total irradiance arising from rotational modulation. For example, irradiance decreases are clearly evident in both data sets in July, August and September of 1988. Neither ERBS nor NOAA-9 can adequately characterize rotational modulation of the sun's total irradiance because of the reduced cadence of 2 - 3

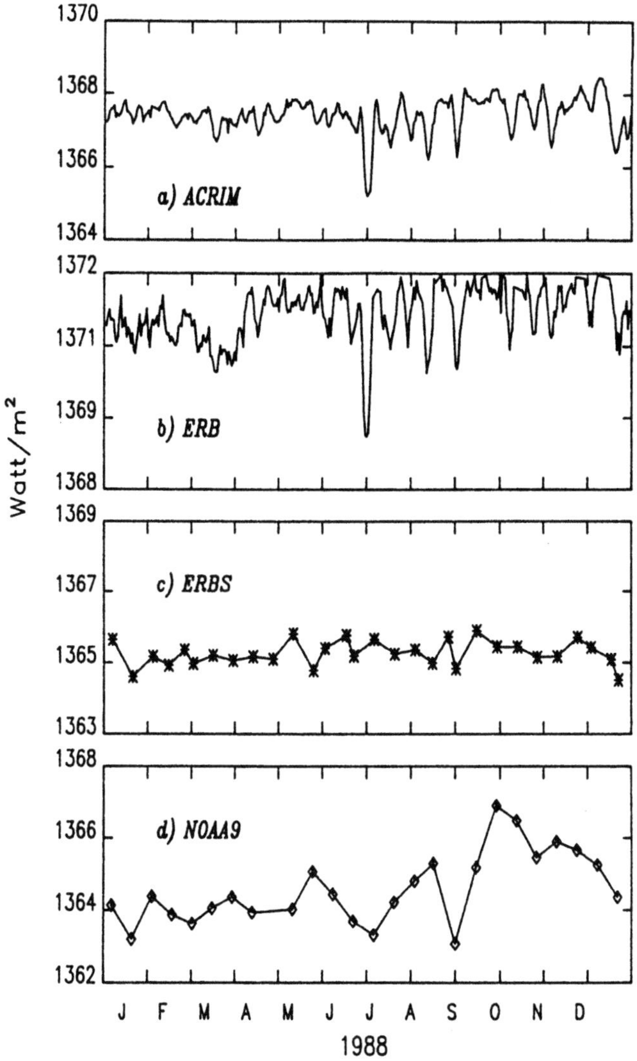

Figure 2. Daily variations in total solar irradiance in 1988, during the ascending phase of solar cycle 22, measured by a) ACRIM, b) ERB, c) ERBS and d) NOAA-9.

measurements per month.

3. Solar surface inhomogeneities

Solar surface inhomogeneities of interest for understanding total irradiance variations are the dark sunspots and bright faculae which occur typically in active regions, grouped together in large scale patterns called active complexes (Gaizauskas et al. 1983). Sunspots are sites of compacted magnetic fields ($B_o \approx$ a few thousand Gauss; Allen, 1973) which have erupted into the photosphere from the convection zone below. Because their concentrated magnetic fields inhibit the upward flow of energy from the convection zone, sunspots are regions of diminished bolometric emission; they are $\approx 33\%$ darker than the background quiet sun (Foukal, 1981).

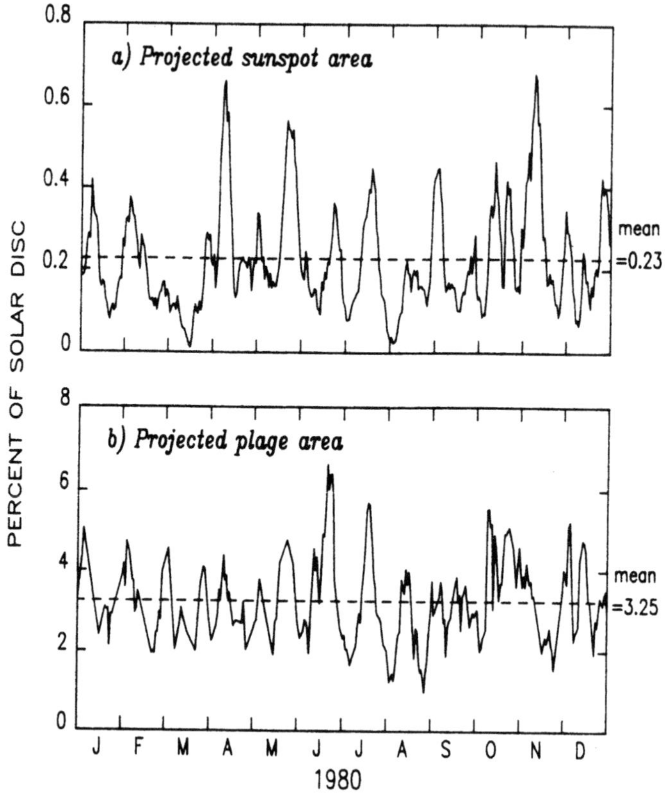

Figure 3. Daily variations during 1980, a time near maximum activity in solar cycle 21, in the fraction of the solar disk covered by a) sunspots, A_s, and b) faculae, A_f, as determined from CaII K plage areas.

Sunspots are often found in groups of ten or so. A sunspot group may cover as much as 0.02% of the solar hemisphere and remain on the solar disk anywhere from a few days to 1.5 months. Surrounding the sunspots are extended regions where the local bolometric emission is enhanced, by a few percent, relative to the background sun. These bright regions, the faculae, are also regions of enhanced magnetic activity ($B_o \approx$ a few hundred Gauss), but of lesser concentration than in sunspots. Faculae cover some ten times more of the disc than do sunspots, and are 2 - 3 times longer lived.

Since sunspots and faculae are both solar magnetic activity phenomena, they occur more frequently and with larger areas at times of maximum solar activity, than at solar minimum when, on certain days, they may be absent entirely from the sun's surface. Figure 3 illustrates the extent to which solar rotation modulates the fractional area of the solar disk covered by both sunspots, As, and faculae, A_f, during a time near maximum solar activity. In this figure, the facular areas were determined from CaII K spectroheliograms, as recorded by the WDC/NOAA.

The emission deficit in sunspots and the emission enhancement in faculae are both functions of wavelength. This is illustrated in Figure 4 which shows the spectra of sunspot and faculae contrasts, compared with the spectral distribution of the sun's radiation. At the visible wavelengths that contribute most to the total solar irradiance, sunspots are an order of magnitude darker than faculae are bright. When viewed at shorter UV wavelengths, sunspots are darker and faculae are brighter than they are in visible light. But, whereas sunspot darkness increases

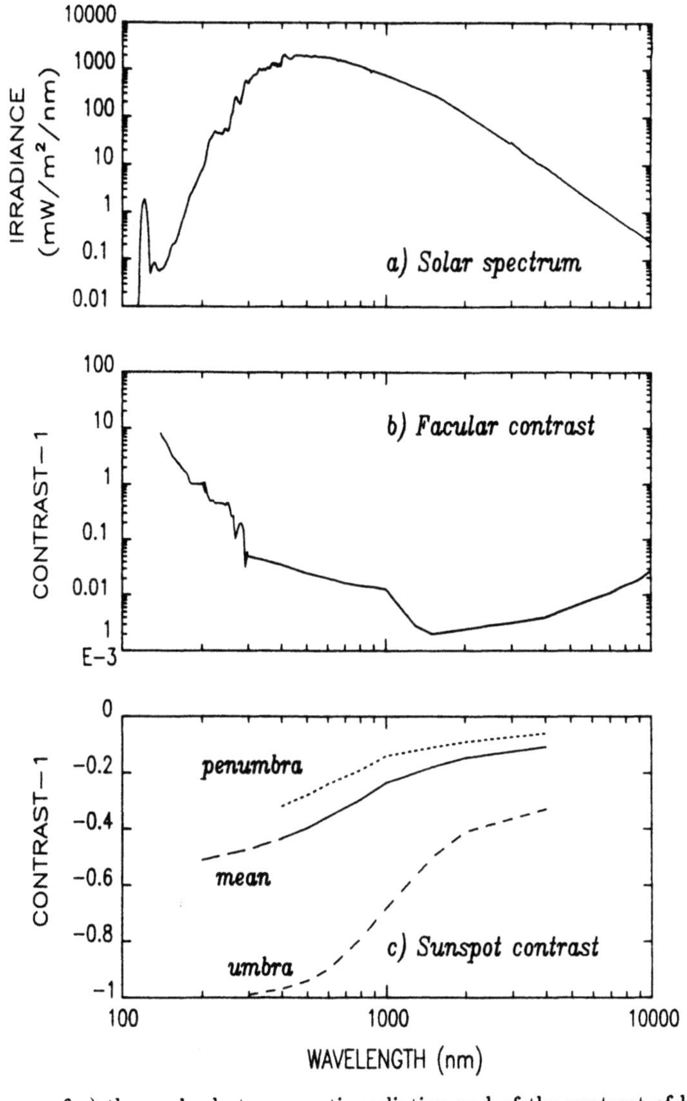

Figure 4. Spectrum of a) the sun's electromagnetic radiation and of the contrast of b) faculae and c) sunspot umbrae and penumbrae (from Lean, 1989).

only moderately from visible wavelengths to 200 nm (from ≈ 30% to ≈ 50%), facular brightness increases dramatically, by two orders of magnitude (from ≈ 2% to ≈ 200%).

4. Effect of sunspots on solar irradiance

The association between short term decreases observed in the sun's total irradiance, such as those seen in Figure 2, and the presence of sunspots on the solar disk was first reported by Hudson et al. (1982). A quantitative estimate of the effect of sunspots on the total irradiance is obtained by evaluating the sunspot blocking function (Foukal, 1981), sometimes called the photometric sunspot index (PSI)

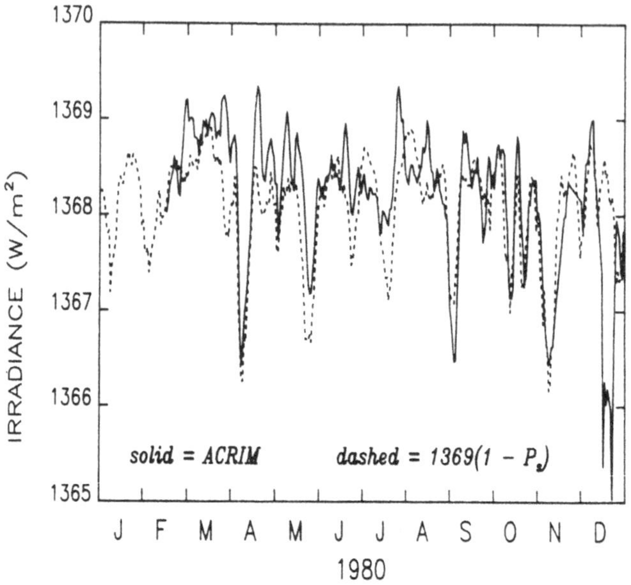

Figure 5. Variations in total irradiance measured by ACRIM (*solid line*), compared with the variations generated by the presence of dark sunspots on the disk, as parameterized by the sunspot blocking function, P_s; for these calculations the quiet sun irradiance is 1369 W m^{-2}.

$$P_s = (C_s\text{-}1)\sum A_i\mu_i(3\mu_i + 2)/5$$

where A_i and μ_i are the area (in units of the solar hemisphere) and heliocentric location of the ith sunspot, obtained from white-light spectroheliograms, $(C_s\text{-}1) = 0.33$ is a measure of the average bolometric emission deficit in sunspots, relative to the background photosphere, and $(3\mu + 2)/5$ is the bolometric center-to-limb variation.

Figure 5 demonstrates the agreement between the calculated P_s function and the observed total irradiance. The coincidence of a number of observed irradiance dips with dips in P_s provides evidence that blocking of the sun's radiation by sunspots does indeed generate solar irradiance variations over solar rotation (≈ 27 days) time scales. The agreement between the two curves in Figure 5 is especially good, in view of the sometimes large uncertainties in the measured sunspot areas, and in the use of a mean bolometric contrast to approximate the true umbral and penumbral darkening.

5. Effect of faculae on solar irradiance

Obtaining a useful parameterization of the contribution of facular brightening to changes in total solar irradiance is less straightforward than the determination of the sunspot effect, as described above. In principle, the same white-light spectroheliograms, from which sunspot areas and locations are obtained, could be used to generate a facular index analogous to the P_s index for sunspots. However, this is not done because, as discussed previously, white-light faculae are only a few percent brighter than the background sun and are more widely dispersed than are sunspots; total facular area is therefore difficult to measure.

Because faculae are not dramatic magnetic phenomenon in white-light spectroheliograms, they were originally considered to be unimportant for understanding total solar irradiance variations. Since then, Foukal and Lean (1986, 1988) have established conclusively that faculae

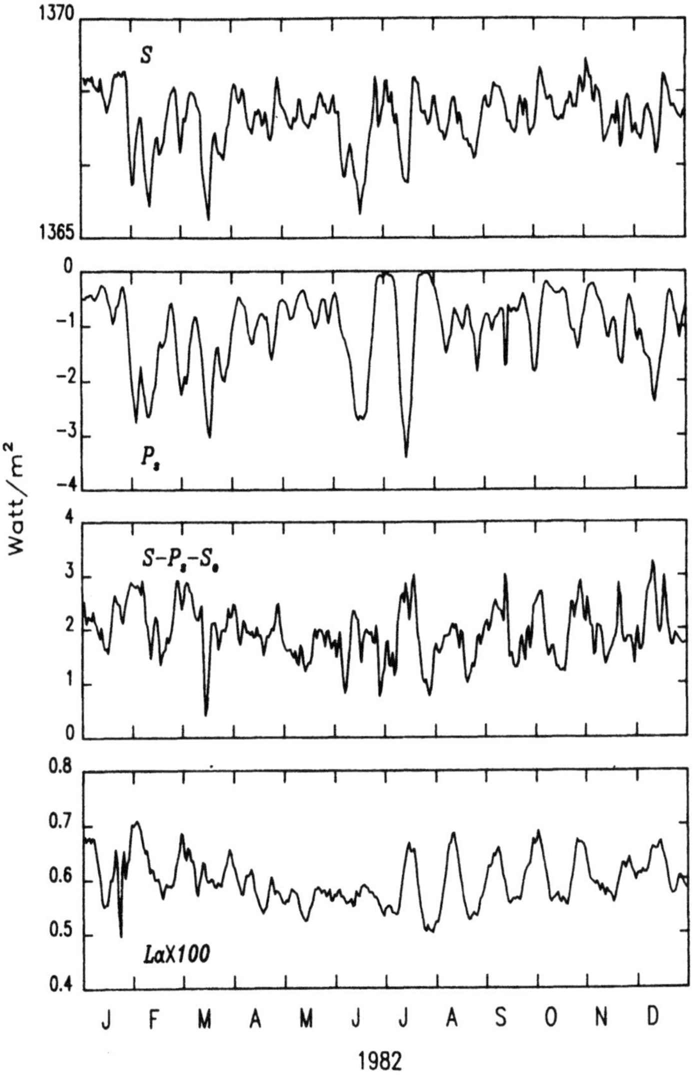

Figure 6. Variations associated with solar rotation during 1982, in S, measured by ACRIM (*top panel*), P_s (*second panel*), $S - P_s - S_o$ (*third panel*) and in the HI Lyman α irradiance at $\lambda 121.6$ nm, measured by the solar spectrometer on the Solar Mesosphere Explorer (SME) satellite (*bottom panel*, Rottman, 1988).

are indeed an important source of total irradiance variations. Instead of trying to obtain information about changes in facular emission from spatially resolved data they used, as a facular proxy, full disk measurements of the sun's spectral UV irradiance. Because faculae/plages are much brighter at UV wavelengths than at visible wavelengths (see Figure 4), changes in UV irradiance at wavelengths shorter than ≈ 250 nm arise primarily from changes in the emission of bright faculae/plages alone; the sunspot effect is minimal at these wavelengths. It was shown statistically that the temporal structure of a residual time series, $S - P_s - S_o$, obtained by

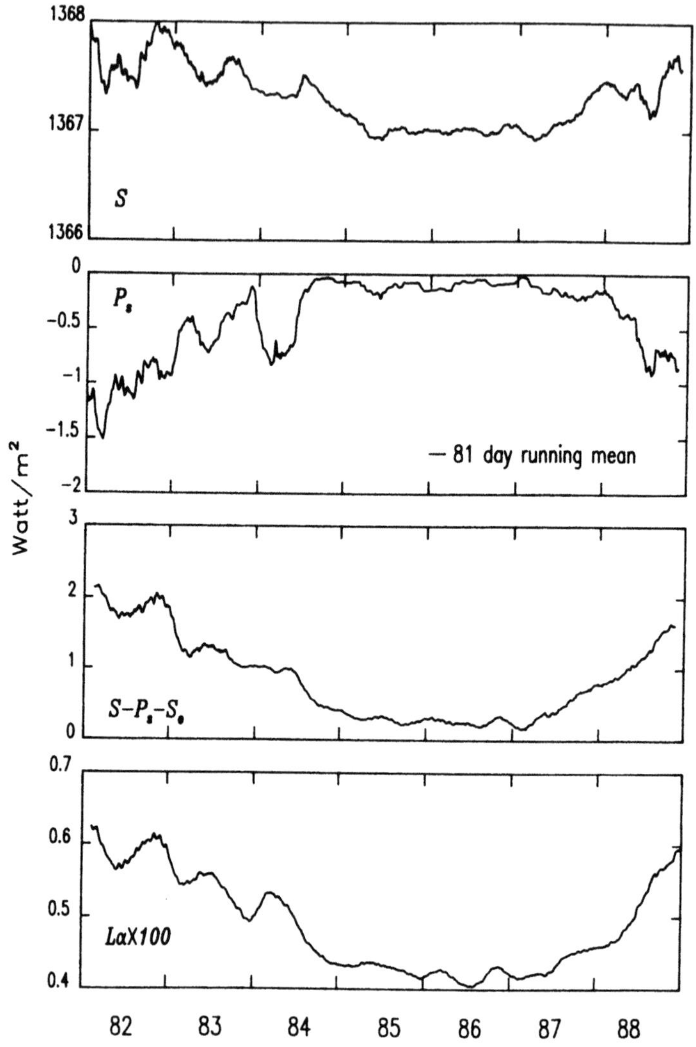

Figure 7. Variations during solar cycle 21 and the ascending phase of solar cycle 22 in 81-day running means of S, measured by ACRIM (*top panel*). P_s (*second panel*), $S - P_s - S_o$ (*third panel*) and the HI Lyman α irradiance at λ121.6 nm, measured by the solar spectrometer on the SME satellite (*bottom panel*).

subtracting the sunspot effect from the measured total irradiance, S, was similar to that of the solar UV irradiance at λ205 nm, but somewhat different from the P_s function itself. This approach is illustrated in Figure 6, where the temporal correspondence over solar rotation time scales between $S - P_s - S_o$ and the solar UV HI Lyman α emission line at λ121.567 nm (used here instead of the λ205 nm irradiance) is evident.

Faculae are now known to contribute to total irradiance variations, not only over solar rotation time scales, but throughout the entire 11-year solar cycle (Foukal and Lean, 1988; Wilson and Hudson, 1988). The temporal correspondence between the total irradiance residuals, $S - P_s - S_o$ and the UV irradiance (*i.e.* a facular proxy) persists, even when the two time series are

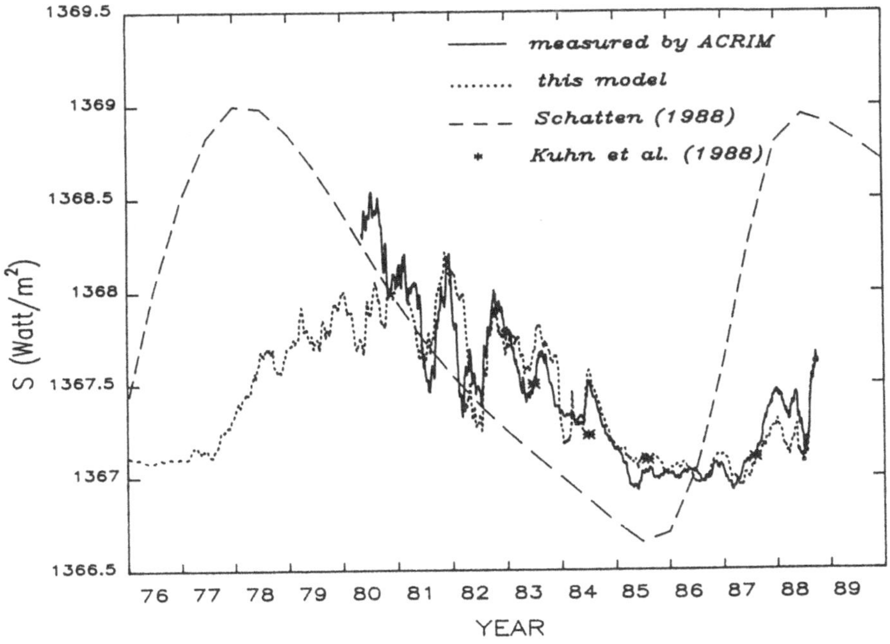

Figure 8. Comparison of observed total solar irradiance variations during solar cycle 21 and the ascending phase of solar cycle 22, with the empirical model of Foukal and Lean (1990), and with other models.

smoothed to remove the rotation modulation. This is seen clearly in Figure 7, where variations in 81-day running mean data from 1982 to 1988 are compared. Intermediate term variations, of duration 3 - 6 months, occur together in both the $S - P_s - S_0$ and Lyman α time series, and reflect the growth, evolution and decay of complexes of activity, which are sites of continual emergence of bright white-light faculae and cospatial chromospheric plage enhanced in emission at all UV wavelengths, including at HI Lyman α.

6. Empirical models of total solar irradiance variations

An empirical model of the sun's total irradiance variations can be constructed by using a calculation of P_s, the sunspot blocking effect, together with a parameterization of the facular effect derived from measured Lyman α irradiances (Foukal and Lean, 1988, 1990). Figure 8 compares such a model with the total irradiance variations measured by ACRIM, and with other models. The empirical model reproduces much of the observed variability in total irradiance on solar rotation, intermediate and solar cycle time scales. Reasons for the differences between the model and the observations in 1980 are unclear; they reflect either changes in ACRIM responsivity during the first months of its mission, or a true solar variation whose source is not associated with sunspots or faculae, and is therefore not reflected in the empirical model, or both of these.

The generally good agreement between the observed and modelled total solar irradiance variations, as shown in Figure 8, suggests that the approach of combining sunspot data with a full disc facular proxy can be used to extrapolate total irradiance variations backwards in time, during solar cycles prior to cycle 21, when no measurements of the total solar irradiance were made. To do this requires a facular proxy other than the HI Lyman α irradiances, which were

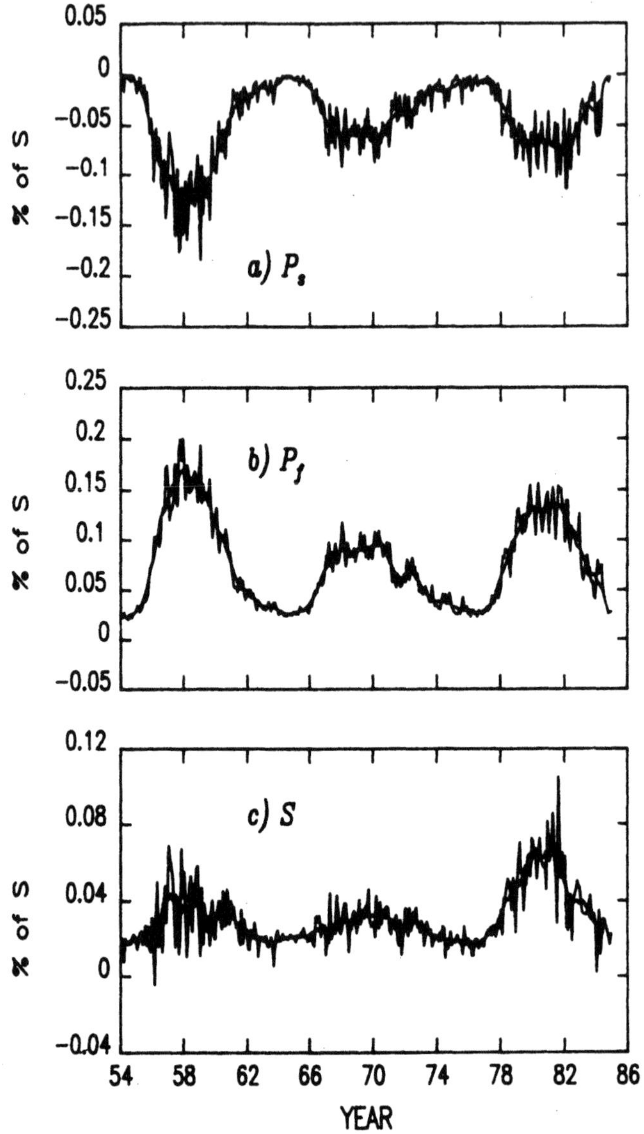

Figure 9. Comparison of a) P_s, the sunspot contribution to the total irradiance with b) P_f, the facular contribution obtained from regression of monthly $F_{10.7\,cm}$ data against ACRIM-P_s, and their cumulative effect on total solar irradiance from 1954 to 1984 (Lean and Foukal, 1988).

not measured prior to 1981. Using, instead, the 10.7 cm radio emission, Lean and Foukal (1988) have reconstructed total irradiance variations in solar cycles 19, 20 and 21, as shown in Figure 9. According to these, and more recent calculations (Foukal and Lean, 1990), the 11-year cycle in the sun's total irradiance was larger during solar cycle 21 than during any previous cycle this century.

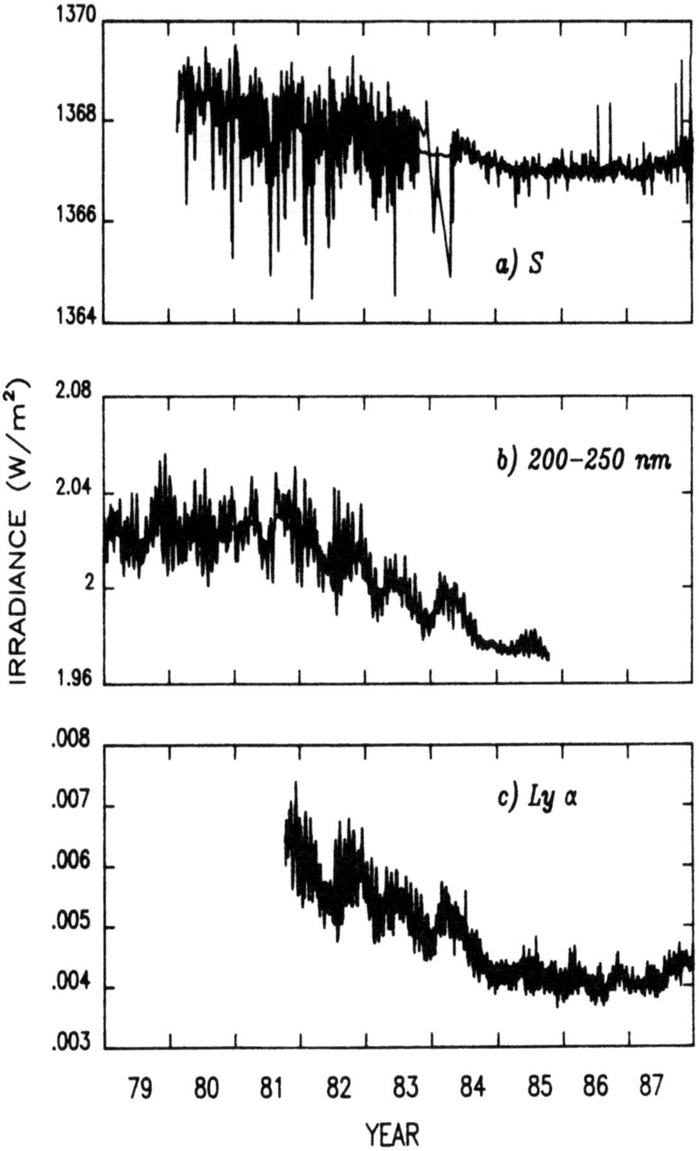

Figure 10. Variations in the sun's emission during the declining phase of solar cycle 21, as evidenced by 81-day running means of a) S, the total irradiance measured by ACRIM and the UV irradiance at b) 200 - 250 nm, determined from the Mg index model of Heath and Schlesinger (1986) and at c) HI Lyman α, measured by SME.

7. Spectral dependance of facular emission

At the same time that the sun's total irradiance was decreasing during the descending phase of solar cycle 21, so too were its emissions at ultraviolet wavelengths. This is illustrated in Figure 10. Although the sun's integrated emission from 200 to 300 nm is just 1.04% of its total radiative output, this radiation is an order of magnitude more variable than the visible emissions

Table 1.
Solar UV and Total Irradiance Variations
1 July 1981 to 30 June 1985
$S = 1\,366.8\ \mathrm{W\,m^{-2}}$....solar min total irradiance $\Delta S = 0.71\ \mathrm{W\,m^{-2}}$....change in S
$\Delta S_c = 1.86\ \mathrm{W\,m^{-2}}$....change in S corrected for sunspot blocking

λ nm	I ($\mathrm{W\,m^{-2}}$)	% of S	ΔI ($\mathrm{W\,m^{-2}}$)	% of ΔS	% of ΔS_c
121.6 (Lyα)	0.004	0.0003	0.0025	0.35	0.13
200 - 250	1.97	0.14	0.059	8.3	3.2
250 - 300	14.0	1.02	0.077	10.8	4.1
300 - 350*	43.11	3.15	0.031	4.4	1.7
350 - 400*	61.76	4.52	0.061	8.6	3.3
sum	120.84	8.83	0.23	32.1	12.3

*estimated from Mg Index model, which assumes that both short and long term changes are dominated by enhanced emission from bright regions.

that dominate the total irradiance. Because of this, changes in solar emissions between 200 and 300 nm contributed 19% of the total irradiance variations from July 1982 to June 1985 (Lean, 1989). It is clear from the data in Table 1 that solar ultraviolet emissions contribute significantly more to variations in the total solar irradiance than they do to its absolute value.

8. Effects other than sunspots and faculae?

Variations in sun-as-a-star UV irradiances and in total irradiance corrected for sunspot blocking are associated with changes in enhanced emission from bright faculae/plage on the solar disc. This is clearly demonstrated for the short term variations generated by the sun's 27-day rotation by calculations (Lean, 1988) using independent measurements of faculae/plage contrasts (*e.g.* Figure 4b) together with scalings of CaII K plage areas (*e.g.* Figure 3b). However, the same calculations extended over the 11-year solar cycle underpredict, by about a factor of two, the measured sun-as-a-star solar cycle variations. This is demonstrated in Figure 11. Brightness variations outside of the largest, contiguous faculae/plage regions (those identified in the scalings from CaII K spectroheliograms) must also be occurring on the longer time scale of the solar cycle. Since this excess brightness source modulates emissions formed in both the non-thermally heated chromosphere and the photosphere, it is possibly generated by smaller, more dispersed regions of enhanced magnetic activity, which are missed in the scalings of the CaII K plage areas. Other mechanisms, such as a solar cycle variation in some global solar parameter, such as temperature, have also been suggested (Willson and Hudson, 1988; Kuhn et al. 1988).

9. Summary

Variations in the sun's total irradiance reflect the growth and evolution of magnetic active regions throughout the sun's activity cycle, with the relative contributions of the dark sunspots

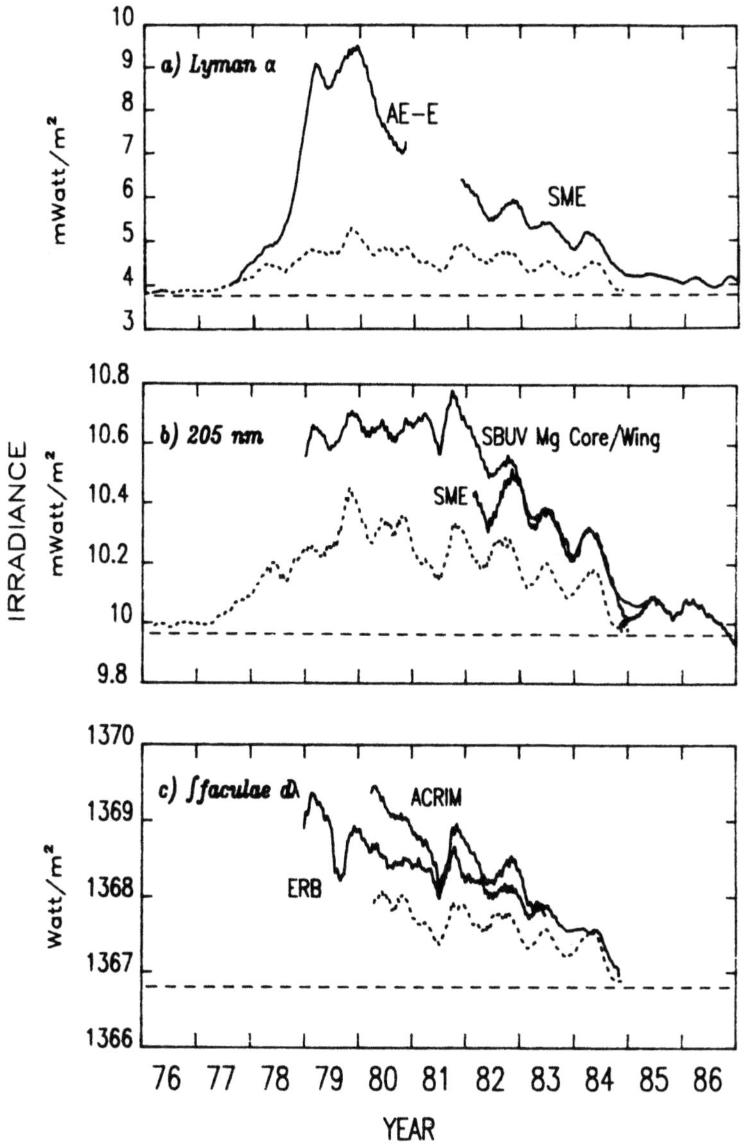

Figure 11. Comparison of 81-day running means of the daily measured variations during solar cycle 21 (*solid lines*) with the calculated variations arising from CaII K plage emission (*dotted lines*) at a) HI Lyman α, b) λ205 nm and c) all wavelengths, as determined by S-P_s-S_o.

and bright faculae/plage dependent on wavelength. Over the 27-day time scale of solar rotation, enhanced emission from bright faculae competes with the emission deficit in sunspots. The signature of facular emission is also seen in total irradiance variations over time scales of 3 - 6 months, the lifetime of the active region complexes which are sites of continuous magnetic flux emergence. The eleven year cycle in total solar irradiance reflects the increase, from minimum to maximum activity, in both the sunspots and faculae on the solar disc. The decrease in total irradiance which has been observed by satellites during the descending phase of solar cycle 21 can be explained by a brightness source which more than compensates the sunspot deficit. This same source of brightness variations in the sun's total irradiance is responsible for changes in the sun-as-a-star in the ultraviolet portion of the spectrum, suggesting that its origin is related to magnetic flux tubes.

References

Allen, C.W.: 1973, *Astrophysical Quantities,* The Athlone Press, University of London

Foukal, P.: 1981, in *The Physics of Sunspots,* eds. L.E. Cram and J.H. Thomas, Sacramento Peak Observatory, Sunspot, New Mexico, USA, p. 391

Foukal, P., Lean, J.: 1986, ApJ, 302, 826

Foukal, P., Lean, J.: 1988, ApJ, 328, 347

Foukal, P., Lean, J.: 1990, Science, 247, 505

Friedman, H.: 1986, *Sun and Earth,* Sci. American Books, New York

Gaizauskas, V., Harvey, K.L., Harvey, J.W. Zwann, C.: 1983, ApJ, 265, 1056

Heath, D.F., Schlesinger, B.M.: 1986, J. Geophys. Res. 91, 8672

Hickey. J.R., Alton, B.M., Kyle, H.L., Hoyt, D.: 1988, Space Sci. Rev. 48, 321

Hoyt, D.: 1990, Conference on Climatic Impact of Solar Variability, NASA/GSFC, 24-27 April

Hudson, H.S.: 1988, ARA&A, 26, 473

Hudson, H.S., Silva, S., Woodard, M., Willson, R.C.: 1982, Solar Phys. 76, 211

Kuhn, J.R., Libbrechht, K.G., Dicke, R.H.: 1988, Science, 242, 908

Lean, J.: 1987, J. Geophys. Res. 92, 839

Lean, J.: 1988, Adv. Space Res. 8(7), 85

Lean, J.: 1989, Science, 244, 197

Lean, J., Foukal, P.: 1988, Science, 240, 906

Lee, R: 1987, in *Solar Radiative Output Variations,* ed. P. Foukal, Cambridge Res. & Instr. Inc., Cambridge, Massachussetts, USA, p. 199

Radick, R.R., Lockwood, G.W., Baliunas, S.L.: 1990, Science, 247, 39

Rottman, G.: 1988, Adv. Space Res. 8(7), 53

Schatten. K.: 1988, Geophys. Res. Lett. 15, 121

Studies in Geophysics: 1982, *Solar Variability, Weather and Climate,* National Academy Press

White, O.R.: 1977, *The Solar Output and Its Variation,* Colorado Assoc. Univ. Press, Boulder

Willson, R.C., Hudson, H.S.: 1988, Nature, 332, 810

OSCILLATIONS & SEISMOLOGICAL DIAGNOSTICS OF SUNSPOTS

Jürgen Staude
Zentralinstitut für Astrophysik
Sonnenobservatorium Einsteinturm, DDR-1561 Potsdam

1. Introduction

Sunspot oscillations have attracted the attention of astronomers for many years. Magneto-atmospheric waves are likely candidates for energy transport and heating in stellar atmospheres, and sunspots provide a unique example of a magnetized stellar atmosphere where up-to-date observational techniques enable us to resolve many details of such periodic disturbances and of the basic atmospheric structure as well. Thus models of both waves and atmospheres can be tested. The fundamentals for a satisfactory theoretical treatment of the generation, propagation, transformation, and dissipation of such waves are not yet completely available. For sunspot umbrae, however, excellent observed data as well as first theoretical models for their interpretation are now available, and a comparison with the properties of ro Ap stars suggests itself.

The present paper cannot give an exhaustive review of all aspects of either observations nor of the theory of waves and oscillations in a magnetized atmosphere. Instead I intend to outline some special aspects of the theme stimulated by the current discussion but, of course, also determined by the individual interests and the point of view of the author. Emphasis will be placed on attempts to explain the observed features of sunspot umbral oscillations by comparison with detailed model calculations. The latter require us to consider waves under the simultaneous influence of compressibility, gravity, and magnetic field in a realistic model of the convective zone and the atmosphere of a large sunspot umbra. Therefore I shall skip studies which pre-suppose simplified conditions such as the thin flux tube approximation or an isothermal atmosphere, although such work can greatly improve our understanding of basic physical processes. The present paper is partly based on two earlier reviews by Zhugzhda et al. (1987) and Staude et al. (1987).

2. Observations

Observational work on oscillations in sunspots has been summarized in several excellent reviews (see, *e.g.* Moore, 1981; Thomas, 1985; Kneer, 1990).

Interest in sunspot oscillations started with the discovery of umbral flashes in the CaII resonance lines (Beckers and Tallant, 1969; Wittmann, 1969) and of running penumbral waves in Hα (Giovanelli, 1972; Zirin and Stein, 1972). Velocity oscillations are also present at photospheric sunspot levels (*umbra:* Beckers and Schultz, 1972; Bhatnagar et al. 1972: *penumbra:* Musman et al. 1976) but the amplitudes are much smaller here than in the chromosphere. Nevertheless the kinetic energy density of the photospheric oscillations can be larger than the energy density in the chromosphere, because the mass density, ρ, is much larger at photospheric levels (Abdelativ et al. 1986). Abdelativ et al. published for the first time a diagnostic (ω - $k_\perp$) diagram for an umbra showing some interesting features, *e.g.* an attenuation of oscillatory power for larger values of frequency, ω, and horizontal wave number, $k_\perp$, to the right hand of a line corresponding to $\omega/k_\perp \approx 25\,\mathrm{km/sec}$. Sometimes the velocity oscillations are connected with oscillations of line intensities at both levels, but the velocity oscillations appear as the more fundamental phenomenon. While the umbral chromospheric oscillations are mainly restricted to a narrow band of periods, P, around 3 min ($100\,\mathrm{sec} \leq P \leq 200\,\mathrm{sec}$), the range of periods is much broader in the umbral photosphere ($2\,\mathrm{min} \leq P \leq 8\,\mathrm{min}$). Strictly speaking, bands of closely packed narrow resonant power peaks around 3 min and 5 min are included. Also in the penumbra the range of significant periods at photospheric levels is much broader than that in the chromosphere. Some authors found oscillations of the photospheric magnetic field (Mogilevsky et al. 1973; Gurman and House, 1981), while others did not (Schultz and White, 1974).

Observations from space with the UVSP instrument on the SMM spacecraft discovered oscillations in velocity and partly in intensity in lines formed in the chromosphere-corona transition region, but only above umbrae (Gurman et al. 1982; Henze et al. 1984). The data clearly show a number of closely packed resonance peaks in the 3 min band. Exactly the same behaviour is observed at chromospheric heights of the umbra. The oscillations in the transition region appear as a continuation of the chromospheric umbral oscillations. Observations of sunspots close to the solar limb (Mattig et al. 1984) have shown that the oscillations in the umbral chromosphere are *longitudinal*, that is to say the plasma displacements caused by the wave take place along the magnetic field.

Simultaneous measurements in several lines are very important because they provide information on the height dependence of amplitudes and phases of both velocity and intensity oscillations. Such data are required in order to test detailed model calculations and to identify the effective wave modes. Giovanelli et al. (1978) were the first to obtain such simultaneous measurements in two or more lines. Later, Kneer et al. (1981), von Uexküll et al. (1983), Lites (1984, 1986a,b), Lites and Thomas (1985), and Thomas et al. (1987) observed umbral oscillations in combinations of lines formed at different heights. The data show characteristic dependences of both amplitudes and phase on the geometrical height, z_o, and on the frequency, ν ($P = 1/\nu = 2\pi/\omega$).

In particular this remark also refers to the difference between the phases of vertical velocity u, ϕ_u, and of intensity oscillations I', ϕ_T, I' being produced by oscillations of thermodynamic quantities such as temperature, T', and mass density, ρ'. The various data are compatible with the assumption of waves propagating upwards in the chromosphere with a phase speed close to the sound speed, c_S, but the detailed behaviour of amplitudes and phases is rather complex, suggesting the existence of several modes with different nodes across the height extent of the umbral atmosphere. Coherence between the oscillations in the lower photosphere and the lower chromosphere has been found for the umbral oscillations in the 3 min and 5 min bands. The 3 min oscillations form a vertical standing wave in the photosphere, while the 5 min oscillations show a small positive phase difference between both heights (Lites and Thomas, 1985). First simultaneous measurements in the photosphere, chromosphere, and transition region of an umbra (Thomas et al. 1987) show oscillations at $P \approx 2.2$ min which are strongly coherent over the whole range of heights, but also horizontally coherent across the entire umbra. Only the power at chromospheric heights is reduced, indicating the presence of a node. At $P \approx 3$ min the oscillations are coherent in chromosphere and transition region, but rather weak in the photosphere. At lower photospheric levels the oscillations in the 5 min band clearly dominate over those in the 3 min band, the latter are hard to detect with increasing depth (see also Balthasar et al. 1986). Compared with the 5 min oscillations in the quiet photosphere outside sunspots, the power of the umbral 5 min oscillations is reduced (Balthasar et al. 1987).

The horizontal extent of an oscillating element is not yet completely clear. Some authors observed coherent oscillations across the whole umbra (Thomas et al. 1984; Lites and Thomas, 1985; Thomas et al. 1987 - see above); other data seem to indicate a subdivision of an umbra into several (2-4) resonators (Turova et al. 1983; Lites, 1986a); Kneer et al. (1981) argued in favour of coherent oscillations across the entire umbral chromosphere, notwithstanding the observed horizontal inhomogeneities of the atmospheric structure. In any case it has now been proved by Lites (1986a) that the closely packed multiple peaks of velocity and intensity power in the 3 min band are indeed cospatial modes with different resonance frequencies. Lites (1986a) found an absence of spatial correlation between the oscillations in the 5 min and the 3 min bands which seems to point out different physical mechanisms or modes for the two period ranges. His data also show an attenuation of higher frequency modes in the chromosphere. In the upper chromosphere the oscillations are sometimes nonlinear ("sawtooth" shape of the waveform at large amplitudes: Lites, 1986b). The same is valid for the transition region, where sometimes sinusoidal waveforms exist (Gurman et al. 1982). In other cases shock-like disturbances are observed (Shine and McWhirter, 1986).

So far we did not consider oscillations with periods larger than those in the 5 min band. In the lower photosphere there seems to exist a deep minimum of power around $P \approx 500$ sec (Balthasar et al. 1986), but again some power for $P \geq 20$ min. Such a time scale points out that umbral bright dots could be representatives of oscillations with these large values of P and could form a third band of umbral oscillations. Berton and Rayole (1985) as well as Gopasyuk (1985), independently of one another, discovered a torsional oscillation of umbral flux tubes with a period of $P \approx 40$ min. Gopasyuk (1982) also gave evidence for a torsion-like oscillation with a much larger period of $P \approx 6$ days.

3. Interpretation of umbral oscillations

The majority of hitherto existing observations of periodic phenomena in sunspots relate to umbral oscillations. Therefore I shall place emphasis mainly on the modelling of waves in a realistic umbral atmosphere permeated by a vertical magnetic field $\underline{B}$. Two basic models have been developed to explain the observed features of such umbral oscillations. Following an earlier work by Uchida and Sakurai (1975) for pure Alfvén waves, Scheuer and Thomas (1981) and Thomas and Scheuer (1982) proposed a model for the resonant trapping of fast-mode, magneto-atmospheric waves in the photosphere and below in the subphotosphere, while the present authors (Zhugzhda et al. 1983, 1984) and Gurman and Leibacher (1984) discussed a model for the resonant trapping of slow magneto-atmospheric waves in the chromosphere of an umbra. Subsequent discussions of these models, including comparisons with new observations (Thomas, 1984; Zhugzhda et al. 1985; Lites, 1986a) have demonstrated strong points, but also shortcomings, in both types of models. Generally, there was agreement that both types of resonators should be excited and interact within a more complex system of coupled resonant layers in the umbra (Zhugzhda, 1984; Thomas, 1984). Recently, an attempt has been made to approach the problem of developing a model for such a more general system of resonators (Zhugzhda et al. 1987). I shall largely follow this approach.

3.1. MAG waves in a vertical magnetic field

For simplicity we consider only small adiabatic disturbances in a stratified atmosphere permeated by a uniform $\underline{B}$, assuming the latter parallel to the gravity $\underline{g}$ (z_0 - direction in a Cartesian coordinate system). From the linearized hydromagnetic equations the equation for pure Alfvén waves is then decoupled; Alfvén waves in an umbral atmosphere have been investigated, *e.g.*, by Nye and Hollweg (1980) and by Zhugzhda and Locans (1982). The remaining two equations describe magneto-atmospheric (or MAG) waves under the influence of gravity (buoyancy), pressure (compressibility), and magnetic forces (Thomas, 1983). The equations simplify for extreme values of the plasma parameter, $\beta = c_S/v_A$ (v_A is the Alfvén speed), *e.g.* for a strong magnetic field $\beta \ll 1$ which is valid in the chromosphere and in the greatest part of the photosphere. Scheuer and Thomas (1981) used a straightforward numerical integration of the 2 coupled MAG equations for special boundary conditions. Such a solution exactly takes into account the relation between the fast-mode and the slow-mode waves, or between the longitudinal and transverse oscillations and its mutual transformation. Transformation is effective around the level $\xi = \omega H / v_A \approx 1$ (H is the pressure scale height) which is situated in the deepest photospheric layers (here also $\beta \approx 1$ is valid) or in the uppermost subphotospheric layers of an umbral model. At larger heights or depths the different modes are decoupled and can be considered separately. In the greatest part of the umbral atmosphere $\beta \ll 1$ is satisfied; moreover, $k_\perp / k_z \ll 1$ ($k_z = \omega / c_S$) is valid for the observed 3 min or 5 min oscillations. In this case the slow mode degenerates to a pure sound wave propagating parallel to $\underline{B}$, while the fast wave exists only as an evanescent mode. At larger subphotospheric depths we have $\beta \gg 1$. Here the fast waves propagate similar to the usual sound waves, that is as a quasi-longitudinal mode. The slow wave, however, is now a quasi-transverse mode. An analytic theory for linear coupling of modes in an isothermal atmosphere has been developed by Zhugzhda and Dzhalilov (1982, 1984a) for vertical and for inclined $\underline{B}$, respectively.

In our approach we used this theory for calculating local coefficients of transformation and reflection of waves incident from below on to the level $\xi \approx 1$; in this way we have justified our simplified treatment which considered the longitudinal mode separately from the other mode. Indeed, our calculations have shown that the fast (acoustic) mode from subphotospheric layers is almost completely 'transformed' into the longitudinal slow mode in the umbral atmosphere above. This means that the longitudinal character of waves is preserved over the whole range of heights, without being disturbed by interaction with the transverse mode (Zhugzhda et al. 1987). The transformation decreases for larger values of ω and $k_\perp$, which could explain the observed attenuation of oscillatory power to the right hand of a line $\omega / k_\perp \approx 25$ km/sec mentioned in Section 2 (Zhugzhda et al. 1987).

3.2 Model assumptions and the method of calculation

For a realistic treatment of the umbral atmospheric structure we used semi-empirical models; the procedures for constructing such models have been described in papers by Staude (1981, 1982, 1985) and summarized in earlier work on umbral oscillations (Zhugzhda et al. 1983, 1984, 1987). The model atmospheres are based on self-consistent solutions of the equations of state, of non-LTE radiative transfer and statistical equilibrium for hydrogen, and of hydrostatic equilibrium. The code gives the height dependence of all required thermodynamic quantities. Subphotospheric layers are calculated by treating convective heat transfer by means of a generalized version of Opik's mixing length theory of cellular convection (Öpik, 1950; Mullan, 1971) as described by Staude (1976, 1978).

Each model describes the basic state of a considered atmosphere (corresponding quantities are indicated by a subscript 0) which has subsequently been subjected to periodic disturbances as described in the preceding subsection. We investigated the behaviour of longitudinal waves in such a model by studying the resonant transmission or the filtering of a broadband flux of wave energy F_0 incident from deeper convective layers. That is, we calculated the coefficient of transmission D of F_0 from the convective zone up to the corona above the umbra, taking into consideration partial reflections at all intermediate heights depending on the atmospheric structure. The resonance frequencies $\omega_i = 2\pi / P_i$ correspond to the maxima in $D(\omega)$. Moreover, the method provides information on the real positions of the boundaries of each resonator and on the height dependence of amplitudes and phases of oscillations of u, T$'$, gas pressure, P', h, etc., thus enabling a direct comparison with observations (h is the vertical plasma displacement, $z = z_0 + h$, the Lagrangian coordinate).

3.3 The two sub-photospheric resonators

In the convective zone below the umbral photosphere two resonators are acting almost independently of each other. The quasi-transverse slow wave appears as oscillatory convection (overstability) along $\underline{B}$ (Kato, 1966; Meyer et al. 1974). Such a type of convection is possible, if $\chi < \nu_m$, where χ and ν_m are the thermal (radiative) and magnetic diffusivities, respectively.

The layer $x = \nu_T$ completely reflects the slow wave upwards and forms the lower boundary of the resonator, while the upper boundary must be assumed at $\xi \approx 1$. Our model calculations (Zhugzhda et al. 1987) resulted in a depth extent of $z \approx 2000$ km for the resonator and a period of $P_0 \approx 20$ min ($\approx$ twice the Alfvén travel time across Δz). The value of P_0 is close to the observed lifetime of bright umbral dots. Therefore we tentatively identify this mode with umbral dots. Non-linear calculations (Knobloch and Weiss, 1984) result in somewhat larger values of P_0 which could still be compatible with observed lifetimes of dots.

The fast wave in the subphotosphere with weak $\underline{B}$ ($\beta \gg 1$) is converted into a quasi-longitudinal acoustic wave. The resonator for this wave differs only slightly from that in the convective zone of the quiet sun with its trapped p-modes. The differences are limited to the upper boundary layer where $\beta \ll 1$, where the Wilson depression is active, and the atmospheric structure in the umbra clearly differs from that in the quiet sun down to a depth of $\Delta z_0 \approx 10^3$ km. For $\lambda_\perp \approx 10^4$ km and $P = 3$ min even up to the bottom of the convective zone (Zhugzhda et al. 1987); the corresponding harmonics have 5 - 10, or even more, nodes over this height distance. Our simple calculations have shown that the observed peaks in the power spectrum of observed photospheric umbral oscillations in the 5 min band can be identified with the response of the lower photosphere to the lowest harmonics P_i, $i = 1, 2, 3, \ldots$, of the subphotospheric acoustic resonator. In the 3 min band, however, the P_i of the fast-wave resonator are so closely packed that they approach the appearance of a continuum; the same is valid for the higher harmonics of the slow wave resonator. This property disagrees with the observations of umbral oscillations in the 3 min band showing several significant and clearly separated peaks at all atmospheric heights. In our opinion, this feature can be ascribed to the action of another resonator, namely the photospheric-chromospheric resonator (see the following section). Only the 'fine structure' splitting of single resonance peaks, sometimes observed in the umbral chromosphere (e.g., Lites and Thomas, 1985), could be explained by the combined effect of both resonators. Formally the spectrum will be determined by the spectrum of the acoustic resonator, multiplied by the coefficient of transmission $D(\omega)$ through the atmospheric layers.

3.4 The chromospheric and photospheric-chromospheric resonator

The existence of a resonator in the umbral chromosphere has been demonstrated by model calculations of Zhugzhda et al. (1983, 1984, 1985) and Gurman and Leibacher (1984). It is formed by the strong reflection of waves from the temperature minimum T_{min} and the chromosphere-corona transition region. The thin transition region is a free boundary which practically oscillates as a whole. At the lower boundary near T_{min} we have tunneling at the acoustic cutoff frequency $\omega_c = \gamma g(2c_S)^{-1}$ (γ is the ratio of specific heats). For frequencies $\omega > \omega_0$ we have little reflection, and therefore a bad cavity, while for $\omega < \omega_0$ there is decreased transmission from below. Nodes exist at T_{min} and at the lower boundary of the photosphere, where the strong gradient of temperature forms a third level of reflection. The scheme described outlines the mechanism of the fundamental chromospheric resonance P_1. Besides this basic period of the chromospheric resonator there exists also a resonance with larger period P_0 of the whole system (photosphere + chromosphere), without a node at T_{min}. Moreover, there are further resonances P_i ($i = 2, 3, \ldots$), with additional nodes. Obviously, the whole umbral atmosphere comprises a system of two resonators, the photospheric-chromospheric resonator and the chromospheric resonator, where the latter forms a subsystem of the other. The whole system of subphotospheric and atmospheric resonators has been outlined in Figure 1.

The values of P_i and the height dependences of amplitudes and phases, including the positions of the nodes, depend on details of the umbral models. In this way we get a method of probing the atmospheric structure. For instance, the periods P_i for each subscript i become larger if the extent of the chromosphere increases. At larger T_{min} the reflection from this intermediate level

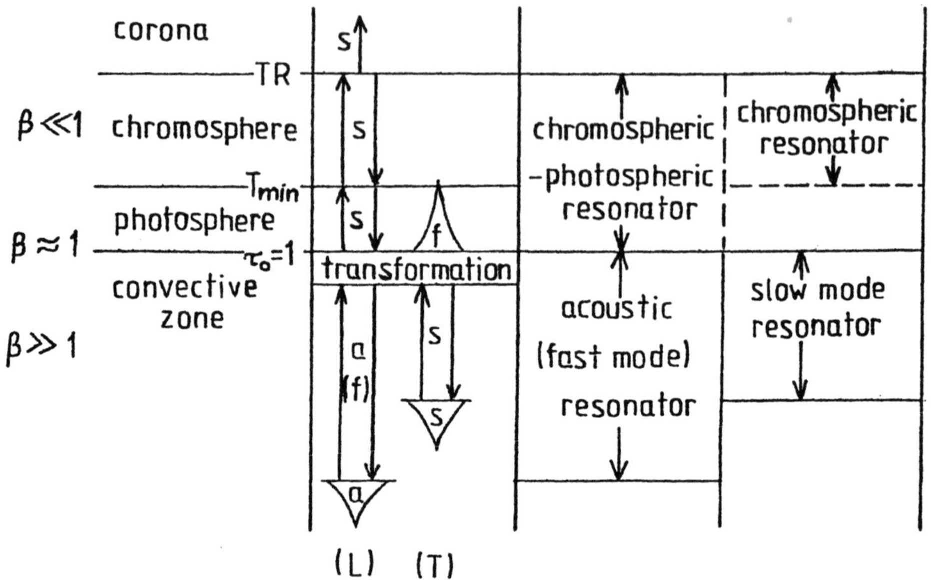

Figure 1: Scheme of the resonators for magneto-atmospheric waves in a sunspot umbra as described in Section 3. The letters *s*, *a* and *f* indicate slow mode, acoustic, and fast mode waves, respectively. (L) means quasi-longitudinal waves, (T) quasi-transverse waves. TR is the transition region, τ_0, the optical depth in the continuum at a wavelength of 500 nm. The other symbols have been defined in the text. Vertical arrows show the direction of wave propagation, triangles, mean evanescent waves.

is less important and the quality of the chromospheric resonator decreases. In our earlier papers it has been shown that it is always possible to select an umbral model atmosphere which can explain a given set of observed oscillatory data. Umbral models which do not differ very much from our 'standard' model SO5 (Staude, 1981; Staude et al. 1983) especially are able to explain the majority of measurements. Oscillations observed in lines from the umbral transition region can be explained by assuming a concentration of such oscillations in cold fine structure elements which occupy only a few percent of the volume in the transition region (Staude et al. 1985).

3.5 Excitation and coupling of the different resonators

In principle we should distinguish between two different types of resonators. An 'ideal' resonator with complete reflection from both boundaries can only be excited from inside. For example, Scheuer and Thomas (1981) considered such a mechanism by assuming a common lower boundary of total reflection or overstable convection for both subphotospheric resonators; its corresponding depth has been arbitrarily prescribed in such a way that a fundamental period $P_0 \approx 3$ min results. Larger periods or closely-packed peaks in the 3 min band as described in Section 3.3 are then impossible, and the physical mechanism which is responsible for such a lower boundary remains unclear. The transmission-type longitudinal resonators, on the other hand, are analogous to a Fabry-Perot interferometer. Reflection from boundaries remains incomplete. It is determined by the structure of the atmosphere or subphotosphere, and we are free from additional assumptions about the driving mechanism for the oscillations and about the boundaries. The probability for excitation of such a 'realistic' resonator is much higher because both inside and outside sources are possible.

Nevertheless, considerations about the source of energy are of some interest for completing the specification of our model. In the earlier papers by Zhugzhda et al. (1983, 1984, 1985) or Gurman and Leibacher (1984), a broad-band ('white') flux of wave energy has been assumed incident on the deepest subphotospheric layers which are still treated in detail in the model of resonant transmission. The subphotospheric acoustic resonator can be excited by various mechanisms which could produce such a broad-band flux, similar to those which are under discussion for the p-modes of the quiet sun. The strong coupling of the acoustic resonator and the photospheric-chromospheric resonator (see Section 3.1) results in a system of coupled longitudinal resonators over a large range of heights, from the convective zone up to the base of the corona. If we are interested only in the chromospheric 3 min oscillations it is even a good approximation to consider separately the atmospheric umbral layers and to assume a white flux incident on the base of the photosphere (Zhugzhda et al. 1987).

The subphotospheric slow wave resonator is nearly independent of the longitudinal oscillations. It could be excited by overshooting from the region of overturning convection, 2000 km to 20 000 km below the umbral photosphere where $x < \nu_m$ (Meyer et al. 1974; Staude, 1978).

4. Penumbral waves

Penumbral waves permit us to investigate wave propagation in a stratified atmosphere permeated by an almost horizontal magnetic field, thus complementing our information on waves in the vertical field of an umbra. While we considered umbral waves at some length, only a few remarks will be added here on penumbral waves. Zhugzhda and Dzhalilov (1984b) proposed a model which is based on their theory of linear transformation of MAG waves. All observed types of penumbral waves can be explained by the transformation of the subphotospheric p-modes (trapped acoustic waves) under the influence of the almost horizontal magnetic field, if the Evershed shear flow is taken into account (the latter effect has already been considered by Cally and Adam, 1983). For instance, the running chromospheric waves could result from the transformation into slow waves in well-defined points of the (ω - $k_\perp$) plane. Moreover, surface fast waves are produced which are similar to those in the model of Nye and Thomas (1974, 1976). Generally speaking, the development of models for penumbral waves in realistic model atmospheres has not yet achieved a high level comparable with that for umbral waves. Because also observations of oscillations in the inhomogeneous penumbra are complicated, hitherto comparisons between theory and observed data are less reliable than those for umbrae.

5. Analogy with rapidly oscillating Ap stars?

The investigation of solar oscillations with periods of a few minutes, mainly of the '5-min' oscillations, has opened up a promising new field of astronomy, *i.e.* 'helioseismology'. We have acquired a method for probing subphotospheric as well as atmospheric layers, thus completing customary spectroscopic diagnostics. Analogous stellar investigations, however, are still in their infancy. So far the rapidly oscillating Ap (ro Ap) stars represent the only main-sequence stars to which the methods of asteroseismology can be applied.

At present, 14 ro Ap stars are known. They belong to the coolest Ap stars, with a spectral type near F0. That is, these stars have at least a thin envelope convective zone which could excite the oscillations. The oscillations have periods in the range 4 - 20 min and usually appear as small-amplitude variations in broad-band photometry. The oscillations of brightness are modulated in phase with the variations of the magnetic field, with maximum amplitudes coinciding with the peak field strength. Kurtz (1982) applied this property for deriving his oblique pulsator model. That means, the oscillations are concentrated in the magnetic pole or spot, while the (dipole) magnetic axis is inclined to the rotational axis of the star. In this model the oscillations are interpreted as high overtone (large n) p-modes of zonal ($m = 0$) nonradial oscillations of low degree ($l = 0, 1, 2$). Only recently, hints at oscillations of radial velocities have been found in some of those stars as well (Matthews et al. 1988; Ando et al. 1988), but the results are still contradictory and difficult to interpret (Weiss and Schneider, 1989).

It is tempting to look for analogies between the oscillations in ro Ap stars and in sunspot umbrae. However, the scantiness of observed data on ro Ap stars, especially of data from line spectroscopy such as radial velocities, makes such a comparison rather difficult. For the same reason several mechanisms have so far been proposed to explain the ro Ap stars. Shibahashi (1983) and Cox (1984) suggested overstable magnetic convection to account for the observed data. This would be analogous to the subphotospheric slow mode resonator of sunspot umbrae, and the quasi-transverse oscillations would be difficult to detect in radial velocity observations. In the original oblique pulsator model of Kurtz (1982) the oscillations are high overtone nonradial p-modes of low degree. This model has been worked out by other authors taking into account additional effects such as the magnetic field and advection (*e.g.*, Dziembowski and Goode, 1985). This mechanism should be compared with the umbral 5 min oscillations. However, in sunspots these oscillations are weakened compared to the quiet sun (perhaps due to redistribution of wave energy over several modes in the magnetic field), apparently in contrast to the behaviour of ro Ap stars. An atmospheric resonator and its coupling with the subphotospheric acoustic resonator have not yet been considered for ro Ap stars. Detailed model calculations for the different magneto-atmospheric wave modes in ro Ap stars would be highly desirable; the predicted spectrum of resonance frequencies could help to identify the mode acting. For such calculations reliable models of the basic thermodynamic structure in the subphotosphere and atmosphere of the ro Ap star are required. However, the seismological diagnostics themselves provide a means for improving the initial models and for getting insight into such an interesting phenomenon as the ro Ap stars.

6. Conclusions

Comparisons between recent observations of umbrae and detailed numerical model calculations have achieved a substantial agreement. This implies that the hitherto developed models for the structure and dynamics of sunspot umbrae seem to reflect essential features of the nature of sunspots, notwithstanding the various simplifying assumptions. Recent analytic calculations for simplified umbral models also confirmed our model of a system of umbral resonators (Campos, 1989; Evans and Roberts, 1990); Campos (1989) only argued in favour of an additional resonator for pure Alfvén waves. The model provides an important tool for sounding subphotospheric and atmospheric layers of sunspots and possibly of other magnetized stellar atmospheres. Further observations of oscillations should try to improve the spatial resolution on the sun and the coverage of the data in space and time. Model calculations should abandon the restrictive assumptions by considering nonlinear and nonadiabatic processes such as radiative dissipations of waves and, moreover, self-consistent calculations of wave propagation and non-LTE line formation are needed to improve the diagnostics.

Acknowledgements: The author wishes to thank the organizers of the Armagh Bicentenary Colloquium for the invitation and support to attend the meeting.

References

Abdelativ, T.E., Lites, B.W., Thomas, J.R.: 1986, ApJ, 311, 1015.

Ando, H., Watanabe, E., Yutani, M., Shimizu, Y., Nishimura, S: 1988, PASJapan, 40, 249.

Balthasar, H., Fangmeier, E., Küveler, G., Wiehr, E.: 1986, Mitteil. Astron. Ges. 65, 155.

Balthasar, H., Küveler, G., Wiehr, E.: 1987, Solar Phys. 112, 37.

Beckers, J.M., Schultz, R.B.: 1972, Solar Phys. 27, 61.

Beckers, J.M., Tallant, P.E.: 1969, Solar Phys. 7, 351.

Berton, R., Rayole, J.: 1985, AA, 152, 219.

Bhatnagar, A., Livingston, W.C., Harvey, J.W.: 1972, Solar Phys. 27, 80.

Campos, L.M.B.C.: 1989, MNRAS, 241, 215.

Cox, J.P.: 1984, ApJ, 280, 220.

Dziembowski, W., Goode, P.R.: 1985, ApJ, 296, L27.

Evans, D.J., Roberts, B.: 1990, ApJ, 348, 346.

Giovanelli, R.G.: 1972, Solar Phys. 27, 71.

Giovanelli, R.G., Harvey, J.W., Livingston, W.C.: 1978, Solar Phys. 58,

Gopasyuk, S.I.: 1982, in: W. Fricke and G. Teleki (eds.), *Sun and Planetary System*, Proc. 6th Europ. Reg. Meeting, p. 125.

Gopasyuk, S.I.: 1985, Izv. Krymsk. Astrofiz. Observ. 73, 9.

Gurman, J.B., House, L.L.: 1981, Solar Phys. 71, 5.

Gurman, J.B., Leibacher, J.W.: 1984, ApJ, 283, 859.

Gurman, J.B., Leibacher, J.W., Shine, R.A., Woodgate, B.E., Henze, W.: 1982, ApJ, 253, 939.

Henze, W., Tandberg-Hanssen, E., Reichmann, E.J., Athay, R.G. : 1984, Solar Phys. 91, 33.

Kato, S.: 1966, ApJ, 143, 893.

Kneer, F.: 1990, Proc. 6th Europ. Meeting on Solar Phys. Publ. Debrecen Heliophys. Obs. Hungary, in press.

Kneer, F., Mattig, W., von Uexküll, M.: 1981, AA, 102, 147.

Knobloch, E., Weiss, N.O.: 1984, MNRAS, 207, 203

Kurtz, D.W.: 1982, MNRAS, 200, 807.

Lites, B.W.: 1984, ApJ, 277, 874.

Lites, B.W.: 1986a, ApJ, 301, 992.

Lites, B.W.: 1986b, ApJ, 301, 1005.

Lites, B.W., Thomas, J.H.: 1985, ApJ, 294, 682.

Matthews, J.M., Wehlau, W.H., Walker, G.A.H., Yang, S.: 1988, ApJ, 324, 1099.

Mattig, W., von Uexküll, M., Kneer, F.: 1984, Proc. 4th Europ. Meeting on Solar Physics, ESA SP-220, 59.

Meyer, F., Schmidt, H.U., Weiss, N.O., Wilson, P.R.: 1974, MNRAS, 169, 35.

Mogilevsky, E.I., Obridko, V.N., Shelting, V.D.: 1973, Radiofizika 16, 1357.

Moore, R.L.: 1981, Space Sci. Rev. 28, 387.

Mullan, D.J.: 1971, MNRAS, 154, 467 .

Musman, S., Nye, A.H., Thomas, J.H.: 1976, ApJ, 206, L175.

Nye, A.R., Hollweg, J.V.: 1980, Solar Phys. 68, 279.

Nye, A.H., Thomas, J.H.: 1974, Solar Phys. 38, 399.

Nye, A.H., Thomas, J.H.: 1976, ApJ, 204, 582.

Öpik, E.J.: 1950, MNRAS, 110, 559.

Scheuer, M.A., Thomas, J.H.: 1981, Solar Phys. 71, 21.

Schultz, R.B., White, O.R.: 1974, Solar Phys. 35, 309.

Shibahashi, R.: 1983, ApJ, 275, L5.

Shine, R.A., McWhirter, R.W.P.: 1986, BAAS, 18, 990.

Staude, J.: 1976, Bull. Astron. Inst. Czechosl. 27, 365.

Staude, J.: 1978, Bull. Astron. Inst. Czechosl. 29, 71.

Staude, J.: 1981, AA, 100, 284.

Staude, J.: 1982, HHI-STP Report No. 14, Berlin, p. 24.

Staude, J.: 1985, Astron. Nachr. 306, 197.

Staude, J. et al.: 1983, Acta Astron. 33, 441.

Staude, J., Zhugzhda, Y.D., Locans, V.: 1985, Solar Phys. 95, 37.

Staude, J., Zhugzhda, Y.D., and Locans, V.: 1987, in L. Hejna and M. Sobotka (eds.), *The Sun*, Proc. 10th ERAM, Vol. 1; Publ. Astron. Inst. Czechosl. Acad. of Sci. 66, 161.

Thomas, J.H.: 1983, Ann. Rev. Fluid Mech. 15, 321.

Thomas, J.H.: 1984, AA, 135, 188.

Thomas, J.H.: 1985, Austral. J. Phys. 38, 811.

Thomas, J.H., Cram, L.E., Nye, A.R.: 1984, ApJ, 285, 368.

Thomas, J.H., Lites, B.W., Gurman, J.B., Ladd, E.F.: 1987, ApJ, 312, 457.

Thomas, J.H., Scheuer, M.A.: 1982, Solar Phys. 79, 19.

Turova, I.P., Teplitskaya, R.B., Kuklin, G.V.: 1983, Solar Phys. 87, 7.

Uchida, Y., Sakurai, T.: 1975, PASJapan 27, 259.

von Uexküll, M., Kneer, F., Mattig, W.: 1983, AA, 123, 263.

Weiss, W.W., and Schneider, H.: 1989, AA, 224, 101.

Wittmann, A.: 1969, Solar Phys. 7, 366.

Zirin, H., Stein, A.: 1972, ApJ, 173, L85.

Zhugzhda, Y.D.: 1984, MNRAS, 207, 731.

Zhugzhda, Y.D., Dzhalilov, N.S.: 1982, AA, 112, 16.

Zhugzhda, Y.D., Dzhalilov, N.S.: 1984a, AA, 132, 45.

Zhugzhda, Y.D., Dzhalilov, N.S.: 1984b, AA, 132, 52.

Zhugzhda, Y.D., Locans, V., Staude, J.: 1983, Solar Phys. 82, 369.

Zhugzhda, Y.D., Locans, V.: 1982, Solar Phys. 76, 77.

Zhugzhda, Y.D., Locans, V., Staude, J.: 1985, AA, 143,

Zhugzhda, Y.D., Locans, V., Staude, J.: 1987, Astron. Nachr. 308, 257.

Zhugzhda, Y.D., Staude, J., Locans, V.: 1984, Solar Phys. 91, 219.

Discussion: Staude

Rodono How much energy is involved in oscillations in comparison to other forms of energy, *e.g.* magnetic?

Staude The energy involved in the oscillations is small compared with other forms of energy such as magnetic energy or the energy flux missing in sunspots.

Rodono Has any systematic frequency drift has been observed with increasing sunspot age?

Staude Frequency drifts with increasing sunspot age have indeed been observed.

van't Veer What can we hope to learn about the inner structure of the Sun and stars in future from this type of observations?

Staude We hope that improved observations of spatial fine structures of the oscillations in umbrae can provide the necessary data for seismological diagnostics of the subphotospheric structure of sunspots, similar to the sounding of the convective zone of the quiet Sun. For stars the lack of reliable observed data makes such an analysis still very complicated, as it has been shown for ro Ap stars, but in the near future the situation will surely improve.

Skumanich Spots have finite horizontal scale and hence limit the horizontal wave numbers to specific values. Your analysis assumes a continuum of wave numbers. What is the effect of going from a continuum to a discrete set? Note that reflections off the lateral "walls" will occur.

Staude In the present numerical modelling the analyis has been restricted to pure 'longitudinal' waves (with displacements along the vertical magnetic field) propagating similar to pure sound waves along the magnetic field; considerations of wave transformation have shown that this could be a reasonable first approximation. Of course, a more detailed analysis of observations such as diagnostic (ω - $k_\perp$) diagrams would require to consider the whole system of magneto-atmospheric wave equations and the lateral boundaries of the spot as well.

Mullan Oscillation periods must change with changes in model temperature. What can the observations say about different T_{eff} in different spots? Does it change much from one spot to another?

Staude So far we did not consider different T_{eff} but different temperature stratifications. The calculations show that the resulting details of the oscillations such as resonance periods depend on the temperature model. It has been shown that the periods increase for increasing height of the transition region, and the periods as well as the quality of the resonator depend on the assumed values of T_{min} as well.

Spruit Are you assuming that all that happens is a wave flux from deep below the umbra being transmitted through a cavity in the atmosphere? If so, how do we know that there is enough wave flux in the relevant range of frequencies? The p-mode power in the normal Sun at these frequencies is known (or upper limits are known). So one can answer this question in principle. Another possibility would be an instability mechanism exciting the oscilliation inside the cavity itself; then one is not limited by the incident wave flux (but the instabiltiy has to be strong enough to overcome the losses due to the finite Q of the cavity).

Staude In the presented analysis nonideal "transmission-type" resonators are considered which in principle can be excited from inside as well as from outside (in contrast to an "ideal" resonator with completely reflecting boundaries which can only be excited from inside). In our case, the energy in the high-frequency tail of the 5 min oscillations of the umbral subphotospheric resonator for the fast mode seems to be sufficient to excite the atmospheric 3 min oscillations.

Spruit The observed oscillations amplitudes are small enough to be treated linearly. We know from helioseismology that there is power in the subphotospheric cavity at 2 min. La Bonte, Duvall *et al* have found that active regions "absorb" about half of the nonradial pulsation energy at horizontal wave numbers larger than 2π over the size of the region.

SURFACE MAPPING OF SLOWLY ROTATING, COOL STARS USING LINE BISECTOR VARIATIONS

Clifford G. Toner and Barry J. LaBonte
Institute for Astronomy
University of Hawaii
2680 Woodlawn Drive
Honolulu HI 96822
USA

1. Introduction

Standard Doppler Imaging techniques (*e.g.* Vogt and Penrod 1983; Rice, Wehlau and Khokhlova 1989) rely on the correspondance between location on the stellar disk and position across a rotationally broadened line profile in order to infer the distribution of features on stellar surfaces. Unfortunately, most F, G, and K stars are slow rotators; when $v \mathrm{Sin}\, i$ decreases to the width of the specific intensity profile for a star, or more correctly the macroturbulence width, standard Doppler Imaging techniques can no longer be applied.

For slow rotators ($v \mathrm{Sin}\, i \leq 10$ km/sec) line bisectors provide an extremely sensitive tool for monitoring the shape of line profiles. This makes bisectors very useful for measuring the small line profile distortions which result when rotation carries a feature across a star's apparent disk (*e.g.* Toner 1988; Toner and Gray 1988; Toner and LaBonte 1990a, 1990b). By applying a technique called "Bisector Mapping" (see Gray 1988, Toner 1988), the observed bisector variations can be used to generate a simple map of the star's surface.

2. Observations and Analysis

We obtained high resolution ($\lambda/\Delta\lambda \approx 10^5$) spectra for ξ Boo A (G8V) and β Com (G0V) using the coude spectrograph and Reticon detector of the 1.2 m telescope at the University of Western Ontario (Gray 1986). For this study we used 60 exposures of ξ Boo A from the 1986 season, and 32 exposures of β Com from 1988. Typical S/N ratios for the data are ≈ 300 to ≈ 400.

A mean line bisector was generated for each exposure by averaging the individual bisectors for 6 - 8 spectral lines. Next, a time series of the velocity span (essentially a measure of the mean bisector slope) for the bisector was produced. Periodogram analyses of the velocity spans show a significant peak at 6.5 days for ξ Boo A, and 11.1 days for β Com. These are the same periods found from modulation of the CaII H and K emission (*e.g.* Baliunas *et al* 1985), and are undoubtedly the rotational periods for these stars.

We also reduced the data using standard Fourier techniques (see Gray 1988) and found an $\approx 5\%$ variation in the mean line broadening for ξ Boo A. β Com, on the other hand, shows no line broadening variation above the noise level.

The equivalent widths of some of the temperature sensitive lines in our data are a temperature diagnostic. For ξ Boo A we used V I (λ 6251.83) and Fe II (λ 6257.56), and for β Com we used Ti I (λ 6258.11 and λ 6261.11) and Fe II (λ 6238.39). Since the strength of the V I and Ti I lines increases with decreasing temperature, while the strength of the Fe II lines decreases, we used the ratios of the two. For ξ Boo A we find that V I/Fe II varies by $\approx 7\%$,and for β Com we find that Ti I/Fe II varies by $\approx 5\%$.

3. Modeling and Results

The observed equivalent width variations allow us to constrain the mean temperature and

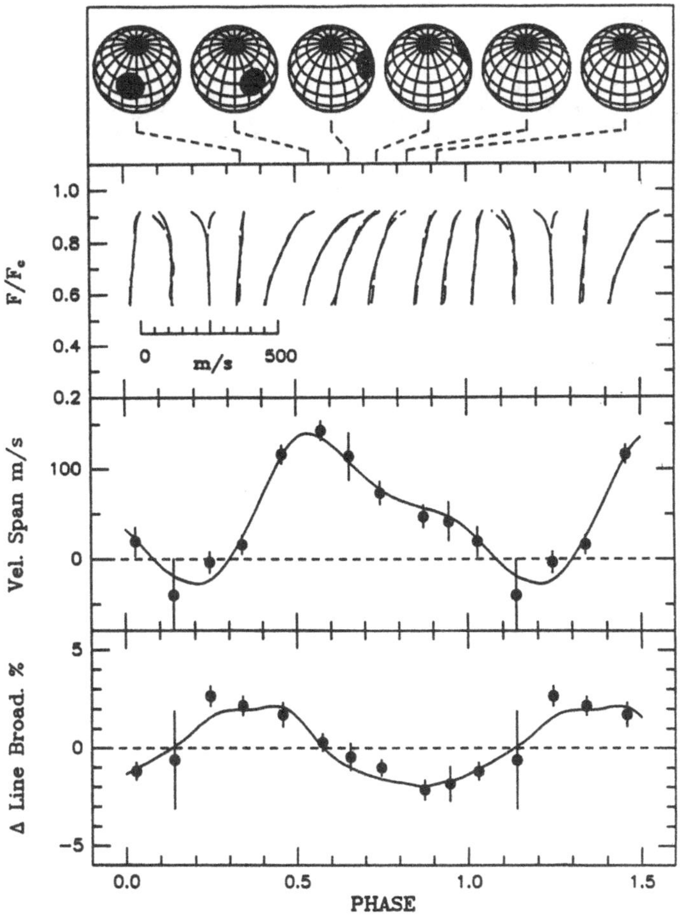

Figure 1. The best tangential flows model for ξ Boo A is shown with the observations. Along the top are pictures of the model star for six representative phases. Below this the observed bisectors (dashed lines) are compared to the simulated ones (solid lines). In the bottom panels the computed velocity spans and line broadening variations (solid curves) are plotted with the observations (dots with error bars).

brightness of the feature, given its size. Numerical experiments with dark, cool features show that a brightness temperature difference alone cannot reproduce the observations. Furthermore, we find that the line bisector and line broadening variations show no correlation with the magnetic sensitivity of the spectral lines. This forces us to turn to velocity fields to explain the observations. Two possible velocity fields have been explored: 1) enhanced granulation velocity dispersion (modeled as a radial flow), and 2) tangential mass flows (analogous to the Evershed and moat flows observed near spots on the Sun).

To model the lines we integrated over the stellar disk. We used a simple two-stream simulation for the granulation (Gray and Toner 1985), and assumed a circular feature. By searching parameter space (*e.g.* spot latitude, spot size. and velocity characteristics) we were able find which combination of parameters best fit the data. The reduced χ^2 was used as the measnre of the goodness of fit.

When assuming tangential flows, we found the following characteristics for the feature on ξ Boo A: areal coverage $10\% \pm 3\%$ of the visible disk, latitude $30^o \pm 4^o$, mean brightness 0.85 ± 0.05

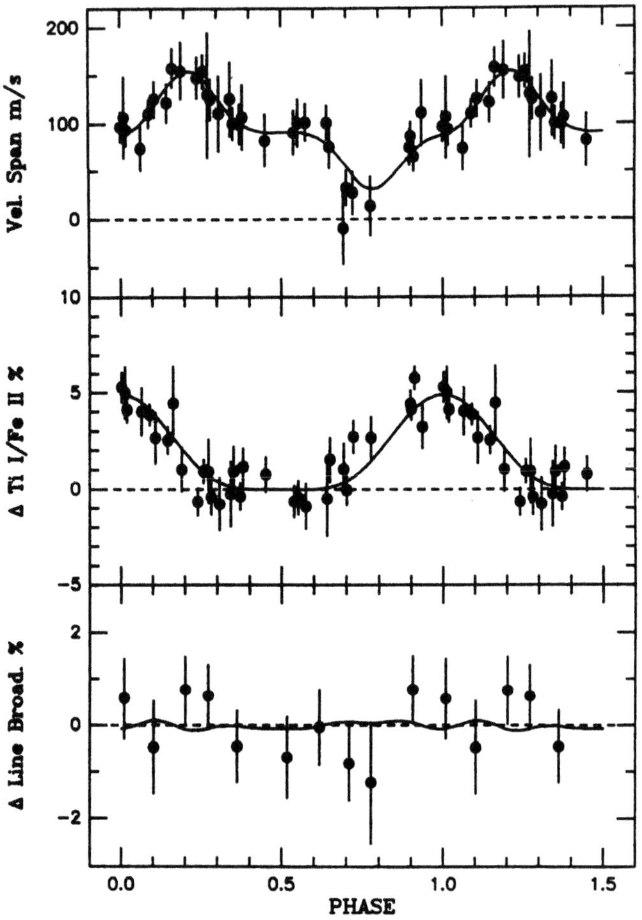

Figure 2. Along the top are the velocity spans, next down the equivalent width ratios, and at the bottom the line broadening. Observations are dots with error bars, and the model calculations

relative to the "quiet" photosphere, mean tangential flow velocities of 8.0 ± 1.5 km/sec, and dispersions about the mean of 8.0 ± 2.0 km/sec. Alternately, when we assumed an enhanced granulation velocity dispersion we found: areal coverage $14\% \pm 4\%$, latitude $45^o \pm 5^o$, mean brightness 0.85 ± 0.05. and granulation velocity dispersions which are enhanced by a factor of ≈ 1.7 relative to the surrounding photosphere. Both models fit the data with adequate precision ($\chi^2 \approx 1$), however the tangential flows model yields marginally lower values of χ^2. The best tangential flows solution is compared to the data in Figure 1.

Only a preliminary analysis, assuming tangential flows, has been completed for the β Com data. The results of this suggest a feature at a latitude of $\approx 40^o$, covering $\approx 10\%$ of the visible disk. The feature is $\approx 15\%$ fainter than the rest of the star, and the mean tangential flow velocities and dispersions are ≈ 2.5 km/sec. This model is compared to the observations in Figure 2.

Acknowledgments: We express our gratitude to D.F. Gray and W. Lockwood for supplying some of the data used in this study. This work was supported by NASA grant NAGW-1698.

References

Baliunas, S.L., Horne, J.H., Porter. A., Duncan, D.K., Frazer, J., Lanning, H., Misch, A., Mueller, J., Noyes, R.W., Soyumer, D., Vaughan, A.H., and Woodard. L. 1985, ApJ, 294, 310.
Gray, D.F. 1986, in Instrumentation and Research Programs for Small Telescopes, IAU Symp. No. 118. ed. J.B. Hearnshaw and P.L. Cottrell (Dordrecht: Reidel), p. 401.
Gray. D.F. 1988, Lectures on Spectral-Line Analysis: F, G, and K Stars (Arva: The Publisher).
Gray, D.F.. and Toner, C.G. 1985, PASP, 97, 553.
Rice, J.B., Wehlau. W.H., and Khokhlova. V.L. 1989, AA, 288, 179.
Toner, C.G. 1988, PhD. Thesis, University of Western Ontario.
Toner, C.G., and Gray, D.F. 1988. ApJ, 334, 1008.
Toner, C.G., and LaBonte, B.J. 1990, in Cool Stars, Stellar Systems and the Sun, ed. G. Wallerstein (San Francisco: ASP Conference Series), p. 161.
Toner, C.G.. and LaBonte, B.J. 1990. ApJ, submitted.
Vogt, S.S., and Penrod, G.D. 1983. PASP, 95, 565.

Discussion: Toner

Pettersen Have you applied your technique to observations of the Sun, to compare with actual direct images?

Toner No, I have not, but I would expect that when viewing the Sun as a star we would not see anything. The activity on the Sun is too low, and requisite non-uniformity in longitude does not exist.

Byrne Can you be more quantitative in your elimination of magnetic field; say by specifying an upper limit to the (B $\times$ filling factor) product.

Toner I have done some very preliminary calculations in which I consider a two component feature (dark and "non-dark"). For ξ Boo A I find that to maintain consistency with the photometic and equivalent width variations one needs a filling factor of at least ≈ 0.35 (*i.e.* 35% of the feature is dark, $I/I_o \approx 0.5 - 0.6$). If I introduce a magnetic field in the dark component only, I find that the field has to be greater than ≈ 1 kG before it is seen. Below this level the velocity fields completely dominate.

Innis Did you take any precautions to eliminate the effects of seeing and guiding on the spectrograph optics illumination?

Toner We use an image slicer with an aperture of $\approx 7 \times 2$ arcsec. This means that practically all of the star's light enters the spectrograph. Guiding corrections are made manually by monitoring the image slicer entrance aperture with a TV camera.

Innis You see no difference between the magnetically sensitive line bisectors and the non-magnetically sensitive line bisectors, and concluded the spot was non-magnetic. But if the spot was dark (*i.e.* gives no light relative to the non-spotted photosphere) would you expect such an effect in the two sets of bisectors?

Toner If the spot were so dark that it gave no light relative to the non-spotted photophere we would not see any variation of the equivalent widths - but we do. This does not rule out the possibility of a few very dark regions scattered throughout the feature which may contain magnetic fields. But if this is the case, these regions would not contribute to the line asymmetry variation. The excess line broadening responsible for the line asymmetry variations originates in a region which is only slightly cooler than the photophere (as evidenced by the equivalent width variations). Comparing the line bisectors for magnetically sensitive and insensitive lines, as well as the line broadening itself, shows no evidence for magnetic fields within the feature.

Pasquini Did you observe any star which does not show bisector velocity?

Toner I have one season of observations of σ Dra, which is a chromospherically inactive star. To within the errors, the line bisectors are constant.

Pasquini Did you try a period search without doing a mean of the bisector?

Toner No, I only computed a periodogram of the velocity spans of the mean bisectors.

van't Veer When you determine the bisectors you take the mean of the different lines. These lines have different depths of formation. What is the effect of this?

Toner To first order all of the line bisectors have the same shape. I have not observed a large enough number of lines to be able to group them by line strength or excitation potential. If we could observe a large number of lines simultaneously (such as with a cross-dispersed Echelle) we could obtain important information on the depth dependence of velocity fields within these features.

(l to r) G. Doyle tells J. Gurman and J. Lean that the weather is not always like this in Irish Summer!

SURFACE BRIGHTNESS CHARACTERISTICS OF W URSAE MAJORIS

Albert P. Linnell
Department of Physics and Astronomy
Michigan State University,
E. Lansing, MI 48824,
USA

1. Introduction

Tbe problem of interpreting photometric data for W Ursae Majoris is exacerbated by unpredictable light curve variability and "anomalies". These anomalies primarily afflict the W-type contact binary systems, of which W UMa is the prototype. It is of interest to note that the W-type systems, in which primary minimum corresponds to eclipse of the less massive component, typically are of spectral type later than F5, are believed to have convective atmospheres, and, based on indications of strong surface activity (Vilhu 1984) may be expected to have photospheric starspots. These light curve instabilities are in contrast to the more predictably placid behavior of the A-type contact binaries, in which primary minimum corresponds to eclipse of the more massive component (Binnendijk 1970), and for which the spectral types are earlier. It is of interest to inquire whether the instabilities can be explicitly simulated by a physically reasonable light synthesis model, and, assuming simulation success, what deductions follow concerning derivation of system model parameters.

The anomalies may be classified into several specific effects. A major one is the O'Connell effect (unequally high consecutive light maxima). An example is in the light curve by Hilditch (1981). During primary minimum the light curve may show a negative slope during totality (Evren, Tunca and Tumer 1979), or positive slope (Baldinelli and Ghedini 1977). The depth of primary minimum is variable (Hamzaoglu 1982), and secondary minimum (Baldinelli and Ghedini). Apparent phase shifts of eclipse ingress or egress may occur (Baldinelli and Ghedini). A prominent effect is delay of light maximum to orbital phase 0.27 (Linnell 1985).

In spite of light curve instabilities, the underlying light curve always remains W-type. The cause of W-type light curves has been contentious (Eaton, Wu, and Rucinski 1980; Smith, 1984; Linnell 1987). This paper adopts the hot secondary model, based on the baroclinic atmosphere model of Webbink (1977). The basic light synthesis package (Linnell 1984) has been augmented with the starspot algorithm by Budding (1977). Tests indicate there is little light curve sensitivity to spot latitude, within the range ($+25^o$, -25°). On the other hand, spot longitude and angular diameter have substantial effects. The simulations reported here adopted a fixed (-1500 K) temperature contrast between spots and the adjacent photosphere, based on a solar analogy. Since the spots will be permitted to drift in longitude, the physical consequences of an encounter between a drifting equatorial spot and the neck connecting the two components are obscure. This consideration, plus an analogy to sunspot latitudes at sunspot maximum, has prompted choice of spot pairs at fixed latitudes ($+25^o$, -25°). If we restrict consideration to cool starspots, only one pair cannot simultaneously produce both thc O'Connell effect and the delay of light maximum to orbital phase 0.27. Most efficient production of the O'Connell effect, for the higher maximum following primary minimum, is for starspot location on the advancing hemisphere of the more massive component. Adoption of the more massive componcnt as the starspot locale is based on the observations of Barden (1985) associating Hα emission with that component exclusively. On the other hand, the delay of light maximum to orbital phase 0.27 results from depression of the egress light curve from primary minimum. This depression requires

Table 1
Starspot Parameters

Parameter	I		V		B	
Latitude (°)	+25	+25	+25	+25	+25	+25
	-25	-25	-25	-25	-25	-25
Longitude (°)	280	70	280	70	280	70
(± 10°)	280	70	280	70	280	70
Angular Radii (°)	15.0	13.0	13.5	13.0	14.3	13.0
(±1°)	15.0	13.0	13.5	13.0	14.3	13.0
Temperature	-1500	-1500	-1500	-1500	-1500	-1500
Contrast (K)	-1500	-1500	-1500	-1500	-1500	-1500

starspots on the trailing face of the more massive component, and displaced in longitude from the quadrature (90°) location. Thus, the simulation was driven to choose a minimum of two spot pairs. The O'Connell effect then becomes a rather delicate balance between the two pairs of competing spots.

A value for the mass ratio is necessary in setting preliminary values of the system parameters. This simulation adopted the cross-correlation spectroscopic value by McLean (1981). The photometric data to be fitted were the UBVRI data of Linnell (1985).

Extensive experimentation was necessary before achieving a final fit in I, V, and B. It was found that varying the longitudes of the spot pairs, while keeping the longitude separation of the pairs constant, can reproduce all of the light curve anomalies reported in the historical literature. However, the light curve effects of spot longitude drift can simulate changes in underlying system parameters. For example, one longitude setting for the spots may indicate an incorrect intercomponent temperature difference, but a shift in longitude erases that indication. This situation implies the necessity of additional criteria to assist in setting spot longitudes, with approximately correct system parameters, followed by modification of system parameters and new simulations to check on spot longitudes. The criteria that were most sensitive were: (1) an absence of observed light curve tilt during total eclipse phases, (2) the amplitude of the O'Connell effect, and (3) the phase displacement of maximum light to orbital phase 0.27. These criteria, together with the fixed spot latitude and temperature contrast from the adjacent photosphere, established fairly restricted values for spot angular sizes (as seen from the star center) and longitudes. The final fit to I, V, and B light curves is excellent. It was a result of empirical parameter adjustments, not differential corrections. The spot parameters are in Table 1 and the system parameters in Table 2. The successful precise representation of the observed light curves

Table 2
System Parameters

Parameter	Value	Parameter	Value
q^* ($q = 1/q^*$)	0.49 ± 0.03	$A_h = A_s$(adopted)	0.50
$\Omega_h = \Omega_c$	5.25 ± 0.03	$b_h = b_c$(adopted)	0.08
i (°)	86.0 ± 0.02	Filling Factor	0.097
T_h(K,pole)(adopted)	6100	Ω (inner critical)	5.3082
T_c(K,pole)	5875±25	Ω (outer critical)	4.7098

with the spectroscopic value of the mass ratio eliminates possible indication of a third object in the system (Whelan, Mochnacki, and Worden 1974). A more extensive paper is in preparation giving full details of this study.

References

Baldinelli, L., Ghedini, S.: 1977. Mem. Soc. Astr. Ital., 48, 83.

Barden, S.C.: 1985, ApJ, 295, 162.

Binnedijk, L.: 1970, Vistas Astr., 12, 217.

Budding, E.: 1977, Ap&SS, 48, 207.

Eaton, J.A., Wu, C.-C., Rucinski, S.M.: 1980, ApJ, 239, 919.

Evren, S., Tunca, Z., Tumer, O.: 1979, IBVS, No. 1577.

Hamzaoglu, M.: 1982, IBVS, No. 2196.

Hilditch, R.W.: 1981, MNRAS, 196, 305.

Linnell, A.P.: 1984, ApJS, 54, 17.

Linnell, A.P.: 1985, ApJS, 57, 611.

Linnell, A.P.: 1987, ApJ, 316, 389.

McLean, B.J.: 1981, MNRAS, 195, 931.

Smith, R.C.: 1984, Q. Jl. R. astr. Soc., 25, 405.

Vilhu, C.: 1984, AA, 133, 117.

Webbink, R.F.: 1977, ApJ, 215, 851.

Whelan, J., Mochnacki, S.W., Worden, S.P.: 1974, MNRAS, 168, 31.

Discussion: Linnell

Maceroni I have done the same type of computations you have presented here with the Wilson-Devinney program. I agree that one can model all the observed distortions by means of spots. However the "inverse" problem, that is to solve a distorted light curve with a spotted model, is much more difficult to solve. My experience is that often the same light curve can be solved equally well allowing for dark and bright spots in different positions. Besides, there exist correlations among spot and system parameters that make it dangerous to perform solutions without indications on the unspotted system.

Linnell The uniqueness problem becomes much worse if you allow both dark and bright spots.

I would find it difficult to justify bright spots. In modelling effects like those discussed here I believe internal physical consistencey is especially important, where we are talking about effects at the level of 0.03 mag or less. For that reason I have a strong preference for my program, which assigns theoretical limb darkening coefficients that vary over the photosphere, to the Wilson-Devinney program that adopts a single limb darkening coefficient for an entire stellar component. I agree that a very large data set would be preferable covering a long enough time interval to disentangle effects of spot drift. I do not feel other data of adequate accuarcy and phase density are available. The criteria used to set spot longitudes for the present data set seem adequate. Empirical tests for determinability of the remaining parameters, (orbital inclination, photosphere potential, polar temperature of primary) indicate that the data set is good enough to constrain them.

Eaton Why do you give more credence to the spectroscopic mass ratio than the photometric mass ratios? As I read in the literature, the two agree reasonably well.

Linnell The introduction of cross-correlation radial velocities largely has eliminated the glaring discrepancies that used to exist between photometric and spectroscopic mass ratios. The descrepancy for W UMa itself was a famous one. My interest here has been to minimize the number of disposable parameters. I believe the credibility of spot longitude effects would have been reduced if I had tried to extract a photometric mass ratio from the data set.

Budding I think Yamasaki put a spot on W UMa in a paper round about 1978. I wonder how your results compared with these.

Linnell I mistrust analyses with small data sets. My impression from the results reported here is that you must have a large and very accurate data set to produce a credible Roche model as well as spot parameters. I have tried to fit only the large data set shown here.

AUTOMATED PHOTOMETRIC TELESCOPES IN THE STUDY OF ACTIVE STARS

Marcello Rodonò
Institute of Astronomy of Catania University,
and
Astrophysical Observatory,
95125 Catania,
Italy

1. Introduction

Several papers at this Armagh Colloquium have already demonstrated the importance of systematic observations of non-repeating variables and the enormous value of multiband simultaneous observations when dealing with such non-stationary phenomena, as those induced by solar-like magnetic activity on stellar atmospheres. The changing level of activity, the seemingly unlimited variety of different patterns that a finite number of compact surface features can make to *"decorate"* stellar surfaces, lead to apparently unresting stars, whose light curves may dramatically change on time scales as short as one week. Occasionally, during the course of flare events, the star luminosity may increase conspicuously with a sudden release of an enormous amount of energy.

Although we actually observe only epidermic phenomena, our present understanding of stellar activity suggests that these phenomena have deep roots penetrating deeply into the stellar interior, which is not accessible to direct observations, so that the study of stellar magnetic activity turns out to be a powerful indirect tool to investigate the internal structure of stars.

The main purpose of my review will be a concise inventory and critical overview of the principal achievements of photometry in disclosing and studying the various facets of stellar activity phenomenology. Their physical interpretations and expected progress from the implementation of novel techniques, such as the operation of fully automated or robotic telesccopes, will also be considered.

2. Theory versus observations

The general perception is that the increased power and sophistication of observations, which have allowed us to progress significantly in the physical description of solar and stellar activity phenomena, have not been paralleled by a correspondingly significant progress in the theoretical modellings and by a clear understanding of those fundamental mechanisms and processes, which may allow us to interpret the available observations.

Theoretical models of stellar activity, almost invariably inspired and guided by the solar analogy, have had the invaluable merit of indicating that the basic ingredient is the presence of a complex magnetic field, which may suppress or enhance the normal quiescent stellar flux at the different atmospheric levels, from the photosphere up to the corona, and on different time scales, from tens of seconds (flaring events), to tens of days (spot formation and evolution), to tens of years (activity cycles). Early papers on the application of the $\alpha - \omega$ dynamo theory to stars were actually dealing with parametrized linear models scaled down from the solar case (cf. Belvedere 1983, 1985, and references therein). Therefore, only differential parameters could be derived. More recent, non linear dynamo models (Gilman 1983a, 1983b) have attempted to progress in the theoretical modelling of stellar activity, but this difficult task has led only to preliminary results due to the complexity of numerical simulation and interpretation of the results.

Quite generally, theoretical studies of stellar activity have indicated which parameters, other than the magnetic field strength, are of importance to constrain and or guide theoretical models. Among them,

- angular rotation velocity;
- surface differential rotation and turbulence;
- depth, and especially structure and dynamics of the convection zone;
- convective and magnetic energy deposition into the outer stellar atmospheres;
- size, location and migration of compact activity features, such as cool spots, plages and coronal features and their covering and/or filling factors;
- time scales of activity features and, more importantly, their non-cyclic and cyclic long-term behaviour.

A comprehensive review of the methodology and results obtained from observations and their comparison with theoretical predictions is beyond the purpose of the present paper, also because several recent review papers presented at specialized international meetings are available (Byrne and Rodonò 1983, Stenflo 1983, Mangeney and Praderie 1984, Zeilik and Gibson 1986, Gondhalekar 1986, Stencel and Linsky 1988, Schroter and Schussler 1987, Linsky and Stencel 1988, Haisch and Rodonò 1989a, 1989b, Mirzoyan et al. 1990). Moreover some aspects are thoroughly reviewed in this volume by Eaton, Collier-Cameron, Neff, Linsky, and Mullan.

3. The role of photometry and APTs (Automatic Photometric Telescopes)

Relevant progress in the study of stellar activity phenomena relies on multiband simultaneous observations that allow us to investigate the activity signatures formed at different atmospheric levels and then to study the formation and evolution of activity features as a function of time. This is not always feasible, so that several fundamental questions requiring long-term photometric and spectroscopic studies still remain to be answered.

The important role played by photometry in stellar activity studies is directly addressed in section 4, where some highlights are presented. Here, I should like to stress that photometry alone offers a low-cost and effective means of investigation for survey and long-term monitorings of active stars. In fact, small aperture (< 1 meter) telescopes, which are available world-wide and may be satisfactorily used for this much needed and fundamental research project, are quite adequate.

The advent of remotely controlled space telescopes in the last two decades and the development of new technology used to operate semi-automatically large ground-based telescopes have triggered the interest in building fully-automated remotely-controlled small telescopes for photometry. Pioneering work in this direction has been done since the late seventies by Russell M. Genet and Louis J. Boyd, who were able to start a joint undertaking involving several Institutions in USA and to establish an APT service in 1985 to provide, upon request, high-quality photometric observations. The APT service concept and operation have been described in detail by Boyd et al. (1986) and Genet et al. (1987). Basically, the most significant aspect behind the APT concept is that of obtaining a homogeneous and systematic data base of high-quality photometry of variable stars, including solar-like active stars, by means of unmanned automated telescopes that can be remotely controlled. The remote access mode allows to run the telescope at low cost, but, more importantly, the automated operation mode leads to observation efficiency (ratio between actual photometric measurement time and telescope total operation time) of

Figure 1. An APT built by *Autoscope Co.* with its typical horseshoe mount. A similar 80-cm diameter telescope is being installed at Catania Observatory.

the order of 60-80 per cent, while the efficiency of manually operated telescopes seldom reaches 30-40 per cent. The presently operated APTs - most of them built by Autoscope Co. (Mesa, Arizona, USA) - have high-luminous f/2 thin primary mirrors with diameters from 25-cm to 75-cm, Cassegrain or Ritchey-Chretien combinations, with resulting f/8 focal ratios, on horseshoe mounts, which are the best suited for automated unmanned operation. An instrument selector, close to the telescope focal plane allows the observations to be performed with a photomultiplier-based photometer for measurements in the U-B-V or u-b-v-y photometric systems, or with interference filters centered at specific narrow or wide wavelength regions, such as $H\alpha$, or with a photodiode-based B-V-R-I photometer. A fast scan CCD camera, also mounted on the instrument selector, is presently used for star centering. A slow scan CCD camera is presently being implemented for imaging and (area) photometry. Future developments include a spectrophotometer, a fiber feed system for off-telescope gravity-independent large instruments and a photopolarimeter.

A large concentration of APTs is foreseen at the Fred G. Whipple Observatory at Mt. Hopkins (Arizona, USA). In addition to the three small APTs (25-40 cm diameter) already in service for several years, four 75-cm are being installed or planned. Overseas undertakings include a nearly completed 80-cm APT for Catania Observatory (Italy) and another 80-cm for Nanango

Observatory (Australia). These overseas telescopes will constitute the first seed of a Global Network of Automated Telescopes (GNAT) for internationally coordinated programs to study active cool stars, as well as target-of-opportunity objects, such as supernovae and other types of variable stars (Hayes and Genet 1989).

Three small-aperture APTs (25-cm diameter) are used by the Optical Communication Group at the Jet Propulsion Laboratory (Pasadena, California, USA) to monitor the atmospheric transmission and cloud cover with the aim of studying the feasibility and performance of an optical communication link between a planetary spacecraft and the Earth (Cowles 1989).

A successful unmanned APT program relies on the following minimal conditions:

- *photometric precision* as close as possible to 1 milli-mag;
- *excellent site* ensuring reasonably good seeing ($\sim 1-2$ *arcsec*) and stable atmospheric trasparency with a high percentage of photometric nights;
- well designed *observation strategy*, as far as it concerns the target selection (including symthomatic types of active and/or variable stars with the widest possible distribution of global parameters, ages, and sky positions) and an operation software capable of taking real-time intelligent decisions, according to target visibility, priority and weather conditions;
- availability for coordinated *multi-band* and *multi-site studies* within a flexible global network capable of responding to last-minute calls for target-of-opportunity observations;
- efficient and reliable *remotely controlled* operation mode and data retrieval.

4. Some highlights of photometric observations

For several decades photometric observations have given substantial contributions in the detection and study of activity signatures on stars with fundamental results on the physics of photospheric starspots and flares (cf. Rodonò 1986a, 1990, and references therein). The establishment of an APT network will certainly add considerable momentum to the photometric study of active stars. The few operative APTs at Mt. Hopkins in a few years have already produced an impressive amount of high-quality data on several newly discovered active stars (Strassmeier and Hall 1988, Strassmeier et al. 1989) and have complemented existing data on several others (cf. Cutispoto, Rodonò, 1990). As an example of the excellent photometric capability of APTs, in Figure 2 (upper panel) the dramatic evolution of UX Ari light curve on time scales of the order of 10 days is shown. The observations were collected at Mt. Hopkins, upon Catania Observatory request, with the 25-cm Phoenix-APT under the *"Rent-a-Star"* service operated by Mike Seeds of the Franklin and Marshall College (Lancaster, PA). There is clear evidence that at least two major spotted areas with different rotational velocities, as well as changes of the mean activity level, are responsible for the remarkable light curve variations. The lower panel of Figure 2 shows the long-term behaviour of UX Ari activity level with bars representing the peak-to peak amplitude of the seasonal photometric wave. These observations will be continued, also with the Catania APT, for the purpose of studying the long-term behaviour of UX Ari and several other selected RS CVn, BY Dra and flare stars.

An exhaustive review of photometric results is beyond the scope of the present paper. On the other hand, specific aspects are discussed elsewhere in this volume. Therefore, I shall confine myself to a few topics, that are of interest to APT research programs, especially to those concerning the study of spots and flares, namely, spot evolution and migration, differential rotation, activity cycles, and flare physics.

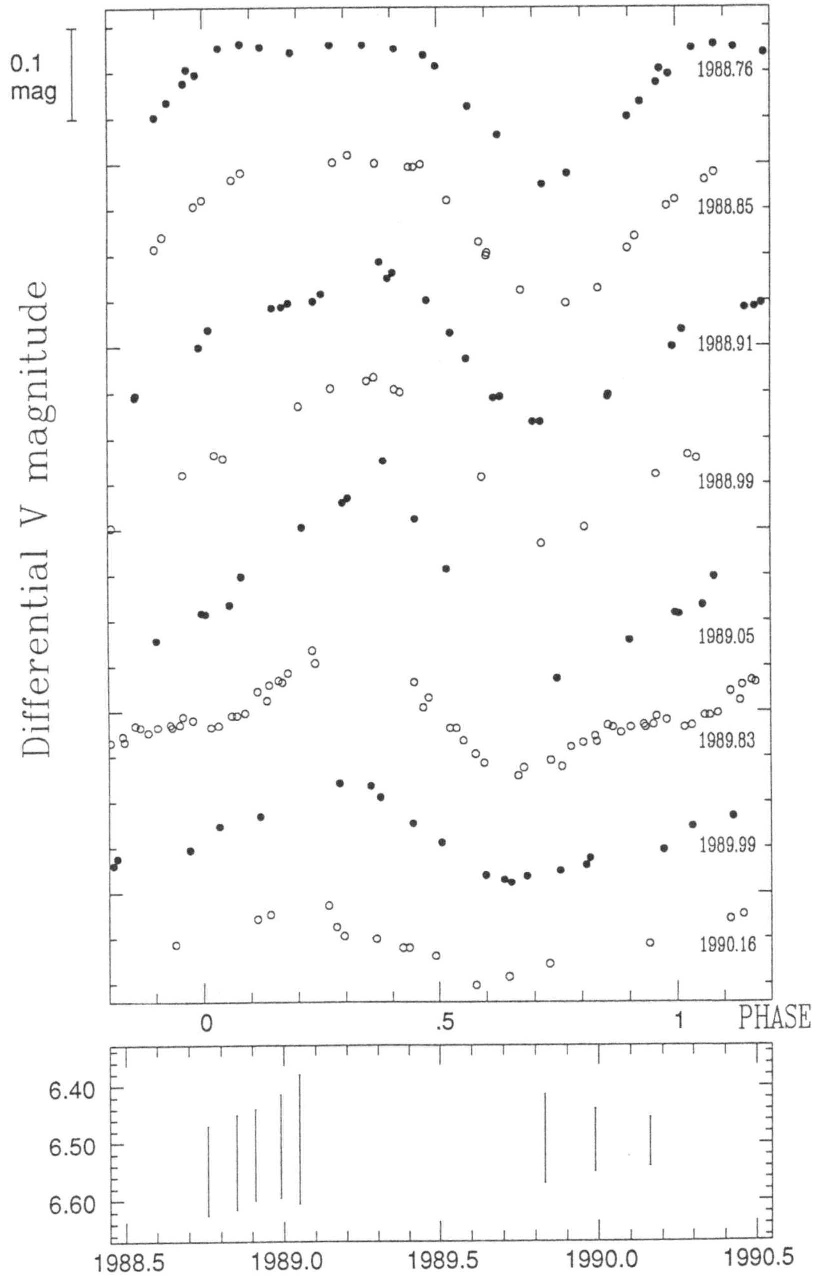

Figure 2. Time-evolution of UX Ari light curve showing the effect of at least two main spotted areas with different rotational velocities (*upper panel*) (from Rodonó and Cutispoto 1990). The observations were collected with the 25-cm unmanned Phoenix APT under the *"Rent-a-Star"* service. The *lower panel* shows the long-term behaviour of UX Ari activity level with bars representing the peak-to-peak amplitude of the seasonal photometric waves.

4.1 Spot migration and differential rotation

As known (cf. Rodonò 1986b; Eaton, 1990), the interpretation of low-amplitude wave-like light curves in terms of spot-induced modulation of the observed flux, as the star rotates, has allowed us to derive spot physical parameters and their surface distribution. Although problems do exist in deriving unique solutions, some spot parameters, such as their huge dimension and longitude location, may be derived with confidence.

Therefore, while the analysis and interpretation of a single light curve may be subject to some uncertainty, by modelling a series of homogeneous light curves, obtained over an extended time interval, it is possible to derive relevant information on latitude and longitude spot migrations and reliable estimates of stellar rotation and differential rotation (see, however, Lanza et al. 1990). This is particularly true for eclipsing binaries because the inclination of the rotation axis can be independently derived, thus avoiding a major source of ambiguity in starspot modelling methods. As an example of systematic studies that can lead to an estimate of differential rotation, the results of spot modelling of a series of light curves of BY Dra, the prototype spotted star discovered by Chugainov (1966), are shown in Figure 3 (from Cutispoto 1987). Two major spotted areas are at least required to reproduce the asymmetric light curves obtained from 1981 to 1987. A low-latitude smaller spot shows equatorwards and faster migration rate than the high-latitude larger spot, which shows polewards migration. From the differential migration rates of the two spot forming regions and their mean latitude location, a differential rotation rate of about one order of magnitude smaller than solar was derived. This result confirms previous data on other spotted stars (cf. Rodonò 1986a, 1986b) that indicate surprisingly small differential rotation rates for very active dwarf stars. Assuming that this result is not an artifact of spot modelling methods, the low rates of surface differential rotation may be due to a strong orbit-rotation tidal coupling that apparently affects at least the outer layers of active stars in close binaries, without suppressing their ability of producing huge active regions. Another possibility to be explored is whether the huge magnetic flux tubes implied by the existence of so large spotted areas may completely control the plasma dynamics to the effect of perturbing or even suppressing differential rotation.

4.2 Spot evolution and activity cycles

Spot modelling of light curves systematically acquired over several years can also address the problem of active region evolution and lifetimes. The observed light curves sometimes show dramatic changes on time scales of the order of ten days to months. These variations may be due to changes of the surface distribution of spots as well as to intrinsic variations of their size and number (cf. Figures 2 and 3). Spot modellings allow us to disentangle intrinsic from topological effects. Most of the short-term light curve changes, however, turns out to be due to spot redistribution over the stellar surface, while spotted areas generally appear to be quite stable structures with lifetimes from several months to years.

The long-term behaviour of activity features and the determination of activity cycles are important questions that systematic photometry with APTs can properly address. For a few stars long-term studies already suggest cyclic behaviour (e.g., Anders et al. 1990, Cutispoto and Rodonò, 1990), but a much longer data base is required. APTs can play a relevant role because nowadays it is not feasible to provide such a large amount of data from a single place, not to mention from a single observer or research group. Clearly, APTs are marking the passage from *dedicated observers* to *robotic observations* of active stars.

The length of activity cycles is an important parameter to constrain or test the theories of stellar activity. As an example, non-linear $\alpha-\omega$ dynamo models (Gilman 1983a, 1983b) predict cyclic behaviour only under specific circumstances. In particular, within a limited range of

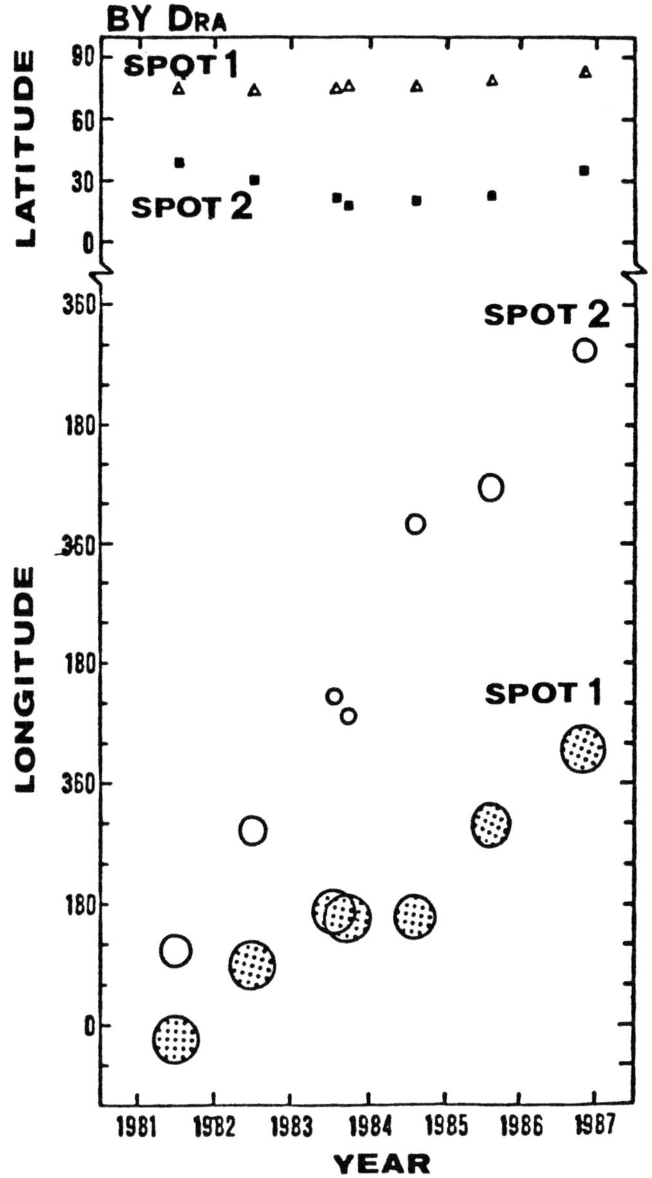

Figure 3. Spot latitude and longitude migrations of spots derived by Cutispoto (1987) from modelling a series of light curves of BY Dra. The low-latitude spot appears to rotate faster than the high-latitude spot. Both equatorwards and polewards migrations are apparent.

electrical conductivity, cyclic dynamos occur if rotation is not too low (because the α-effect would be too small) or too high (because fast rotation reduces convection). On the other hand, if electrical conductivity is very low, the dynamo will not operate at all (Figure 4). Therefore, very fast rotation or very short rotation periods would not allow cyclic behaviour to occur. This prediction was supported by the results of Wilson's Ca II line flux monitoring (see Vaughan 1983) indicating that no cyclic behaviour was found for stars with rotation period shorter than

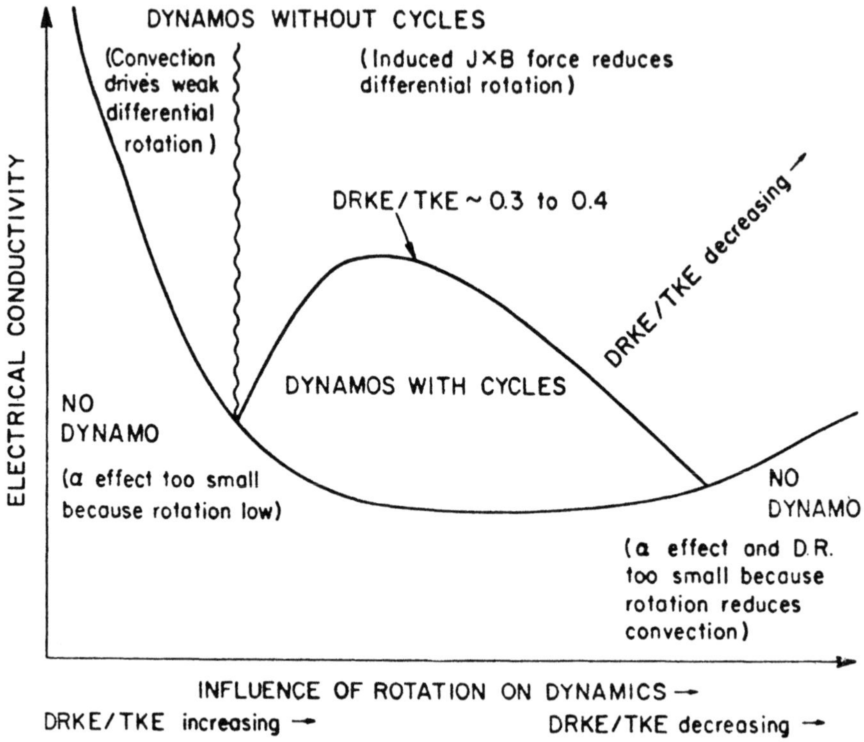

Figure 4. Schematic diagram for dynamos expected from global non-linear models as functions of conductivity and influence of rotation on the plasma dynamics (from Gilman 1983a). Very slow and very fast rotating stars should not allow cyclic dynamos to develop.

about 20 days. But photometric data on several stars with rotation period of the order of a few days, such as BY Dra, II Peg and V 711 Tau, do indicate cyclic activity with time scales of the order of 1-10 years (Figure 5). This question clearly requires additional theoretical and observational efforts. The Catania APT aims at collecting at least a 10 year high-precision photometric data base with the purpose of addressing also the question of spot evolution and activity cycle timescales.

The duration of activity cycles may have much wider astrophysical implications, as suggested by the results of Maceroni et al. (1990), who found that the cycle periods for different types of variables, including single and binary stars, cluster around 6 years.

4.3 Spot-plage spatial correlation

Another area that would strongly benefit from systematic APT network observations is the still open question of spot-plage spatial correlation, i.e., whether bright chromospheric and transition region plages overlie photospheric cool spots, as on the Sun. Both positive and negative well documented evidence has been presented (Rodonò et al. 1986, 1987, Gondoin 1986, Butler et al. 1987, Panagi and Doyle 1987, Andrews et al. 1988, Doyle et al. 1989). Most of these studies are based on concurrent photometric and IUE spectroscopic studies.

Obviously, the latter observations cannot be carried out systematically over extended intervals of time, although a few targets (AR Lac, II Peg, V711 Tau) have been observed continuously

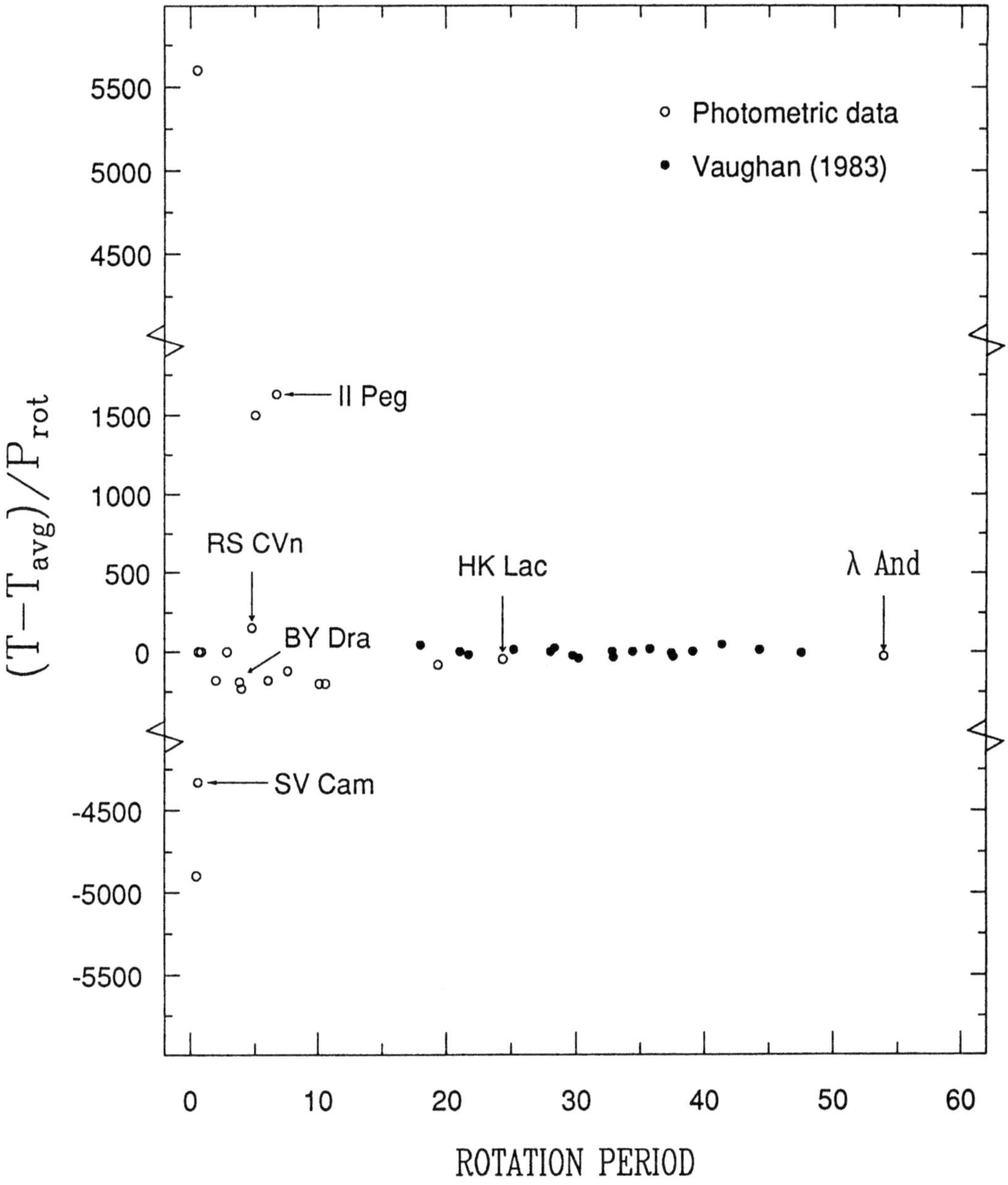

Figure 5. An updated version of Vaughan (1983) diagram (*dots*) of mean cycle timescale (T_{avg} = 10 yr), divided by P_{rot}, as a function P_{rot} with the addition of photometric results (*open circles*). It is evident that also stars with rotation period shorter than about 20 days do show long-term activity cycles.

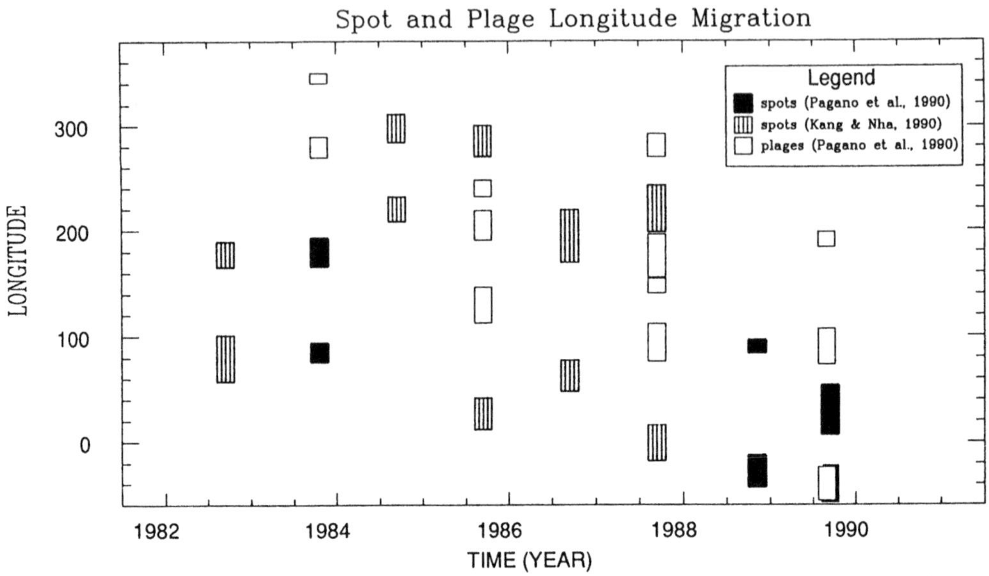

Figure 6. Longitude migration of spots (*dashed* and *filled rectangles*) and plages (*open rectangles*) on the atmosphere of AR Lac derived from optical photometry and UV Doppler imaging, respectively.

over one or two complete orbits (Walter et al. 1987, Neff et al. 1989, Rodonò et al 1987, Andrews et al. 1988). Therefore, it is essential to carry out systematic photometric monitoring of those stars that have been selected for IUE studies, so that the spot configuration at the time of IUE or ground based spectroscopy is known. As an example, the paper by Pagano et al. (1990) at this colloquium primarily deals with Doppler imaging of plages on AR Lac from 1983 to 1989. Only limited photometry was concurrently obtained in 1983 and 1989 by the same group, while another paper at this meeting by Kang and Nha (1990) presents independent light curves and spot modelling from 1982 to 1987. In Figure 6, the results of both papers are combined. It is apparent that one of the two spots, that are required to reproduce the observed light curves, does not seem to be associated with any of the plages derived from the Doppler imaging study, except in 1989, while the other spot longitude is often rather close to the longitude of one of the plages. Unfortunately, no reliable information on spot and especially, plage latitudes can be derived to address more stringently the question of spot-plage spatial correlation. Figure 6 clearly demonstrates that if a photometric network had provided systematic photometry of AR Lac, whose unfavourable period (almost exactly equal to two days) makes it unsuitable for complete coverage from one single observatory in a short interval of time, a more conclusive test on spot-plage spatial correlation would have been possible. In this context it is important to bear in mind that on the Sun the spot-plage spatial correlation does not seem very tight: sunspots appear to lead associated plages by an amount that is directly proportional to plage areas (Dobson-Hockey and Radick 1986).

4.4 Stellar Flares

Stellar flares were the first phenomena that suggested the possibility of solar-like activity to occur on stars. At the recent IAU Colloquium No. 104 (Haisch and Rodonò 1989, 1989b) our present knowledge of these sudden, rapid and intense phenomena was discussed with particular emphasis on similarities and differences between solar and stellar flares. Their unpredictable times of occurrence require continuous uninterrupted monitoring, that makes flare stars one of the most suitable type of variables for APT network observations. Large collections of data already exist and have been used to derive statistical properties of flares and flare stars (cf. Shakhovskaya 1989, and references therein). More recent studies are dealing with the physical nature of flares, their origin and development. Solar and stellar flares involve almost simultaneously the entire stellar atmosphere from photospheric up to coronal levels. Therefore, co-ordinated multi-band and multi-site observations with ground-based and space telescopes are required (cf. Byrne 1989). A simple inspection of observed light curves (Cristaldi and Rodonò 1970, 1973; Moffett 1974) clearly shows how complex the flux and time-evolution of flares might be. Therefore, flare physics studies require accurate multi-wavelength observations of many flare events with high time and spectral resolution. These observations are now becoming available (cf. Rodonò 1990) and fundamental insights into the flare plasma physics are at hands (cf. Houdebine et al. 1990a), but many more observations are needed. As a matter of facts, the most recent observations of stellar flares have rised new questions that need to be answered. In particular:

- Is the simple loop solar flare model applicable to stellar flares (cf. Haisch 1989, Mullan 1989)?. In other words, is the solar case a valid paradigm for the stellar activity scenario?
- Can purely thermal mechanisms, that successfully interpret the general behaviour of optical spectra, also account for the observed short-lived flares with time scales sometimes shorter than 1 s (Beskin et al. 1989a, 1989b; Katsova et al. 1981)?
- What is the nature of the neglected oscillatory behaviour of optical and radio fluxes close to or during flares (Rodonò 1974, Gibson 1983)?
- The flare total energy budgets need considerable revision after the observations of Rodonò et al. (1989) and Rodonò and Cutispoto (1988) revealed a huge amount of missing energy in the infrared K-band (about ten times larger than the energy emitted in the optical region) and Houdebine et al. (1990b) showed spectroscopic evidence for a high-velocity mass ejection event at the onset of an intense flare on AD Leo involving kinetic energy and mass ejection 500 and 40 times larger, respectively, than for Coronal Mass Ejection (CME) events on the Sun.
- Is there any spatial and/or physical correlation between spot and flare activity in stars? Systematic studies are sorely lacking.
- How is flare activity affected by the interacting environment in BY Dra, RS CVn and Algol-type close binary systems (cf. Linsky 1983, Catalano 1990)?

Lastly, systematic and homogeneous data sets need to be collected to improve the presently available statistics on stellar flare activity. The available studies are based on inhomogeneous data sets independently collected at different observatories: a clearly unsatisfactory situation.

5. Conclusion

At least an entire page conclusion could be brought to the reader attention by summarizing the arguments presented in this paper, but I prefer to stress the main theme I have tried to demonstrate, i.e. that nowadays automated photometric (and spectroscopic) telescopes, possibly unmanned, constitute one important essential means of securing the much needed collection of long-term systematic data, that are essential for addressing the several open questions concerning solar-like magnetic activity on stars.

Acknowledgements.

It is my pleasure to thank very warmly Brendan Byrne and Dermott Mullan, Co-Chairmen of the Armagh Observatory Bicentenary Colloquium, for inviting me to serve on the Scientific

Organizing Committee and to present this contribution, the Director, Mart de Groot, and the Armagh Observatory Staff for providing a pleasant and scientifically fruitful environment assisted by incredibly beautiful weather, that destroyed many jokes about the ill-famed *"Irish weather"*.

Solar and stellar activity researches at Catania University and Astrophysical Observatory are supported by the Italian Ministry for Universities and Scientific and Technical Research (***MURST***), the National Research Council (*CNR*) and the Italian Space Agency (*ASI*), that are gratefully acknowledged for their generous financial support.

Finally, I should like to thank Giuseppe Cutispoto and Isabella Pagano for their kind assistance in preparing the illustrations included in the text.

References

Anders, G.J., Coates, D.W., Thompson, K., Innis, J.L..: 1990, this volume.

Andrews, A.D., et al.: 1988, A&A, 204, 177.

Belvedere, G.: 1983, in Activity in Red-Dwarf Stars, IAU Coll. No. 71, P.B. Byrne and M. Rodonò (Eds.), Reidel, Dordrecht, Holland, p. 579.

Belvedere, G.: 1985, Solar Phys. 100, 363.

Beskin, G.M., et al.: 1989a, in Solar and Stellar Flares, Poster Papers, B.M. Haisch, M. Rodonò (Eds.), Publ. Catania Astrophys. Obs., Special Volume, p. 95.

Beskin, G.M., Plakhotnichenko, V.L., Pustil'nik, L.A., Shvartsman, V. F., Gershberg, R.E.: 1989b, in Solar and Stellar Flares, Poster Papers, B.M. Haisch, M. Rodonò (Eds.), Publ. Catania Astrophys. Obs., Special Volume, p. 99.

Boyd, L.J., Genet, R.M., Hall, D.S.: 1986, PASP, 98, 615.

Butler, C.J., et al.: 1987, A&A, 174, 139.

Byrne, P.B.: 1989, in Solar and Stellar Flares, IAU Coll. No. 104, B.M. Haisch, M. Rodonò (Eds.), Solar Phys. 121, 61.

Byrne, P.B., Rodonò, M. (Eds.): 1983, Activity in Red-Dwarf Stars, IAU Coll. No.71, Reidel, Dordrecht, Holland.

Catalano, S.: 1990, in NATO-ASI on Active Close Binaries, C. Ibanoglu, I. Yavuz (Eds.), Kluwer Acad. Publ., Dordrecht, Holland, in press.

Chugainov, P.F.: 1966, Comm. 27 IAU, IBVS No. 122.

Cowles, K.: 1989, in Remote Access Automatic Telescopes, D.S. Hayes, R. M. Genet (Eds.), Fairborn Press, Mesa (Arizona), p. 27.

Cristaldi, S., Rodonò, M.: 1970, A&AS, 2, 223.

Cristaldi, S., Rodonò, M.: 1973, A&AS 10, 47.

Cutispoto, G.: 1987, PhD Thesis, University of Catania.

Cutispoto, G., Rodonò, M.: 1990, this volume.

Dobson-Hochey, A.K., Radick, R.R.: 1986, in Fourth Cambridge Workshop on Cool Stars, Stellar Systems, and the Sun, M. Zeilik, D.M. Gibson (Eds.), Springer-Verlag, Berlin, Germany.

Doyle, J.G., Butler, C.J., Byrne, P.B., Rodonò, M., Swank, J., Fowles, W.: 1989, A&A, 223, 219.

Eaton, J.A.: 1990, this volume.

Genet, R.M., Boyd, L.J., Kissell,K.E., Crawford,D.L., Hall, D.S., Hayes, D.S., Baliunas, S.L.: 1987, PASP, 99, 660.

Gibson, D.M.: 1983, in Activity in Red-Dwarf Stars, IAU Coll. No. 71, P.B. Byrne, M. Rodonò (Eds.),Reidel, Dordrecht, Holland, p. 273.

Gilman, P.A.: 1983a, in Solar and Stellar Magnetic Fields: Origin and Coronal Effects, IAU Symp. No. 102, J.O. Stelflo (Ed.), Reidel, Dordrecht, Holland, p. 247.

Gilman, P.A.: 1983b, ApJS, 53, 243.

Gondhalekar, P.M. (Ed.): 1986, Flares: Solar and Stellar, Astron. Astrophys. Workshops, Rutherford Appleton Laboratory, RAL-86-085, Chilton, Didcot, U.K.

Gondoin, P.: 1986, A&A, 160, 73.

Haisch, B.M.: 1989, in Solar and Stellar Flares, IAU Coll. No. 71, B.M. Haisch, M. Rodonò (Eds.), Solar Phys. 121, 3.

Haisch, B.M., Rodonò, M. (Eds.): 1989a, Solar and Stellar Flares, IAU Coll. No.104, Solar Phys. 121, Kluwer Ac. Publ., Dordrecht, Holland.

Haisch, B.M., Rodonò, M. (Eds.): 1989b, Solar and Stellar Flares, Poster Papers, Publ. Catania Astrophys. Obs., Special Volume.

Hayes, D.S., Genet, R.M. (Eds.): 1989, in Remote Access Automatic Telescopes, Fairborn Press, Mesa (Arizona, USA).

Houdebine, E.R., Butler, C.J., Panagi, P., Rodonò, M., Foing, B.H.: 1990a, A&A,, in press.

Houdebine, E.R, Rodonò, M., Foing, B.H.: 1990b, A&A, in press.

Kang, Y.W., Nha, Il-Seong: 1990, this volume.

Katsova, M.M., Kosovichev, A.G., Livshits, M.A.: 1981, Astrofizika 17, 285.

Lanza, A.L., Rodonò, M., Zappalà, A.R.: 1990, this volume.

Linsky, J.L.: 1983, in Third Cambridge Workshop on Cool Stars, Stellar Systems, and the Sun, S.L. Baliunas, L. Hartmann (Eds.), Springer- Verlag, Berlin, Germany.

Linsky, J.L., Stencel, R.E. (Eds.): 1987, in Fifth Cambridge Workshop on Cool Stars, Stellar Systems, and the Sun, Springer-Verlag, Berlin, Germany.

Maceroni, C., Bianchini, A., Rodonò, M., Van't Veer, F., Vio, R.: 1990, A&A,, in press (see also this volume).

Mangeney, A., Praderie, F. (Eds.): 1984, Space Research Prospects in Stellar Activity and Variability, Obs. Paris-Meudon, France.

Mirzoyan, L.V., Pettersen, B.R., et al. (Eds.): 1990, Flare Stars in Star Clusters, Associations, and Solar Vicinity, IAU Symp. No. 137, Kluwer Ac.Publ., Dordrecht, Holland.

Moffett, T.J.: 1974, ApJS, 29, 1.

Mullan, D.J.: 1989, in Solar and Stellar Flares, IAU Coll. No. 71, B.M.Haisch, M.Rodonò (Eds.), Solar Phys. 121, 239.

Neff, J.E., Walter, F.M., Rodonò, M., Linsky, J.L.: 1989, A&A, 215, 79.

Pagano, I., Rodonò, M., Neff, J.E.: 1990, this volume.

Panagi, P., Doyle, J.G.: 1987, in Fifth Cambridge Workshop on Cool Stars, Stellar Systems, and the Sun, J.L.Linsky, R.M.Stencel (Eds.), Springer-Verlag, Berlin, Germany.

Rodonò, M.: 1974, A&A, 32. 337.

Rodonò, M.: 1986a, in The Atmospheres of M-Dwarfs, H.L.Johnson, F.Querci (Eds.), CNRS-NASA Monograph Series, NASA SP-492, p. 409.

Rodonò, M.: 1986b, in Highlights of Astronomy, J.-P. Swings (Ed.), Reidel, Dordrecht, Holland, p. 429.

Rodonò, M.: 1990, in Flare Stars in Star Clusters, Associations and Solar Vicinity, IAU Symp. No. 137, L.V. Mirzoyan et al. (Eds.), Kluwer Acad. Publ., Dordrecht, Holland, p. 371.

Rodonò, M., Cutispoto, G.: 1988, in Activity in Cool Stars Envelopes, O.Havnes et al. (Eds.), Kluwer Ac.Publ., Dordrecht, Holland, p.163.

Rodonò, M., Cutispoto, G.: 1990, work in progress.

Rodonò, M., et al.: 1986, A&A, 165, 135.

Rodonò, M., et al.: 1987, A&A, 176, 267.

Rodonò, M., et al.: 1989, in Solar and Stellar Flares, Poster Papers, B.M. Haisch, M. Rodonò (Eds.), Publ. Catania Astrophys. Obs., Special Volume, p. 53.

Schroter, E.H., Schussler, M. (Eds.): 1987, 5th Eur. Solar Meeting on Solar and Stellar Physics, Springer-Verlag, Berlin, Germany.

Shakhovskaya, N.I.: 1989, in Solar and Stellar Flares, IAU Coll. No. 104, B.M. Haisch, M. Rodonò (Eds.), Solar Phys. 121, 375.

Stenflo, J.O. (Ed.): 1983, Solar and Stellar Magnetic Fields: Origin and Coronal Effects, IAU Symp. No. 102, Reidel, Dordrecht, Holland.

Strassmeier. K.G., Hall, D.S.: 1988, ApJS, 67, 453.

Strassmeier, K.G., Hall, D.S., Boyd, L.J., Genet, R.M.: 1989, ApJS, 69, 141.

Vaughan, A.H.: 1983, in Solar and Stellar Magnetic Fields: Origin and Coronal Effects, IAU Symp. No. 102, J. O. Stelflo (Ed.), Reidel, Dordrecht, Holland, p. 113.

Walter, F.M., Neff, J.E., Gibson, D.M., Linsky, J.L., Rodonò, M., Gary, D.E., Butler, C.J.: 1987, A&A, 186, 241.

Zeilik, M., Gibson, D.M. (Eds.): 1986, in Fourth Cambridge Workshop on Cool Stars, Stellar Systems, and the Sun, Springer-Verlag, Berlin, Germany.

Discussion: Rodonó

Strassmeier With current APT techniques there is no way to do mmag photometry. One of the reasons is that one must use a very large diaphragm, normally 1 arc min, in order to find and centre the star. The external standard deviation for the 25 cm Fairborn APT is 0.01 mag., while it is a bit better for the 40 cm Vanderbilt APT, 0.007 mag. What do you expect for the 80 cm Catania APT?

Rodono A well-designed observing strategy to account for the various sources of systematic and accidental errors would allow us, at least in principle, to approach the mmag limit and, occasionally, to beat it.

Young Concerning the putative connection or non-connection between dark spots and bright plages, it appears that you associate a large difference in longitude between such features as evidence that they are not connected. However, at the risk of irritating our hosts, I should point out that if the spots and the plages are at high latitude they can differ in longitude by arbitrarily large amounts while still being close together on the surface.

Rodono Yes, indeed. However, in the particular case I have shown, light curve spot modellings indicate spot latitudes from 60 to 0 degrees. Unfortunately, no accurate plage latitudes can be derived from spectral imaging of UV emission lines.

Linnell The delineation of phenomena that occur on time scales of minutes or hours sets conditions on observing technique that are important for the operation of APTs. It is desirable to follow a given object for extended time periods, picking it up near the horizon and following it until it sets. As little time as possible needs to be spent on the comparison star. The target needs to be maintained centred as accurately as possible in the diaphragm. This in turn means that the telescope pointing control needs to allow for differential refraction.

Rodono The present APTs are designed to observe sequentially many variable stars on one night, including comparison, check and standard stars and specific objects for extinction measurements. As I have shown, extended observations of active stars are necessary for specific purpose, e.g. detection and multi-band studies of flares. Hence APT tracking capability over extended periods of time should be implemented, as we have planned for the Catania APT.

Zeilik What is the magnitude limit of the presently operating APT's?

Rodono Because of the limited telescope aperture the presently operating APTs are not adequate to address some of the fundamental questions concerning active stars. However, the present 0.01 mag. limit (or even better sometimes) is adequate for long-term variability studies, as delineated in my talk.

Zeilik What is your view on archiving data from APTs?

Rodono At the last APT workshop in Tucson last February, some ideas on APT data archiving were spread around. Most of the attendees did support the general concept.

Linsky I wish to support Mike Zeilik's statement that the data obtained with APT's should be archived and be available to everyone after a proprietary period. The archiving problem is not difficult compared to that of archiving data from satellites (which is done routinely) and that from imaging and echelle spectroscopy. The Strasbourg data centre and other data centres are probably willing to archive the APT data sets.

Rodono I strongly support an easy-access APT archive, as well as other source photometry, after a proprietary period. This is presently done by individual observers on a voluntary basis by sending their data to the IAU Comm. 27 depository of variable star observations. I believe that also for ground-based data the space data policy, as far as data rights, archiving and retrieval are concerned, will inevitably be followed in the future. It would be a bad service to science to leave precious photometric data buried in Observatory archives all over the world.

Strassmeier Four years of photoelectric UBV data of 100 stars, obtained with the prototype Fairborn APT, have been sent to NSSDC and to the CDS in Strasbourg.

Rodono I think that this is the correct way to address the question of data archiving.

MODELLING INHOMOGENEITIES ON B AND A TYPE MAGNETIC PECULIAR STARS

John D. Landstreet
Department of Astronomy,
The University of Western Ontario
London, Ontario N6A 3K7,
Canada

About 10% of main sequence A and B stars ($7500\,K \leq T_{eff} \leq 20\,000\,K$)have large ($\approx 3\,10^2$ - $3\,10^4$ G), coherent (roughly dipolar) magnetic fields. Many of the magnetic Ap and Bp stars have quite inhomogeneous distributions of such trace elements as He, Si, Ca, Ti, Cr, and rare earths over their surfaces, and mean abundances which depart strongly from solar ratios. The magnetic field structure is often highly inclined to the rotation axis, and so the mean field modulus and the mean line-of-sight field component vary as the star rotates. Furthermore, the inhomogeneously distributed elements often seem to take the dipole axis (rather than the rotation axis) as a rough axis of symmetry, and so as the star rotates, the line strengths and profiles vary. There is at present no strong evidence that either the magnetic fields or the abundance distributions change intrinsically on short time scales ($\tau \leq 10^2$ years), and so the observed variations of the magnetic field components and spectral line profiles with rotation may be used to derive models of magnetic field geometry and chemical abundance distributions by methods analogous to those used to analyse structures on cool stars.

The occurrence of the magnetic peculiar stars, and of a variety of other chemically anomalous B, A and F stars (Am & Fm stars, λ Boo stars, Ap HgMn stars, non-magnetic He-weak stars, *etc.*) is important because these stars reveal the existence of powerful chemical segregation processes, which operate in the envelopes, atmospheres, and perhaps, winds of early-type stars. These processes probably also operate in "normal" B-A-F stars. Understanding them is very important for correctly interpreting observed chemical abundances, for appreciating the extent to which a particular atmospheric abundance pattern may be produced by initial conditions, by stellar evolution, or by chemical segregation processes. As is the case for cool stars, mapping of inhomogeneous structures is expected to help elucidate the physical mechanisms producing the inhomogeneities.

I have been involved (together with D. Bohlender and T. Lanz) in an effort to obtain low resolution maps of a few magnetic Ap stars in which the magnetic field is large enough to strongly influence line profiles. In this case we may hope to obtain well-constrained maps of both the magnetic field geometry and the surface distribution of several elements.

For this mapping project, I have developed a line synthesis code that (like some codes used to study magnetic fields in the Sun or cool stars, e.g. Wittmann 1974 and Basri and Marcy 1988) takes account of Zeeman splitting of spectral lines and of radiative transfer of polarized light. This code assumes a starting magnetic field geometry (specified as a superposition of colinear dipole, quadrupole, and octupole components) and an abundance distribution geometry (axisymmetric about the magnetic axis, with a maximum resolution of 30° in colatitude) for all elements contributing to an observed spectral absorption feature. Theoretical profiles are computed for all the rotational phases for which spectra are available, for up to three different spectral lines, and compared to the observed spectra. The abundance distribution may be modified by the program to search for the model giving the best fit to the observed profiles, or for fitting magnetic field parameters the model is modified manually. Model parameters are

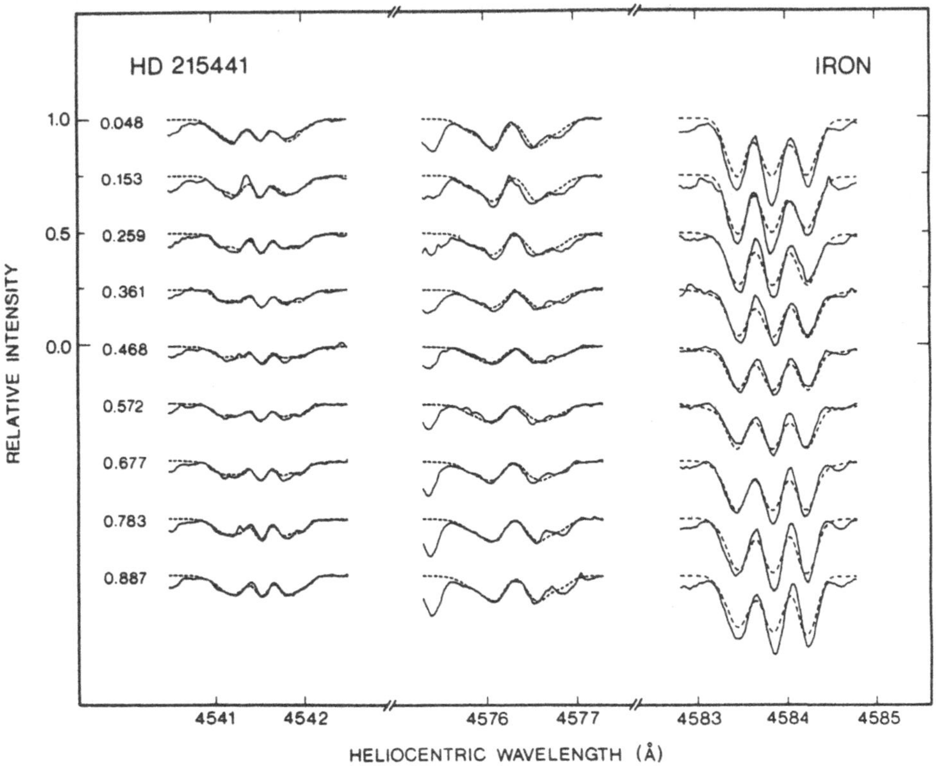

Figure 1. Profiles observed (solid lines) and calculated (dashed) for three different iron lines at nine different rotational phases for HD 215441. See text for discussion. (©1989, The American Astronomical Society. Used by permission.)

iterated until acceptable agreement with observations is obtained.

An example of the results of this fitting process is shown in Figure 1, which illustrates the fit obtained for three lines of iron for the Bp star HD215441 (Landstreet *et al* 1989). Observed line profiles are shown as solid lines, models are dashed. Rotational phases through the 9.49 day rotation phase of the star are shown along the left margin. In this star the lines are widened mainly by the Zeeman effect of a field of about 35 kG. Spectrum variations are visible as changes in the depths of spectral lines. There are clearly some problems with the model of line formation; weak lines are fit well but theoretical profiles of the strong line are not deep enough.

Models have been published so far for the Ap SrCrEu star 53 Cam (Landstreet 1988), the Ap Si star, HD215441 (Landstreet *et al* 1989), and the He-strong star HD64740 (Bohlender and Landstreet 1990). Further modelling is in progress on the Ap SrCrEu star CS Vir, several Ap Si stars (with T. Lanz) and the He-strong stars σ Ori E and HD37776 (with D. Bohlender).

The magnetic field structure over the stellar surface can be reasonably well approximated by the low order multipole expansion used (in wild contrast to the situation for cool stars!). The field structures are usually dominated by the dipole component, but higher multipoles, especially the quadrupole, are important. In effect it seems that most magnetic Ap and Bp stars have unequal field strengths at the two magnetic poles. Some chemical elements (H, Fe) are normally uniformly abundant, while others (He, Ca, Ti, Cr, rare earths) may, or may not, have abundance contrasts of up to 2-dex, usually with a high abundance at one magnetic pole and a low abundance at the other. Furthermore, some elements concentrate at one pole and some at the other. In 53 Cam, for example, Ti is concentrated near the strong (negative) pole, while Ca concentrates near the weak pole. Cr, Fe, and some rare earths are fairly uniformly distributed. In CS Vir, in contrast, Cr is abundant near one pole, some rare earths are found near the other, and Ti and Fe are fairly uniform.

Babel and Michaud (1990) have recently discussed the physics involved in producing such large abundance contrasts. They conclude that the observed abundance patterns may arise from the diffusion of trace elements under the combined influences of gravity and radiation pressure, in a weak ($\approx 10^{-13}\,M_\odot$/year) stellar wind which is strongly modulated over the stellar surface by the magnetic field.

References

Babel, J. and Michaud, G. 1990, ApJ in press.

Basri, G. and Marcy, G.W. 1988, ApJ, 330, 274.

Bohlender, D.A. and Landstreet, J.D. 1990, ApJ, in press.

Landstreet, J.D. 1988, ApJ, 326, 967.

Landstreet, J.D., Barker, P.K., Bohlender, D.A. and Jewison, M.S. 1989, ApJ, 344, 876.

Wittmann, A. 1974, Solar Phys, 35, 11.

Discussion: Landstreet

Linsky A good test of whether the fields are nearly dipolar in these stars would be a simultanous measurement of the magnetic flux using the circular polarization and line broadening (Robinson) techniques. Has this test been done?

Landstreet The models I have made are constrained simultaneously by the available polarisation data and by line profiles and are therefore consistent with both. Models based solely on polarimetry at present are very poorly constrained, and do not provide much check on more elaborate models made from all the data. However, at least in the case of HD215441, there is not much room or any extra magnetic flux not seen by polarimetric observations: the mean longitudinal field in this star can reach 20 kG, more than half the value of 35 kG seen in the line splitting (Borra and Landstreet 1978, Ap. J. 222, 226 and Landstreet *et al* 1989, Ap. J. 344, 876).

Eaton What, if anything, does the distribution of Ap stars with spectral type tell us about the spectral type at which subsurface convection becomes important in A stars?

Landstreet This is not yet very clear. The known magnetic Ap stars do not occur at surface temperatures as low as those ($T_{eff} \approx 7500$ K) for which deep convection begins, but the drop-off in frequency between about A0 (where they are $\approx 10\%$ of A stars) and late A (where they are $\approx 1\%$ of all stars) cannot be explained by changes in convection. Futhermore, diffusion below the convection zone should be able to produce some atmospheric chemical peculiarities even in F stars, where no magnetic Fp stars are known. A further discussion of this question will be found in my paper in the proceedings of the 1990 Helsinki meeting on Cool Stars.

Staude Specialists in dynamo theory suggested that there should be a preferential excitation of magnetic multipoles which are either only symmetric or only antisymmetric with respect to the equator of rotation. Therefore one should expect magnetic field distributions the axis of which either coincides with the axis of rotation or which is perpendicular to it ('orthogonal rotator'). Do your results contradict such a picture?

Landstreet The star HD215441 (Babcock's star) provides a clear example of the difficulty of reconciling the requirement for equatorial symmetry or anti-symmetry with the observations. Strenuous efforts (Borra and Landstreet 1978, Ap. J. 22, 226) to find an acceptable symmetric or antisymmetric magnetic model for this star were not successful , while a magnetic field roughly dipolar and inclined to the rotation axis by about 35° describes the available data well. I do not think the field of HD215441 has the desired symmetry, and so it seems unlikely that a dynamo with this requirement functions in it.

Strassmeier What is the actual difference between your method and Zeeman Doppler Imaging?

Landstreet In Zeeman Doppler Imaging, one tries to use high resolution polarimetry through the modelled lines, as well as high resolution intensity profiles, as constraints on the model. With this kind of extra constraint, more precise magnetic models should be possible. My programme already calculates both intensity and polarization data, and when polarimetry becomes available it can easily be incorporated into the modelling process.

SURFACE INHOMOGENEITIES ON CAPELLA

M.M. Katsova
Sternberg State Astronomical Institute
Moscow State University
119899 Moscow V-234, USSR

1. Introduction

Capella is one of the most interesting stellar objects, where we are faced with such astrophysical problems as stellar activity and the influence of binarity effects on activity and stellar evolution. Usually the name Capella refers to the double-line spectroscopic binary, α Aur, with $P_{orb} = 104.0214$ day. Now it is clear that the F-star secondary component with rotational period $P_F \approx 8$-10 day is a fast rotator compared to the primary G-star, *i.e.* the system has a strong asynchronism.

13 years ago we proposed that the F-star secondary component is mainly responsible for the EUV and X-ray emission (Katsova and Livshits 1978). Its outer atmosphere can be outflowing. Our concept of the higher activity level on the F-star has got direct support from high-resolution UV observations carried out by Ayres and Linsky (1980) with the IUE satellite.

2. Observations

It is well known that the IR He I λ 10830Å line is a good index of chromospheric activity. We searched Capella in this line during several seasons (Aug - Sep 1980, Feb 1982 and Aug 1985 - Apr 1988) (Katsova and Shcherbakov 1983; Shcherbakov *et al*, 1990). For the first two years we used high-sensitivity infra-chromatic film. Then, from 1985, the spectra were obtained using the Helsinki University Observatory CCD imaging system. It is installed on coude spectrograph of the 2.6 m telescope of the Crimean Astrophysical Observatory.

Analysis of the first two seasons observational data gave evidence for 8 - 10 day variability of the equivalent width (EW) of the HeI line, which we assumed to be related to axial rotation of the secondary F-component (Katsova and Shcherbakov 1983). At the second epoch we obtained high dispersion CCD near-IR spectra of Capella over several orbital periods (Shcherbakov *et al* 1990). Their analysis showed unexpected results.

Firstly, the EW(HeI) varies with orbital period. It reaches its maximum at phase 0.5 (at one of the elongations), and a 3 times smaller minimum value occurs at phase 0.0. Fig. 1 shows EW(HeI) for different phases of the orbital period. Secondly, the coincidence is found of the radial velocity curves of the HeI and photospheric lines. It is clear that the HeI line is formed on the G-star (primary) which we consider to be the less active component.

We therefore have the following observational results.

- The HeI absorption source is connected with the G-star
- Only one of the hemispheres of the G-star is an absorption source in the HeI line and the other hemisphere does not show this effect.
- There is evidence for variation of the EW of helium absorption with the period of axial rotation of the F-star *i.e.* about 8 - 10 day.

Thus, we observe the HeI λ 10830Å radiation source on the G-star. modulated by the rotation of

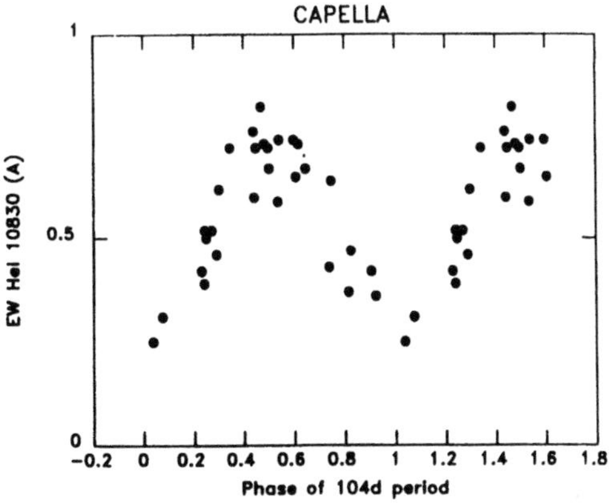

Figure 1

the F-star. These three strange facts stimulated the development of the following model.

3. The model

The physical parameters of the source, which is emitting lines of ions at $T_i \geq 10^5$ K as well as X-ray emission, are close to those obtained by Katsova (1985) in the homogeneous model of the outer atmosphere of the secondary, F-component of Capella. Temperatures of the coronal loop plasma in this model ($T = 10 - 50\,10^6$ K) are close to typical values for the primary active components of the RS CVn binaries.

Apparently, an active region on the F-star is a source of an enhanced stellar wind. This arises from the high level of coronal heating and, possibly, from the formation of open configurations in the magnetic fields. This process can be related to reconnection of magnetic field, *i.e.* similar to those in main phase of a solar flare with the subsequent formation of a streamer. Thus, from one of the active longitudes of the F-star a strong outflow of gas can occur with a characteristic velocity close to parabolic (about 350 km/sec). The stellar wind from the F-star then reaches the Lagrangian point L_1, and part of the flow is captured by the gravitation field of the G-star.

We considered particle trajectories in a binary system with strong non-synchronism in the framework of the two-dimensional problem (Kruszewski 1984). The usual expression for gravitational potential is

$$\psi = -GM_1/R_1 - GM_2/R_2 - \tfrac{1}{2}\omega_*{}^2 R_1 + \omega_{bin}{}^2 M_2 A/(M_1 + M_2)$$

where ω_* is the angular velocity of the star's rotation, $\omega_{bin}{}^2 = G(M_1 + M_2)/A^3$ is the angular velocity of its orbital motion in the binary system and A is the distance between the stars' centers. The force acting on the line connecting the centers of the stars is $F = -\nabla\psi$. Then we introduce dimensionless masses $m_1 = M_1/(M_1 + M_2)$ and $m_2 = M_2/(M_1 + M_2)$ and the degree of non-synchronism is measured with the non-synchronism parameter, f, defined as $\omega_* = (1 + f)\,\omega_{bin}$. Here the plane of orbital motion only is considered and the polar coordinate system is used, $\xi = x/A$. Finally. we obtain

$$F = -m_1/\xi_1{}^2 + m_2/\xi_2{}^2 + (1 + f)^2\xi - m_2 = 0$$

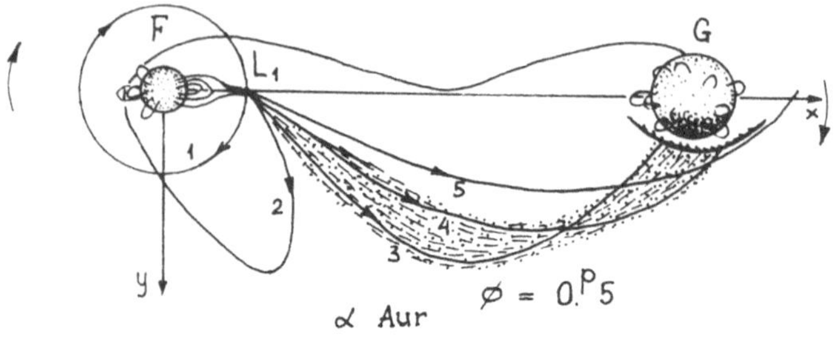

Figure 2

where ξ is the radial coordinate of the point considered, which lies on the star's equator, and $\xi_2 = 1 - \xi_1$. As far as Capella is concerned, $\omega_*/\omega_{bin} \approx 10$, so then $\xi^3 = m_1/(1+f)^2 \approx 1/200$. This permits the determination of the position of the Lagrangian point L_1 and then the trajectories of particles arriving in this point can be found.

The stellar wind from the G-star gives the radial component of the velocity and fast rotation of this star provides the tangential component of velocity of the outflowing matter. In such a case, the stellar wind plasma coming to the neighbourhood of the L_1-point, goes below an angle of about 45° to the line connecting the stars' centers.

A calculation provided by N. Shakura considers 7 cases of velocity parameters v_x and v_y. For their case 3 the particles fall directly on the G-star and a region of shock wave formation results in front of the G-star between angles 48° and 36°. This flux of matter, infalling on the G-star, results in the helium absorption and emission in the low-temperature UV lines such as OI, SiII, *etc.* There are at least two possible mechanisms for this. Firstly, this stream can form a shock wave in the corona of the G-star, and the X-ray emission arising behind the shock front can lead to excitation of the helium 2^3S level. This is the photorecombination mechanism considered in Katsova and Shcherbakov (1983). A second possibility is the overshoot of a gas-dynamic perturbation directly into the chromosphere, where gas heats up to $T = 2 - 3\,10^4$ K.

In any case this scheme, illustrated in Fig. 2, allows us to explain most of the stranger observational features on Capella. Here it is important that the behaviour of the helium spot on the G-star traces those processes, which occur on the F-star. In particular, modulation of the helium absorption by the 10 day period can be related to the stream from active region on the F-star coming to the L_1-point at each revolution. Thus, this picture makes clear why only one of the hemispheres demonstrates the helium spot.

In this connection it is of interest to analyse the radio emission from Capella. One part of the radio flux (0.18 - 0.55 mJy) measured at 6 and 2 cm by Drake an Linsky (1986) is associated with thermal bremsstrahlung from the F-star corona. For $T \approx 4\,10^6$ K a radio flux of about 0.2 mJy corresponds to an emission measure $EM \approx 2 - 3\,10^{53}\,cm^3$. Available data do not yet allow us to conclude about the properties of the radio emission, *e.g.* plausible variability of a cm-radio emission on a timescale close to the axial rotation period of the F-star. The second part of radio flux at longer wavelengths can occur through radiation of the shock wave located near the G-star. In this case weak variations on a timescale close to the 104 day orbital period of the binary system might be observed.

A full version of this contribution will be published in the Soviet Astronomy.

References

Ayres, T.R., Linsky, J.L.: 1980, ApJ, 241, 279

Drake, S.A., Linsky, J.L.: 1986, AJ, 91, 602

Katsova, M.M.: 1985, Sov. Astron., 29(6)

Katsova, M.M., Livshits M.A.: 1978, Sov. Astron. 22, 208

Katsova, M.M., Shcherbakov, A.G.: 1983, Sov. Astron. 27, 153

Kruszewski, A.: 1964, Acta Astron. 14, 231

Shcherbakov, A.G., Tuominen, I., Jetsu, L., Katsova, M.M., Poutanen, M.: 1990, AA, in press

Discussion: Katsova

van't Veer What is the mass ratio?

Katsova The mass ratio is about unity (1.07).

van't Veer Do you have an explanation for the asynchronisation?

Katsova No, I have not.

Linsky Is the He I absorption due to excitation of the 2^3S level by (a) recombination following X-ray photoionization or (b) collisional excitations in the shock of temperatures near 50,000 K? If your model uses explanation (a), is the amount of He I absorption consistent with the observed X-ray luminosity?

Katsova 1. We discuss both possibilities. Here the radio observations at the different wavelengths can help to determine the size of the source.
Yes, the HeI absorption is consistent with the X-ray luminosity.

Young The flow calculations that you show offer a nice explanation for selective heating of the leading hemisphere of one star. However, in many systems in which both components are active (*e.g.* AR Lac, ER Vul and others) we observe that the leading hemispheres of both components are hotter. In such cases your mechanism may not be appropriate as an explanation. Do you have a proposed mechanism for such cases?

Katsova I think that in the case of Capella we are dealing with a unique binary system with a strong asynchronism. Our considerations describe only such a case. Your question does require special attention, however and will be a good topic for further investigation.

Multi-Site Spectroscopic Networks for the Study of Late-Type Stars

Bernard H. Foing and the MUSICOS team
Space Science Department of ESA,
ESTEC, 2200AG Noordwijk, NL
and
Institut d'Astrophysique Spatiale,
BP 10, 91371 Verrieres le Buisson, France

1. The need for Multi-site spectroscopy

We organised in June 1988 a national workshop in France about the scientific use of multi-site spectroscopy (see Catala and Foing 1988). The workshop was attended by 50 participants, from different fields of solar and stellar physics. The interest for continuous spectroscopy was recognised for various topics for the study of asteroseismology, stellar rotational modulation, surface inhomogeneities, for stellar activity short term phenomena,or Doppler imaging of surface structures. Also the study of wind variability, flare patrol, or eclipse imaging was stressed. Additionally, a joint network of high resolution spectroscopy and photometry would give a simultaneous support to continuous satellite observations. The need for multi-site continuous coverage (taking as example networks of solar seismology) was especially recognised for the asteroseismology programs e.g. of OB, Be, delta Scu, fast rotating B stars and solar type stars (see Dappen 1988, Catala and Foing 1988).

2. The MUSICOS project

MUSICOS is an international project started in 1988 which aims at developing a multi-site network of high resolution spectrographs. This project has received support from many scientists, from European countries, America and Asia so far. The goal of Musicos is to facilitate multi-site spectroscopic observations, first by setting up an organisation helping the coordination of observations from existing instruments at different sites, then by designing, developping and installing similar spectrometers in well-chosen sites around the world, for which part of the time would be devoted to multi-site observations.

Given the positive impact of the first MUSICOS workshop and the interest shown by several other European teams , we have decided to get started on the practical and technical aspects of the project as soon as 1989. In a first phase, we are organising multi-site campaigns using resident instruments on various telescopes around the world and a transportable fiber-fed mono-order spectrometer (ISIS, developed at Meudon Observatory and used until 1988 at the 1.93 m telescope at OHP). In a second phase, we shall continue in 199 the design and development of an cross echelle fiber-fed spectrometer, and perform tests in the laboratory and at telescope on this prototype to assess the scientific performances of this fiber-fed spectrograph. In a third phase, we shall develop an instrument that can be proposed for duplication to different foreign collaborators. This will allow at long term multi-site campaigns with identical instruments, and to reach the limits in quality, sensitivity and homogeneity of ground based continuous spectroscopy that are set by asteroseismology requirements.

3. Requirements for asteroseismology and stellar activity

Asteroseismology requires high S/N, and the possibility of observing a large number of photospheric lines, was stressed , together with the specific requirement of a high spectral spectral stability and possibility of very accurate velocity calibration. The need for prereduction of data at each site, in order to control in real time the instrumental parameters (quality of night, drifts) was made clear at the workshop.

For solar-type oscillations, the velocity signal is much reduced (of the order of 15 cm/s per stronger solar mode). An observing continuity better than 60 percent in order to decrease the effect of parasite sidelobes and aliasing in the power spectrum, and to increase the signal to noise and resolution on the oscillation modes was requested, which argues for multi-site network. Classical slit spectrographs using one or a few lines are not stable and efficient enough. Resonance cell techniques have a good stability but can apply only to 0 or 1st magnitude stars with 4m class telescopes, which are oversubscribed and make difficult a continuous observing from several sites. For having access to fainter stars from later-type closest stars or from other classes in the H-R diagram, the multi-line concept may be relevant in the future. For seismology velocity measurements, the MUSICOS spectrometer at 30000 resolution only with a 2m telescope would give on a 3 magnitude star a S/N of 100 in 1 mn, which coupled to the measurement of 400 lines relatively free of blending available on the spectral domain in the cross echelle mode would in principle allow to reach an oscillation detection of 3m/s per mn exposure, and thus a noise of 5cm/s for a continuous 72 hours observation.

Stellar activity requires also the observation of different lines simultaneously, for instance: photospheric lines to be used for magnetic measurements; different unblended lines for Doppler imaging of spots; chromospheric lines such as the Ca II H and K or infrared triplet allow to probe different chromospheric heights; in particular the H alpha line is a very sensitive indicator of extended active structures; Balmer lines give very important temperature, density and dynamic diagnostics for flare studies; both H Balmer line and strong resonance lines such as Na or Ca II can probe circumstellar absorbing material transiting over the disk; also high excitation lines such as He I allow to follow more energetic phenomena or variations of EUV or X radiation.

4. Technical design of MUSICOS instrument

From the scientific requirements after the MUSICOS workshop it was realised that a large range of scientific programmes require a spectral resolution 10000, 30000, and 80000. The main requirements for stellar activity studies and asteroseismology is the availability of large spectral domain in order to measure a number of photospheric lines in solar-type stars, and the stringent constraint for spectral stability down to the 10m/s range. The spectrograph can be fed by an optical fiber (Felenbok, 1988) which makes the instrument transportable and adaptable on another telescope.

The current design of the instrument include the following specifications: a spectral resolution of 30000 minimum, a spectral domain 3900-8700 A that can be covered in two successive exposures (blue and red), an efficiency allowing a S/N better than 200 in 1 hour for a star of 6th magnitude with a 2m telescope and an astronomical CCD; a Thomson CCD 1024x1024 (19 μ pixel) proposed as a detector; a fiber feeding; the possibility to acquire 2 spectra simultaneously. Initially, until october 1989, the selected fiber (spectran SG820) had a core of 110 μ, which allowed to input 1.5 arcsec with an efficiency of 85 percent in the spectrograph for a 2m telescope opened at f/5.5. An image slicer, of Bowen-Walraven type, would thus have been needed to give finally on the detector 2.5 layers of width 50 μ, superposed on a total height of 315 μ, with an efficiency better than 90 percent. At present, the selected fiber has a core of 50 μ, which corresponds to 2 arcsecs for an entrance aperture f/2.5. Its chromatic transmission is near 90 percent for a large domain in the visible and in the red, and remains better than 70 percent at 3900 A. Thus, an image slicer is not anymore necessary. The cross dispersion will be given by a FLINT rectangle prism working in double pass at deviation minimum. The whole is articulated around an objective (from the market). A slight tilt of the grating, and a lateral mirror will allow a comfortable mounting of the detector cryostat. This set up has the advantage of being very compact, luminous and unexpensive.

5. The MUSICOS december 1989 observing campaign

For this first broad MUSICOS campaign at the end of 1989, three programs were chosen among those which really require multi-site spectroscopic observations:
1) Short-period spectroscopic variations in Be stars
2) Corotating stream structures in the winds of PMS Herbig Ae stars
3) Doppler Imaging and flare monitoring of RSCVn-type active stars

Initially, the sites, telescopes and observers to be involved in this campaign, were the following: Mauna Kea 2.2m UH, 3.6m CFHT, Kitt Peak MacMath, La Silla 1.4m CAT, France 1.5m OHP, Crimea 2.6m Shajn, China 2.16m Xinglong.

After the general announcement sent by J. Butler through the Multi wavelength IAU Working group, and also the IAU circular for our multi-site continuous spectroscopy campaign from 8 to 17 december 1989 (MUSICOS 89 campaign), we received a large observing support (ground based spectroscopic or photometric, or satellite observations) and finally 17 telescopes , including IUE, were pointed towards our targets . In addition to resident spectrographs at OHP, Crimea, CFH, ESO, Lick, Mc Math , we used the two existing versions of the ISIS fiber-fed spectrograph in mono-order mode (one of them used until 1988 at OHP), and specially transported for the MUSICOS campaign on the 2.2m University of Hawaii telescope and on the 2.16m newly installed telescope at the Chinese XingLong station.

Final participants for the december 89 MUSICOS campaign included: B.H.Foing, (ESA/IAS), co-PI MUSICOS, coordination HR1099; C.Catala (DESPA) , co-PI MUSICOS, coordination AB Aur; and the participants in the SPECTROSCOPIC campaign at Hawaii: T. Simon (UH), J. Baudrand, J.G Cuby, (Meudon), Installation of ISIS bis spectrograph on 2.2m U. Hawaii; P.Felenbok, J.Czarny (Meudon) for observations at Canada France Hawaii 3.6m; U.S.A.:J. Neff GSFC and J.Avellar, for NSO observations at MacMath Kitt Peak ; Chile: S.Jankov (IAS) for remote control CES/CAT observations from ESO Garching; S.Char (IAS), A.M. Lagrange et al (IAP) for 1m Ca spectrophotometry at ESO La Silla; IUE: T. Ayres (Boulder) for IUE observations; Brazil: R. de la Reza, C.Torres (Brazil National Observatory); France: A.M. Hubert, H. Hubert, M. Floquet (Meudon) with spectro Aurelie OHP; Austria: J. Hron,H.Maitzen (Wien) for 1.5m Schoepfl; USSR: I. Tuominen, Huovelin (Helsinki) P.Petrov, A. Scherbakov, I. Savanov (Crimea Obs), for observations with Coude Spectrograph on Crimea Shajn telescope; China: Li Qi Bin, Huang Lin, Jiang Shi-Yang, Zhao Disheng (Beijing AO), C.Catala, J.Guerin, M.Dreux, B.Foing installation of ISIS instrument at Xinglong Obs.2.1 m telescope; India: K.K. Ghosh Vainu Bappu Observatory.
The participants in the PHOTOMETRIC campaign were: Chile, G. Cutispoto (Catania), photometry ESO La Silla; Mexico, San Pedro martir: Alvarez (UNAM), photometry ; Turkey :V. Keskin, C. Akan, S. Evren, Ege University UBV 48cm; Greece: J.Seidarakis, S.Mavridis; Australia: Dr. Page (Mt Tambourine observatory); US: Dr Brown (Manitoba Glenlea Obs.).

The first results from the MUSICOS 89 campaign have been presented in the 2nd MUSICOS workshop (Catala and Foing, Eds, 1990). On the RS CVn-type HR 1099, this includes a complete phase coverage allowing a Doppler imaging of photospheric spots. Quasi-simultaneously, we observed modulation of CaII K profile due to chromospheric plage regions. At least two exceptional white-light flares were detected photometrically and by their remarkable spectral signatures in H alpha. The 14 December flare starting at 13:00 UT was followed from China and Catania, and the 15 December flare starting at 2:00 UT was followed from ESO, Hawaii, China and Catania (Foing et al, 1990). We estimated the energy budget for these two events, the first of this importance detected on HR 1099. The interpretation of a filament ejection occurring over a magnetic arcade and the magnetic energy budget are discussed in Foing et al (1991). These observations bring a new light and a puzzle about energy transport mechanisms in stellar flares.

6. MUSICOS Project organisation and collaborations

There is in France a MUSICOS project group composed of C. Catala (Meudon) and B.H.Foing (ESA/IAS) as Principal Investigators, P. Felenbok, A.M. Hubert , J. Czarny (Meudon), J.M. Le Contel and E. Fossat (Nice). J. Baudrand (Meudon) is the instrument project manager. A group of 30 associated scientists in France is involved in the project. Also contacts were established with European groups in Italy (in Trieste and Catania), in Scandinavia (in Upssala, Helsinki, and Aarhus), in Spain, in United Kingdom (at Armagh). Associate countries such as US (in Hawaii, in Boulder, Goddard, and with the SYNOP group), in USSR (in Crimea) and in China (at Beijing Astronomical observatory) wish to participate to the multi-site project. A group of international co-investigators from institutes providing funding, hardware, observing site or manpower support to the project is being set, together with an international group of Associated scientists interested in the further scientific use of the network.

The MUSICOS project will be open to any input and collaboration during the development phase , and through the observing proposals to the community during the operational phase. The philosophy of the MUSICOS project is to associate the groups interested also on the scientific return , and this is the reason why we start already collaborative campaigns to learn how making, reducing , analysing the results from multisite data. Asteroseismology will require continuous coverage and a good control on the stability and the quality of the measurements. Also, some programs require special observational strategy and might be interleaved with other programs or require service observing.

7. MUSICOS Planning and Perspectives

Since 1982, with our collaborators , we have participated to Multi-Site Multi-Wavelength Observing campaign, giving the collaboration framework for operations and scientific analysis. At the end of 1989, we organised an observing campaigns including transport of ISIS and ISIS-bis instruments in complementary sites (Xing Long, Hawaii). In january-mars 1990, we worked on the preliminary reduction and analysis of the results from this campaign. A MUSICOS workshop organised at Meudon on 27-30 March allowed presentation of these results, further discussion and future organisation and realisation of the MUSICOS project (see Catala and Foing 1990). 1989 was devoted also to the design of the spectrograph. In 1990, the current ISIS bis spectrograph will be transformed into a cross echelle spectrograph . At the end of 1990, instrument tests and qualification of this MUSICOS prototype are to take place at Pic du Midi Observatory . Next MUSICOS campaigns with existing instruments would include sites in Hawaii, ESO, OHP, Pic du Midi, Canarias, Crimea, Kitt Peak, AAT, China etc). 1991 should see the final development of the MUSICOS spectrometer model to be duplicated by the participant countries. In 1991-1992, we plan the installation in remote sites (such as Hawaii, Xinglong, Canarias), and the start of the full network operations. When the network is operational, a Multi-Site Guest Observer program will be offered to the community.

Multi-site multi-frequency campaigns in coordination with satellite continuous observations will be also a driver for the MUSICOS operations. A multi-site spectrophotometric network will complement several space missions such as EVRIS on Soviet Mars 94 probe, PRISMA currently under study at ESA for asteroseismoloy, or Hubble Space Telescope, ROSAT, and future UV and X-ray observatories for stellar activity or microvariability programmes. Also, moon-based astronomy (either from the future lunar base or its precursor missions) will provide very favourable viewing conditions for photometric asteroseismology and multi-frequency stellar activity studies, that can be complemented by global networks.

The perspective of a global network should also include the possibility of a node in Antarctica, allowing exceptional sites and continuous viewing possibilities. Robotic facilities may be envisaged in support to the courageous scientists spending one continuous polar observing night, or the PhD students in a hurry for a thesis done in 3 days and 3 nights.

References

Catala, C. and Foing , B.H., editors, 1988, *Proceedings 1st MUSICOS Workshop on MUlti SIte COntinuous Spectroscopy,* (Meudon Observatory Publications).

Catala, C. and Foing , B.H., editors, 1990, *Proceedings 2nd MUSICOS Workshop on MUlti SIte COntinuous Spectroscopy,* (Meudon Observatory Publications).

Dappen, W., 1988 , in *Proceedings 1st MUSICOS Workshop on MUlti SIte COntinuous Spectroscopy,* eds.C.Catala, B.Foing, (Meudon Observatory Publications), p.11.

Felenbok, P., 1988 , in *Proceedings 1st MUSICOS Workshop on MUlti SIte COntinuous Spectroscopy,* eds.C.Catala, B.Foing, (Meudon Observatory Publications), p.123.

Foing, B.H. et al, 1990 , in *Proceedings 2nd MUSICOS Workshop on MUlti SIte COntinuous Spectroscopy,* eds.C.Catala, B.Foing, (Meudon Observatory Publications), p.117.

Foing, B.H., Char, S., Jankov, S., Catala, C., Zhai, D.S. et al, 1991 , *Astron. Ast.* , ().
submitted

FLARING AND QUIESCENT X-RAY EMISSION FROM THE RS CVn SYSTEM II Peg

G. Tagliaferri and N.E. White
EXOSAT Observatory, European Space Agency
ESTEC, Netherlands

1. Introduction

The RS CVn system II Peg (= HD224085) is a 6.7 day single-line spectroscopic binary with a K2IV-V primary and an unobserved companion. It is at a distance of $d = 29 \pm 5$ pc. Orbital photometric variations are explained by large-scale dark photospheric spots on the surface of the primary. The shape and amplitude of the orbital light curve vary on timescales of months from sinusoidal, to asymmetric, to constant, and indicate short term changes in the star spots (see Doyle (1988) and references therein).

The X-ray luminosity of II Peg, $4\,10^{31}$ erg s^{-1}, and its temperature, $\sim 10^7$ K, are similar to those observed from other members of the RS CVn class (Walter et al. 1980, Swank et al. 1981). A spectral survey of RS CVn systems, performed with the EINSTEIN SSS instrument, shows in the 0.5 - 4.5 keV band two distinct temperature components (Swank et al. 1981). This result has been confirmed by spectra of Capella and σ^2 CrB obtained with the EXOSAT grating spectrometer in the 0.05 - 2 keV band (Lemen et al. 1989). The failure to detect an eclipse in the 2 - 6 keV EXOSAT light curves of AR Lac and TY Pyx, while it was detected in the 0.05 - 2 keV light curves, shows that the hot component comes from an extended region and the cooler component from a more compact region, similar in size to solar active regions (White et al. 1990; Culhane et al. 1990).

2. The Observations

The observations of II Peg by the EXOSAT X-ray observatory (White and Peacock 1988) consisted of a single observation in 1984 August and seven observations in 1985 July. The

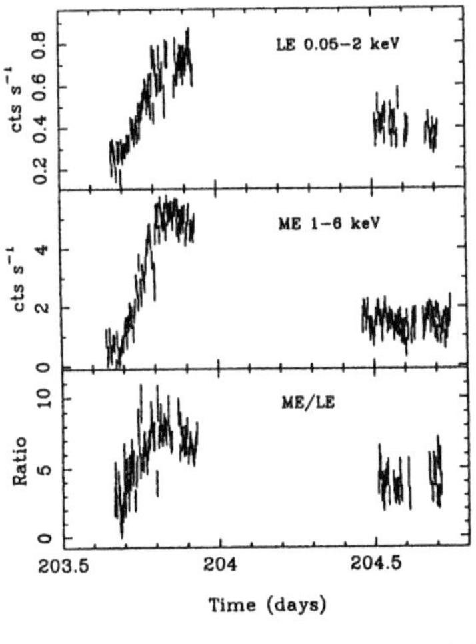

Figure 1. The strong X-ray flare detected on July 22, 1985 both by LE and ME. *Bottom panel* shows the hardness ratio, accumulation time is 300 sec. The gap in coverage is caused by an EXOSAT perigee passage.

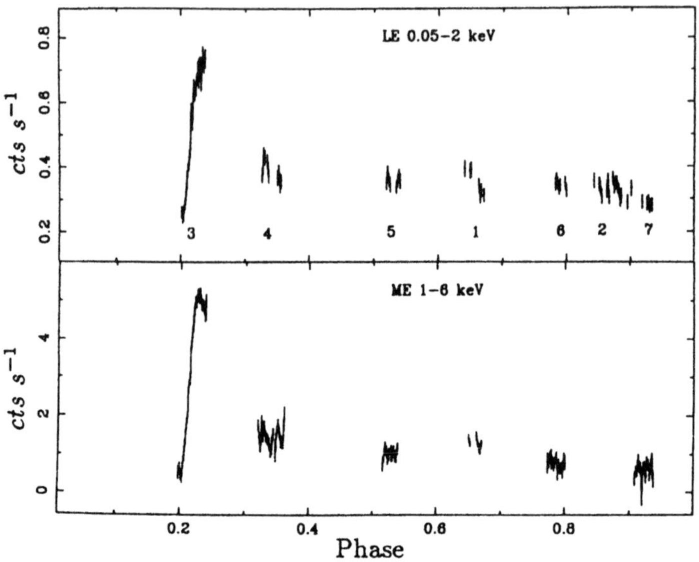

Figure 2. The folded light curve of II Peg as recorded by the ME and LE on the period July 18-27, 1985. In the *top panel* the observations are numbered so that their relative phasing can be seen.

seven observations made in July 1985 sampled 1.5 orbital cycles of the system and were quasi-simultaneous with 16 IUE observations (Doyle 1988). The full analysis of the EXOSAT + IUE data are in progress (Tagliaferri et al. 1990). Here we present preliminary results from the EXOSAT data.

The first observation on August 12, 1984 lasted 12.5 hr. The ME (1 - 6 keV) and CMA (0.05 - 2 keV) count rates were constant for the first $\sim$ 5 hr, then a gradual increase in flux began that continued until the end of the observation. The ME count rate increased by a factor of $\sim$ 3, whereas in the CMA the count rate increased by only 0.25.

The most notable feature in the July 1985 light curves is a flare during the July 22 observation (Figure 1). In the ME the count rate increases by a factor $\sim$ 10, whereas in the LE it increased by only a factor 3. The rise time of the flare is one hour in the ME, but in the LE it is longer and did not reach a peak when the observation ended. When II Peg was again observed 12 hours later, the ME and LE fluxes were still a factor of 2 above the quiescent value observed prior to the flare (Figure 2). The ME and LE folded light curves are shown in Figure 2 (Period and epoch are from Vogt 1981).

Two background subtracted ME spectra were accumulated from the 1984 August observation, covering the interval when the flux was stable, and one when the flux was increasing. A thermal model (Mewe et al. 1985) was fit to the spectra and in both cases a single temperature gave an acceptable fit. The temperature increased from 2.2 ± 0.5 to $2.9 \pm 0.3\, 10^7$ K as the flux increased (all uncertainties are at 90% confidence), with only a small increase in the emission measure (EM) from 1.7 ± 0.3 to $2.0 \pm 0.25\, 10^{53}\, \mathrm{cm}^{-3}$.

The ME spectra accumulated for the July 1985 observations are also well fitted by a single temperature model. For the July 22 observation a spectrum was accumulated during the rise and one during the peak of the flare. We used as background a pre-flare spectrum that included the quiescent source plus background, to isolate only the component responsible for the flare. The best-fit temperatures are 4.38 ± 1.0 and $3.7 \pm 0.3\, 10^7$ K and the EM's are 2.9 ± 0.5 and $7.5 \pm 0.5\, 10^{53}\, \mathrm{cm}^{-3}$ respectively. In the observation made 12.5 hr later, the temperature had decreased to 2.3 ± 0.2 keV. In the following four observations covering an interval of 5 days, the temperature continues to decrease, reaching $1.68 \pm 0.30\,10^7$ K similar to the pre-flare value.

The ME is only sensitive to temperatures $\geq 10^7$ K, while the LE is sensitive to temperatures $> 5\,10^5$ K. The large energy range spanned by the LE (3-Lex and Al/P filters) and ME combination of spectra increases the sensitivity to the two temperature plasma model reported from these systems. However the actual sensitivity of the LE is critically determined by the interstellar absorption N_H, while the ME is insensitive to values of $N_H < 10^{20}\,\mathrm{H\,cm^{-2}}$, and in most cases the single temperature model still provides an acceptable fit to the data. The best fitting value of N_H ranges between $< 0.5\,10^{19}$ and $\sim 1\,10^{19}\,\mathrm{H\,cm^{-2}}$. Adding a second component to set upper limits to the presence of a low T component is only useful if the N_H is fixed. We used the value of $5\,10^{18}\,\mathrm{H\,cm^{-2}}$ from Paresce (1981).

The best fit temperature of the soft component varied between 3.5 and $8\,10^6$ K, in good agreement with the range of (4 - $8\,10^6$ K) reported by Swank et al. (1981) for a sample of RS CVn systems. The best-fit values of the harder component vary between 17 - $30\,10^7$ K. Swank et al. (1981) found 20 - $100\,10^7$ K.

The two models (a single temperature plasma plus a variable amount of N_H, or a two temperature plasma plus a fixed amount of N_H) give equally acceptable results in three cases. In the other four cases, the two temperature model gives a significant improvement of the χ^2. The flare spectra from July 22 do not require a second component.

3. Discussion

The flare has a peak luminosity of $1.1\,10^{31}\,\mathrm{erg}\,s^{-1}$ and a total energy release of $> 1.6\,10^{35}$ erg in the 0.05 - 10 keV band. These values are much higher than the values detected in compact and two-ribbon solar flares or detected in flare stars (*e.g.* Table 8 in Pallavicini et al. 1989), but are not unusual for RS CVn and Algol-like systems. For instance, similar values are reported for a flare from Algol (White et al. 1986).

For the II Peg flare, we determine a value of $\tau = 40\,000$ sec with an intrinsic error of $\sim 25\%$. Following White et al. (1986), we can obtain the loop height H as a function of $\alpha^2 N_L$ in the two extreme hypothesis of pure radiation cooling or conduction cooling, where N_L is the number of loops and α the ratio between the loop radius and the loop height.

In the case of pure radiation cooling, we obtain an electron density of $n_e = 1.9\,10^{10}\,\mathrm{cm^{-3}}$, a volume of $2.0\,10^{33}\,\mathrm{cm^3}$, and a loop height $H = 5.9\,10^{10}(\alpha^2 N_L)^{-1/3}$ cm. Assuming $\alpha = 0.1$, as typical for the solar case, we derive a loop height $H = 2.7\,10^{11} N_L^{-1/3}\,\mathrm{cm} \simeq 2 N_L^{1/3}\,R_*$ (Vogt 1981). A single loop flare defines the maximum height, $H \simeq 2\mathrm{R}_*$, with a footpoint area, $A \simeq (\alpha H)^2\pi \simeq 2.3\,10^{21}\,\mathrm{cm^2}$, about $4\,10^{-2}$ of the stellar surface.

The conduction cooling time is given by $\tau_c \simeq 1.18\,10^{-10}\,n_e L_c^2/T^{2.5}$ sec (Culhane et al. 1970), where L_c is a characteristic length for the temperature gradient (often identified with L). Assuming that $L_c = L$, we obtain $H = 1\,10^{12}\,\alpha\,N_L$ cm. The *minimum* loop height is given where the two values of H are equal ($\tau_c = \tau_r$). For the II Peg flare this occurs when $\alpha^2 N_L \simeq 0.12$ and gives a minimum loop height of $\sim 1\,10^{11}$ cm. If $\alpha \simeq 0.1$, then N_L has to be ≥ 12. For values of $\alpha^2 N_L \geq 0.12$ conduction cooling dominates, while for lower values, radiation cooling dominates. Conductivity can be inhibited by up to a factor of 10 by constriction of the loop legs (Antiochos and Sturrock 1978) and the derived values of H can be lower by up to a factor of 100. For $\tau_c = \tau_r$ this increases $\alpha^2 N_L$ to ~ 4 and decreases the minimum loop height to $4\,10^{10}$ cm.

The pressure in the flaring loop is $p = 2nkT \simeq 200\,\mathrm{dyne\,cm^{-2}}$ and from the confinement condition $B^2/8\pi > p$, we estimate a lower limit to the magnetic field in the flare region of $B \geq 70$ Gauss. The thermal energy at the peak is given by $E = 3nkTV$ erg, and, from the density and volume inferred from the radiative cooling time, we obtain a total energy of $\simeq 6\,10^{35}$ erg, comparable to the total energy radiated in X-rays during the flare.

Comparison with solar flares shows that this flare is unusual. The characteristic sizes and time scales are substantially larger. On the other hand the gas pressure and magnetic field involved are not.

References

Antiochos, S.K., Sturrock, P.A.: 1978, ApJ, 220, 1127

Culhane. J.L., Vesecky, J.L., Phillips. K.J.H.: 1970, Solar Phys. 15, 394

Culhane, J.L., White, N.E., Shafer, R.A., Parmar, A.N.: 1990, MNRAS, 243, 424

Doyle, J.G.: 1988, A&A, 192, 281

Lemen, J.R., Mewe. R., Schrijver, C.J., Fludra, A.: 1989, ApJ, 341, 474

Mewe, R., Gronenschild, E.H.B.M., van den Oord, G.H.J.: 1985, A&AS, 62, 197

Pallavicini, R., Tagliaferri, G., Stella, L.: 1989, A&A, 228, 403

Paresce, F.: 1984, AJ, 89, 1022

Swank, J.H., White, N.E., Holt, S.S., Becker, R.H.: 1981, ApJ, 246, 214

Tagliaferri, G., White, N., Doyle, G., Hassal, B., Swank, J., Culhane, J.L.: 1990, in preparation

Walter, F.M., Cash, W., Charles, P.A., Bowyer, C.S.: 1980, ApJ, 236, 212

White, N.E., et al.: 1986, ApJ, 301. 262

White, N.E., Peacock, A.: 1988, Mem. Soc. Astron. Ital. 59, 7

White, N.E., Shafer, R.A., Horne, K., Parmar, A.N., Culhane, J.L.: 1990, ApJ, 350, 776

Discussion: Tagliaferri

Gurman From the SMM experience, if the solar flare analogy is of any use, it has been found that the soft X-ray emitting loops must be highly filamented in order to reproduce the observed cooling times. Also, perhaps more relevantly, it appears that the largest SMM soft X-ray flares require a two-temperature solution for differential emission measure; the temperatures are similar to yours.

SUMMARY AND CONCLUSIONS

D.J. Mullan
Bartol Research Institute, University of Delaware
Newark, DE 19716, USA.

Introduction

In this summary of the Colloquium, Section 1 discusses the physical properties of spots on the Sun, beginning at the photosphere, where observational information is most complete. Although the Colloquium title refers to inhomogeneities on the "surfaces" of cool stars, the physical structure of magnetic inhomogeneities cannot be discussed without also referring to vertical extensions above and below the visible photosphere. Therefore, the properties of magnetic inhomogeneities at other levels of the solar atmosphere form an integral part of our understanding of the physical processes occurring inside a magnetic flux tube. Sub-sections within Section 1 summarize information on vertical extensions of inhomogeneities.

In Section 2, analogous information is summarized for spots on other cool stars. The arrangement is parallel to that used in Section 1 for discussing sunspots. The choice of a parallel structure for Sections 1 and 2 is intentional: during the Colloquium, the "solar-stellar connection" was referred to in rather disparaging terms. The parallels between Sections 2 and 1 enable us to present a critical evaluation of the "solar-stellar connection".

In Section 3, theoretical work on spot models is summarized: the most serious lack at present centers around a consistent discussion of the energy equation. This lack means that predictions of T_{eff} in a starspot are uncertain. Observationally, the most notable lack at present is *total ignorance of how the bolometric luminosity L_{bol} of a spotted star behaves when a spot is on its surface, relative to its luminosity when the surface is free of spots:* if the solar analogy is any guide (with net excess emission during sunspot maximum), L_{bol} should be greatest when a star is maximally spotted.

In Section 4, spot lifetimes are discussed. On the basis of material presented at the Colloquium, it appears that the lifetimes of spots on the sun and cool stars may be quantitatively interpreted in terms of diffusive decay, with diffusivities determined by convective turbulence. The existence of quantitative agreement between observed and expected lifetimes is remarkable in view of the phenomenological nature of most convective models. Since convective processes also seem to play a crucial role in determining the dimensions of the smallest permissible sunspots, we refer in Section 5 to recent attempts to model stellar convection more realistically. The occurrence of convection in the photosphere does not mean that convective effects are important only at the surface: rather, these effects extend into the corona. The question of the minimum scale-size for inhomogeneities in a stellar atmosphere is discussed briefly: the minimum scales may be orders of magnitude smaller than anything discussed at the Colloquium.

A word about notation: the abbreviation TP in association with an author's name indicates a reference to material presented at the Colloquium which should appear elsewhere in These Proceeedings.

1. Properties of spots and associated inhomogeneities in the sun

In this Section and the next, the discussion starts at the photospheric level and proceeds upwards.

1.1. Properties of spots in the photosphere.

1.1.1 Positions. As far as the heliographic longitudes of sunspots are concerned, there appear to be no serious restrictions. Although there have been discussions in the literature of "preferred longitudes" for active regions and large flares (Bai, 1988), the evidence for "preferred longitudes" of spots is not very strong. However, there are hints of some sporadic clustering in certain regions ("nests") (Maltby: TP).

As far as the heliographic latitudes of spots are concerned, spots are confined to relatively low latitude, between about 10° and 30°. For future reference, we note that the reason for spots to be confined to low latitudes is usually attributed to the effects of latitudinal differential rotation (Babcock, 1961): in the Sun, the sense of the differential rotation is that Ω_{eq} exceeds Ω_{pole}. The effect of this is to cause poloidal flux to be converted most efficiently to toroidal flux at low latitudes: therefore, the buoyancy effects (which increase with increasing field strength) responsible for bringing intense toroidal flux ropes to the surface and create spot pairs, are most pronounced at low latitudes.

1.1.2 Dimensions. The smallest spots (the so-called "pores") have diameters of 1 - 2 arcsec (about 1000 km) (Bray and Loughhead, 1979 [hereafter BL], p. 70). It is important to note that the scale-sizes of the smallest spots are very similar to the preferred horizontal scale-sizes of photospheric granules. Since granules are believed to be the fundamental "unit cells" of convection near the solar surface, the fact that spots never exist with dimensions smaller than those of a granule suggests that spots owe their existence to interference with convection: whatever the process is which reduces the efficiency of convective heat transfer, it must operate over a horizontal scale-size which is large enough to interfere with a complete granule before a characteristic extended dark patch appears.

In terms of heliographic angular coordinates, the minimum diameters of spots are about 0.1^o. Spot diameters extend to larger scales, although the frequency of spots falls off steeply towards larger sizes. Maltby (TP) refers to a spot with an umbral diameter in excess of 10 arcsec (i.e. radius of 3600 km) as a "large spot". The largest group of sunspots ever recorded occupied 0.00613 of the visible hemisphere (BL:229): the diameter of an equivalent circle (having equal area to the group) would be about 10 heliographic degrees.

1.1.3 Radiant intensities: temperatures. The ratio R_{up} of umbral intensity I_{umb} to the intensity of the adjacent photosphere I_{ph} depends on wavelength and on distance from disk center. At disk center, R_{up} ranges from 0.6 - 0.7 at $\lambda \approx 2\mu$ to 0.02 at $\lambda \approx 4000$Å (Maltby, TP). Corrections for scattered light must be made particularly carefully at the shortest wavelengths: earlier works had suggested that at 4000Å, R_{up} might be as large as 0.25 (BL:113), but the recent work now indicates that some 90% of the measured light in those works was in fact due to scattering in the instrument and in the Earth's atmosphere. Confidence in scattered light corrections has been gained by observations of a transit of Mercury in 1973.

As a spot approaches the limb, R_{up} decreases at all wavelengths, with larger decreases at longer wavelengths (Maltby, TP): *i.e.* the umbra becomes darker relative to the ambient photosphere. In terms of radiative equilibrium atmospheres, this suggests that the temperature in the umbra has a steeper gradient relative to optical depth than in the photosphere.

Combining the darkness of a spot with its area, the maximum decrease in solar luminosity due to spots is of order $0^m.002$, or 2 - 3 W/m^2 at Earth orbit, where the "solar constant" is about 1370 W/m^2 (Lean, TP). Surrounding spots, facular areas of the photosphere emit excess radiation: although their emission levels are only 10% or so of the missing levels in spots, they occupy typically 10 times more area than the spots. Thus, the net effect of spots and faculae in bolometric luminosity is almost a cancellation: in fact, there is observed to be a net facular excess, and the sun's luminosity is maximum when sunspots areas are maximum (Lean, TP).

The difference in brightness temperature ΔT_b between a sunspot umbra and the photosphere is about 2000 K at wavelengths ranging from $0.4\,\mu$ to $\lambda > 2\,\mu$ (Maltby, TP). (In the older literature, it was suspected that values of ΔT_b as high as 2400 K might exist in the largest spots [BL:112]: however, recent data suggest that there is no correlation between spot temperature and spot area (Maltby, TP).) The difference in effective temperatures between umbra and photosphere is also $\Delta T_{eff} \approx 2000\,K$: thus, $T_{eff}(spot) \approx 3800\,K$. This is an average over the umbra. However, umbrae are not uniformly dark (Skumanich, TP): brighter regions ("umbral dots"; "umbral granulation" [BL:84]) can be seen at various locations. The temperatures and other physical characteristics of the umbral dots are currently uncertain, as is their physical origin. A correct interpretation of the umbral dots may eventually provide information about the magnetic structure underneath the visible surface of a spot: for example, is the flux rope uniform at great depths (perhaps reaching to the base of the convection zone), or does it become shredded into separate flux ropes below the visible surface? Or does the lifetime of the dots really correspond to the Alfven crossing time in the convective layer (Weiss et al. 1990)?

1.1.4 Magnetic fields. Near the central axis of an umbra, the field is vertical, with strength of 2000-4000 G (Maltby, TP). Farther from the axis, the vertical component decreases monotonically, while a radial component at first grows in size, and subsequently also decreases (Skumanich, TP). A correlation between axial field strength and spot area has not been definitively established (Maltby, TP). For spots which have well developed circular symmetry, and diameters of L_{sp}, the observed vertical and radial field distributions can be fitted satisfactorily by a "buried dipole", centered at a depth of order L_{sp} below the visible surface (Skumanich, TP). This is a general feature of fields which we shall revert to below: a feature which has a surface scale of L is associated with a vertical extent of order L below and above the surface.

The size of a spot is determined by the magnetic flux in the rope Φ and the field strength B. As regards the magnetic flux in a particular rope, the factors which control this are at present obscure: the action of a dynamo in the deep sub-surface gas must be a key factor, but the theory of dynamo operation in the sun is not sufficiently well developed to allow predictions of the maximum Φ which the sun can generate in an individual rope. As regards the field strength, B is apparently determined by lateral force balance with photospheric gas, $B_{eq} \approx 8\pi\Delta p$, where Δp is the difference in gas pressures at a particular geometrical depth outside the spot and inside (Skumanich, TP). Thus, spot size need have nothing directly to do with convection cell sizes (except for pores), and one should not argue that spots on other stars will scale in size according to convection cell sizes.

1.1.5 Lifetimes. Pores survive for a few hours [BL:70], whereas the largest spots may survive for a few months. The lifetime of a spot group is related to its area A (in units of millionths of the visible hemisphere) by $\tau\,(\text{days}) \approx 0.1\,A$ [BL:229]: if this also applies to individual spots, we note that the decay of a spot of linear dimension $L \sim A^{0.5}$ may be related to a diffusivity $\kappa \sim L^2/\tau$ which is the same for all spots. With the above expression for τ, we find $\kappa \approx 10^{12.5}\,cm^2\,sec^{-1}$ for spots and pores. Interpretation of these quantities will be discussed in Section 4. Note, however, that some sunspots live longer than average, suggesting that the diffusive decay can at times be impeded (Spruit, TP), perhaps by organized flow patterns around the spot. (For a discussion of various reasons for the existence of organized flows around spots, see Mullan, 1983 and references therein.)

1.2. Properties of spots at the temperature minimum.

According to brightness measurements of the sun in the ultraviolet continuum, the temperature minimum above a spot occurs at 3200 K (Gurman, TP). According to one particular analysis of oscillations in the chromospheric cavity above an umbra, the temperature minimum is 3000 K (Staude, TP). Thus, R_T, the ratio of T_{min} to T_{eff}, is $\approx 0.8 - 0.85$: this is higher than the value for the photosphere, where R_T is closer to 0.7. The possibility of thermal bifurcation in the solar atmosphere (Ayres, 1981) suggests that certain regions of the atmosphere may be much cooler (by at least 900 K) than elsewhere if molecular cooling is effective (however see Athay and Dere,

1990). In a spot, molecule formation is even more efficient than in the photosphere, and so the possibility of extreme cooling in certain parts of an umbra may also need to be considered.

1.3. Properties of spots in the chromosphere.

In the low to middle chromosphere, emission from surrounding plage occupies typically 10 times more area than the spot. Nevertheless, spots remain visible as dark areas in CaII K spectroheliograms, indicating that flux tube spreading is not so excessive at levels of 1000 - 1500 km above the photosphere as to "canopy" the spot from view. Higher up, at the level of Lyman α formation, spreading of flux tubes causes plages to occupy 40% more area than at the level of Ca K (Lean, TP). In the sun, spots are always accompanied by plage, although the inverse is not necessarily true. The question as to whether or not starspots are always accompanied by plage will be examined below (Section 2.4)).

In spectra of the Ca II K-line and the Mg II K-line observed above umbrae, the emission core is found to be narrow and singly peaked at line center (Gurman, TP). This contrasts with the doubly peaked broad emission which is observed in plages and in the quiet sun (although in the latter, there may be single peaks shifted redward or blueward of line center). The peak intensity over an umbra may exceed that in plage. In small umbrae, the formation of Ca K and Mg K emission peaks is influenced not only by local collisional processes, but also by horizontal influx of radiation from the surrounding photosphere. In large umbrae, the radiation terms are of less importance. (*A fortiori*, in starspots where dimensions may be much larger than the largest sunspot, effects of photospheric radiation in K- and k-line formation must be even less important.)

Oscillations are observed in the chromospheric portions of umbrae of essentially all spots (Staude, TP). Times series of umbral spectra obtained in chromospheric lines (*e.g.* CaIIλ8542Å) reveal a sharp peak at periods of 170 seconds in both velocity and intensity, and the phase relation between velocity and intensity for these oscillations is distinctly different from that in the quiet sun (Lites, 1988, and references therein). The 5 minute p-mode oscillations in the photosphere do not seem to drive the 170 second oscillations in the chromosphere: in fact, for high-order p-modes, 40 - 50 percent of the incident power on a spot is absorbed (Braun et al. 1988). Although the r.m.s. velocity amplitude of the chromospheric oscillations is small at the level of formation of CaIIλ8542Å (0.26 km/sec), this is not necessarily true at higher levels: when the oscillations propagate upwards to the level of HeIλ10830Å formation, amplitudes of 11 km/sec are seen, and the oscillations may then substantially enhance the radiative losses from these levels of the sunspot atmosphere.

Analysis of the 170 second oscillations which are peculiar to sunspots provides a powerful probe of the thermal structure of an umbra (Staude, TP). The oscillations in all likelihood represent the normal modes of a magnetized atmosphere. The eigenfrequencies are sensitive not only to the local chromospheric properties but also to the run of temperature with depth in the convection zone below the visible surface: these properties determine the transmission of various modes into the overlying chromosphere. Thus, these oscillations hold the promise not only of determining which modes are permitted in the chromosphere, but also of quantifying how subphotospheric convection is altered in the presence of a strong magnetic field. This is essential information if spots are to be modelled consistently.

1.4. Properties of spots in the transition region (TR).

The properties of spots as seen in TR emission lines appear to vary with solar cycle. On the one hand, SMM images in CIV lines (close to solar maximum) show no clear signature of spots (Gurman, TP), suggesting that the gas pressure over a spot at 10^5K is comparable to that outside the spot. On the other hand, when Skylab data were obtained close to solar minimum, observations in a variety of TR lines, formed at temperatures ranging from 10^5 K to 10^6 K, indicate that areas directly over sunspot umbrae are the brightest features (by a factor of 10) in an active region: moreover, analysis of density-sensitive line ratios suggested that densities over umbrae are significantly less than in the surrounding plage (Foukal et al. 1975). From these data, it was concluded that the temperature gradient in the TR over an umbra is shallower by an order of magnitude than in the quiet sun.

SMM polarimetry indicated fields of order 1 kG at heights $z = 4000$ km (Gurman, TP): at such heights, the ambient gas pressures are so low that the field must be force-free.

Gas motions in the TR over sunspot umbrae are characterized by downflows and reduced turbulence. Downflows as large as 200 km/sec have been reported in rocket data (the HRTS instrument), although SMM data have not definitively seen analogs of these flows (Gurman, TP). As a measure of the reduced turbulence in the TR over umbrae, we may refer to the widths of TR lines observed by SMM: the widths of CIV lines over umbrae are about 15 km/sec (FWHM), compared with about 25 km/sec in quiet sun (Gurman, TP). The reduction in line width can be interpreted as a reduction in the dimensionality of turbulence from 3 (in field-free regions) to 1 (when motions are confined along a flux tube).

1.5. Properties of spots in the corona.

As a corollary of a remark which was made above concerning sub-surface dipoles associated with spots, it is in general true that, given typical solar boundary conditions, magnetic structures of dimension L on the surface must have features extending to distances of order L both above and below the visible surface. Thus, spots are associated with coronal loops with lengths $L = 10^{9-10}$ cm, i.e. $(0.01 - 0.1)R_*$. Conversely, if a coronal structure is seen to extend to altitude L, the feet of the flux tube in all likelihood occupy a patch of surface with diameter of order L.

Although magnetic fields are not usually measurable in the solar corona, sunspot loops are an exception. Field strengths on the coronal portions of sunspot loops are derivable from microwave polarization: the values extend to several hundred gauss (Lang, 1983). Radio data also yield densities on the coronal portions of sunspot loops: the values are of order 10^9 cm^{-3} (Lang, 1983), i.e. comparable to those in the ambient low corona. (This is consistent with equality of pressures in the TR.) If we use 300 G as a typical value for B on a sunspot coronal loop, Alfven speeds v_A in the coronal portions of spot loops are found to be $\approx 3\,10^9$ cm sec^{-1}. This is considerably larger than values which are often assumed for coronal Alfven speeds: however, the latter are usually based on mere guesses for "average" field strengths and "average" densities, whereas the estimate here is based on direct measurements of the coronal parameters. The time required for an Alfven wave with the above speed to cross a sunspot loop of length $L = 10^{9-10}$ cm is 0.3 - 3 seconds. Analogous time-scales for stellar loops will be discussed below (Section 2.3).

2. Properties of spots and associated inhomogeneities in cool stars

The discussion here parallels that in Section 1, with comments on the connection between solar and stellar phenomena.

2.1. Properties of spots in stellar photospheres.

Extraction of properties of spots on other cool stars has become quite reliable since high signal/noise multicolor photometry and Doppler imaging have been combined to yield complementary information (Eaton, TP: Cameron, TP). The extraction of spot positions and sizes from photometric data is particularly complete in the case of eclipsing systems, and quantitative criteria can be derived to enable maximal extraction of information without over-interpretation (Zeilik, TP). As a result, the solutions for spot properties are much less subject to problems of non-uniqueness than was the case a few years ago.

2.1.1 Positions. In certain stars, spots appear to cluster around certain "active longitudes", but in other stars, there seems to be no clear preference for one longitude over another. Thus, similar to the Sun, there may be tendencies for stellar spots to have preferred longitudes, but these tendencies are not universally in force.

In latitude, stellar spots are seen at essentially all values, including the vicinity of the poles. This behavior is in striking contrast to the Sun. However, in one particular star for which sufficient data are now available, a crucial aspect of the star has been determined reliably for the first time, namely, the latitudinal differential rotation. The result is also in striking contrast to the solar case, not only as regards the magnitude of the differential rotation, but especially in its sign: Ω_{eq} has been found to be *less* than Ω_{pole}, (Strassmeier et al. TP), *i.e.* the polar regions rotate more rapidly than the equatorial regions. In such a star, field lines would be wrapped more quickly near the pole than near the equator. Hence the conversion of poloidal to toroidal flux would proceed most efficiently at *high* latitudes. If Babcock's (1961) scenario of the solar dynamo were to be applied to such a star, spots would be expected to be buoyed to the surface predominantly at much higher latitudes than in the sun. Therefore, the existence of high latitude spots on a star does not necessarily indicate that the spot-generating processes are physically different from those at work in the Sun.

2.1.2 Dimensions. The smallest spot which can be reliably detected on a star has an angular diameter of about 4° (Zeilik, TP). There appears to be no reported upper limit on stellar spot diameter. In terms of angular size, starspots therefore overlap with the distribution of spots on the Sun.

In the case of stellar spots, photometric data alone cannot distinguish between a single large spot and an aggregate of small spots. However, on the basis of the discussion above on the vertical extent of flux ropes (Section 1.5), it can be argued that small spots (no matter how many there are on the surface of a star, or how much of the surface area they occupy) will be capable of producing only slight extensions vertically upwards into the corona, whereas a single large spot with L of order R_* will be accompanied by a coronal loop extending to a height h of order R_* above the surface. Evidence from UV and X-rays (see Section 2.3 below) indicates that in fact, spotted stars exhibit extensive coronal counterparts ($h \sim R_*$): this suggests that what appears in the photometry as a darkening is probably *not* composed of a collection of small spots, but is more likely to be attributable to a single large spot. This inference indicates the advantage of multi-wavelength studies of inhomogeneities in cool stars.

2.1.3 Radiant intensities: temperatures. The earliest searches for color changes between spotted and unspotted hemispheres of cool stars were performed in the B and V bands. However, these bands are at wavelengths which are too short to be useful for determining temperature differences between a star with T_{ph} of, say, 5000 K and a spot with $T_{sp} = 4000$ K: in B and V, the spot is essentially black. Multicolor photometry at longer wavelengths is essential if spot temperatures are to be measurable. Such data indicate temperature differences $\Delta T = T_{ph} - T_{sp}$ of 700 - 1900 K with uncertainties of at least 100 K (Eaton, TP; Zeilik, TP). However, observational selection may favor detection of $\Delta T \approx 1100$ K (Byrne, TP). The stellar values of ΔT appear somewhat smaller than ΔT_{eff} for sunspot umbrae. However, there may be no serious inconsistency: whereas the solar figure refers to temperature differences between spatially resolved photosphere and umbra, the stellar results represent a weighted average over umbra and penumbra. In view of this, it must be regarded as rather surprising that values of ΔT as large as 1900 K (i.e. almost equal to the solar value) have ever been detected (but see Eaton, TP): if penumbrae contribute at all to the stellar data, the values of ΔT quoted above must be regarded as lower limits to the true values. Thus, the umbra of a spot on an M dwarf with $T_{eff} = 4000$ K might have T_{eff} less than 2000 K, and a temperature minimum of less than 1600 K. Molecular equilibria in such a spot might be highly unusual. And in view of the high densities in a dwarf (which are orders of magnitude larger than those in a supergiant, where astrophysical dust grain formation is usually discussed), there might even be a chance for dust grains to condense out of the gas. The interaction between such grains and the stellar corona can be discussed by analogy with the solar case (Mullan, 1977): the grains would probably be rapidly destroyed by sputtering.

2.1.4 Magnetic fields. There are no clean measurements of fields in umbrae of starspots: the fields which are extracted from spectra are weighted averages over the visible disk, and are heavily biassed against the dark umbrae. The largest reported fields in cool stars (up to 5-6 kG: Saar, 1987) are therefore lower limits on the strengths in the umbrae. Hence, starspot umbrae have fields which are definitely stronger than in the Sun. However, a correlation between photospheric gas pressure and stellar field strengths (Saar, 1987) suggests that lateral force balance exists in the photospheric regions of spots on both the Sun and stars (Skumanich, TP). This is an important conclusion in the context of starspot modelling.

2.1.5 Lifetimes. Spots on stars have a range of lifetimes. For example, in AB Dor, the lifetime is several years (Innis, TP), and there may be even longer lived spots for which evolutionary effects are too slow to have been noticed so far. At the other extreme, a spot in UX Ari is reported to have lasted no more than one month (Vogt and Hatzes, TP). In AB Dor, the spot dimensions can be estimated to be about 2×10^{10} cm, while in UX Ari, the spot diameter is of order 6×10^{10} cm. If these lifetimes are interpreted in terms of diffusive decay, the equivalent diffusivities are $\kappa \approx 10^{12.5}$ (AB Dor) and $10^{15}\,\mathrm{cm^2\,sec^{-1}}$ (UX Ari). The lower of these numbers coincides remarkably well with the solar value estimated above, while the value for UX Ari is definitely larger than solar: if the UX Ari diffusivity were to occur in the sun, a spot with diameter 30 arcsec (a "large" spot) would survive for only one hour!

2.2. Properties of starspots in chromosphere and transition region.

Doppler imaging techniques in the upper atmospheres of spotted stars are cruder than in the optical, and there are fewer constraints on the interpretation: solutions are therefore not usually unique (Neff, TP). The solutions which do exist suggest that in certain RS CVn systems, cool spots exist covering several tens of percent of a hemisphere, analogous to what has been deduced on the photosphere. However, the chromospheric data also suggest that the leading hemisphere of both stars in binaries is brighter than the trailing hemisphere (Neff, TP; Katsova, TP), and one hemisphere has larger line widths than the other. Moreover, there appear to be large systematic downflows (of order 100 km/sec) over parts of the surface. This raises the concern that rotation may not be the only, or even the dominant, mechanism for broadening the spectral lines in such stars: if this conclusion is correct, Doppler imaging will be compromised. Difficulties of interpretation arise also because of the binary nature of these systems: gas flows between components may serious significantly distort the spectra (Katsova, TP).

2.3. Properties of starspots in coronal emission.

Various lines of evidence point to the existence of spatially extended structures (with scales up to a stellar radius, R_*) in cool star coronae. For example, the absence of hot X-ray eclipses in RS CVn systems suggests emission from regions at least as large as the radius of the K subgiant (Linsky, TP). And analyses of X-ray flare light curves from a variety of giants and dwarfs suggests loop lengths of at least a few tenths of R_* (Haisch, 1983; Tagligerri, TP; Kellett, TP; Cheng, TP). Moreover, microwave emission from RS CVn systems requires source dimensions of order R_* to interpret the observed fluxes (Linsky, TP). Density-sensitive line data from RS CVn systems, however, suggest also the existence of smaller loops, with dimensions less than $0.1\,R_*$ (Landini et al. TP). Thus, as far a dimensions are concerned, stellar coronal loops range in size from less than $0.1R_*$ to R_*. This range overlaps with the solar range, although it certainly extends to the large side of the solar range.

The existence of large coronal structures suggests (see Section 2.1.2) that the surface inhomogeneities in these stars consist of single large flux ropes.

Densities in coronal loops in spotted stars have been derived for loops which have flared: for RS CVn stars, they range from $10^{10.4}$ to $10^{11.5}\,cm^{-3}$ (Landini et al. TP; Kellett, TP). These are comparable with values for solar flaring loops, $10^{10.5-10.8}\,cm^{-3}$.

Coronal fields in RS CVn systems can be derived from microwave polarization: values of 220 G are typical (Linsky, TP). Lower limits on magnetic field strengths in coronal loops can be obtained if loops confine the X-ray plasma. Here also, values of 70 - 550 G have been reported in RS CVn stars and in flare stars (Haisch, 1983; Kellett, TP; Tagliferri, TP). These values are comparable to values observed in sunspot loops: however, the latter are much closer to the surface, and therefore the surface fields at the feetpoints of stellar coronal loops may be larger than the solar value by factors as large as 10 (Mullan, 1984).

Combining the larger fields and larger spot sizes in cool stars, we find that the magnetic fluxes in starspots Φ may exceed solar values by several orders of magnitude. This may be viewed as a measure of how much more efficient dynamo action is in spotted stars.

Using the loop sizes, densities, and field strengths quoted above for stellar loops, we find that the Alfven crossing time is of order 50 - 100 seconds. This is interestingly close to periodicities reported in low-amplitude optical variability in flare stars and correlated with X-ray luminosity (Andrews, TP). Individual loops in flare stars may be so large as to dominate the coronal emission, and free-free emission from such a loop could contribute detectably to the optical luminosity of the star: in such a case, Alfven waves bouncing back and forth on the loop might modulate the emission measure of the loop by altering the volume or by precipitating trapped electrons each time the wave packet bounces. In either case, there is the possibility of imposing a small-amplitude modulation on the optical luminosity. Forthcoming pointed X-ray observations of Andrews' target stars will help in determining if these periodicities are in fact associated with coronal loops: if it can be shown that they are, this will provide valuable information for theories of coronal loop heating.

A feature of certain cool star coronae which seems at first sight quite different from the Sun is the presence of cool material (cold enough and dense enough to scatter Hα detectably) at distances of a few R_* above the surface (Foing et al. TP; Young, TP). In AB Dor, the mass of the cool hydrogen in the atmopshere at any one time may be as large as 10^{19} gm. Cool material (with temperatures of no more than 2×10^4 K) is also present in the wind from a cool dwarf (Mullan et al. 1989). Although there are no reports of solar wind gas as cool as the above values, material in the solar wind can be significantly cooler than normal if it is thermally isolated from ambient (by, e.g. magnetic shielding). It is possible that this process occurs more extremely (aided by radiative cooling in the denser wind) in rapidly rotating cool stars where efficient dynamos ensure copious magnetic structures (Mullan et al. 1989).

2.4. Are starspots correlated with plages and coronal loops?

Is the tight correlation of spots and plages which occurs in the Sun also seen in cool stars? The evidence is mixed. Among the reports at this Colloquium, solar-like behavior (i.e. maximum emission close to spots) has been reported for Ca K emission in AB Dor (Innis, TP), for Mg k emission in CC Eri (Byrne et al. TP), and for microwave emission in AB Dor (Budding et al. TP). However, in CC Eri, emission lines formed in the transition region ($T = 10^5$K) show no spatial correlation with the spots (Byrne et al. TP), perhaps because of flux-rope spreading. And Hα emission in AB Dor is not correlated with spot location: however, in this case, Hα is contaminated by absorption features in the corona, and this complicates the interpretation. Also in contrast to solar spots (where LiI lines are greatly strengthened compared to the photosphere), there is no evidence for such strengthening in a sample of cool spotted stars (Pallavicini et al.

TP). However, before we infer from this that non-solar behavior is at work, we must remember that, in the sun, chromospheric emission from plages causes LiI lines to be filled in by up to 50% (Giampapa, 1984): in view of this, it is possible that when one integrates over the entire visible hemisphere of a spotted star, the weighted average of plages (with LiI lines filled in chromospherically) and starspots (with deep LiI lines due to cooling) is such as to cancel each other. The possibility of cancellation between spots and plages is not at all implausible when we recall that in the sun, the bolometric effects of spots and plages (faculae) are almost exactly equal and opposite (Section 1.1.3 above).

2.5. Solar-stellar connection

Overall, the evidence presented in this Section suggests that solar analogies still provide useful guides for interpreting stellar phenomena.

3. Spot models

The process of constructing a model for a cool magnetic spot in the convection zone near the surface of a star is sufficiently complicated that no complete depth-dependent and radial-dependent model yet exists.

3.1. Momentum equation

If one insists only that the momentum equation be satisfied, progress can be made if one specifes the magnetic field distribution (Skumanich, TP). For example, a dipole placed at a certain depth below the visible surface gives rise to a determinate field distribution which in turn, in a magnetohydrostatic model, leads to a determinate run of pressure inside the spot as a function of r and z. Below the surface, in depth ranges where the field B may be considered as uniform, lateral force balance ensures that the gas pressure inside the spot is less than that outside by an amount $\Delta p = B^2/8\pi$. If the flux rope is completely evacuated, lateral force balance requires that the field strength at a level where the undisturbed photosphere has gas pressure p_{ps} is $B = \sqrt{8\pi p_{ps}}$. Saar (1987) has reported that this in fact yields an acceptable scaling for measured field strengths in a sample of cool stars.

Above the surface, where gas pressures become much smaller than $B^2/8\pi$, the field must become essentially force-free.

3.2. Energy equation.

In order to model the cooling in a spot, an energy equation is necessary. Reduction in the efficiency of convective heat transport is almost certainly an essential component in spot structure: during the stage when a new pore is forming near an old one, the granulation between the pores is observed to become disturbed and the intergranular lanes become darker (BL:68). This suggests that a rising flux tube interferes with the normal granule flows.

The earliest models of convection in the sun were one-dimensional mixing length models with purely vertical motions. In a vertical magnetic flux tube, vertical motions are in no way inhibited, and as a result, there exists no plausible way to modify the model of the photosphere in order to obtain a model of a spot. (Reductions of the mixing length parameter have been proposed, but these are purely *ad hoc.*) Thus, in order to make any progress in constructing a spot model, at least two dimensions must be incorporated in order to allow for overturning (i.e. horizontal flows): magnetic interference with the horizontal flows provides a possible physical mechanism for reduction in convective efficiency in a vertical flux tube.

Convection cannot be inhibited entirely in a magnetic field because the electrical conductivity never becomes infinitely large: there is always some residual convection. Interactions between the residual convection and the strong field in an umbra are likely sources of MHD waves, and these may contribute to spot cooling also.

In the context of Boussinesq convection in cells (i.e. incorporating horizontal exchange of gas between upward and downward flows), umbral models were constructed some years ago incorporating not only the reduction in convective efficiency due to magnetic effects (allowing for finite conductivity in a partially ionized gas transverse to the field) but also incorporating the generation of Alfven waves by the residual convection (Mullan, 1974a, b). These models suggest that fields in spots on cool stars may be as large as 10 - 20 kG, and that the cooling may be as large as $\Delta T = 2000\,\mathrm{K}$. Neither of these conclusions is inconsistent with the properties of starspots as summarized at this Colloquium.

More recently, a 2-D compressible study of magnetoconvection (Weiss et al. 1990) has found that the properties of convection in an umbra depend on the ratio ζ of magnetic to thermal diffusivity. If ζ exceeds unity at all depths, the magnetic field diffuses from the gas faster than heat does, and so the gas behaves essentially as if it were nonmagnetic: convection occurs by overturning. Conversely, if ζ is less than unity at all depths, magnetic effects interfere significantly with convective heat transfer, and overturning is no longer permitted: instead, the convection becomes oscillatory. In a real sunspot, the value of ζ probably changes from less than unity at the top to greater than unity at great depths: this leads to complicated couplings between inhibited convection near the surface and overturning flows at depth.

Quantification of the extent to which magnetic fields reduce the efficiency of convective heat transfer in the sun may eventually be achieved by modelling of chromospheric oscillations in umbrae (Staude, TP).

3.3. Effects of a spot on stellar structure and luminosity

Although the problem of creating a physically consistent model of a spot has not yet been solved satisfactorily, it is worth considering the inverse problem. The inverse problem of spot cooling can be stated as follows: given that a cool spot (i.e. an insulating plug) has appeared on the surface of a star, what effect does it have on the structure of the star as a whole? The answer depends on the time-scale over which the plug (spot) exists (Spruit, TP). In steady state, where the spot lasts at least 10^7 years in the sun, the plug causes heat to pile up beneath it, but the excess temperature rises only to about 1 degree: the diffusivity of the turbulent convection is so large that even this small temperature difference suffices to transport the entire missing flux horizontally from the base of the spot. In this case, the blocked energy reappears as a slight excess of emission from a large area of surface around the spot. Even with a large sunspot, the appearance of a detectable "bright ring" around the spot is very unlikely.

Steady state is not relevant in the case of a spot which appears and disappears on time-scales of a few months. In the latter case, when a spot appears, the luminosity of the star falls by $-\Delta L$, and during the lifetime τ of the spot, an excess energy $\Delta E = \Delta L \times \tau$ is deposited in the convection zone. The total thermal energy U of the convection zone is very large, of order $L \times 10^5 yrs$. Hence, for a spot which blocks $10^{-3}L$ of the star's luminosity L for a time of a few months, the excess energy deposited in the convection zone is only one part in 10^8 of its thermal energy. Such a small change can be absorbed in the thermal reservoir with only minute changes in structure: the star should decrease in brightness by ΔL as long as the spot lasts (Spruit, TP).

However, the above discussion neglects the effects of faculae. In the sun, faculae are observed to contribute an excess of radiation which not only compensates for the flux blocked by spots, but even overcompensates, so that the sun emits more luminosity when it is maximally spotted (Lean, TP). Faculae may be composed of flux tubes which are smaller in diameter than a typical granule, and are therefore unable to impede convection in the manner of a pore. With Wilson depressions of a few hundred kilometers, faculae may provide the means for the Sun to emit radiation from regions which are not normally visible beneath the surface, thereby enhancing the radiative efficiency (Spruit, TP).

It is important to ask the question: *How is the bolometric luminosity of a cool star altered (if at all) between epochs when spots are present and absent?* In the course of the Colloquium, no answer was given to this question. *In view of what is now known about the solar irradiance as a function of spottedness, this question must be regarded as one of highest priority in starspot physics.*

4. What determines spot lifetimes?

Since Cowling's work on the decay of sunspot fields almost 40 years ago, it has been known that the dissolution of a sunspot cannot be due to ohmic dissipation in stationary material: the time-scales for such dissipation are much longer than the observed lifetimes (BL:270). In view of the proportionality betwen spot areas and lifetimes observed in the sun, it is more natural to ascribe the decay of a spot to a diffusive process associated with turbulent convection $\kappa_t \sim \lambda v$ (where λ is a mean free path and v is an r.m.s. turbulent velocity). Near the surface of the sun, $v \approx 1 - 3\,\mathrm{km/sec}$. The choice of λ is not immediately obvious: the local pressure scale height H_p is 100 - 300 km, while the preferred scale of granules L_g is about 1000 km. Setting λ equal to H_p or L_g, we find $\kappa_t \approx 10^{12-13.5}\,\mathrm{cm^2\,sec^{-1}}$. With increasing depth into the convection zone, models suggest that κ_t will increase by perhaps another order of magnitude (Spruit, TP).

The diffusivities inferred from lifetimes of spots on the surface of the sun fall exactly in the above range. Convective conditions in the K2 dwarf AB Dor are expected to be not greatly different from solar: the fact that spot lifetimes in AB Dor also yield L^2/τ equal to solar therefore suggests turbulent diffusive decay. And in UX Ari, the reduced gravity (by a factor of almost 10) in the subgiant star means H_p is also larger by 10, and the reduced densities in the photosphere may also mean somewhat larger r.m.s. convection speeds: thus, even a value of $10^{15}\,\mathrm{cm^2\,sec^{-1}}$ may not be inconsistent with turbulent decay.

It must be regarded as rather remarkable that turbulent decay reproduces the observed lifetimes of spots quantitatively so well. After all, convection models are quite crude; moreover, the turbulent time-scales should be appropriate for purely gasdynamic structures immersed in the convection. One could imagine that the field in a sunspot might be strong enough that the spot would be able to resist the incessant "pummelling" by the eddies in the surrounding convection zone, and defend itself against these eddies. This scenario would be even more likely if spots rely on organized flow patterns for their existence: such flows have been proposed in the past for various reasons (see Mullan, 1983) although some observers doubt their existence (Maltby, TP). It may be that these flows exist around the largest and most long-lived spots, shielding them (for a time, at least) from the convective eddies.

5. Is there a minimum size for inhomogeneities?

In the photosphere of the sun, the length-scale associated with diffusivity appears to be related to convection scales, as is the minimum size of a spot (pore). Hence, in order to understand sunspots in their entirety, it is necessary to have an understanding of convection in the solar context.

It was mentioned above that, if a physically plausible model of a sunspot is ever to be constructed, the convection model must incorporate at least two spatial dimensions. Recent modelling of solar/stellar convection has progressed to the point where the codes are three-dimensional, and fully compressible. Results from these codes suggest that the local pressure scale-height is a relevant quantity in the vertical structuring of convective flows (Hossain and Mullan, 1990): horizontal scales are of the same order as the vertical. Thus, since H_p is a few hundred kilometers just beneath the visible photosphere, convection cell sizes in the photosphere should also be a few hundred kilometers: the observed granules are somewhat larger than this, but terrestrial atmospheric effects may be obscuring the true size distribution of granules. The preference for H_p in the vertical direction may be associated with sound wave trapping.

The effects of granules in creating inhomogeneities are not confined solely to the photosphere, but extend into the overlying corona. The smallest known inhomogeneities in the solar atmosphere have scale-sizes of 1 km in the corona: if these are associated with coronal heating (Mullan, 1990), their analogs should also exist in cool stars.

References

Athay, R.G. Dere, K.P. 1990, ApJ 358, 710.

Ayres, T. 1981, ApJ, 244, 1064.

Babcock, H.W. 1961, ApJ, 133, 572.

Bai, T. 1988, ApJ, 328, 860.

Braun, D.C. et al. 1988, ApJ, 335, 1015.

Bray, R.J. Loughhead, R.E. 1979, *Sunspots* Dover, New York.

Foukal, P.V. et al. 1974, ApJ, 193, L143.

Giampapa, M. 1984, ApJ, 277, 235.

Haisch, B.M. 1983, in *Activity in Red Dwarf Stars,* eds. P.B. Byrne and M. Rodono (Reidel), p. 255.

Hossain, M., Mullan, D.J. 1990, ApJ, 354, L33.

Lang, K. 1983, in *Activity in Red Dwarf Stars,* eds. P.B. Byrne, M. Rodono, Reidel, p. 331.

Lites, B.W. 1988, ApJ, 334, 1054.

Mullan, D.J. 1974b, ApJ, 187, 621

Mullan, D.J. 1974b, ApJ, 192, 149.

Mullan, D.J. 1977, AA, 61, 369.

Mullan, D.J. 1983, in *Activity in Red Dwarf Stars,* eds. P.B. Byrne and M. Rodono (Reidel), p. 527.

Mullan, D.J. 1984, ApJ, 279, 746.

Mullan, D.J. 1990, AA, 232, 520.

Mullan, D.J. et al. 1989, ApJ, 339, L33.

Saar, S.H. 1987, in *Cool Stars, Stellar Systems, and the Sun,* eds. J. Linsky, R.E. Stencel (Springer), p. 10.

Weiss, N.O. et al. 1990, MNRAS, 245, 434.

Discussion: Mullan

Byrne I believe that we can say something on the matter of changes in bolometric flux associated with large spots on stars. In the case of the 1986 spot configuration on II Peg JHKL observations have been made (Lazaro and Arevalo, AA 1990) to complement the UBVRI data from our group. These show a very small amplitude decreasing with increasing wavelength, with a maximum amplitude in R or I-band.

Linsky I suggest that, in your assessment of the validity and usefulness of the solar-stellar connection, you make a distinction between physics and phenomenology. The physics, of course, must be the same on the Sun and stars, but the phenomena may differ due to different boundary conditions. For example, the magnetic flux in the photospheres of the RS CVn and dMe stars is 10-100 times that of the Sun, as I mentioned in my talk. This can lead to phenomena like very large starspots and large rates of coronal heating that are well beyond the range of solar phenomena.

Poster

Papers

THE AUGUST 1989 SPOT LIGHTCURVE OF II Peg AND RELATED FLARE ACTIVITY

S. Avgoloupis,[1] J.G. Doyle,[2] L.N. Mavridis,[3] J.H. Seiradakis[1] and M. Mathioudakis[2]
[1]Department of Astronomy, University of Thessaloniki, GR-54006, Thessaloniki, Greece
[2]Armagh Observatory, Armagh BT61 9DG, N. Ireland
[3]University of Thessaloniki, GR-54006, Thessaloniki, Greece

1. Introduction

The RS CVn stars are a group of active binaries in which at least one component shows evidence of spots and other phenomena associated with stellar activity. II Peg (= HD224085) is a single-line spectroscopic binary of spectral type K2-3 V-IV and a period of approximately 6.72 days. It shows wavelike variations usually interpreted as due to cool spots. Based on IUE observations II Peg appears to have a very high rate of activity, flaring perhaps every 10 hr (Doyle *et al.* 1983). However in the optical regime no flare on this star had been reported before. In the present study we present 'quiescent' state variations of II Peg in August 1989. We also show that its rate of flaring activity in the U band is very high, this high activity been associated with the 'quiescent' state spot variations.

2. Observations and Data Reduction

The photoelectric observations of II Peg were carried out between 2 and 17 August 1989 at Stephanion Observatory, Greece, using the 30 in Cassegrain reflector of the Department of Geodetic Astronomy of the University of Thessaloniki. The telescope, photometric system and

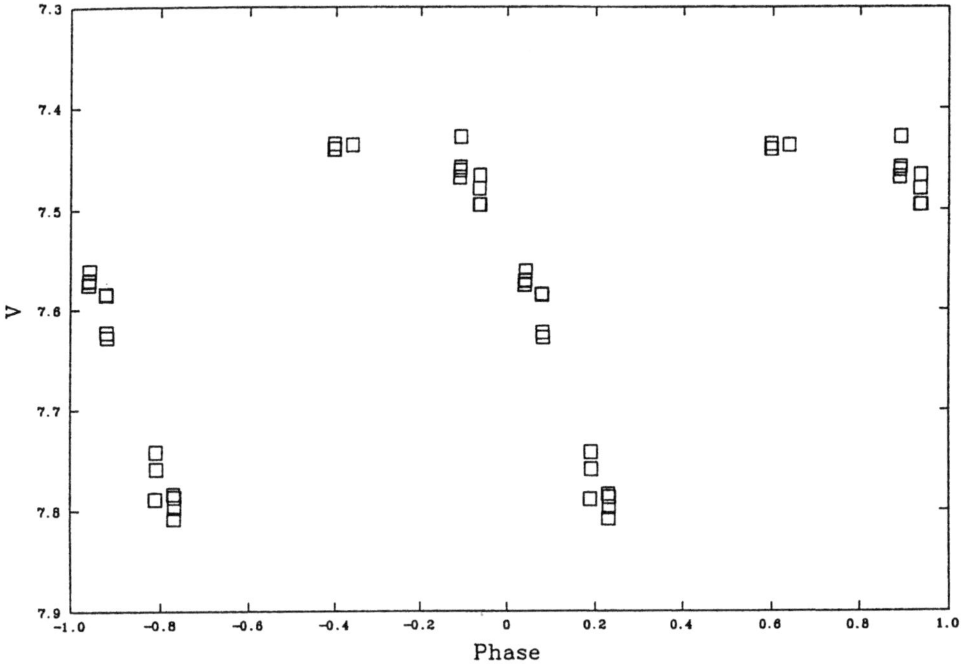

Figure 1. The 'quiescent' state V magnitude light-curve of II Peg; JD0 = 2442021.7246, $P = 6.724464$ days.

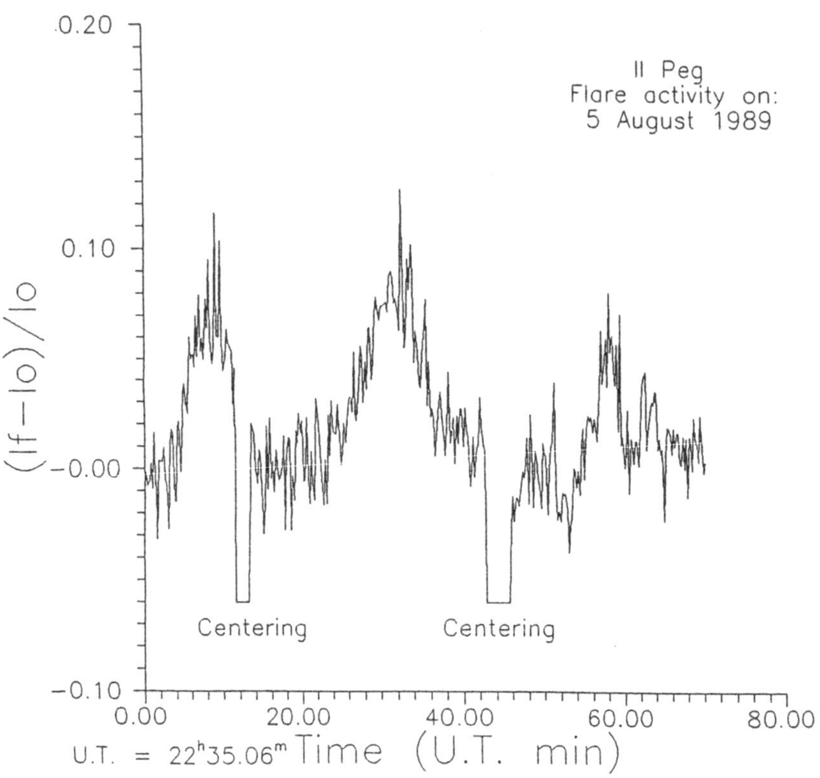

Figure 2. U band flares on II Peg on 5 August 1989 during an interval of 71 min, showing the high activity level of the star in the optical region

the observing procedure has been described by Mavridis *et al.* (1982). The equations for transformation of the instrumental system to the UBV system were derived from observations of standard stars. Part of the observations were carried out during a co-ordinated, multiwavelength campaign (Doyle, 1989). Our observations of II Peg consist of *(a)* continuous monitoring in the U band (2-8 and 15-17 August 1989) and *(b)* three-color photometry in the UBV system (3-14 August 1989). Some of our results during the international campaign have been presented in previous works (Doyle *et al.* 1990).

3. Results and Discussion

The V magnitude lightcurve of II Peg with an amplitude $\Delta V \approx 0.4$ is presented in Fig. 1. Phases were calculated using the ephemeris of Raveendran *et al.* (1981) *viz.* JD = 2442021.7246 + 6.724464 E. Observations of II Peg have proved especially prolific during the last few years. A few years ago its light-curve typical amplitude was about 0.12 mag. (Andrews *et al.* 1988). After 1985 the star became fainter and its light curve amplitude increased to about 0.4 mag. Our 1989 data confirm the results of Byrne (1986) and Cutispoto *et al.* (1987) regarding amplitude and phase of light minimum. Thus following the two-spot model of Andrews *et al.* (1988) we note that the position and contribution of Spot 1 to the star's light-curve has remained rather stable between 1981 and 1989, whereas Spot 2 seems to be responsible for the observed variations in the light curve. The mean V magnitude of the star has not changed much during that interval. On the contrary, Spot 2 (around phase 0.2) has become larger and/or cooler, resulting in the large amplitude of the light-curve observed after 1985. The relative large variations in the (U-B) and (B-V) colors are definite. This compares with no variation in (B-V) and a small one in (U-B) in 1981 (Rodono *et al.* 1986).

Between 2 and 17 August, 1989 we observed seven flaring events during a total monitoring time of $19^h 41^m$, one of which was of unprecedentedly large amplitude (Doyle *et al.* 1990). The lightcurves for some of those flares are presented in Fig. 2. All flares occured between phases 0.5-0.05 during a monitoring time of $13^h 41^m$, that is at phases close to Spot 1, which at the time was the less dominant feature on the surface of the star.

References

Andrews, A.D., Rodonó, M., Linsky, J.L., Brown, A., Butler, C.J., Catalano, S., Scaltriti, F., Busso, M., I-Seong Nha., Oh, J.Y., Henry, M.C.D., Hopkins, J.L., Landis, H.J., Engelbrektson, S.: 1988, A&A, 204, 177

Byrne, P.B.: 1986, IBVS, No. 2951

Cutispoto, G., Leto, G., Pagano, I., Santageti, G., Ventura, R.: 1987, IBVS, No. 3034

Doyle, J.G.: 1989, IBVS, No. 3250

Doyle, J.G., Kellett, B.J., Byrne, P.B., Avgoloupis, S., Mavridis, L.N., Seiradakis, J.H., Bromage, G.E., Tsuru, T., Makishima, K., McHardy, I.M.: 1990, MNRAS, submitted

Doyle, J.G., Byrne, P.B., van den Oord, G.H.J.: 1989, A&A, 224, 153

Mavridis, L.N., Asteriadis, G., Mahmoud, F.M.: 1982, in *Compendium in Astronomy*, eds. E. Mariopoulos, P.S. Theocaris, L.N. Mavridis, Reidel, Dodrecht, p. 253.

Raveendran, A.V., Mohin, S., Mekkaden, M.V.: 1981, MNRAS, 196, 289

Rodonó, M., Cutispoto, G., Pazzani, V., Catalano, S., Byrne, P.B., Doyle, J.G., Butler, C.J., Andrews, A.D., Blanco, C., Marilli, E., Linsky, J.L., Scaltriti, F., Busso, M., Cellino, A., Hopkins, J.L., Okazaki, A., Hayashi, S.S., Zeilik, M., Helston, R., Henson, G., Smith, P., Simon, T.: 1986, A&A, 165, 135

SPOT CHARACTERISTICS OF CONTACT BINARIES

S.A. Bell, R.W. Hilditch and P.P. Rainger
University Observatory, Buchanan Gardens
St. Andrews, Fife, KY16 9LZ, Scotland

Three V light curves of BX And have been obtained with the TPT in the interval 1985-1988 as well as J and K Photometry from UKIRT and radial velocity data from the INT during 1987. A full analysis of these data and two other light curves obtained during 1976 (Samec *et al.* 1989) and 1981/82 (Rovithis and Rovithis-Livaniou 1984) suggest that BX And is a marginal-contact binary with a hot spot on the secondary component (Bell *et al.* 1990a). The parameters describing the spot itself are its radius and over-temperature defined as the excess temperature of the spot over that of the surrounding photosphere. The spot has been assumed to be circular for this analysis and has very sharply defined edges in the temperature domain such that across the boundary of the spot the temperature increases as a step function from that of the photosphere to that of the spot itself. The spot is ≈ 1200 K hotter than the surrounding photosphere with a diameter of $\approx 72°$ (20% of the star) centred at the substellar point. A schematic diagram of the system configuration is shown in Fig. 1.

With the exception of the 1981/82 data, the spot size ranges from 30° to 38° and the spot over-temperature from 1100 K to 2200 K. The solutions for the 1985 TPT data and the UKIRT data should be regarded with some caution as the light curves are not well defined on either side of secondary minimum. The 1986 and 1988 TPT data and that taken during 1976 do show good agreement. However, it has proved impossible to obtain a satisfactory solution to the 1981/82 data which reflects a larger anomalous luminosity in the ingress to and egress from secondary minimum. This could be interpreted as a larger and/or hotter spot caused by enhanced mass transfer, an explanation which may be supported by the scatter in the photoelectric (O - C)

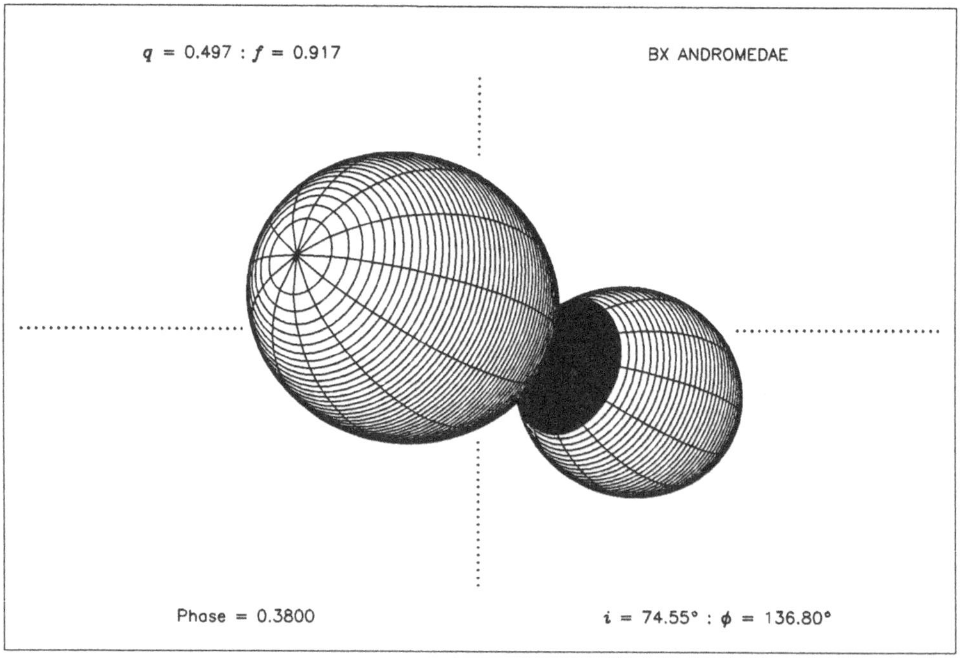

Figure 1. A schematic diagram of BX And at $\varphi = 0.38$.

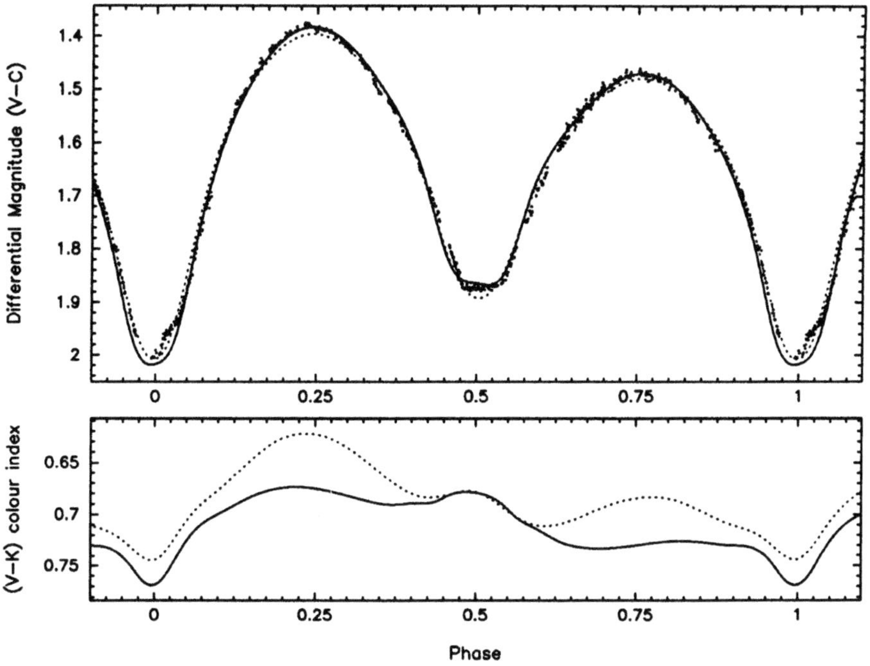

Figure 2. Spot modelling of the light curve of AG Vir.

residuals at that time. It seems likely that the period variations exhibited by this system can be linked with mass transfer between the components which subsequently have a direct effect on the spot radius and/or temperature and ultimately on the shape of the light curve. New photometric and spectroscopic observations of the eclipsing binary AG Vir are also presented. Medium resolution spectroscopy has allowed the measurement of velocities for the secondary component for the first time. A full analysis of the spectroscopic and photometric data has been made (Beli *et al.* 1990b) which suggests that the system is in either a marginal state of contact or a deep contact configuration depending on the type of spot model invoked. It is difficult to see why a hot spot should exist on either component perpendicular to the line of centres of the stars as there is no conclusive evidence for ongoing period variations which would support the presence of mass transfer between the components and there is no room in the system for an accretion stream. Cool spots appear to be the more likely option, since the system is composed of two rapidly-rotating stars with convective envelopes and substantial magnetic activity can be expected of which cool starspots are the most readily observed manifestation (*e.g.* Rucinski 1986).

The 1989 TPT V observations of AG Vir and generated models for hot and cool spots located on the primary component are plotted in the upper half of Fig. 2. The lower half of Fig. 2 shows the *predicted* (V - K) colours for these models. In both sections of the diagram the solid line refers to the deep contact/cool spot model and the dotted line to the marginal contact/hot spot model. The corresponding models for spots on the secondary component are very closely similar. Doppler imaging techniques may indicate the true location of the spot and simultaneous infrared and optical photometry will establish the temperature of the spot.

It would appear that BX And is a good example of how an area of enhanced luminosity driven by mass transfer can influence the shape of the light curve. On the other hand, AG Vir constitutes an excellent example of the expected manifestations of surface spot activity on a light curve. It also demonstrates the ease with which the spot phenomenon can be invoked to explain the appearance of a light curve and to yield conflicting results.

References

Bell, S.A., Rainger, P.P., Hill, G., Hilditch, R.W.: 1990a, MNRAS, 244, 328

Bell, S.A., Rainger, P.P., Hilditch, R.W.: 1990b, MNRAS, in press

Rovithis, P., Rovithis-Livaniou, H.: 1984, Ap&SS, 105, 171

Rucinski, S.M.: 1986, in *Instrumentation and research programmes for small telescopes*, IAU Symp. No. 118, ed. Hearnshaw, J.B. & Cottrell, P.L., Reidel, Dordrecht, Holland, p. 159

Samec, R.G., Fuller, R.E., Kaitchuck, R.H.: 1989, AJ, 97, 1159

(l to r) S. Bell explains to A. Linnell that the Scottish Summer weather is not normally so good either!

OBSERVATIONS OF AB Dor AT 8.4 GHz AND INTERPRETATION

E. Budding[1], A. Burgess[2], S. Chan[3] and O.B. Slee[3]
[1]Carter Observatory, Wellington, NZ
[2]Department of Astrophysics, University of Sydney, Australia
[3]Australia Telescope National Facility, Epping, NSW, Australia

1. Introduction

AB Dor (=HD36705) is a relatively rapidly rotating star at a distance of 27 ± 7 pc. The rotation period is 0.51 day. Absence of any clear indications of a radial velocity curve, together with a detected high lithium abundance, provide compelling reasons to believe that the star is indeed single. It is probably a young object (age $\sim 7\,10^7$yr), in the final stages of core contraction down to the Main Sequence.

The star has a history of 'activity', starting with early reports on chromospheric emissions, and continuing with various photometric studies showing 'waves' of a kind associated with RS CVn. Available space limits a full citation list here, but for a recent review see *e.g.* Innis *et al.* (1988). Comparisons can also be made with the radiometric data at 8.4 GHz of Slee *et al.* (1986).

The star has usually shown photometric evidence of at least one major spot, or spot group and more commonly the picture is interpretable in terms of two large ($\gtrsim$ 10 deg) 'spots', separated by $\sim$180 deg, so that the optical light curve tends to show a double-humped pattern. This pattern has the absence of very good periodicity which is associated with the 'migrating waves' of RS CVn stars and indeed it is not necessarily always there (*cf.* Anders, 1990).

Anders also inferred the presence of some general background maculation. This point is not obvious, however, since the Sun redistributes the total outflow of radiation so that the mean unspotted photosphere tends to slowly brighten during periods of increased sunspot number (Hudson, 1988).

2. Results and main points summary

Observations of AB Dor were made using the 64 m antenna of the Australia Telescope National Facility at Parkes, starting on the evening of 10 Jan 1990, and continuing, with $\sim$ 11 hr runs through that night and each of the four successive ones. This took place as part of an international campaign organized by A. Cameron Collier.

With one exception, the nights were generally good and show definite emission at 8.4 GHz. There appears to be a double humped trend, with peaks of about 15 mJy centered at phases of around 0.0 and 0.5, and an amplitude of about 10 mJy. There is an indication that the second maximum peaks somewhat lower than that of phase $\sim$0. The ephemeris is that used by Anders (1990). This pattern in the background flux appears to follow, in an inverted sense, roughly contemporaneous optical photometry (Anders, private communication; Banks *et al.* 1990) which show depressions, probably due to large spotted regions of the photosphere, at similar, though centered somewhat later (by about 0.1), phases.

The field lines associated with the regions of photospheric field concentration (the natural explanation for the maculation effects) will create extended loop structures above, where hot plasma may be magnetically confined — by analogy with the solar corona. The sources of the double humped microwave background radiation can be interpreted in such a scenario. The loops are essentially present, though in different projections, at all phases. The evidence of this run indicates some preference for a second harmonic term in the sectorial distribution of the underlying field structure.

At about 08:00 UT on 14 Jan a remarkable flare was observed. The distinct characteristic sudden rise of the event (rise time <3 min) is noticeable. This impulsive feature invites comparisons with phenomenologically similar solar flare instabilities, though the AB Dor event (solid angle $\sim 10^{-14}$ that of the Sun) has to be much stronger than even the strongest solar flares.

The sudden rise indicates an acceleration mechanism capable of accelerating electrons to energies $\sim$ 1 MeV in a minute or so. Such processes are observed during flares in the lower solar corona, as evidenced by fast rise times in both microwave and hard (> 30 KeV) X-rays, and may originate in a similar confined source region (Dulk, 1987). The main differences must lie in the much higher field strengths, which can be expected for AB Dor in view of the enhanced scale of general 'activity' diagnostics, as well as the numbers of electrons involved in the burst. A more detailed model and discussion of this flare will be presented elsewhere.

References

Anders, G.: 1990, IBVS No 3437

Banks, T., Kilmartin, P., Budding, E.: 1990, *Jan 1990 Photometry of AB Dor*, preprint

Dulk, G.A.: 1987, in *Cool Stars, Stellar Systems, and the Sun*, eds. J.L. Linsky and R.E. Stencel (Springer-Verlag), p. 72

Hudson, H.S.: 1988, ARA&A, 26, 173

Slee, O.B. *et al.* (6 authors): 1986, PASA, 6, 312

OPTICAL PHOTOMETRY & UV SPECTROSCOPY OF CC Eri IN Nov 1989

P.B. Byrne[1], D.J. Agnew[1], G. Cutispoto[2], D.W. Kilkenny[3], J.E. Neff[4], P.M. Panagi[1]

[1]Armagh Observatory, Armagh BT61 9DG, N. Ireland
[2]Osservatorio Astrofisico di Catania, Catania I-95125, Italy
[3]SAAO, P.O. Box 9, Cape Town 7945, S. Africa
[4]NASA Goddard Space Flight Center, Maryland MD20771, USA

1. Introduction

CC Eri (= HD16157) is a 1.56 day spectroscopic binary with a mass ratio ≈ 2 (Evans 1959). The primary is either K7Ve or M0Ve and corotates with the orbital motion. Therefore it is the third fastest rotating solar neighbourhood M dwarf known. It is a flare star (Busko and Torres 1976) and a BY Dra variable (Evans 1959). It is a powerful X-ray source with an X-ray luminosity, $\log L_X \approx 29.67\ \mathrm{erg\,s^{-1}}$, making it one of the two X-ray brightest solar neighbourhood M dwarfs.

In spite of its many interesting properties, CC Eri is a relatively neglected object observationally. Its ultraviolet spectrum has been observed only once by the IUE satellite (Byrne *et al* 1980) and then only a single long-wavelength (≈ 1900 - 3200 Å) spectrum was obtained. During 2 - 4 November 1989 we mounted an intensive campaign of UV and optical spectroscopy and optical photometry over almost one-and-a-half rotations of CC Eri. Here we report on the results of the UV spectroscopy and the optical photometry.

2. Observations

The optical photometry was carried out at the South African Astronomical Observatory's 1 m telescope at Sutherland during November and at the European Southern Observatory's 50 cm telescope at La Silla during early December 1989. The resulting light curve is shown in the top panel of Fig. 1. A low amplitude modulation is visible in V with a peak-to-peak variation of $\approx 0^m.05$.

UV spectroscopy was obtained over a 48 hr interval on 2 - 4 November 1989 using both the long-wavelength (≈ 1900 - 3200 Å; henceforth LWR) and short-wavelength (≈ 1150 - 1950 Å; henceforth SWP) cameras in low spectroscopic resolution mode (R ≈ 300). Two spectra were taken in each image by positioning the star, first at one end of the spectrograph slit, and then at the other. The latter were reduced using the IUEDR computer programme on the Armagh Observatory's STARLINK computer system. The fluxes of the principal emission lines were derived by fitting each with a gaussian profile of instrumental width in the manner described by Byrne and Doyle (1989)

3. Results

A selection of line fluxes are plotted in Fig. 1 using the ephemeris JD = 2447129.52934 + 1.56145E. Comparing the line flux variations with the broadband V light curve we note the following.

1. There is good evidence of modulation of the MgII resonance lines in antiphase with the optical. To aid in this comparison we have fitted a mean curve to the V variation and scaled the inverse of this to the MgII variation.

2. There is a much larger scatter in the individual measurements of the MgII flux than would be expected from measurement errors alone (≈ 10%).

3. The mean MgII line flux at Earth is $\approx 4.3\,10^{-12}$ erg cm^{-2} s^{-1}, comparable to that found in 1979 by Byrne *et al* (1980).

4. The contrast in the MgII (max-to-min) is much larger, $\approx$ 40%, than in V, $\approx$ 5%.

5. The evidence for modulation of the higher temperature lines, *i.e.* SiII, CII and CIV resonance lines, is much weaker. Any variation here is dominated by scatter, which is $\approx$ 50%, much greater than typical errors of measurement ($\approx$ 10 - 20%).

6. There is evidence for two discrete flares, marked as such, in the CIV line flux plot. The corresponding points in the CII and SiII line fluxes are also marked.

7. The mean CIV line flux at Earth is $\approx 4.3\,10^{-12}$ erg cm^{-2} s^{-1}, excluding the two flare points.

4. Discussion

1. The observed mean CIV and MgII line fluxes, when extrapolated to the distance of CC Eri ($d \approx 11.4$ pc), make it the most luminous solar neighbourhood dMe star in these lines yet measured (*cf.* Byrne and McKay 1990).

2. The observed MgII variation with phase confirms earlier results that chromospheric emission is correlated in a general sense with optical spots. However, the lesser modulation of the broadband optical illustrates that the area coverage in the two is not related on a one-to-one basis.

3. The lack of detectible modulation in CIV due to the presence of a large scatter, possibly due to low level flaring, is in keeping with previous results for the stars BY Dra and AU Mic (Butler *et al* 1987).

4. The two flares radiated $2.7\,10^{31}$ erg and $1.6\,10^{31}$ erg in the CIV line alone. These are large flares by the standards of IUE observations of dMe stars. Using the scaling between CIV luminosity and bolometric luminosity derived for solar flares by Bruner and McWhirter (1988), we estimate the energy radiated by the entire outer atmosphere ($4.0 \leq \log T \leq 8.0$) of these flares as $\approx 10^{36}$ erg.

References

Bruner, M.E., McWhirter, P.: 1988, ApJ, 326, 1002

Busko, I.C., Torres, C.A.O.: 1976, IBVS, No. 1186

Butler, C.J., Doyle, J.G., Andrews, A.D., Byrne, P.B., Linsky, J.L., Bornmann, P.L., Rodonò, M., Pazzani, V., Simon, T.: 1987, A&A, 174, 139

Byrne, P.B., Butler, C.J., Andrews, A.D.: 1980, Irish AJ, 14, 219

Byrne, P.B., Doyle, J.G.: 1989, A&A, 208, 159

Byrne, P.B., McKay, D.: 1990, A&A, 227, 490

Evans, D.S.: 1959, MNRAS, 119, 526

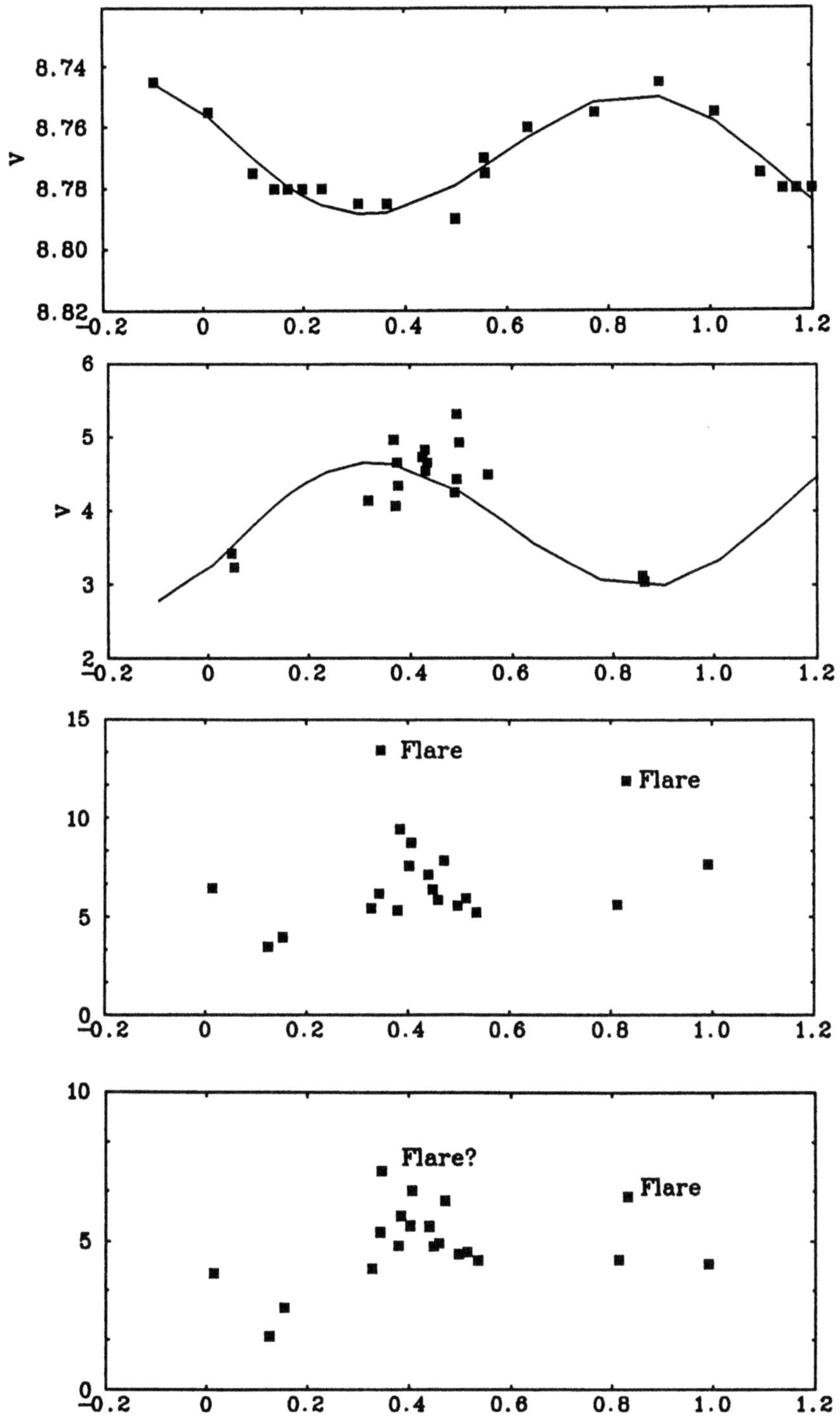

Figure 1: Variation in V (*top panel*), MgII (*2nd panel*), CIV (*3rd panel*) and CII (*4th panel*) *vs* phase for CC Eri in Nov 1989. The solid line in the second panel is explained in the text.

NUMERICAL SIMULATIONS OF FLARES ON LATE-TYPE STARS: HYDRODYNAMICS AND X-RAY SPECTRA

Chung-Chieh Cheng[1] and Roberto Pallavicini[2]
[1]Naval Research Laboratory, Washington, D.C. 20375, U.S.A.
[2]Osservatorio Astrofisico di Arcetri, Florence, Italy

1. Introduction

Recently, many flares on dMe stars have been observed with good time and spectral coverage in X-rays by EXOSAT (for a review, see Pallavicini, Tagliaferri, and Stella 1990). The X-ray emission measure of these flares is much larger than that of solar flares, ranging from 10^{50} to more than 10^{53} cm^{-3}. To understand the diverse physical parameters of the observed stellar flares, we have made a series of numerical simulations of flare loop models with different values of loop size, flare energy input, and initial loop conditions. The purpose was to study in detail the physics of flare hydrodynamic processes, such as the chromospheric evaporation, the temporal evolution of temperature and density in the flaring loop, and to determine how the hydrodynamic evolution varies in parameter space. In addition, the hydrodynamic results were used to calculate the expected X-ray spectra and X-ray light curves. The results were compared with the X-ray observations of stellar flares from EXOSAT.

2. Hydrodynamic Results

The numerical simulations of the flare hydrodynamics were done with the NRL one-dimensional flux tube code (Cheng et al. 1983; Cheng, Karpen, and Doschek, 1984). We have calculated nine loop models, with different loop sizes ($2 - 8 \times 10^9$ cm), energy input functions, and initial loop conditions. The flare energy is applied at the top of the loop at a constant input rate for 300 sec before an exponential decay with a decay constant t_d. For the model dMe star, we have assumed a stellar gravity twice that of the Sun and a chromosphere at a temperature of around 9 000 K. The initial loop model is obtained by solving the full set of fluid equations with a steady heating input to maintain the required initial loop conditions in an equilibrium state.

We find that, although the basic hydrodynamic results are similar, there are important quantitative differences in their evolutions. When the flare energy input is applied at the top of the loop, the temperature rapidly increases and a conduction front moves toward the chromosphere and reaches it within a few seconds. The large conductive flux heats the chromosphere and generates a large pressure pulse which causes the gas to expand upward. The velocity of the chromospheric evaporation increases rapidly (Figure 1a), and the density of the loop increases while the location of the transition region is pushed downward. There is also formation of a chromospheric condensation. This, however, has a relatively small amplitude, due to the large gravity and high chromospheric pressure scale height in the stellar atmosphere.

As soon as the energy input stops in the flare decay phase, the loop temperature drops immediately, due to large radiative losses. However, the loop will evolve differently according to whether there is energy input or not in the decay phase. If the energy input is stopped at the flare maximum, the flaring plasma cools very fast and suffers some condensational instabilities. The cooled flare plasma will drop down toward the loop footpoints with high velocity on the order of several hundreds km per sec. On the other hand, if there is still energy input and the flare is allowed to cool gradually, then no condensational instability will occur. The flare loop gradually returns to its initial state (Figure 1b). The different evolutionary tracks can be used

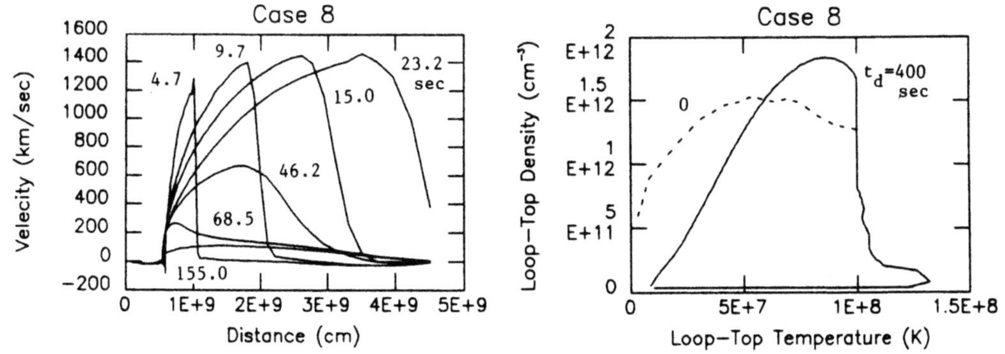

Figure 1. *a.* Evaporation velocity *(left) b.* Evolutionary tracks for two cases $t_d = 0$ and 400 sec *(right)*.

as a diagnostic for the dynamics in the flare cooling phase.

3. X-Ray Spectra and Predicted EXOSAT Light Curves

From the hydrodynamic calculations, we have calculated the X-ray spectra as a function of time during the flare evolution, by folding the Raymond-Smith radiation code with the output from the hydro-code. The spectra around 1.85Å(FeXXIV, XXV) and 3.2Å(CaXVIII, XIX) are very useful. The line ratios in these wavelength regions can be used to determine the temperature of the emitting plasma. The large evaporation velocity of more than 1000 km sec $^{-1}$ will be manifested by a large blue-shifted component in the line profiles of MgXI, XII, CaXVII, XIX, XX, and FeXXI, XXII, XXIV, XXV, which could be observed by future X-ray missions such as the AXAF.

From the X-ray spectra, we can calculate the temporal change of the total X-ray flux in the (1 - 10 KeV) interval, which can be converted into the expected count rate for the ME channel of EXOSAT. Figure 2a shows the predicted X-ray light curves for Model 8 if the flare was observed by the EXOSAT. The aspect ratio of the loop used for the models ranges from 0.2 to 0.3. These light curves are similar to those observed in stellar flares by EXOSAT.

4. Discussion

One of the important observational results for stellar flares is the large X-ray EM ranging from 10^{50} to more than $10^{53}\,\mathrm{cm}^{-3}$, and the empirical linear relationship between the EM and the estimated total flare energy (Figure 2b). Questions have been raised whether a very large loop of the size of the stellar radius is required to produce the large observed EM. The calculated total X-ray EM for different models at the time of flare maximum are compared with the observations in Figure 2b. The calculated total flare energy is the spatially integrated energy deposited in the loop during the entire evolution of the flare. Model 5 has the same length as Model 8 (*i.e.* $8\,10^9$ cm - large loop), but with 10 times less energy input. Model 9 has the same length as Model 3 (*i.e.* $2\,10^9$ cm - short loop), but with 10 times more energy input. Figure 2b shows that it is possible to produce the fairly large EM ($\sim 10^{52}\,\mathrm{cm}^{-3}$) with fairly small loops of total length of only $2\,10^9$ cm (Model 9). The same EM could also be produced by a larger loop (Model 5). That is, medium-sized flares can be produced by short loops with a large energy input. It is not necessary that the loops be as large as previous simulations seem to indicate (Reale et al. 1988). For really large EM, however, a large loop with a large energy input is required (Model 8). It might be surmised that a greater energy input, say, by a factor of 100 more than that for Model 3 might be able to evaporate enough material to increase its EM by two orders of magnitude. However, from the series of simulations, we find that the amount of

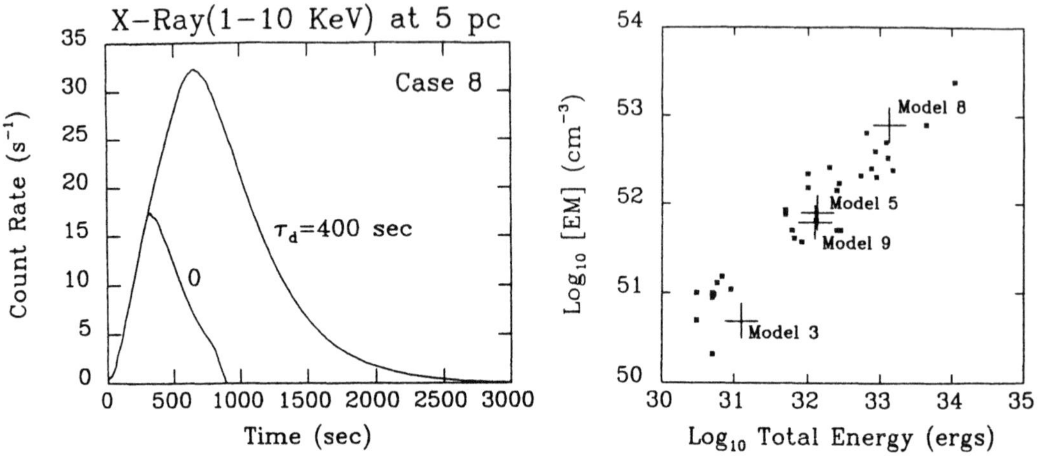

Figure 2. *a.* Predicted X-ray count rate *(left)* and *b.* Predicted *EM* compared with observations *(right).*

evaporated material does not vary linearly with the energy input. The reason is that as more energy is deposited in the loop, the initial evaporation velocity will increase, resulting in a faster initial increase of loop density. As soon as the density is increased, the evaporation material will encounter more resistance, and the evaporation velocity rapidly decreases, as we have seen in previous figures. Also, radiative energy losses will increase rapidly, and less energy is then available for driving the chromospheric evaporation. The processes eventually becomes very inefficient in supplying materials to the loop. For physically realistic situations, a large EM of more than $10^{53}\,cm^{-3}$ requires a large energy input in a large loop, of the size of $8\,10^{9}\,cm$.

References

Cheng, C.-C. et al.: 1983, ApJ, 265, 1090

Cheng, C.-C., Karpen, J.T., Doschek, G.A.: 1984, ApJ, 286, 787

Pallavicini, R., Tagliaferri, G., Stella, L.: 1990, A&A, 228, 403

Reale, F. et al.: 1988, ApJ, 328, 256

NON-LOCAL THERMAL CONDUCTION IN SOLAR AND STELLAR CORONAL LOOPS

A. Ciaravella[1], G. Peres[2] and S. Serio[1,3]

[1]IAIF/CNR, Palermo, Italy
[2]Osservatorio Astrofisico di Catania, Italy
[3]Instituto di Astronomia, Universitá di Palermo, Italy

1. Introduction

Stationary models of loop-like structures in solar and stellar coronae which have so far appeared in literature (for a list *cf.* Ciaravella et al. 1990a) have used Spitzer's formulation of thermal conductivity in ionized plasmas (Spitzer, 1962, and references therein). This is based on the assumption that the electron mean free path is much less than the temperature scale height, $L_T = T_e / |\nabla T_e|$, and that the plasma velocity distribution is everywhere nearly Maxwellian. Since the hypothesis of local thermal equilibrium becomes invalid already at relatively large L_T *i.e.* of the order of 10^3 electron mean free paths (Gray and Kilkenny, 1980), and because of the steep temperature gradients present in the chromosphere-corona transition zone (Shoub, 1983), non-local thermal transport might be important. Non-local heat flux has been shown to be important during the early phases of flares, when the initial temperature rise has not yet been accompanied by a substantial increase in plasma density (Karpen and De Vore, 1987; Peres et al. 1987; Karpen et al. 1989). Since heat transport plays a crucial role in determining the coronal thermal structure and therefore also the stratification of density, it is very important to understand whether the Spitzer theory, which has been generally used to compute the thermal flux in models of coronae, is an appropriate description of the actual physics.

2. Static Model

We have explored the effect of non-local thermal conduction using the non-local description of the thermal flux by Luciani, Mora and Virmont (1983).

Our model assumes that the coronal plasma is magnetically confined inside a circular loop of semi-length L of constant cross-section, subject to heat deposition, and mirror symmetric with respect to its apex. The basic equations of this description describe hydrostatic equilibrium, the energy balance among the non-local conductive flux, a phenomenological heating term, assumed to be uniform along the loop field lines and the radiative losses according to the optically thin radiative losses provided by Raymond (1976).

The energy equation is an integro-differential one, in which the link between heat flux and temperature is not direct, at variance with the Spitzer formulation. Hence the models are computed iteratively. We first integrate the differential equation of energy balance assuming Spitzer conductivity. We then compute the non-local flux according to Eq. 1, and use the new flux in the energy equation. The calculation is repeated until we find convergence, using the criterion that the maximum fractional difference between the differential emission measure in two consecutive iterations be below a given threshold (typically 0.1).

3. Results

We have computed models of static coronal loops for several values of loop length and coronal pressure, and for two values of surface gravity, corresponding to solar gravity $g_\odot$ and to gravity on M dwarf stars such as Prox Cen, *i.e.* $10\,g_\odot$. Representative results obtained for both values

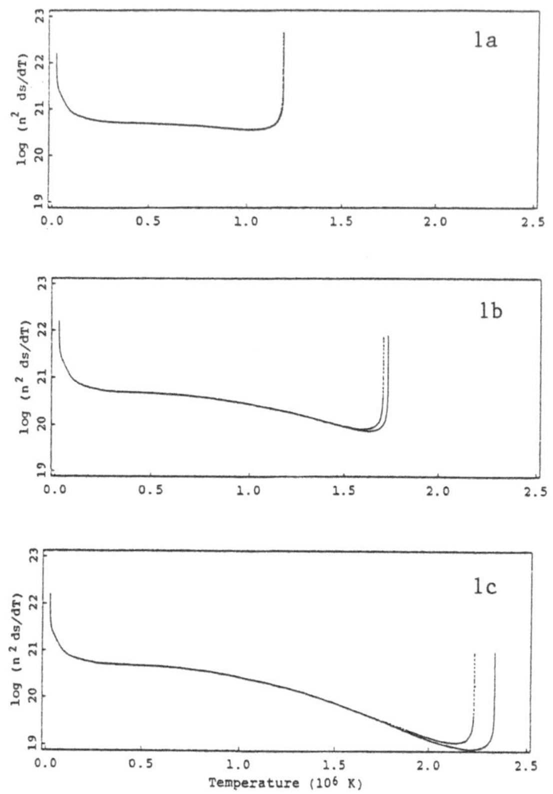

Figure 1. Differential emission measure *vs.* temperature for models of static coronal loops with solar gravity, base pressure 0.1 dyne cm^{-2} and loop semilength $6\,10^9$ cm (*Panel a*), $2\,10^{10}$ cm (*Panel b*) and $5\,10^{10}$ cm (*Panel c*). The initial profile of the iteration, *i.e.* the one computed using Spitzer conductivity, is represented by a *dashed line*, the final profile, derived with non-local conductivity, by a *solid line*. Non local effects increase going from *a* to *c*.

of gravity and showing relevant effects of non-local thermal conductivity on the differential emission measure of the loops are shown in the figures.

A wider set of results, reported in Ciaravella et al. (1990a, b) show that non-local effects become important for the most extended and tenuous loops and for the smallest pressure scale heights. Surface gravity can also play a crucial role. In fact, for loops of a given length and base pressure, effects of non-local thermal conduction are more relevant on stars with higher surface gravity.

We therefore conclude that effects of non-local thermal conduction are important for the longest and lowest pressure coronal loops, while the Spitzer-Harm treatment is substantially valid for small (and high pressure) loops. For more details *cf.* Ciaravella et al. (1990a, b)

We wish to point out, also, that in a few cases (*i.e.* large loops at low pressures, especially in the M star case), even our approach is not completely appropriate, since the effective mean free path at the loop apex becomes larger than the loop dimensions. In this case a more adequate calculation of the loop structure would require a more fundamental approach, such as the use of the Fokker-Planck equation.

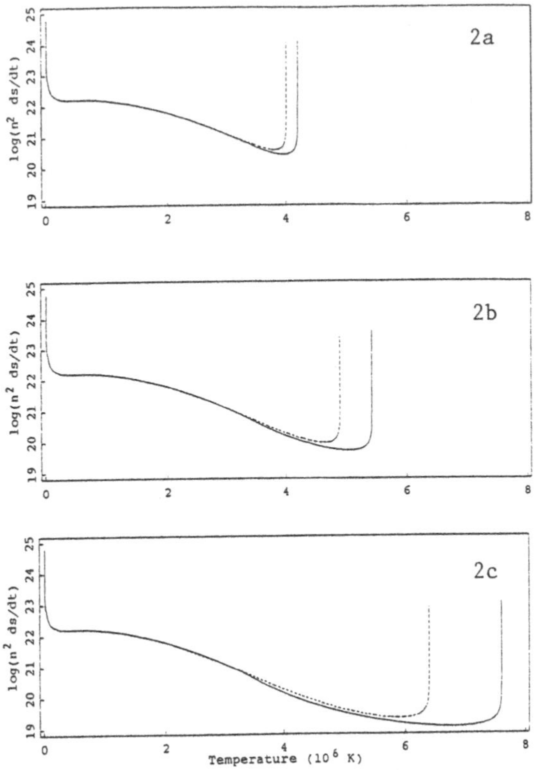

Figure 2. Same as Figure 1, for stellar gravity equal to ten times solar, base pressure 3 dyne cm^{-2} and loop semilength of 10^{10} cm (*Panel a*), 2 10^{10} cm (*Panel b*), 5 10^{10} cm (Panel c). Non-local effects increase going from *a* to *c*.

References

Ciaravella, A., Peres, G., Serio, S.: 1990a, Solar Phys. submitted

Ciaravella, A., Peres, G., Serio, S.: 1990b, in preparation

Gray, D.R., Kilkenny, J.D.: 1980, Plasma Phys. 22, 81

Karpen, J.T., DeVore, C.R.: 1987, ApJ, 320, 904

Karpen, J.T., Cheng, C-C., Doschek, G.A., DeVore, C.R.: 1989, ApJ, 338, 1184

Luciani, J.F., Mora, P., Virmont, J.: 1983, Phys. Rev. Lett. 51, 1664

Peres, G., Rosner, R., Serio, S.: 1987, Il Nuovo Cimento 99B(1), 29

Raymond, J.C., Cox, D.P., Smith, B.W.: 1976, ApJ, 204, 290

Spitzer, L.: 1962, *Physics of Fully Ionized Gases,* Interscience, New York, p. 143.

PHASE DEPENDENT VARIATIONS OF THE Hα LINE IN CHROMOSPHERICALLY ACTIVE STARS

G. Cutispoto[1] and R. Pallavicini[2]

[1]Osservatorio Astrofisico di Catania, Italy
[2]Osservatorio Astrofisico di Arcetri, Florence, Italy

1. Introduction

Emission in the core of the Hα line is believed to originate from active regions in the chromospheres of late-type stars. On the Sun, chromospheric plages are often, although not exclusively, associated with spots and both indicate the emergence of magnetic flux at the stellar surface. If spots and plages are distributed non-uniformly over the stellar disk one would expect to see rotational modulation in broad-band photometric data and in Hα. Moreover, if plages are closely associated with spots, the variations of Hα line emission should be in antiphase with the photometric variations (*i.e.* maximum Hα emission should occur at the minimum of the photometric light curve). Previous observations have sometimes confirmed the above predictions (*e.g.* Nations and Ramsey 1980). More often, however, the observations have provided contradictory results, which have been attributed either to intrinsic short-term variability in Hα or to a complex distribution of spots and plages over the stellar surface.

In order to investigate whether Hα variations are indeed correlated with photometric variations, we have carried out simultaneous photometric and spectroscopic observations of four active stars using instruments at ESO. For all programme stars we report significant photometric and spectroscopic variations that can be attributed to the transit of spots and plages across the stellar disk. For one star (AB Dor) we report extremely large changes in the Hα profile from deep absorption to emission. We also show that Hα emission was varying in phase, and not in antiphase, with the photometric variations.

2. Observations

Four stars were monitored simultaneously in the Hα line and in broad-band UBV$(RI)_c$ filters using two different telescopes at the European Southern Observatory, La Silla, Chile. The programme stars are all well-known "spotted" stars as indicated by their photometric variations. They are also known to show pronounced chromospheric activity as measured in the CaII H and K lines, in UV lines and in Hα. The four stars are HD22468 (= HR1099), HD32918 (= YY Men), HD36705 (= AB Dor) and HD81410 (= IL Hya).

2.1 Photometry

Photometric observations were carried out over a 3-week period in December 1989. They started simultaneously with the spectroscopic run but lasted two weeks longer. We used the 50 cm ESO telescope equipped with a single-channel photometer and standard UBV $(RI)_c$ filters. On each night the photometric data consisted of a series of observations of the programme stars as well as comparison and check stars.

All stars showed clear photometric variations with amplitudes of ≈ 0.1 magnitudes for HR1099 and AB Dor and ≈ 0.2 magnitudes for YY Men and IL Hya. Figure 1 shows the derived light curve in the V-band for AB Dor. The ephemerides derived from the photometric variations were used to determine the phases at which the spectroscopic data were obtained. The latter are marked as S on the light curve of Figure 1.

The phase coverage of the spectroscopic observations was good for HR1099 and AB Dor

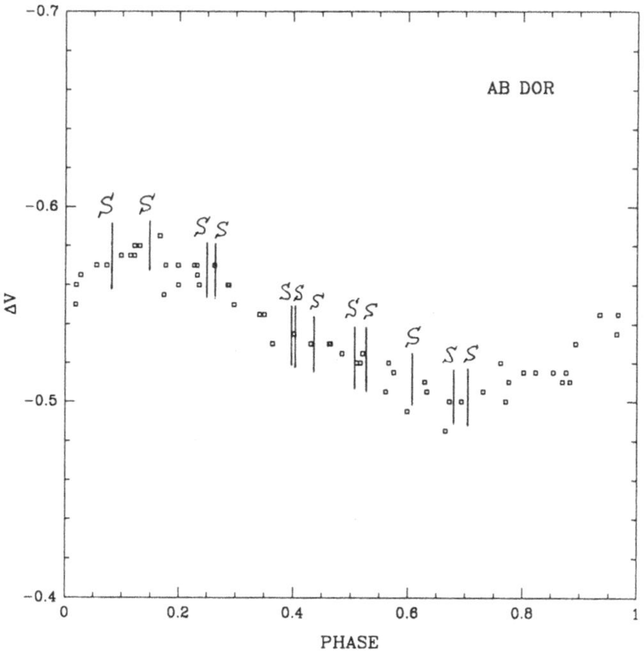

Figure 1. Light curve of AB Dor in the V-band. The phases at which spectroscopic observations were obtained are marked by *S*.

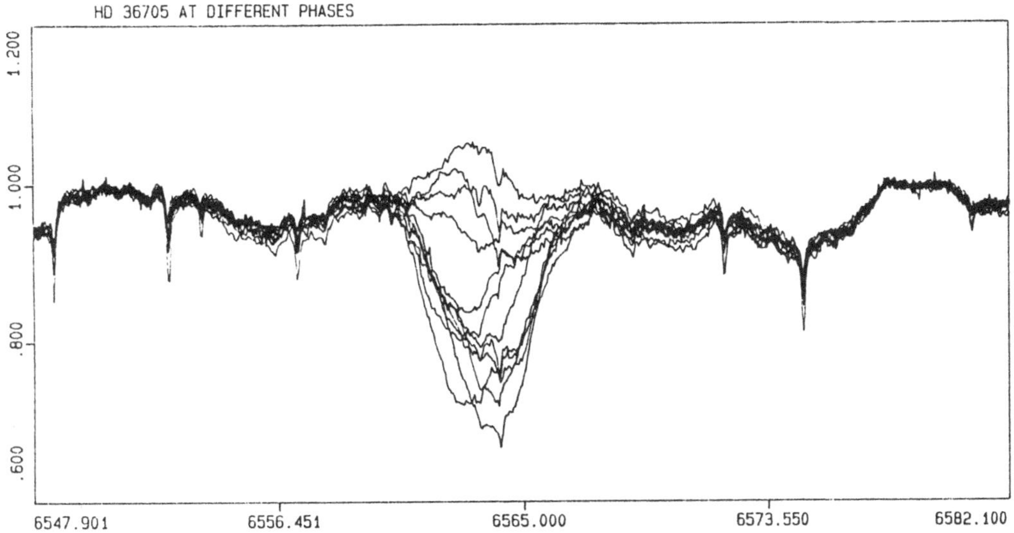

Figure 2. Hα profiles of AB Dor at different phases. Deep absorption occurs at the time of the maximum spot visibility, while the profile goes into emission around the maximum of the photometric light curve.

covering both the maximum and the minimum of the photometric light curve. For YY Men and

IL Hya, which have a longer rotational period, the phase coverage was more limited but still sufficient to cover a photometric variation of ≈ 0.1 magnitudes in V light.

2.2 Spectroscopy

Spectroscopic observations in the H$\alpha\lambda$6563Å line were obtained on five consecutive nights (8-13 Dec 1989) simultaneously with the start of the photometric observations. The instrument used was the Coudé Echelle Spectrometer fed by the 1.4 m CAT telescope. We used the short-camera and a cooled CCD detector at a resolving power R = 50 000.

Significant changes with phase were observed in the profile of the Hα line, which were most likely produced by surface inhomogeneities. Except for AB Dor, the observed profile variations did not change the general appearance of the Hα line, which was in emission in HR1099 and YY Men and in absorption (partially filled) in IL Hya. The most striking variations were observed for AB Dor for which the Hα profile changed from deep absorption to emission at different phases (see Figure 2).

3. Phase-dependent variations in AB Dor

Twelve Hα spectra were obtained for AB Dor which covered all phases from the optical maximum to the minimum. During this period, the Hα profile changed gradually, showing that the observed variations were not chaotic but due instead to the star rotation.

The most interesting aspect of these observations is that, contrary to expectations, the variations of the Hα line emission were *in phase* with the photometric light curve, *i.e.* Hα emission was at maximum at the time of minimum spot visibility. Detailed comparison of the photometric and spectroscopic data shows in fact that the Hα profile was in absorption at the minimum of the photometric light curve. and reversed to emission at the time of minimum spot visibility.

The results obtained for AB Dor contrast with the simple picture of a single hot plage closely associated in space with a cool spot. Rather they suggest a more complex distribution of hot and cool regions over the surface of the star, the bright plage regions being the dominant features on the hemisphere of the star opposite to that dominated by the dark spots. Alternatively, the deep absorption profiles observed at the time of maximum spot visibility could be produced by the passage of dense clouds of absorbing material such as those found for AB Dor by Collier Cameron and Robinson (1989). If this is the case, our observations provide evidence that these prominence-like corotating Hα clouds are preferentially associated with the spotted hemisphere of the star.

References

Collier Cameron A., Robinson. R.D.: 1989, MNRAS, 236, 57

Nations, H.L., Ramsey, L.W.: 1980, AJ, 85, 1086

LONG-TERM VARIABILITY OF ACTIVE STARS

G. Cutispoto[1], M. Rodonó[1,2]
[1]Osservatorio Astrofisico di Catania, v.le A.Doria 6, I-95125 Catania, Italy
[1,2]Instituto di Astronomia, v.le A.Doria 6, I-95125 Catania, Italy

1. Introduction

Cyclical variations of activity indicators such as the total area covered by photospheric spots (covering factor), spot latitudes and/or longitudes and strength of chomospheric emission lines are expected in active stars by analogy with the well known 11-year solar cycle. The lack of spatial resolution and the very high incidence of binary systems, however, can make the stellar data much more difficult to collect and interpret than for the Sun.

The most significant previous studies on the long-term behaviour of active systems have been made by studying the Harvard archival plate collection (Phillips and Hartmann, 1978; Hartmann *et al.* 1979; Hartmann *et al.* 1981) and by monitoring the CaII K emission line strength (Wilson, 1978; Vaughan and Preston, 1980; Vaughan, 1980; Vaughan *et al.* 1981). In this paper we attempt to investigate the photospheric activity level by collecting accurate photometric observations, so that this study could be considered as a natural extension of the previous works made by using the Harvard plates. The major advantage of photometry with respect to the photographic techniques is the precision of the data, that can allow us to distinguish possible short term superimposed on long-term variability, and to define the latter much better. The major disadvantage is that for only a very few stars a long-term photometric database exist (see, for instance, Rodonó, 1986; Strassmeier, 1990; Innes *et al.* 1988). In order to produce significant results long-term photometric studies require a strong observational effort, as the continuous monitoring of active stars obtained by using the Automatic Photoelectric Telescopes (Strassmeier *et al.* 1989). The data presented in this paper have been collected by the authors at Catania Astrophysical Observatory, at ESO and by using the APT service observations with the 25 cm Phoenix telescope. Additional data have been taken from the literature. In Table 1 the stars monitored by our group more intensively are listed.

Table 1. Programme stars monitored by our group.

HD 1835	=	BE Cet	HD 5303	=	CF Tuc	HD 8357	=	AR Psc
HD 8358	=	BI Cet	HD 8435	=	BC Phe	HD 16157	=	CC Eri
HD 21242	=	UX Ari	HD 22468	=	V711 Tau	HD 26337	=	EI Eri
HD 32918	=	YY Men	HD 34198	=	UU Lep	HD 36705	=	AB Dor
HD 37824	=	V1149 Ori	HD 37847	=	TW Lep	HD 39917	=	SZ Pic
HD 45088	=	OU Gem	HD 61245	=	V344 Pup	HD 71071	=	LU Hya
HD 81410	=	IL Hya	HD 82558	=	LQ Hya	HD 86005		
HD 491816	=	LR Hya	HD 98712	=	SZ Crt	HD 101309	=	V829 Cen
HD 102077	=	V838 Cen	HD 106225	=	HU Vir	HD 127535	=	V841 Cen
HD 139084	=	V343 Nor	HD 155555	=	V824 Ara	HD 175742	=	V775 Her
HD 197481	=	AU Mic	HD 199178	=	V1794 Cyg	HD 214479	=	FK Aqr
HD 224085	=	II Peg	HDE 234677	=	BY Dra			

2. Results and Discussion

In Fig. 1 we present the long-term variability of 5 selected stars for which a significant amount of data is available. The behaviour of the individual stars are now discussed.

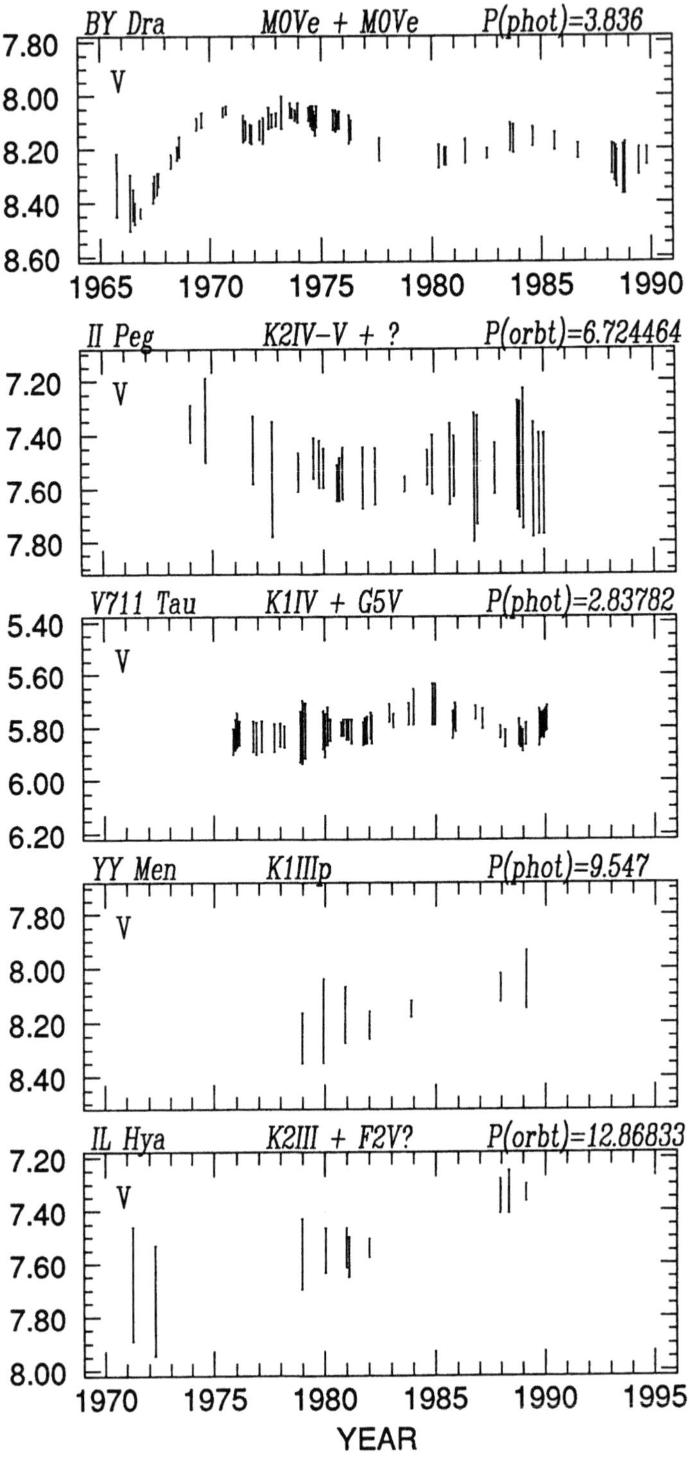

Figure 1. Long-term variability of BY Dra, II Peg, V711 Tau, YY Men and IL Hya.

BY Dra is the prototype of M-dwarf spotted stars (M0Ve + M0Ve; $P_{phot} =$ 3.836 day) and has been observed intensively by several authors, so that an almost complete photometric history since 1965 is available. The data in Fig. 1 clearly show the presence of long-term variability. Taking into account the photographic (Phillips and Hartmann, 1978) and our photoelectric data, a cycle period of about 50 years could be confirmed, but there is also evidence of shorter cyclic fluctuations with relative minima separated by about 8.5 year.

II Peg is a single-line spectroscopic binary (K2IV-V + ?; ,$P_{orb} =$ 6.724464 day) often classified as an intermediate case between the RS CVn and BY Dra-type systems (Rodonó *et al.* 1980). We note that the luminosity decrease after 1968 (Hartmann *et al.* 1979) is confirmed. Indeed, the mean level of II Peg appears to fade up to about 1984. Subsequently, its peak-to-peak amplitude has been very large, indicating a strong asymmetry in spot distribution, and the mean luminosity, with small fluctuation, indicates a rather constant brightness very similar to the 1945-51 level.

V711 Tau is a bright non-eclipsing RS CVn binary (K1IV + G5V; $P_{phot} =$ 2.83782 day), which has shown an almost constant mean brightness from late 1975 to 1980; then its mean magnitude decreased and a luminosity maximum occurred in 1984-85. The latest data show a positive trend that might lead to another maximum within a few years. We would like to point out that the detailed behaviour of the long term of II Peg, BY Dra and V711 Tau variability we have just described would have escaped detection by using photographic techniques.

Finally, the observations of the FK Com star YY Men (K1IIIp; $P_{phot} =$ 9.547 day) and of the RS CVn system IL Hya (K2III + F2V?; $P_{orb} =$ 12.86833 day) have shown considerable increases of their mean luminosities and sizeable changes in their wave amplitudes. We believe that long-term observations will show the occurrence of cyclic variability, whose length cannot be estimated from the present data sets.

References

Hartmann, L., Londoño, C., Phillips, M.J.: 1979, ApJ, 229, 183

Hartmann, L., Bopp, B.W., Dussault, M., Noah, P.V., Klimke, A.: 1981, ApJ, 249, 662

Innis, J.L., Thompson, K., Coates, D.W., Lloyd Evans, T.: 1988, MNRAS, 235, 1411

Phillips, M.J., Hartmann, L.: 1978, ApJ, 224, 182

Rodonó, M., Romeo, G., Strazzulla, G.: 1980, in *2nd Europ. IUE Conf.* ESA SP-157, p. 55

Rodonó, M.: 1986, in *The M-type Stars*, CNRS-NASA, eds. H.R. Johnson and F.R. Querci, NASA SP-492, p. 409

Strassmeier, K.G., Hall, D.S., Boyd, L.J.,Genet, R.M.: 1989, ApJS, 69, 141

Strassmeier, K.G.: 1990, ApJ, 348, 682

Vaughan, A.H.: 1980, PASP, 92, 392

Vaughan, A.H., Preston, G.W.: 1980, PASP, 92, 385

Vaughan, A.H., Baliunas, S.L., Middlekoop, F., Hartmann, L., Mihalas, D., Noyes, R.W., Preston, G.W.: 1981, ApJ, 250, 276 Wilson, O.C.: 1978, ApJ, 226, 379

VILNIUS PHOTOMETRY OF THE TYPICAL AND SUSPECTED CHROMOSPHERICALLY ACTIVE STARS

V. Dadonas, J. Sperauskas
Vilnius University Observatory, Ciurlionio 29
232009 Vilnius, Lithuania

1. Introduction

As is well known, the presence of starspots moving through the stellar surface produces periodic variations in light. This photospheric activity is characteristic of almost all chromospherically active stars. Starspots can cover up to 20% of the total stellar surface area and produce broad-band light variations.

This paper deals with photometric activity in four groups of chromospherically active stars: BY Dra, RS CVn, W UMa, and FK Com. Bopp and Espenak (1977) defined BY Dra variables as late-type dwarf stars (dKe-dMe) with strong Ca II (and often Balmer) emission and low-amplitude light variability with periods of a few days. RS CVn stars are a class of the similar short- and long-period, regular detached binary system consisting of an F-G type primary star and a late G-K secondary (Hall, 1976), or a white dwarf secondary (Fekel *et al.* 1986). Hall (1976) suggested that W UMa binaries may also be chromospherically active, later discussed by Eaton *et al.* (1980). FK Com stars (Bopp and Stencel, 1981) appear quite like these other groups and are rapidly rotating G-K giants.

We report differential UPXYZVS photometry of six typical spotted stars and photometry of eight suspected chromospherically active stars. Using Fourier transform and least-squares methods we derived photometric periods, amplitudes of the variability and maximum brightness in V bandpass for each variable.

2. Data analysis and results

The observations reported here were made with the 48 cm reflector in August 1989, on Maidanak (Uzbekistan).

The first aim of the analysis was to determine the parameters of variability. The best known way to describe these is through Fourier analysis. We prepared a period-finding program based on four procedures. First of all, we smooth the noisy observational data, not necessarily equally spaced or uniformly weighted, by a polynomial smoothing spline. After this, new data equally spaced in time are selected from the smooth curve. The second step is Fourier transformation; we find the most likely frequency f of the variability. Then, by considering a single harmonic component, by which we mean a single co-sinusoidal of harmonic amplitude A, frequency f, phase constant φ, and mean level m_0,

$$m(t) = m_0 + A \cos(2\pi f t + \varphi)$$

we use the least-squares method for the computation of the parameters of the first component of variability. Finally, we remove the contribution of this component from the initial data, forming the residual spectrum. Next, if the stopping condition is not met, we repeat the procedures with residuals but in this case with two components in least-squares, and so on.

Table 1 contains the data for well-known spotted stars from our program and the derived parameters of variability. In Table 2 we present the data for suspected chromospherically active stars.

Table 1. Basic parameters and results for typical spoted stars

Name	Classified	Spectral Type	P_{orb} (days)	P_{phtm} (days)	ΔV_{max}	V_{max}	T_{conj} (2,440,000 +)	Comparison star	Results from our observations		
									P_{phtm} (days)	ΔV	V_{max}
BY Dra (HD 234677)	BY Dra	K4V/K7.5V	5.975112	3.827	≈0.2	8.07	3792.1056	BD+51°2410	3.771 ±0.036	0.067 ±0.008	8.109
VW Cep (HD 197433)	W UMa	G5V/K0V	0.278319	0.2783136	≈0.4	7.23	4176.6161	BD+75°764 BD+75°765	0.291252 ±0.017642	0.282 ±0.014	7.320 August 20, 1989
									0.185280 ±0.027910	0.062 ±0.019	
									0.264138 ±0.004632	0.291 ±0.013	7.314 August 29, 1989
									0.263732 ±0.071762	0.059 ±0.014	
HD 199178	FK Com	G5III-IV	–	3.337	≈0.06	7.27	4395.7000	BD+43°3780	3.325 ±0.016	0.152 ±0.011	7.122
HK Lac (HD 209813)	RS CVn	FIV/K0III	24.4284	24.461	0.25	6.52	5.024	BD+46°3611 BD+46°3711	23.714 ±3.767	0.101 ±0.012	6.906
									11.637 ±0.878	0.083 ±0.012	
KZ And (HD 218738)	BY Dra	dK2/dK1	3.032867	3.03	0.08	7.98	2371.641	BD+46°3969 BD+46°3964	2.978 ±0.079	0.069 ±0.011	7.901
II Peg (HD 224085)	RS CVn	K2-3V-IV	6.724183	6.718	0.43	7.2	3030.239	BD+28°4648 HD 224016	6.759 ±0.080	0.371 ±0.017	7.314

References: for BY Dra, HK Lac, KZ And, and II Peg - Strassmeier et al. (1988)
for HD 199178 - Kaluzny (1984), Huovelin et al. (1987)
for VW Cep - Glownia (1988), Hill (1989)

Table 2. Suspected chromospherically active stars

Name	Classified	Spectral Type		Name	Classifield	Spectral Type
HD1326	flare [1]	dM2.5		HD 19122	CA? [3]	G5
HD6680	RS CVn? [2]	F5		HD 193891	CA? [3]	K0
HD 129674	RS CVn? [2]	F3		HD 200740	CA? [3]	K0
HD 166620	CA?	K2V		HD 222143	CA? [3]	G5

Refs: 1. Pettersen (1976); 2. Fleming *et al.* (1990); 3. Strassmeier *et al.* (1988)

In all chromospherically active stars we find clear evidence of starspot existence. The results of data analysis indicate photometric periodicities variations and multiple periodicities for two program stars. Such behaviour may be explained best by differential rotation. Active areas may drift in longitude, therefore producing a phase shift, as well as drift in latitude and, therefore, rotate with a different rate; they may grow and decay, which will also be reflected in the rotational period. Data from one season alone are usually insufficient to distinguish unambiguously between differential rotation and evolution of active areas.

Unfortunately, we find no evidence for variability during the period of our observations for the suspected chromospherically active stars. However, a null result is also a result.

References

Bopp, B.W., Espenak, F.: 1977, AJ, 82, 916

Bopp, B.W., Stencel, R.E.: 1981, ApJ, 247, L131

Eaton, J.A., Wu, C.C., Rucinski, S.M.: 1980, ApJ, 239, 919

Fekel, F.C., Moffett, T.J., Henry, G.W.: 1986, ApJS, 60, 551

Fleming, T.A., Gioia, I.M., Maccaro, T.: 1990, AJ, 98, 692

Glownia, Z.: 1988, IBVS, No. 3258

Hall, D.S.: 1976, in IAU Coll. 29, *Multiple Periodic Variable Stars,* ed. W.S. Fitch (Dordrecht: Reidel), p. 287

Hill, G.: 1989, A&A, 218, 141

Huovelin, J., Piirola, V., Vilhu, O., Efimov, Y.S., Shakovskoy, N.M.: 1987, A&A, 176, 83

Kaluzny, J.: 1984, IBVS, No.2627

Pettersen, B.R.: 1976, *Catalogue of Flare Star Data,* Inst. of Theor. Ap., Blindern, Oslo, Report No. 46

Strassmeier, K.G., Hall, D.S., Zeilik, M., Nelson, E., Eker, Z., Fekel, F.C.: 1988, A&AS, 72, 291

THE UV SPECTRUM OF THE RS CVn BINARY SZ PSc

E. Danezis,[1] E. Antonopoulou,[1] M. Mathioudakis,[2] and E. Theodossiou[1]

[1]Section of Astrophysics, Astronomy and Mechanics, Department of Physics, Panepistimiopolis, GR-15783 Zografos, Athens, Greece.
[2]Armagh Observatory, Armagh BT61 9DG, N. Ireland

1. Introduction

SZ Psc is a typical RS CVn type binary system which shows all the main characteristics of the group (*e.g.* CaII H & K emission, wave-like variations outside eclipses). It also shows a very unusual, variable behavior of the Hα line (Bopp 1981) and large period variations ($dP/dt = 6 \pm 0.5\,10^{-8}$ days/day; Jahate *et al.* 1976).

In this paper we discuss two IUE spectra of SZ Psc in order to have a better understanding of the system. Similar work has been done for some others RS CVn systems like UX Ari (Simon and Linsky 1980), HR1099, II Peg, AR Lac (Byrne *et al.* 1982), λ And (Baliunas *et al.* 1984), σ Gem (Ayres *et al.* 1984).

2. Observations

This work is based on two IUE spectra, *viz.* SWP22958, a low resolution spectrum covering the wavelength region from 1235 to 1950 Å taken by Fernandez and Gimenez at VILSPA, and

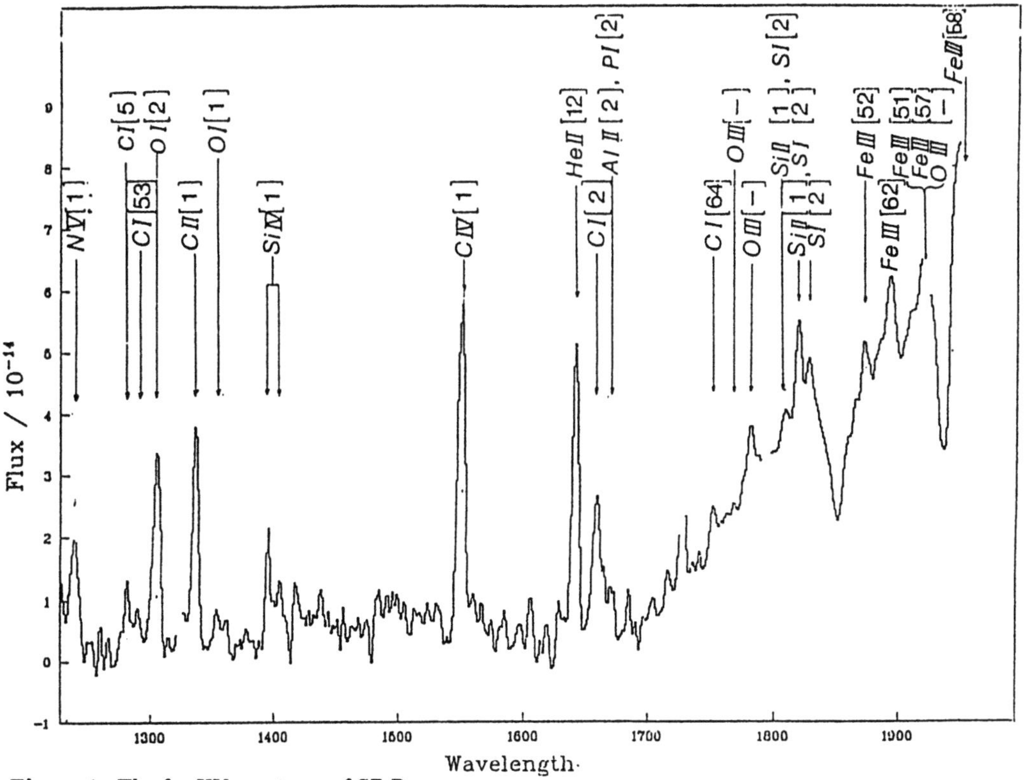

Figure 1. The far UV spectrum of SZ Psc

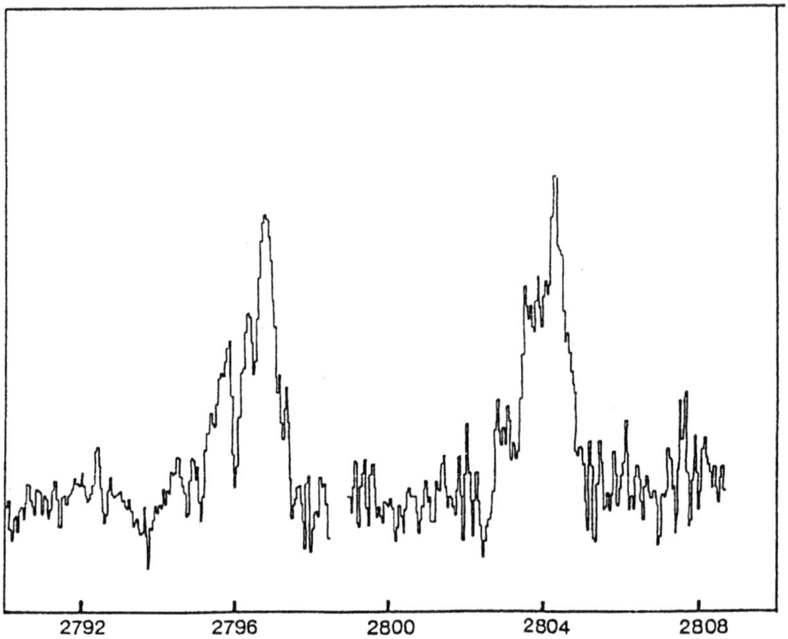

Figure 2. The structure of Mg II doublet

LWR9679, a high resolution spectrum taken by Linsky and Simon in NASA GSFC. The spectra were extracted from de-archived IUE images from the World Data Center (RAL) using the IUEDR package on STARLINK (Gidding 1983).

3. Description of the spectra

3.1 The far UV spectrum SWP22958

This spectrum exhibits chromospheric (*e.g.* CI, OI) and transition region emission lines (*e.g.* CIV, SiIV, NV) (Fig. 1). It was taken on 9 May 1984 and the estimated phase using the ephemeris by Hall and Kreiner (1980) is $\phi = 0.724$ (JD0 = 2442308.946; $P = 3.9658663$ day). The far UV spectrum is crowded by emission lines of CI, OI, SiII, HeII, AlII, PI, SI and FeIII. The most important region is between 1700 and 1950 Å which is dominated by emission lines of SiII and FeIII.

3.2 The UV spectrum LWR9679

This spectrum was taken on 10 January 1981 and the estimated phase is $\phi = 0.362$. The most remarkable feature of this spectrum is the emission of the MgII [1] resonance doublet characterized by a P Cygni profile (Fig. 2). Two emission and one absorption components are present in the blue side of each principal emission lines. The structure of this feature is an indication of the existence of a circumstellar shell around the system, in keeping with Bopp's (1981) suggestion as a possible explanation for the variability of the Hα profiles of SZ Psc.

Acknowledgements: Research at Armagh Observatory is grant-aided by the Dept. of Education for N. Ireland. We acknowledge the support provided by the Starlink Project which is founded by the UK SERC.

References

Ayres, T.R., Simon, T., Linsky, J.L.: 1984, ApJ, 279, 197

Baliunas, S.L., Guinan, E.F., Dupree, A.K.: 1984, ApJ, 282, 733

Bopp, B.W.: 1981, AJ, 86, 771

Byrne, P.B., Butler, C.J., Andrews, A.D., Linsky, J.L., Simon, T., Marstad, N., Rodono, M., Blanko, C., Catalano, S., Marilli, E.: 1982, Proc 3rd Europ. IUE Conf. Madrid.

Giddings, J.R.: 1983, IUEDR, STARLINK User Note No. 37 Rutherford Appleton Lab.

Hall, S.D., Kreiner, J.M.: 1980, Acta Astron. 30, 387

Jakate, S., Bakos, G.A., Fernie, J.D., Heard, J.F: 1976, AJ, 81, 250

Simon, T., Linsky, J.L.: 1980, ApJ, 239, 911

SHORT TIME-SCALE CHANGES IN THE V-BAND LIGHT-CURVE OF II Peg: FLARE, COMPANION OR PROMINENCE?

J.G. Doyle,[1] G.E. Bromage,[2] A.C. Cameron,[3] D.W. Kilkenny,[4] J. Krzesinski[5] H.M. Murphy,[1] J.E. Neff,[6] G. Pajdosz,[5] F. van Wyk[4]
[1]Armagh Observatory, Armagh BT61 9DG, N. Ireland
[2]Astrophysics Div., Rutherford Appleton Lab., Chilton, Didcot, Oxon., England
[3]Astronomy Center, University of Sussex, Falmer, Brighton BN1 0QH, England
[4]South African Astronomical Obser., PO Box 9, Observatory 7935, South Africa
[5]Dept. of Astronomy, Pedagogical University, Krakow PL 30-084, Poland
[6]NASA/Goddard Space Flight Center, Bld. 21, Greenbelt, MD 20771, USA

1. Introduction

II Peg (= HD224085) is a single-lined spectroscopic binary with a photometric period of approximately 6.7 days. It is classified as an RS CVn-type star, although its companion has yet to be observed. Here, we report on the reduction of optical photometric observations taken with the Fine Error Sensor (FES) onboard the International Ultraviolet Explorer (IUE) satellite in August 1989, and its comparison with nearly-simultaneous ground-based photometry.

2. Observational Data

During a continuous 48 hr period of IUE observations beginning August 14, 1989, a total of 32 FES measurements were made. The FES onboard IUE can be used to give V magnitudes to an absolute uncertainty of ≈ 0.05 but with a relative uncertainty substantially better. The magnitudes were derived using a program given by Barylak (1989).

Ground-based photometric observations were from two sources. Firstly, monitoring carried

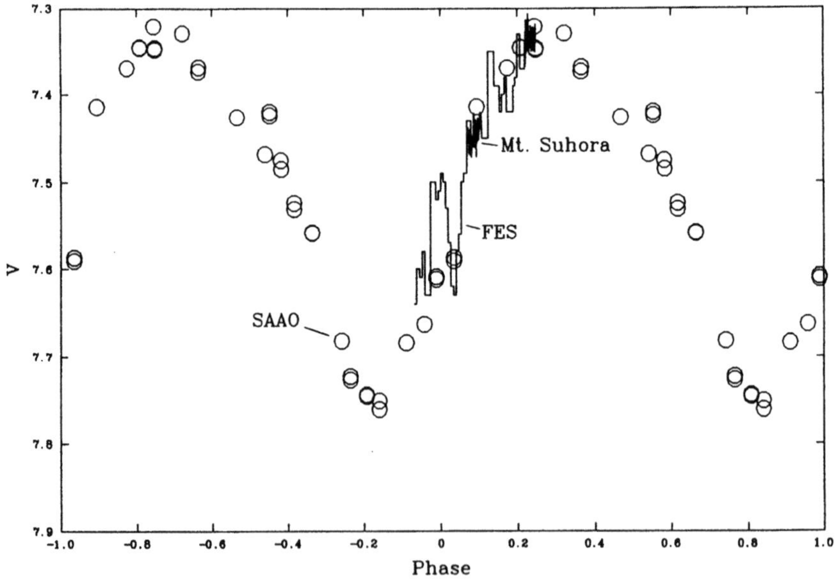

Figure 1. V-band photometry for II Peg in August 1989 from SAAO, FES and Mt. Suhora. Note that the SAAO data points at phases 0.99 and 0.035 were taken one and four weeks respectively after the FES and/or Mt. Suhora data

out in yellow and blue filters (these being close to the Johnson UBV system) at Mt. Suhora Observatory (Poland) using a 60 cm Cassegrain telescope equipped with a double-channel photometer. The monitoring period was 15 August 20:28 to 16 August 04:00 UT and 16 August 20:26 to 17 August 02:51 UT. On all occasions 30 sec integration times were used. The comparison star chosen was BD +28°4648, with the standard deviation in both filters being about ± 0.006 mag.

The second data set comprised of UBV(RI)$_{KC}$ measurements made either once or twice per night for most nights over a two month period beginning mid-July 1989 at the South African Astronomical Observatory on the 50 cm telescope, with a single channel photometer. The comparison star chosen was again BD +28°4667.

3. Results and Discussion

In Fig. 1 we show the V-band data for II Peg as a function of phase, using the ephemeris of Vogt (1981), *i.e.* JD = 2443033.47 + 6.72422E. The overall agreement between the three datasets was very good and no zero point shifts were required. Just after phase 0.0 in the FES data there is a 0.15 mag. "dip", showing a symmetrical decrease and increase lasting a total of ≈ 8 hr. These FES and Mt. Suhora data are shown in more detail in Fig. 2. Three possible explanations for this "dip" are:

- *A prominence:* the "dip" could be the result of an eclipse due to a cloud of neutral hydrogen high in the corona. Such cool clouds of gas have been observed on rapidly rotating K & M dwarfs, *e.g.* Robinson & Cameron (1986) and Doyle & Cameron (1990). An indication of cool dense material being present high in the atmosphere of II Peg during a flare has been reported previously by Doyle *et al.* (1989). However, a major problem with this interpretation is that, although prominences have been observed in absorption lines such as Hα, CaII H & K, *etc.*, no such feature has been dense enough to be optically thick in the visible continuum.

- *A companion:* The secondary in the II Peg system has yet to be observed, but one possible interpretation of the photometric "dip" is that it is due to an eclipse by the companion since phase 0.0 is defined as the time when the primary is furthest from the observer. Taking the effective temperature of the primary to be ≈ 4700 K and a radius of 2.7 $R_\odot$, and assuming the eclipse is total and ignoring limb-darkening, we find the radius of the secondary to be ≈ 1 $R_\odot$. This value is not critically dependent upon the surface brightness of the secondary. The phase difference between the first and fourth contact is ≈ 20°, hence the orbital separation between the components would be $(R_1 + R_2)/\sin 10$, *i.e.* ≈ 20 $R_\odot$. Furthermore, we find the mass of the secondary to be ≈ 0.6 $M_\odot$, the primary ≈ 2 $M_\odot$ and a luminosity ratio between the components of ≈ 0.05.

- *A flare:* An alternative explanation is that the increase in the FES mag. beginning at phase 0.95 is due to a large flare, with the flare decay beginning at phase 0.01 and ending at phase 0.04. In this interpretation, the rising part of the apparent "dip" is simply the underlying spot light-curve. Using the V-band flux calibration of Bessell (1979), implies a V-band flare energy of ≈ $1.2\,10^{36}$ erg. Other large energetic flares have been observed on II Peg previously, *e.g.* Doyle *et al.* (1989) estimated a total radiative energy of $2.4\,10^{36}$ erg for a flare observed with IUE. However, if this is a flare, there is no corresponding increase in the ultraviolet lines.

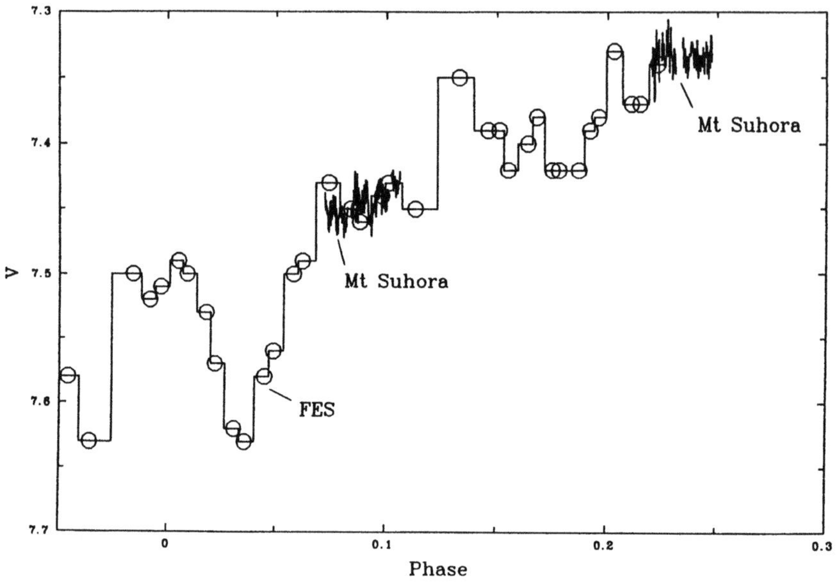

Figure 2. An enlarged plot around phase 0.0 showing the V-band photometry obtained 14 - 17 August 1989 for II Peg from FES and Mt. Suhora

References

Barylak, M.: 1989, ESA IUE Newsletter, 33, 20

Bessell, M.S.: 1979, PASP, 91, 589

Doyle, J.G., Byrne, P.B., van den Oord, G.H.J.: 1989, A&A, 224, 153

Doyle, J.G., Cameron, A.C.: 1990, MNRAS, 244, 291

Robinson, R.D., Cameron, A.C.: 1986, PASA, 6, 306

Vogt, S.S.: 1981, ApJ, 247, 975

PROPOSED UPGRADE OF THE McMATH SOLAR/STELLAR TELESCOPE TO A 4 m APERTURE

M. S. Giampapa. W. C. Livingston. and D. R. Rabin
National Optical Astronomy Observatories
National Solar Observatory
P.O. Box 26732, Tucson, Arizona 85726-6732, USA

It appears feasible to convert the present 1.6 m aperture McMath Telescope to one of 4 m aperture at relatively low cost (≈ US$7M). The building structure would remain unchanged, and most instruments would require only slight modification. Only the mirrors and their support systems would be replaced.

For solar research in the infrared (1 - 20 μm), a large aperture is needed to obtain flux levels and angular resolution comparable to what is available at visible wavelengths. Observing the 12 μm atomic emission lines of MgI, which permit highly sensitive measurements of magnetic field strength in the high photosphere, a 4 m McMath Telescope would have a diffraction limit of 0.75 arc sec. Magnetic fields in the low photosphere would be directly measured with Zeeman sensitive lines near the opacity minimum at 1.6 μm. Vibration-rotation bands of CO, and other molecules, would be used to probe sunspots and the ill-understood temperature minimum.

In solar-stellar oscillation studies, a 4 m aperture means a probable limiting magnitude of $V = 6 - 7$ for observations by Doppler spectroscopy. Therefore, solar-type stars characterized by ages in the range $1 - 5\,10^9$ yr could be observed. Stellar convection deduced from line asymmetry requires spectral resolutions of $\Delta\lambda / \lambda = (1\text{-}2)\,10^5$ and $S/N \approx 10^3$. Such spectra could be acquired in an integration time of 1.5 hr for $V = 8.5$, sufficient to reach solar-type stars in the Hyades. In general, a telescope of this aperture which can be dedicated to solar-stellar synoptic investigations could make significant contributions to the kind of research discussed in this volume.

We visualize the present 2 m heliostat being replaced by a 6 m cell which is held by a modified alt-az tracking system. We show in Fig. 1a the present McMath heliostat which has a diameter of 2.0 m. The 6 m tracking mirror which would replace it is conceptually illustrated in Fig. 1b. Pointing is assured by a laser 'meter-stick' in the servo feedback loop. Wind protection is obtained by a wind-fence of variable height. Both the tracking and image forming mirrors would be of solid aluminium, Kanigen coated, and under active control. Metal mirrors permit rapid temperature adjustment of their surfaces and thus improved telescope seeing. In the IR adaptive control may be achievable. The present performance of the telescope in the visible will not be compromised.

A key aspect of this proposal is its low cost. Economy follows from the use of structures already in place, a developed site, and the use of metal mirrors. Practicality follows from recent advances in the active control of thin mirrors. Cost is estimated at US$7M.

The NOAO is operated by AURA Inc., under cooperative agreement with the National Science Foundation.

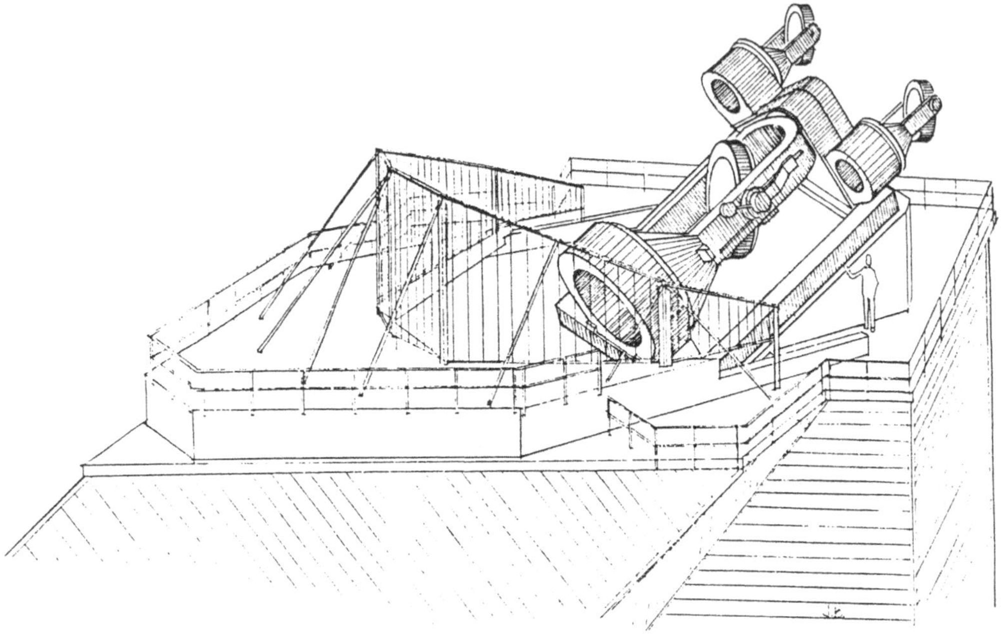

Figure 1a. The present 2.0 m heliostat and wind-screen of the NSO McMath telescope along with the East and West Auxiliary mirrors. The current primary is 1.6 m in diameter.

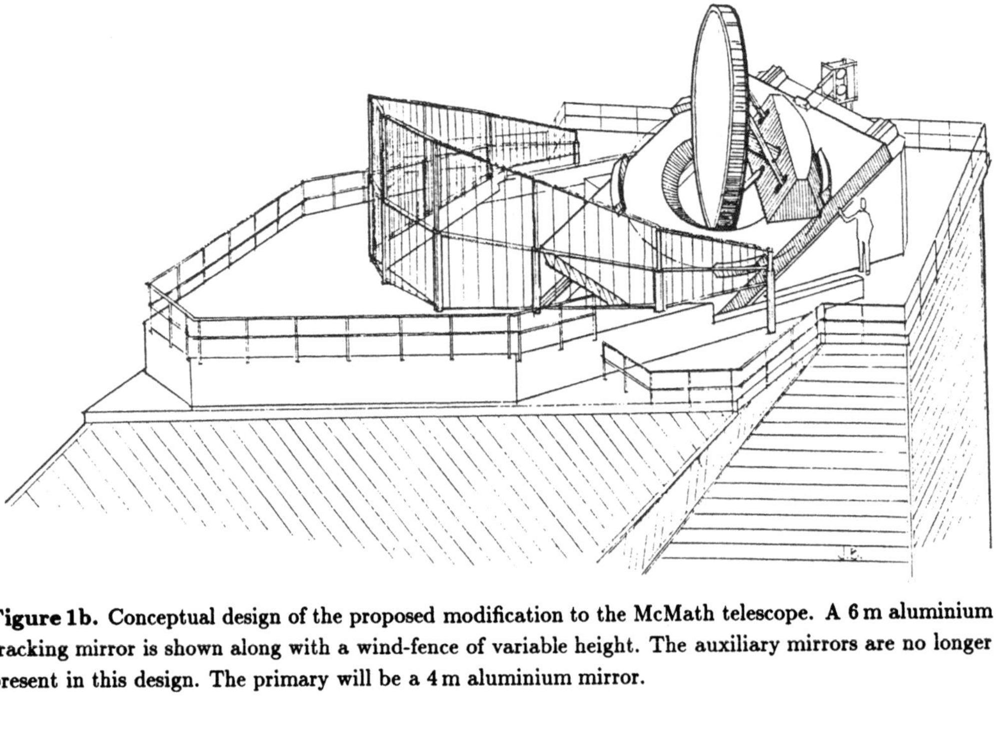

Figure 1b. Conceptual design of the proposed modification to the McMath telescope. A 6 m aluminium tracking mirror is shown along with a wind-fence of variable height. The auxiliary mirrors are no longer present in this design. The primary will be a 4 m aluminium mirror.

SURFACE INHOMOGENEITY ON T Tau STARS AND THE STRUCTURE OF THEIR STELLAR WINDS.

V.P. Grinin, A.S. Mitskevich, N.A. Volchkova
Crimean Astrophysical Observatory, Crimea, P.O. Nauchny, 334413 USSR

1. Introduction

The problem of interpretation of emission lines observed in the spectra of T Tauri stars is the most contradictory problem in the physics of these stars. For example, no contemporary models are able to explain the two-component profiles with blue-shifted absorption, often observed in Hα and other lines. Recently we have shown (Grinin and Mitskevich (GM), 1990a,c) that such profiles can be explained in framework of spherically-symmetrical gas outflow from a star, if its peripheral regions are expanding with deceleration and the wind is not a continuous medium, but composed of separate blobs. The latter means, that the process of matter outflow from the star is eruptive: only a part of stellar surface is involved simultaneously in the wind formation.

2. Stochastic Description of Stellar Wind

We used here a simplified approach. First of all, we have computed the velocity field and excitation and ionization states of atoms and ions in stellar wind within the standard approach of continuous medium ("Basic" model hereinafter). Then, using a random number generator (see GM, 1990c for details) we keep the volumes filled in with gas (blobs) with filling factor q. The velocity field in the Basic model is computed on the assumption that the gas motion is determined by two forces: the gravity of the star and an acceleration force, $F(r) \propto \exp[-(r - R_*)/L]$, where R_* is the stellar radius and L is the dumping length.

The calculation of rate equations in this Basic model is realized on the basis of algorithm described in GM (1990b). The temperature of the star and of the stellar wind (assumed to be isothermal) are 4500 K and 8000 K repectively (GM, 1988, 1990b, Natta *et al.* 1989 and Hartmann *et al.* 1990). Other parameters are stellar radius, $R_* = 3\,R\odot$, stellar mass, $M_* = 2\,M\odot$ and the dumping length $L = 0.8\,R_*$.

It is noteworthy that the CaII IR triplet lines in spectra of T Tau stars are thermalized and that used to be assessed as important evidence in favour of their chromospheric origin (Shanin *et al.* 1975, Herbig and Soderblom 1980). Our computations (GM, 1988, 1990b) have shown that the most probable region of calcium line formation is the base of stellar wind. Therefore the lines of the IR triplet should be sensitive to the surface inhomogeneity of stars.

It is seen from Fig. 1 that in all three cases the triplet lines are narrower than Hα. This confirms our conclusion (GM, 1990b) that the main part of the calcium emission in T Tau stars is formed in the region of wind acceleration, where the gas has not yet reach maximum velocity.

3. The Results of Computations

In the case of a spherically symmetric stellar wind the optically thick CaII IR triplet lines will have P Cygni-type profiles. However the observed profiles do not have absorption components (Fig. 1). This means that the matter outflow from the stars is highly inhomogeneous over the surface. This important property is illustrated by the theoretical profiles, computed in the framework of the stochastic model (Fig. 2).

As it is seen from Fig. 2, theoretical IR triplet line profiles do not show any absorption when $q \leq 0.1$, whereas the blobs are optically thick in the IR triplet lines. The calculations show that in the adopted models the IR triplet is thermalized in agreement with observations.

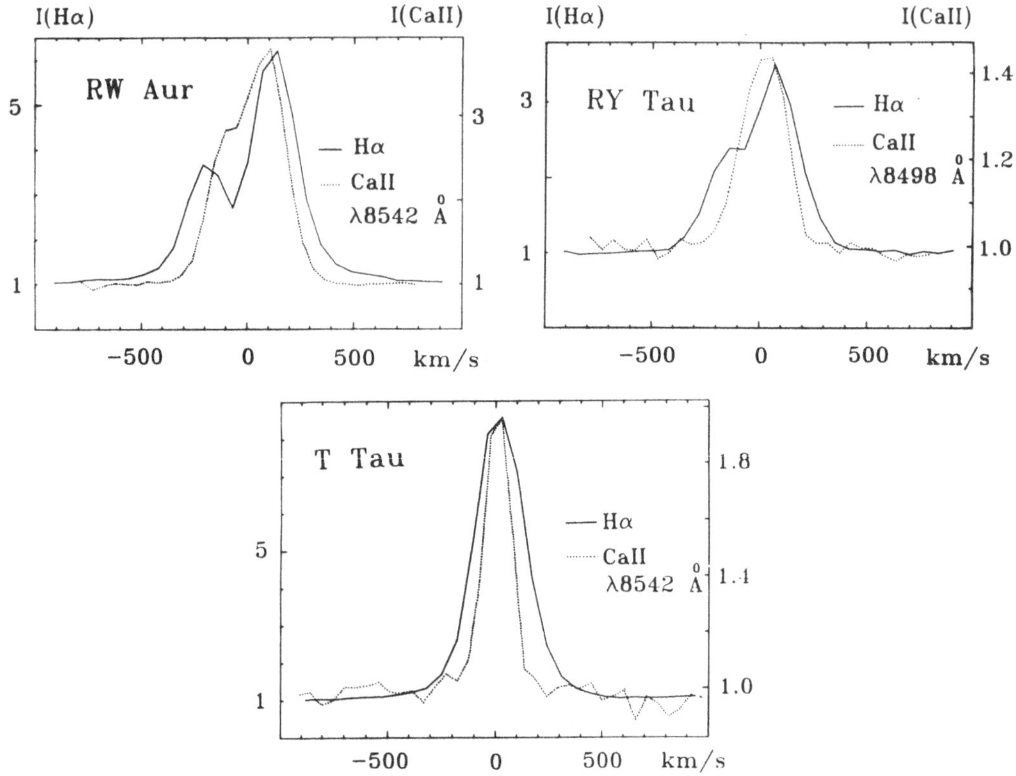

Figure 1. The IR triplet and Hα line profiles for three classical T Tauri stars: RW Aur, RY Tau and T Tau. The observations were carried out by the authors in 1990 at the 2.6 m telescope with the CCD matrix. Spectral resolution in the IR triplet region is 1.5Å.

4. Conculsion

Thus, the matter outflow from T Tauri stars is characterized by extreme inhomogeneity: at each instant of time only < 0.1 of the whole area of the stellar surface is loosing matter. From the viewpoint of MHD-wind models (Hartmann *et al.* 1982) it means that there exist plages on the stellar surface (*i.e.* the spots with maximal magnetic field intensity and open configuration) where intensive matter outflow occurs.

References

Grinin, V.P., Mitskevich, A.S.: 1988. Izv. Krim. Astrof. Obs. 78, 28

Grinin, V.P., Mitskevich, A.S.: 1990a, IAU Symp. No. 137, eds. L.V. Mirzoyan *et al.* (Dordrecht: Reidel) p. 343

Grinin, V.P., Mitskevich, A.S.: 1990b, Afz, in press

Grinin, V.P., Mitskevich, A.S.: 1990c, in preperation

Hartmann, L., Edwards, S., Avrett, E.H.: 1982, ApJ, 261, 279

Hartmann, L., Calvet, N., Avrett, E.H., Loeser, R.: 1990, ApJ, 349, 168

Herbig, G.H., Soderblom, D.R.: 1980, ApJ, 242, 628

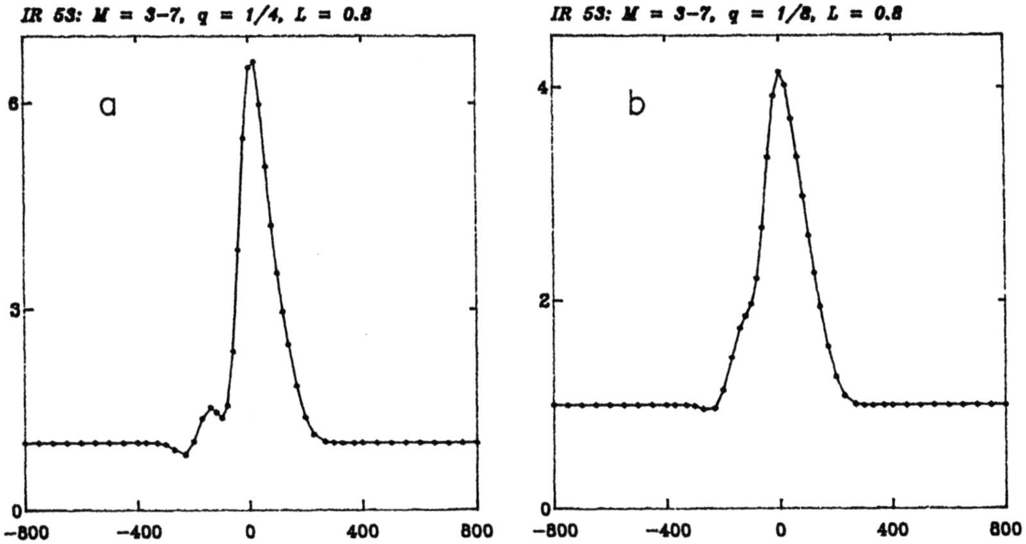

Figure 2. Theoretical line profiles of the IR triplet component λ 8542 Å in a stochastic model with $L = 0.8$ and with filling factors, $a q = 1/4; b) q = 1/8$. The mass loss rate in the Basic model is $3\,10^{-7}\,M_\odot\,yr^{-1}$. The real value of the mass loss rate is q times less. The profiles are smoothed by a gaussian filter with FWHM = 1.5 Å.

Natta, A., Giovanardi, C., Palla, F.: 1988, ApJ, 332, 931

Shanin, G.I., Shevchenko, V.S., Shcherbakov, A.G.: 1975, IAU Symp. No. 67, eds. V.E. Sherwood and L. Plaut (Dordrecht: Reidel), p. 117

ON THE DYNAMICS OF THE EMERGING PRE-FLARE MAGNETIC CONFIGURATION

Haik A. Harutyunian, Vladimir S. Hayrapetyan
Byurakan Astrophysical Observatory
Armenian Academy of Sciences
378433 ARMENIA
USSR

1. Introduction

The theoretical description of flare activity in the Sun and dMe stars as it is generally accepted, is based on magnetic field reconnection. It is known that this approach is a fruitful one for the description of solar flare processes. Hayrapetyan *et al* (1988) have suggested a pinch model for the transformation of magnetic energy during stellar flares. If for reconnection models a resistive instability is required, the the pinch model is based on the action of the instability under conditions of infinite conductivity. This is a very effective source of energy, which enables us to explain the general observational data (Hayrapetyan and Nikoghossian, 1989). Thus, a new problem arises in the upper layers of stellar atmospheres of the formation of closed magnetic configurations with low β which can exhibit pinching and pre-flare reconnection effects. This is the aim of the present paper.

2. Precursor as a flare trigger

Pre-flare radiation of the Sun and dMe stars gives very important physical information about flare origins. Kocharov (1987) has shown that there is a good temporal correlation between the maxima of solar precursors and flare events. Preliminary analysis of some correlations for stellar flares was performed by Shevchenko (1973). Recent observational data suggest that pre-flare processes serve as a trigger for flare energy release.

It is known that both pre-flare and flare processes occur at the same chromospheric heights and the time between their maxima amounts to 500 - 600 sec. Our approach is based on the possibility of existence of magnetic field with low β in emerging flux tubes. An arcade can be unstable with respect to fast reconnection due to a large amplitude perturbation. This perturbation can create a neutral point near the bottom of the stretched configuration as shown in Aly (1989). Due to the reconnection process the magnetic energy release will occur near the bottom of the arcade and a separate closed toroidal magnetic configuration will be formed (see Fig. 1). Quantitatively the value of energy released is in agreement with observed precursor energies. Soft and hard X-ray data of solar pre-flare events yield information about reconnection process (Machado *et al* 1986). The separated magnetic field mentioned above is unstable with respect to the transformation of magnetic energy into thermal and kinetic plasma energy by the pinch effect. The magnetic force lines and current intensity direction are presented in Fig. 2. The MHD instability of magnetic toroid is caused by the azimuthal component of the magnetic field and the Z-component of the current density flowing on the toroidal surface. It is the consequence of very small plasma density in the toroid in comparison with the atmospheric plasma density.

Hayrapetyan *et al* (1988) have presented a model of magnetic toroidal compression to Z-axis in the basis of an ideal one-fluid 2-D MHD. The calculations show two stages of the process: slow compression with sound velocity and fast collapse to the Z-axis with Alfven velocity. In the last stage m = 0 Rayleigh-Taylor instabilities develop, which lead to dense high-temperature

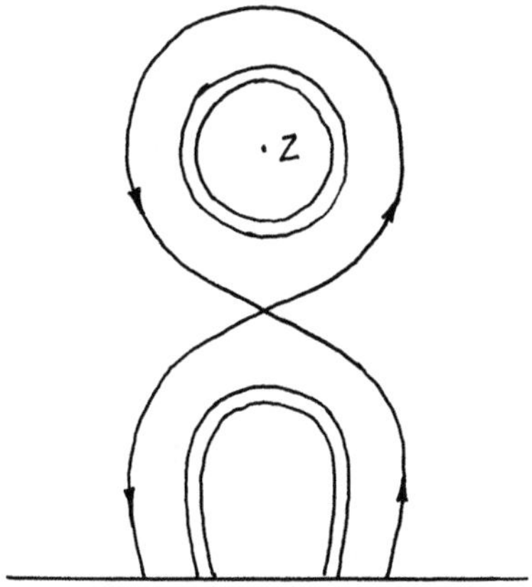

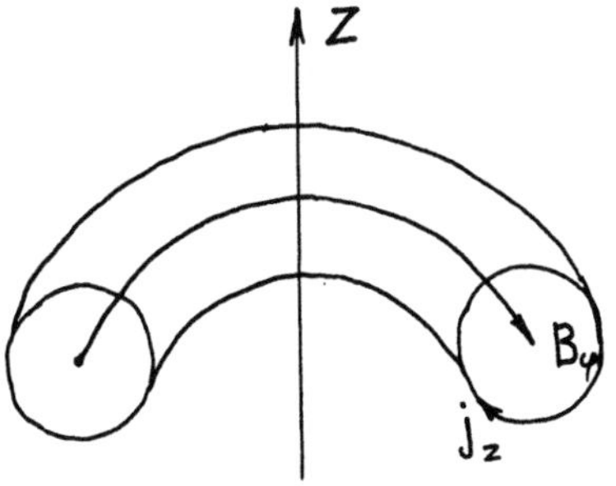

plasma formation along Z-axis (hot spots), serving as the source of impulsive X-ray radiation.

3. Reorientation of the toroidal magnetic field

Let us now consider the dynamics of the separated magnetic toroid. The behaviour of this system is determined by buoyancy instability in the gravitationally stratified stellar atmosphere. We have performed numerical calculations for the magnetic toroid dynamics when its Z-axis is situated as shown in Fig 2. Calculations show that if the toroid main radius, R, is comparable with the pressure height scale, Λ, and is equal to 10^9 cm and for an initial height of toroid, $h = R$ then the reorientation time is ≈ 300 sec. This time is close to the time of sound stage of magnetic toroid compression. With decreasing of ratio R/Λ the reorientation time is increasing and for values $R/\Lambda = 0.5$, $R = 10^8$ cm amounts to 500 sec. During reorientation the magnetic toroid rises due to buoyancy up to 2 initial heights. This means that the region of flare energy release will be situated higher than pre-flare and the plasma pinch column will be formed along the vertical direction in relation to the photospheric level.

4. Conclusion

Thus, we have considered one of the possible ways of flare magnetic structure formation due to fast reconnection process, which is accompanied by pre-flare energy release.

References

Aly, J.J.: 1989, in IAU Coll. No. 104. "Solar and stellar flares", Poster papers. eds. B.M. Haisch, M. Rodono, publ. Catania Astrophys. Obs., p. 259

Hayrapetyan V.S., Nikoghossian, A.G., Vikhrev, V.V.: 1988, in Proc. of "Reconnection in space plasma", Potsdam, GDR. ESA SP-285, Vol. 2. p. 163

Hayrapetyan V.S., Nikoghossian, A.G.: 1989, in IAU Symp. No. 137, "Flare stars in star clusters. associations and solar vicinity" eds. L.V. Mirzoyan, B.R. Pettersen, M.K. Tsvetkov, Kluwer

Kocharov, G.E.: 1987, Itogy nauki i tekhniki, Astronomia (Russ.) Vol. 32, p. 43

Machado, M.E., Orwig, L., Antonucci, E.: 1986, Adv. Space Res. 6, No. 6

Shevchenko, G.G.: 1973, Astron. tsirk. (Russ.), No. 792, p. 2

SPOTS OF AR LACERTAE

Young Woon Kang[1] and Il-Seong Nha[2]
[1] Dept. of Earth Science, King Sejong Univ., Seoul 133-747, Korea
[2] Yonsei University Observatory, Seoul 120-749, Korea

1. Introduction

Kron (1947) was the first one who proposed huge dark patches on the surface of the primary component to explain the intrinsic light variation of AR Lac. Kang and Wilson (1989, hereafter KW) analyzed the six available light curves of AR Lac between 1972 and 1981 to determine starspot parameters and their migration.

Successive observations have been made at Yonsei University Observatory every year during 1982-1987 (Nha, 1990). Phase coverage each year in B and V passbands for this difficult star (because its orbital period is nearly two days) is fairly complete. This enables us to extend our redetermination of the spot parameters with the method applied to earlier observations by KW.

2. Solution with spot parameters

In order to analyze the light curves of AR Lac for starspot parameters, the normal point was weighted according to the number of observations used to create it.

In this study we adopt the light curves of 1983 as a reference for the determination of non-spot parameters, because these curves apparently lack the so-called wave-like distortion showing a clear symmetry. We adjusted the mass ratio of the system, temperature of star 1 ($T1$), and potentials (Ω), gravity darkening coefficients (g), albedo (A), and limb darkening coefficients (x) of both stars. The nonspot parameters obtained are given in Table 1, and these were kept fixed for remaining five epochs, because we are mainly interested in the contribution of the spot parameters to the light curve fitting and in the spot migration with two-spot model. The adjusted spot parameters found for multicolour light curves in five epochs are tabulated in Table 2. The fits to the light curves for six epochs, including the 1983 epoch without spot parameters, are presented in Figure 1 for V only.

3. Spot Migration Period

To investigate the spot migration, the longitudes of the spots by the data in Table 2 with two

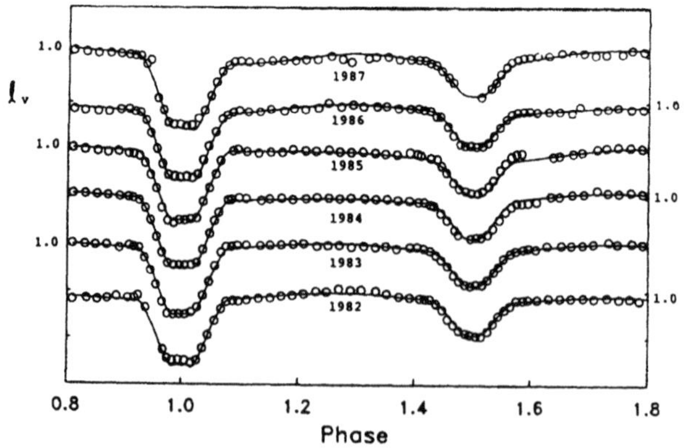

Figure 1. V light curves of AR Lac at six epochs with our fitted model curves.

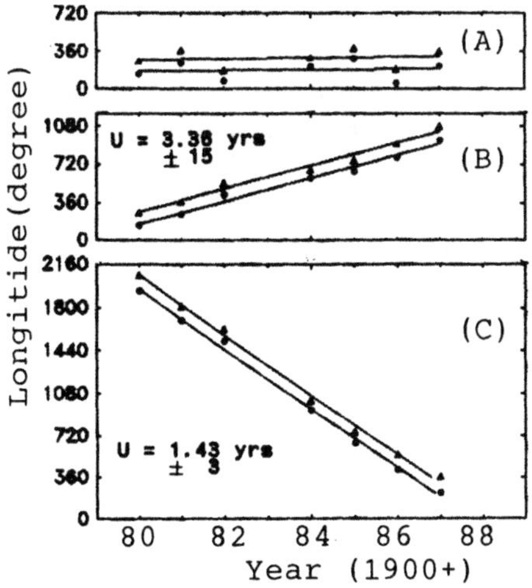

Figure 2. Longitudes of two spots (△ and •) in each year. (A) represents the longitudes in Table 2, (B) the increasing longitudes, and (C) the decreasing longitudes with time.

Table 1. Adjusted parameter of AR Lac at 1983 epoch.

	i	T	g	A	Ω	x(V)	x(B)	M_2/M_1
Star 1	87.0	5368 K	0.30	0.936	6.862	0.117	0.105	1.022
		± 125	± 3	± 129	± 51	± 21	± 19	± 21
Star 2	87.0	4550 K	0.155	0.421	4.374	0.838	0.750	
		± 125	± 15	± 85	± 36	± 49	± 35	

Table 2. Adjusted spot parameters of AR Lac for five different epochs.

	1982	1984	1985	1986	1987
Spot 1					
Latitude	20°	20	20	20	20
Longitude	177°	297	26	194	359
r_{spot}	11°	13	15	24	16
T_{spot}/T	0.82	0.80	0.80	0.93	0.80
Spot 2					
Latitude	20°	20	20	20	20
Longitude	79°	220	285	62	220
r_{spot}	22°	11	15	14	22
T_{spot}/T	0.95	0.80	0.80	0.85	0.82

previous years of KW are plotted against time in Figure 2A which shows no indication of spot migration or maybe too fast migration, if there were any. Figure 2B is made for spots changing the direction in an increasing longitude in time adding one cycle (360 deg) to 1982 - 1983 and two cycles (720 deg) to 1986 and 1987. Figure 2C shows, on the other hand, the spots of decreasing longitude subtracting one cycle for 1981 and five cycles for 1987.

The periods of spot migration 3.36 yr and 1.43 yr are deduced for increasing longitude (Fig. 2B) and for decreasing (Fig. 2C), respectively. These are short compared with that of 10 yr by KW, but closely match 2.03 yr deduced by Kim's (1989) wave minimum analysis if we assume the migration to be retrograde.

All of these attempts with our intensive photometry of seven continuous years allows us, unfortunately, to derive no conclusion concerning spot migration on AR Lac.

References

Kang, Y. W., Wilson, R. E., 1989, Astron. J., 97, 848.

Kim, C. H., 1989, Ph.D. dissertation, Yonsei University, Seoul.

Kron G. E., 1947, Publ. Astron. Soc. Pacific. 59. 261.

Nha, I.-S., 1990. in press.

MODELLING OF THE LARGE X-RAY FLARE ON II Peg OBSERVED WITH *GINGA*

B.J. Kellett, Rutherford Appleton Laboratory, UK
J.G. Doyle, Armagh Observatory, Armagh, N. Ireland
G.E. Bromage, Rutherford Appleton Laboratory, UK
P.B. Byrne, Armagh Observatory, Armagh, N. Ireland

1. Introduction

The UK-Japanese Large Area Counter (LAC) on the X-ray Satellite *GINGA* (Turner *et al.* 1989) observed the RS CVn system II Peg in August 1989. At around 02:00 UT on 17 August, 1989, a very large X-ray flare was detected with *Ginga* and ground-based Johnson U-band monitoring (Doyle *et al.* 1990). Because of the very good sensitivity and good background rejection of the LAC, we have excellent statistics of the decay of this flare. This enables us to test more sophisticated flare models on the data. Initially, we followed the method of van den Oord and Mewe (1989), who analysed the *EXOSAT* Algol flare (White *et al.* 1986). They developed a model which they called "quasi-static" cooling. Fig. 1 shows a typical model fit to our II Peg data. This fit assumes no additional heating took place during the flare decay. Clearly, from Fig. 1, whilst the temperature can be modelled quite well for reasonable input constraints, the run of emission measure (EM) data is not very well fit by the model. This leads to the suspicion that it is the *static* part of the theory that is causing the differences we see; *i.e.* the model assumes a static loop, and perhaps we are observing an expanding loop.

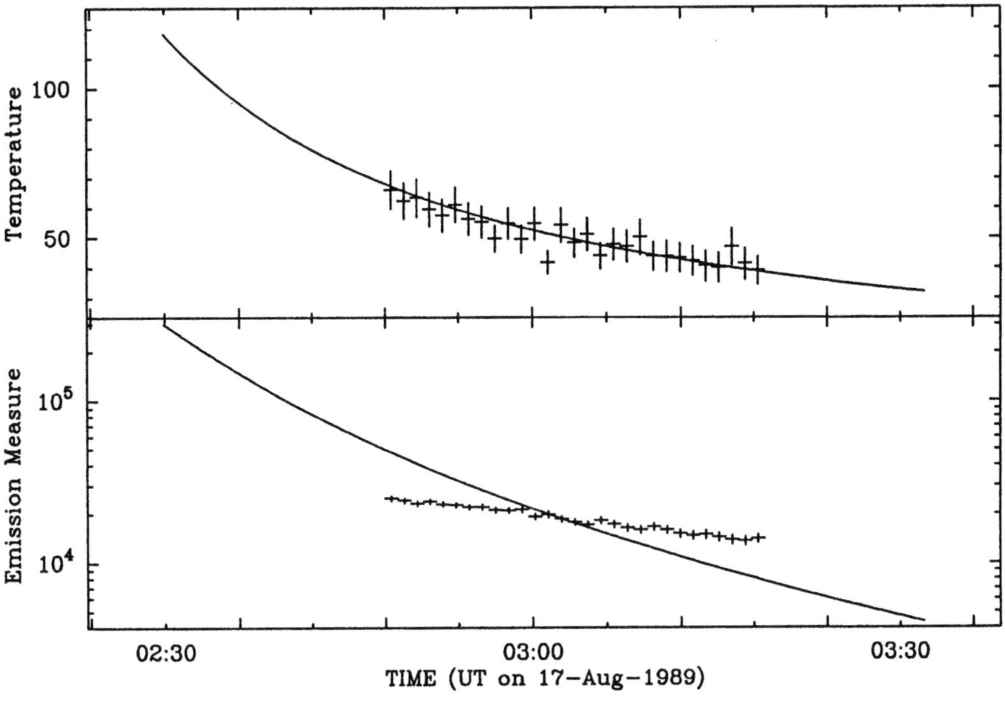

Figure 1. *GINGA* X-ray data on the II Peg X-ray flare. The data points are 64 second integrations. The temperature is in units of 10^6 K, and Emission Measure is in $10^{50}\,cm^{-3}$. The solid line shows a typical "quasi-static" model fit.

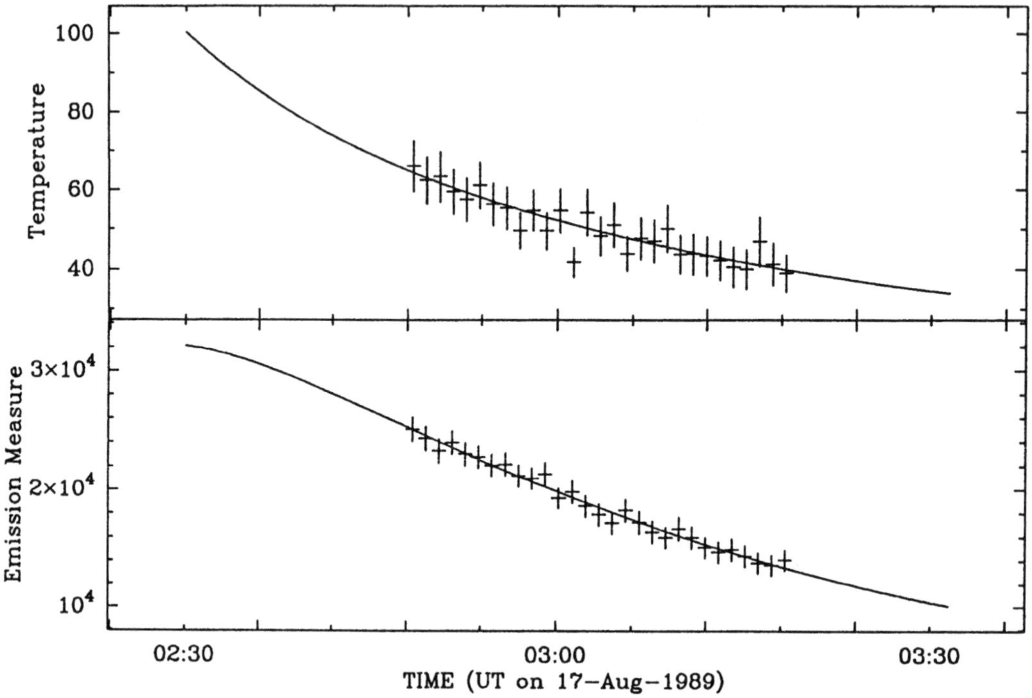

Figure 2. *GINGA* X-ray data on the II Peg X-ray flare. As for Figure 1, but for the modified model, with Poletto *et al.* volume expansion.

2. Modelling

To test this hypothesis, we have developed a simple extension of the quasi-static model, where we use the predicted temperature and density decays given by the quasi-static model, but instead of a rigid (static) loop, we adopt an expanding loop. We use the model of Poletto *et al.* (1986), to describe the volume expansion. This model was developed to study solar and stellar "two-ribbon" flares, and assumes magnetic reconnection in an arcade of loops. Since we are only modifying the *volume* part of the original quasi-static model, we still get the same temperature decay with time, but the EM decay now contains two effects, the "quasi-static" density *decay* and the "magnetic-reconnection" volume *expansion.* This allows the (T, EM) data to be fitted very well as shown in Fig. 2.

3. Discussion

The above modelling is clearly a simplified representation of the complex stellar flare event, and requires further investigation in the future, together with application to similar large flares which we hope to observe with future X-ray instrumentation (*e.g. ROSAT*). One immediate question to ask is whether this extension of the "quasi-static" model actually invalidates that model's assumptions. The main assumption that goes into the quasi-static model is that the pressure in the loop stays uniform. Applying the "standard" single-loop model to our data results in a pressure estimate of $\sim 12\,000$ dyn. This is derived from the equation for the gas pressure, $P = 2n_e kT$, and is perhaps rather a high value. The loop is assumed to be magnetically confined, and in this case would require a field of at least ~ 550 G. The equivalent values for the Algol flare are 5 000 dyn and > 200 G. If the magnetic field strength in the II Peg flare was actually significantly less than 550 G, it is conceivable that the loop would be forced to expand, whilst the pressure within the loop remained approximately uniform; thereby not violating the model constraints. Very sophisticated numerical models do exist for the study of solar flares, but applying them to the much simpler stellar flare data available is not very easy since we do not know the spatial structure of the underlying active region (for example). But with the advent of *Ginga* (and now *ROSAT*), X-ray observations of much higher sensitivity will become available, and the need for better stellar flare models will increase.

References:

Doyle, J.G., Kellett, B.J., Avgoloupis, S., Byrne, P.B., Tsuru, T., Bromage, G.E., Seiradakis, J.H., Pye, J.P., Barstow, M.A., McHardy, I.M., Makishima, K., & Mavridis, L.N.: 1990, (in preparation)

Poletto, G., Pallavicini, R., Kopp, R.A.: 1988, A&A, 201, 93

Turner, M.J.L., Thomas, H.D., Patchett, B.E., Reading, D.H., Makishima, K., Ohashi, T., Dotani, T., Hayashida, K., Inoue, H., Kondo, H., Koyama, K., Mitsuda, M., Ogawara, Y., Takano, S., Awaki, H., Tawara, Y., Nakamura, N.: 1989, PASJ, 41, 345

van den Oord, G.H.J., Mewe, R.: 1989, A&A, 213, 245

White, N.E., Culhane, J.L., Parmar, A.N., Kellett, B.J., Kahn, S., van den Oord, G.H.J, Kuijpers, J.: 1986, ApJ, 301, 262

LIGHT CURVE ANALYSIS OF STARS WITH MORE THAN ONE SPOT

Diana P. Kjurkchieva
Department of Physics, Higher Pedagogical Institute, Shoumen, Bulgaria

1. Introduction

The analysis of light curves with more than one mimimum is normally done by numerical modelling. This approach, however, posseses an important disadvantage, *i.e.* ambiguity in the geometric parameters of the configuration determined. We propose a method for analysis of such light curves on the basis of an analytic solution of the respective direct problem. Such an approach has already been developed for light curves with one mimimum (Kjurkchieva, 1989, Paper I).

The aim of our analysis is the determination of the parameters, i, the inclination of the rotational axis, α_k, the angular size of the k-th spot, λ_k, its phase of maximum visibility and β_k, the polar distance of its centre. The solution of the foregoing problem depends on the light curve shape, mainly on the phase difference $\Delta\lambda = |\lambda_2 - \lambda_1|$ of the spots.

2. Case A

In Case A the light curve possesses two symmetric mimima separated by flat sections (Fig. 1). The condition $\varepsilon_1 + \varepsilon_2 < \Delta\lambda$ is fulfilled and the analysis goes by analogy to the case with one spot (Paper I) in the following order:

1 Measurment of four parameters for each mimimum: *(a)* the phases λ_k of the deepest points of the mimima, *(b)* the phase distances ϕ_k from the points where the rate of change of the light curve slope changes sign, to the phase of minimum, λ_k, *(c)* the phase distances ε_k from the end of the flat section to the phases of minimum, λ_k and *(d)* the brightness differences $\Delta m_k = m(\varepsilon_k) - m(\gamma_k)$ where

$$\gamma_k = \frac{1}{2\pi} \arccos\left(\cos\varepsilon_k + \cos\phi_k\right)/2$$

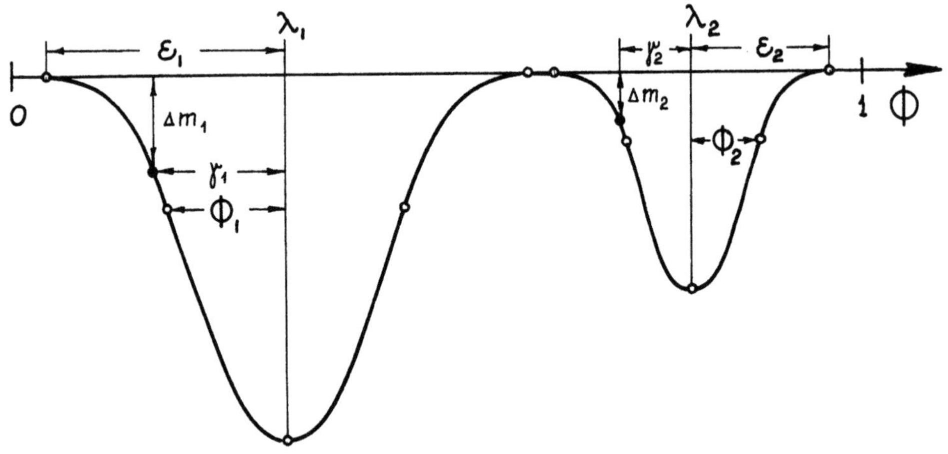

Figure 1. *Case A*

2 Determination of the temperature parameters for each spot by the Vogt method (Vogt, 1981), *i.e.* temperatures T^{st} and T_k^{sp}, surface brightness intensities I_{0k}^{sp} and I_0^{st} and the limb-darkening coefficients u_k^{sp} and u^{st}.

3 Calculation of the geometric parameters similarly to Paper I, *viz.* *(a)* the angular spot sizes α_k from the equations

$$\mathrm{dex}(\Delta \mathrm{m}_k/2.5 - 1) = 0.5E_k(2 - 3\cos\alpha_k + \cos^2\alpha_k) + 3(F_k - E_k)(\alpha_k - \sin\alpha_k \cos\alpha_k)$$

where

$$E_k = (I_{0k}^{sp} u_k^{sp}/I_0^{st} - 1)/(3 - u^{st}) \text{ and } F_k = (I_{0k}^{sp}/I_0^{st} - 1)/(3 - u^{st}$$

(b) The polar distance β_k and the rotation axis inclination i from the equations:

$$i_{ik} = \beta_{2k} = \tfrac{1}{2}\arccos(-\sin\alpha_k - \tfrac{1+N_k}{M_k}) + \tfrac{1}{2}\arccos(\sin\alpha_k - \tfrac{1+N_k}{M_k})$$

where

$$M_k = (\cos\phi_k - \cos\varepsilon_k)/2 \text{ and } N_k = (\cos\phi_k + \cos\varepsilon_k)/2.$$

The case with two spots gives a criterion for a choice between both pairs (i_{1k}, β_{1k}) and (i_{2k}, β_{2k}): obviously these are those pairs that have equal or nearly equal values of the inclination i. So the determination of the geometric parameters of the configuration becomes completely unambiguous.

3. Case B

In Case B the light curve is smooth, contains two clear, nearly symmetric mimima that partially superpose and a flat section (Fig. 2).

In this case the condition, $Max(\varepsilon_1, \varepsilon_2) < \Delta\lambda < \varepsilon_1 + \varepsilon_2$ is fulfilled and the locations of the deepest parts of the mimima coincide with the phases of maximum visibility of the spots. Then the light curve analysis can be made similarly to Case A. For that purpose the parameters $\lambda_k, \phi_k, \varepsilon_k$ and $\Delta \mathrm{m}_k$ could be measured on the nonoverlaping branches of the two mimima.

4. Case C

In Case C the second mimimum of the light curve is poorly expressed and distorted (Fig. 3). In this case the condition $Min(\varepsilon_1, \varepsilon_2) < \Delta\lambda < Max(\varepsilon_1, \varepsilon_2)$ is fulfilled. Then the parameters

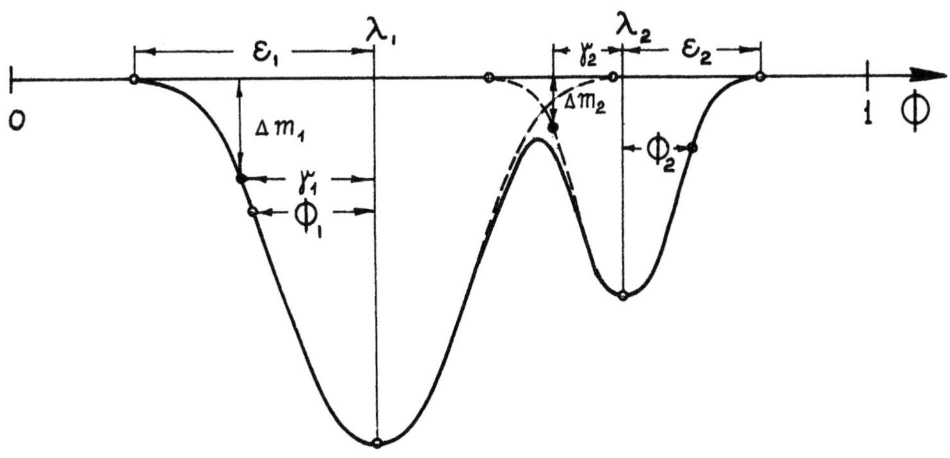

Figure 2. *Case B*

λ_1, ε_1, ϕ_1 and Δm_1 could be measured on this nonoverlaping branch that has more symmetric top and is due to the one spot only. In order to determine λ_2 it is necessary to build its symmetric half with respect to the phase λ_1 and to find the phase λ_2 at which the distance between this symmetric branch m_1 and the total curve m_{tot} is maximum. The parameters ϕ_2, ε_2 and Δm_2 might be measured on the curve $m_2 = m_{tot} - m_1$.

The cases A, B and C allow simple generalizations for light curves with more than two mimima.

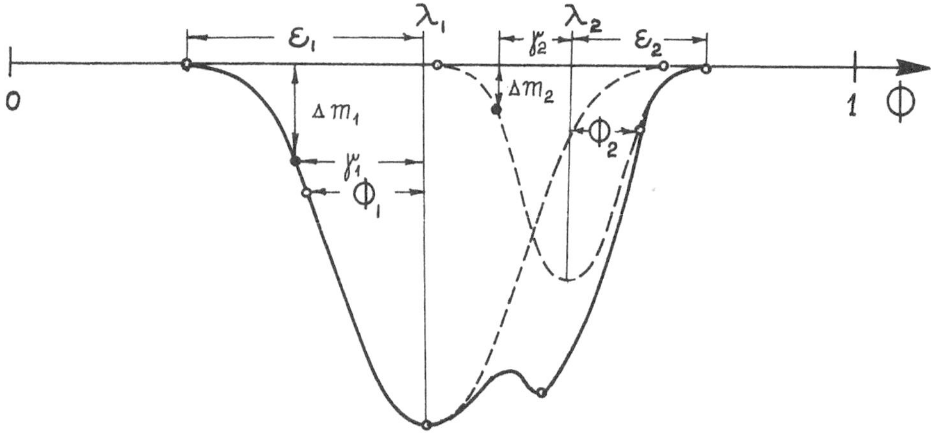

Figure 3. *Case C*

References

Kjurkchieva, D.: 1989, A&SS, 155, 125 Vogt, S.: 1981, ApJ, 250, 327

D. Kjurkchieva reads all about it!

A SEMI-ANALYTICAL MODEL OF STELLAR FLARES

R.A. Kopp[1] and G. Poletto[2]
[1]Los Alamos National Laboratory, Los Alamos, NM 87545, USA
[2]Osservatorio Astrofisico di Arcetri, Largo Enrico Fermi 5, 50125 Firenze, Italy

1. Introduction

Observations of stellar flares in X-rays have been made extensively only in the last few years, by the Einstein and EXOSAT observatories. However, the limited number of observed events and the lack of spatial resolution in the data have not stimulated a thorough analysis of stellar flares, which consequently remain poorly understood phenomena. The hydrodynamics of the flaring coronal loop plasma, for instance, has been studied by only a few authors. The work of Reale *et al.* (1988), who applied the numerical hydro-code calculations previously used in solar flare studies to analyse a flare on Prox Cen, represents, so far, nearly the only attempt to extend methods used in solar physics to stellar flare research. On the other hand, the lack of any information on the spatial characteristics of stellar events suggests dropping such sophisticated numerical schemes in favour of a simplified technique capable of predicting the time dependence of the only observable quantities, namely, the spatially averaged plasma properties.

In the following we propose a simple "point" model which includes thermal conduction, chromospheric evaporation, radiative losses, and gravitational loop draining, and which gives the temporal evolution of temperature, density, and velocity averaged over the flaring loop. The model allows a rapid survey to be made of flare-loop parameter space: we show that the high flare emission measures (EM) can be interpreted within the framework of the model, although it is not possible to evaluate uniquely the physical parameters of the flaring region.

2. The Flare Model

Our model hypothesizes a sudden energy release in a rigid flux tube and describes the successive behavior of the loop plasma. By integrating the partial differential equations for mass and energy conservation along a loop, whose pressure is assumed to be spatially uniform, and equating the downward conductive flux to the upward enthalpy flux at the loop base, we get a set of ordinary differential equations for the average loop pressure, $P(t)$, and density, $\rho(t)$:

$$\frac{dP}{dt} = \frac{(\gamma P v)}{L} + \frac{(\gamma-1)}{L}[(F_c - F_{c0}) + (F_r - F_{r0})]$$

and

$$\frac{d\rho}{dt} = \frac{\rho v}{L}$$

where γ is the specific heat ratio, $v + [(\gamma - 1)/\gamma](F_c - F_{c0})/P - v_{ff}$ is the evaporation velocity minus the gravitational free-fall velocity, and L is the loop semi-length; F_c, F_r, and F_{c0}, F_{r0} represent the conductive and radiative fluxes at times t and $t = 0$ (pre-flare atmosphere), respectively.

The adequacy of the model to reproduce realistically the flaring loop plasma behavior has been tested by comparing its predictions against those from full hydro-code calculations, commonly adopted in solar flare modelling. In order to make this comparison we have selected the "standard model" of Pallavicini *et al.* (1983). Results from the "point" model agree quite closely with the results of Pallavicini *et al.* throughout the decay phase of the flare, for a loop with identical boundary and initial conditions (Kopp and Poletto, 1990).

3. Application to Stellar Flares and Conclusions

Stellar flares cover a broad range of X-ray energies and emission measures. On M dwarf stars' EM's are about 10^3 times larger than solar flare EM's, ranging from $\approx 10^{51}$ to a few times $10^{53}\,\mathrm{cm}^{-3}$ (Pallavicini *et al.* 1990). Our point model has the capability of determining values for the flare parameters which are compatible with these high emission measures. To this end we evaluated the pressure and density profiles resulting from different energy inputs, for loops with an aspect ratio of 0.1 and semi-lengths, $L = 0.5$, 1, and $2\,10^{10}\,\mathrm{cm}$. Observational results suggest that initial conditions may vary from "low" values of $T = 3\,10^6\,\mathrm{K}$, $N_e = 1.2\,10^{10}\,\mathrm{cm}^{-3}$ to "high" values of $T = 1\,10^7\,\mathrm{K}$, $N = 1.2\,10^{11}\,\mathrm{cm}^{-3}$. A check on the influence of initial conditions on the results shows, however, that the subsequent flare-plasma temperature and density values vary by less than a factor of 2 over this range of initial conditions. In all our simulations the preflare state was eventually re-established.

Fig. 1 gives the maximum EM as a function of the flare energy input, for the three loop semi-lengths. These results show that:
1) EM values between 10^{51} and $10^{53}\,\mathrm{cm}^{-3}$ can be easily explained by our choice of initial conditions and loop geometries;
2) the flare energy input is, for the largest flares, a few hundred times larger than in typical solar flares - see Pallavicini *et al.*'s (1983) "standard model";
3) large flares can originate only in large loops, because a large energy input in a small loop results in a catastrophic loop disappearance driven by large radiative losses. "Intermediate" flares, however, can be explained by a variety of parameter values. Observations of flare electron densities, via the density sensitive line ratio technique, may help solve this ambiguity.

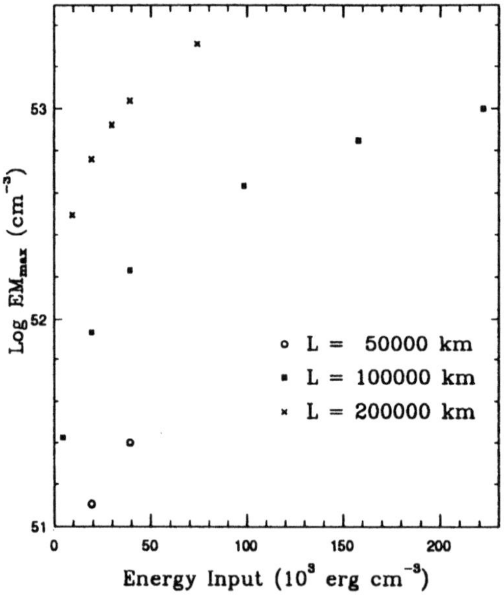

Figure 1 Maximum emission measure versus energy input for different loop semi-lengths, L. Gravity is assumed equal to $2g_\odot$.

Acknowledgements

R.A. Kopp is grateful to Gruppo Nazionale di Astronomia of CNR and to Dipartimento di Astronomia e Scienza dello Spazio of the Universitá di Firenze for support and hospitality. The work of G. Poletto has been supported by the Italian Space Agency.

References

Kopp, R.A., Poletto, G.: 1990, in *Mechanisms of Chromospheric and Coronal Heating,* P. Ulmschneider. Ed., in press.

Pallavicini, R., Peres, G., Serio, S., Vaiana, G., Acton, L., Leibacher, J., Rosner, R.: 1983, ApJ, 270, 270

Pallavicini, R., Tagliaferri, G., Stella, L.: 1990, A&A, 191, 109

Reale, F., Peres, G., Serio, S., Rosner, R., Schmitt, J.H.M.M.: 1988, ApJ, 328, 256

SUNSPOT GROUP BRAKING AND PHOTOMETRIC PERIOD VARIATIONS IN SPOTTED STARS

A.F. Lanza[1], M. Rodonó[1,2], R.A. Zappalá[2]

[1]Istituto di Astronomia dell'Universitá degli Studi
[2]Osservatorio Astrofisico, Cittá Universitaria
Viale A.Doria 6, 95125 Catania, Italy

1. Introduction

The variations of photometric periods in BY Dra stars and of the migration rate of the distortion wave in RS CVn systems have been attributed to latitudinal dependent stellar differential rotation (Rodonó, 1981, Catalano, 1983, Rodonó, 1986, Busso *et al.* 1984). Moreover the values of $\frac{\Delta\Omega}{\Omega} = 10^{-3} - 10^{-2}$ appear to be one or two orders of magnitude less than solar ($\frac{\Delta\Omega_\odot}{\Omega_\odot} = 0.2$) and $\frac{\Delta\Omega}{\Omega}$ shows a general tendency for a decrease toward increasing Ω. The last evidence is in contrast with current rotation theories which predict an increase of $\frac{\Delta\Omega}{\Omega}$ with Ω and with the dynamo theory which requires higher $\frac{\Delta\Omega}{\Omega}$ in order to explain the observed activity levels.

Zappalá and Zuccarello (1989, 1990) studied the rotational angular velocity of sunspot groups as a function of their age, pointing out that the angular rotation velocity, Ω_s of the groups younger than 10 days is on the average significantly higher than for groups older than one synodic rotation. The average angular velocity difference is a function of time only and does not show any significant dependence on latitude or group area. Moreover the proportion of young and old groups varies along the solar activity cycle. In particular, the ratio of the number of young groups to the number of old groups is higher during spot minimum phase. These selection effects might affect the solar rotation determination in the sense that on the average the measured angular velocity is higher during minima than close to maxima (Zappalá et al. 1990). A relevant consequence of the above discussed phenomenology is that the observed solar rotation law $\Delta\Omega(t)$ can be used together with the spot groups distribution function to calculate the photometric period variations which would result by observing the Sun as a star. If the behavior of a spotted star is similar, variations of the photometrically determined rotation period should occur because of the changing proportion of young and old groups along the activity cycle and

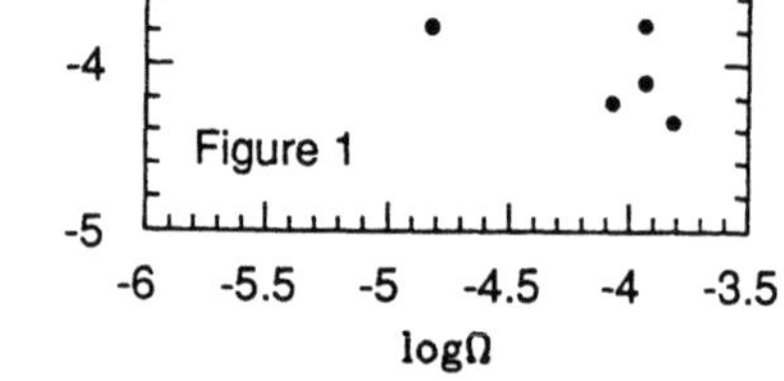

Figure 1. The observed maximum values of $\log[\Delta\Omega/\Omega]$ for a sample of RS CVn systems are plotted versus the orbital angular velocity, Ω. The value of $\log[\Delta P/P]$ computed for the Sun (see text) is labelled.

Table 1. List of the studied RS CVn systems

Star name	SN*	References
RU Cnc	65	Busso *et al.*: 1984, A&A, 135, 255
CQ Aur	48	Busso *et al.*: 1984, A&A, 135, 255
SV Cam	53	Unpublished data.
VV Mon	54	Busso *et al.*: 1985, A&A, 148, 29
SS Boo	98	Busso *et al.*: 1985, A&A, 148, 29
UV Psc	12	Busso *et al.*: 1986, A&A, 156, 106
UX Ari	26	Busso *et al.*: 1986, A&A, 156, 106
BY Dra	126	Vogt: 1981, ApJ, 250, 338
RS CVn	89	Blanco *et al.*: 1982, A&A, 106,311
RT And	163	Milano *et al.*: 1982, A&A, 103, 57
WY Cnc	68	Milano 1981, *Photometric and Spectroscopic Binary Systems,* eds. E.B. Carling and Z. Kopal, 331
CG Cyg	142	Milano: 1981, *ibidem*
XY UMa	69	Milano: 1981, *ibidem*
SZ Psc	164	Blanco *et al.*: 1983, IAU Coll. 71, 395
RT Lac	153	Scharlemann: 1982, ApJ, 253, 298
AR Lac	155	Kurutac *et al.*: 1981, A&SS, 77, 325
HK Lac	154	Catalano: 1983, IAU Coll. 71, 330
V711 Tau	27	Catalano: 1983, *ibidem*
IM Peg	159	Eaton *et al.*: 1983, A&SS, 89, 53
LX Per	24	Ibanoglu *et al.*: 1983, A&SS, 95, 409
SS Cam	56	Hall: 1976, IAU Coll. 29, 287
HR4665	83	Strassmeier *et al.*: 1988, A&AS, 72, 329
Z Her	117	Evren *et al.*: 1982, A&SS, 87, 51
MM Her	118	Sowell: 1983, A&SS, 90, 421
HR7275	132	Fried *et al.*: 1982, A&SS, 83, 181
V741 Tau	29	Evren *et al.*: 1986, A&SS, 120, 97

The main source of the $\frac{\Delta P}{P}$ of the active stars was *A Catalogue of Chromospherically Active Binary Stars* by Strassmeier *et al.* A&AS, 72, 291 (1988) and references therein. We give below the list of the refereces for each star in our sample. *SN is the Strassmeier *et al.* (1988) catalogue number.

the different decay times of their angular velocities. These variations are indeed osservable as variations of the photometric period because of the latitude independence of $\Delta\Omega(t)$. In Fig. 1, we report the observed $\log\frac{\Delta P}{P}$ *vs.* $\log P$ for a sample of active stars (see Table 1) together with the computed value for the Sun. The agreement between solar and stellar values gives credit to the proposed hypothesis and to angular velocity differences among starspot groups of different ages. Present stellar observations cannot yet answer this question.

We have developed a theoretical model aimed at explaining the solar phenomenology and to interpret the active star behaviour.

2. Theoretical approach

Following Schussler (1987), the spot group angular velocities give a tool to probe the internal angular velocity of the Sun. Present models of solar rotation (Pidatella *et al.* 1986) are not in agreement with the hints obtained from sunspot groups dynamics, that is with an increase of Ω of about 2-3% within a depth of about (1-2) 10^7 m (Gilman and Foukal, 1979, Schussler, 1987). Numerical modelling of these layers (Gilman and Foukal, 1979) is very difficult due to

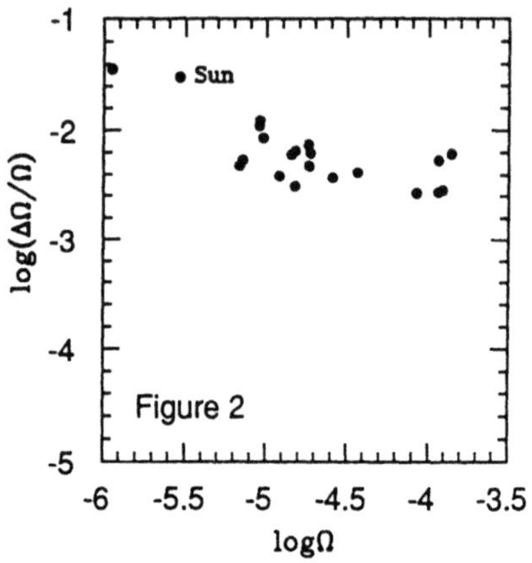

Figure 2. The computed values of the maximum $\log[\Delta\Omega/\Omega]$ *vs.* $\log\Omega$. The Sun is labelled.

the strong non-adiabaticity and the high resolution required. Therefore we have developed an analytic model base on mean stationary continuity, momentum and energy equations. We can solve analytically the problem of the external convective layer rotation by assuming a spherical system of coordinates with its origin in the center of the star and polar axis directed along Ω, and the following simplifying assumptions: *a)* to neglect the tidal interaction (Scharlemann, 1982) and the Lorentz force and the drag force due to magnetic flux tubes (Foukal and Jokipii, 1975); *b)* for the non-rotating case ($\Omega = 0$), the velocity field $u = u_0$ is purely radial and has no azimuthal dependence: $u_0 = u(r,\theta)\hat{}r; c)$ the rotation-induced perturbations on all scalar quantities (such as ρ, T, p, dynamical turbulent viscosity, η, superadiabatic gradient $\Delta\nabla T$, which depend only on r) are assumed to be negligible, both for $\Omega = 0$ and $\Omega \neq 0$ rotation regimes.

In conclusion we find for the external layer:

$$\Omega(r,\theta) = \Omega(r_0,\theta)(\tfrac{r_0}{r}) \tag{1}$$

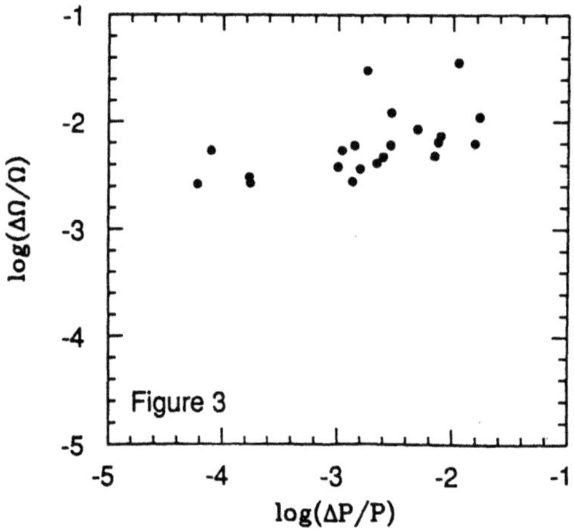

Figure 3. The computed values of $\log\frac{\Delta\Omega}{\Omega}$ *vs.* observed $\log[\Delta P/P]$.

where $\Omega(r_0, \theta)$ is the angular velocity at the base of the layer $r = r_0$. If d is the total depth of the layer the maximum difference of angular velocity across it, is:

$$\frac{\Delta\Omega}{\Omega} \sim \frac{d}{R} \qquad (2)$$

Note that $\Delta\Omega$ is latitude independent in agreement with the evidence given by solar observations.

3. Results and discussion

We have estimated the value of d for the spotted components of the sample of RS CVn systems listed in Table 1 and whose physical parameters are known. We have assumed d to be equal to the distance between the top of the convection zone and the level where $\frac{8z}{v_c(z)} = \frac{1}{2\Omega}$; z is the depth and $v_c(z)$, the convective velocity at depth z computed by a classical mixing length model (Belvedere *et al.* 1980). We report in Fig. 2 the estimated $\frac{\Delta\Omega}{\Omega}$ *vs* Ω for active stars. The orders of magnitude of the computed values are the same as the observed ones and, also, the general trend is the same, except for high Ω values. In order to allow a direct comparison we report in Fig. 3 the computed $\frac{\Delta\Omega}{\Omega}$ *vs* the observed $\frac{\Delta P}{P}$. It should be borne in mind that the computed values are upper limits because the observed values are weighted averages depending on the spot age distribution function and on their decay times. We emphasize that, in spite of its pure speculative nature, the proposed model has the advantage of reconciling the current theory of stellar differential rotation with observations. Nevertheless an important consequence follows: the stellar activity centers do not significantly migrate in latitude along the activity cycle. A new possible scenario for stellar activity cycles appears. We have to assume not only the existence of active longitudes to explain some stellar observation (Busso *et al.* 1984) but also preferred latitudes or long time invariant latitude distribution of starspot groups.

We have also considered the role of small variations of stellar radius along the activity cycle. For the Sun the radius variation is ~ 0.1 arcsec, so that $\frac{\Delta R}{R} \leq 10^{-4}$ (Delache *et al.* 1988). This phenomenon may take place on other magnetically active stars (Bianchini, 1990 and references therein) with the same order of magnitude for the amplitude of the solar radius variation. In any case the associated variations of the moment of inertia of the convection zone is smaller by two orders of magnitude to explain the observed photometric period variation except, perhaps, for systems with $\frac{\Delta P}{P} \sim 10^{-4}$. We present the above ideas in order to stimulate further research and debate on this field.

References

Belvedere, G.,釣Paternó, L., Roxburgh, I.W.: 1980, A&A, 91, 356

Bianchini, A.: 1990, AJ, in press

Busso, M., Scaltriti, F., Blanco, C., Catalano, S., Marilli, E., Pazzani, V., Rodonó, M.: 1984, A&A, 135, 255

Catalano, S.: 1983, in *Activity in Red Dwarf Stars,* Eds. P.B. Byrne and M. Rodonó, Reidel, Dordrecht, p. 343

Delanche, Ph., Laclare, F., Sadsaoud, H.: 1988, in *Advances in Helio- and Asteroseismology,* Eds. J. Christensen-Dalsgaard and S. Frandsen, Reidel, Dordrecht, p. 223

Foukal, P., Jokipii, J.R.: 1975, ApJ, 199, L71

Gilman, P.A., Foukal, P.V.: 1979, ApJ, 229, 1179

Pidatella, R.M., Stix, M., Belvedere, G., Paternó, L.: 1986, A&A, 156, p. 22

Rodonó,: 1981, in *Photometric and Spectroscopic Binary Systems,* Eds. E.B. Carling and Z. Kopal, Reidel, Dordrecht, p. 285

Rodonó M.: 1986, in *Cool Stars, Stellar Systems and The Sun,* Eds. M. Zeilik and D.M. Gibson, Springer-Verlag, Berlin, p. 475

Scharlemann, E.T.: 1982, ApJ, 253, 298

Schussler, M.: 1987, *The Internal Solar Angular Velocity,* Eds. B.R. Durney and S. Sofia, Reidel, Dordrecht, p. 303

Zappalá, R.A., Zuccarello, F.: 1989, A&A, 214, 369

Zappalá, R.A., Zuccarello, F.: 1990, A&A, in press

Zappalá, R.A., Zuccarello, F., Lanza, A.F.: 1990, in preparation

STATISTICS OF MAGNETIC CYCLES IN LATE-TYPE SINGLE AND CLOSE BINARY STARS

C. Maceroni[1], A. Bianchini[2], M. Rodonó[3], F. van 't Veer[4], R. Vio[5]
[1] Osservatorio Astronomico, viale del Parco Mellini 84, I-00136 Roma, Italy
[2] Osservatorio Astronomico, vicolo dell'Osservatorio 5, I-35122 Padova, Italy
[3] Istituto d'Astronomia, via A. Doria 6, I-95125 Catania, Italy
[4] Institut d'Astrophysique, CNRS, 98 bis boulevard Arago, F-75014 Paris, France
[5] Dipartimento d'Astronomia, vicolo dell'Osservatorio 5, I-35122 Padova, Italy

1. Introduction

Magnetic cycles have been detected in late-type single stars and close binary components. In this paper we briefly report on the results obtained applying modern statistical methods to the study of the cycle period distribution and of its correlation with other fundamental stellar properties. A more complete treatment can be found in Maceroni et al. (1990, Paper I.)

2. The samples and their data

2.1 Single stars

The data on magnetic cycles come from Wilson's survey of chromospheric emission, as recently reviewed by Baliunas (1988). Cycles were detected in 60 out of 99 stars monitored for more than 20 years. The same stars also have known rotation periods (P_{rot}), which were directly measured (from spot induced light modulation) or inferred (by means of a calibration P_{rot} *vs.* activity level established by Noyes *et al.* 1984), and known spectral class (from Baliunas, 1988).

2.2 Binary stars.

2.2.1 RS CVn binaries: Our main source of data was the literature referred to in the Strassmeier *et al.* (1988) catalogue. The cycle periods were obtained by studying the photometric wave motion and amplitude (Busso *et al.* 1986 and references therein), or from photometric variability. The difficulties of distinguishing the periodicities effectively related to magnetic cycles reduced the original sample of 168 systems to 21 objects, which show a cyclic behaviour with periods from 3.5 to 60 years (this sample is richer by 30% than that presented in Paper I). The available data are in this case P_{cyc}, P_{rot}, the spectral type, and the absolute elements from the spectro-photometric solution.

2.2.2 Cataclysmic variables (CV) and symbiotic stars: Cycle periods for 19 systems were obtained by Bianchini (1988, 1990), assuming that the solar-like magnetic cycles affect the often unobservable late-type secondaries. On the hypothesis that a solar-like cycle produces a slight modulation of the stellar radius ($\Delta R/R \simeq 10^{-4}$, Warner, 1988) P_{cyc} can be measured from: a) the long term variations of the luminosity in old novae and nova-like variables, b) the interval between outbursts of dwarf novae, c) orbital period variations. The periods derived by Bianchini range from 3.4 to 29 years.

2.2.3 W UMa contact binaries: Cycles may be detected from orbital period variations or from photometric variability (cyclic appearance of spotted areas). We found just two references to reliable spot cycle detections (Bergeat *et al.* 1971, Guinan *et al.* 1988), while period studies are much more numerous. A very long record of period changes (90 years) for 70 contact systems was carried on at Katowice Observatory and presented by Glownia (1986a,b). However, a statistical test verifying the alternation of shortening and lengthening of period (Van 't Veer 1990), expected as a result of solar-like magnetic cycles, gave negative results. The cycles, if present, are masked by period jumps of random sign produced by other phenomena (mass exchange/loss? internal

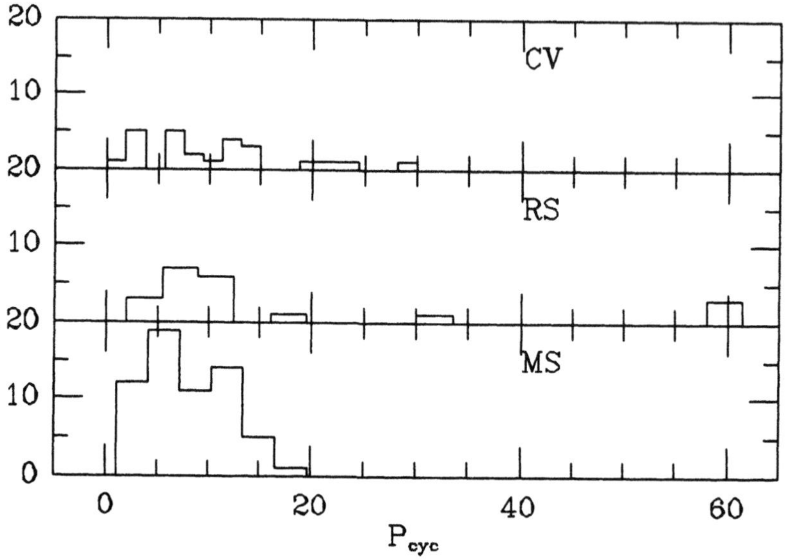

Figure 1. Cycle period histograms for CV's, RS CVn and single (MS) stars.

motions?). As a consequence, only the two systems (44i Boo and VW Cep) with photoeletric variability could be included in the statistical analysis (in the RS CVn figures).

3. The statistical analysis

The P_{cyc} histograms for the various samples were obtained by means of the standard kernel method which reduces the subjectivity introduced by the choice of bin size and left boundary (Fig. 1, see also Paper I).

One-sample and two-sample Kolmogorov-Smirnov tests (90% confidence level) confirm what is suggested by eye inspection, *viz.* all samples have non-uniform distributions, and, at the chosen confidence level, the same distribution. This allows us to add the sample together to get what we will call the total sample (Fig. 2).

The distributions go from the unimodal one of RS CVn's to the multimodal one of CV's. A common feature is, however, a peak for cycle periods around 6 yrs.

Double star samples are unfortunately too small for a meaningful analysis. For the single stars we could instead test the statistical significance of the second peak, performing Monte Carlo simulations in the hypothesis that the "parent sample" had a unimodal distribution, obtained by means of the empirical models of Johnson (Shapiro *et al.* 1981, see Paper I). The result is that the bimodality cannot be safely assessed. Exactly the same result is obtained for the total sample.

In the single star sample the cycle periods become longer for later spectral types. The median of the distribution is gradually shifted to higher values from F to K stars (Fig. 3).

Surprisingly enough there is no correlation between P_{cyc} and P_{rot}. This result is found for every sample and is indirectly confirmed by the fact that the cycle period distribution results to be the same for objects (*i.e.* single stars and CV secondary components) whose rotational velocities differ by a factor of ≈ 100. However, though the linear correlation coefficient is always very small (< 0.4), the analysis of the MS subsamples shows that the dispersion around the rather loose positive correlation P_{cyc} - P_{rot} is appreciably smaller for later spectral types. This

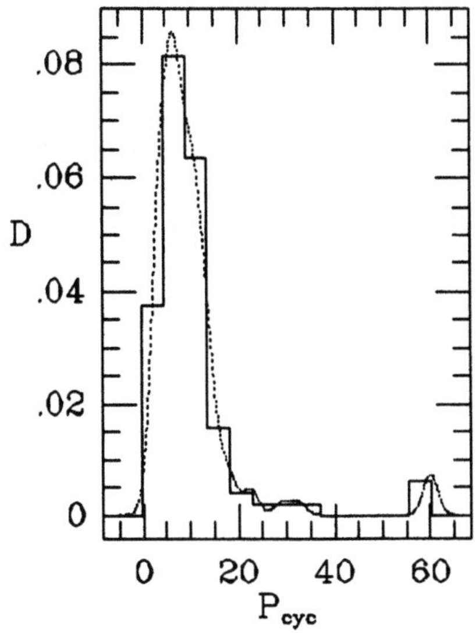

Figure 2. Probability Density Estimate of cycle period distribution for the total sample: Histogram (*continuous line*), kernal estimate (*dotted line*).

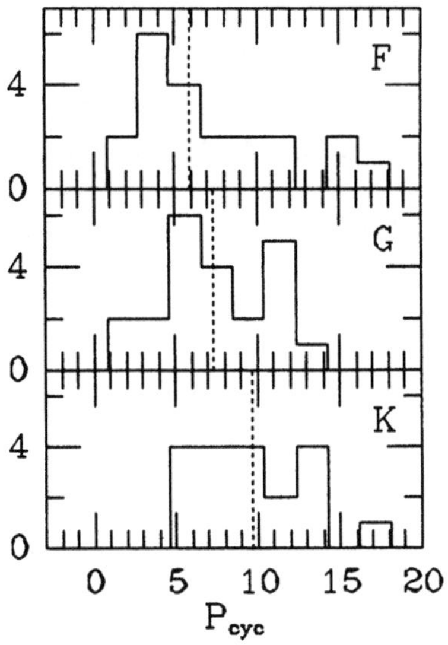

Figure 3 Cycle period histograms for F, G and K single stars. The *vertical dashed lines* indicate the medians of the distributions.

result does not apply to binaries.

4. Discussion and conclusions

Some of our results, and particularly those relevant to data trends, may be biased by the limited timespan of the observations. Long periods, close to this time limit, are certainly less reliable than the short ones. This effect is probably more serious for the single star sample, whose records extend to about 20 years. Close binary observations, though less homogeneous, may date back to several decades.

We believe, however, that the existence of the 6 yr peak is quite reliable; maybe future work on richer samples will assess also the reliability of the 11 yr peak, and even perhaps reveal a multimodal behaviour on which nothing can be said at the present moment.

The only possibility to improve the statistics of single stars is to wait until a longer data record can be accumulated. Close binary samples, instead, yield a second possibility: cycle detection is in their case indirect, but based on photometric effects or period measurements. A great amount of data already exists, stored in photometric or plate archives and only needs patient screening.

References

Baliunas, S.L.: 1988, *Formation and evolution of low mass stars,* eds. A.K Dupree and M.T.V.T. Lago, (Kluwer, Boston), p. 319

Bergeat, J., Lunel, M., Sibille. F., Van 't Veer, F.: 1972, A&A, 17, 215

Bianchini, A.: 1988, IBVS, No. 3136

Bianchini, A.: 1990, AJ, 99, 1941

Busso, M., Scaltriti, F., Cellino, A.: 1986, A&A, 156, 106

Glownia, Z.: 1986a, Thesis, N. Copernicus Astron. Center, Warsaw, Poland

Glownia, Z.: 1986b, Adv. Sp. Sci. Res. 6, 211

Guinan, E.F., Mc Cook, G.P., Bradstreet, D.H.; 1986, BAAS, 18, 977

Maceroni, C., Bianchini, A., Rodonó, M. , Van 't Veer, F., Vio, R.: 1990, A&A, Paper I, in press

Noyes, R.W., Hartmann, L.W., Baliunas, S.L., Duncan, D.K., Vaughan, A.H.: 1984, ApJ, 279, 763

Silverman, B.W.: 1986, in *Density Estimation for Statistics and Data Analysis,* Chapman and Hall.

Shapiro, S.S., Gross, A.J.; 1981, *Statistical Modelling Techniques,* Marcel Dekker, Inc., p. 194

Strassmeier, K.G., Hall, D.S., Zeilik, M., Nelson, E., Eker, Z., Fekel, F.C.: 1988, A&AS, 72, 291

Van 't Veer, F.: 1990, in preparation

Warner, B.: 1988, Nature, 336, 129

ON AN UNUSUAL CORRELATION FOR FLARE STARS

A.L. Mirzoyan
Byurakan Astrophysical Observatory, Armenia, USSR

1. Introduction

The distribution of flare stars on the (magnitude, colour) diagram reflects their evolutionary status. Consideration of the (V, B-V) diagram for the Pleiades flare stars by Mirzoyan and Mnatsakanian (1990) has shown, their flare activity increases with the decreasing V and increasing B-V. In agreement with this Jones (1981) has found that the flare stars situated above main sequence on this diagram are more active than stars below it.

In this paper we discuss the problem on the basis of more complete observational data.

2. Results and Discussion

The (V, B-V) diagram for 273 Pleiades flare stars derived from photographic photometry carried out by Chavushian and Garibjanian (1975) is presented in Fig. 1. The portion of galactic flare stars among them cannot exceed 10% (Mirzoyan *et al.* 1988)and so they cannot change the general picture qualitatively. As the dispersion of stars on this diagram is quite large we give mean data for groups of flare stars that have shown k flares ($k = 1, 2, ...6, \geq 7$). The numbers near the circles correspond to the magnitudes k (*below*) and the numbers of flare stars in groups (*above*). The flare activity, which is characterized by the magnitude k, increases almost regularly to the right lower corner of the diagram. Fig. 1 testifies to our conclusion that flare activity increases with decreasing V and increasing (B-V) (Mirzoyan and Mnatsakanian, 1990). A similar picture is observed on the (V, U-B) diagram of the Pleiades flare stars.

To understand the nature of these correlations we consider the dependance of flare activity on V and B-V, separately (Figs. 2 and 3).

The remarkable dependance of the flare activity (k) on B-V (Fig. 3) can be explained by the fact that the redder a flare star, the higher its flare activity. It seems very improbable that this is the result of observational selection.

It is more difficult to explain the decreasing flare activity with V (Fig. 2). It is possible that here observational selection plays some role, in the sense that, for weaker flare stars the

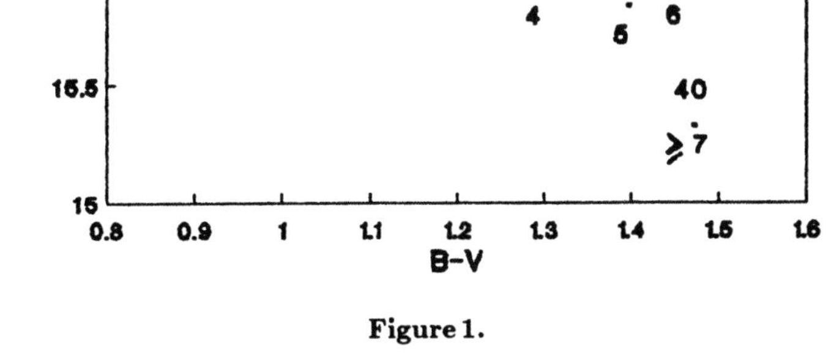

Figure 1.

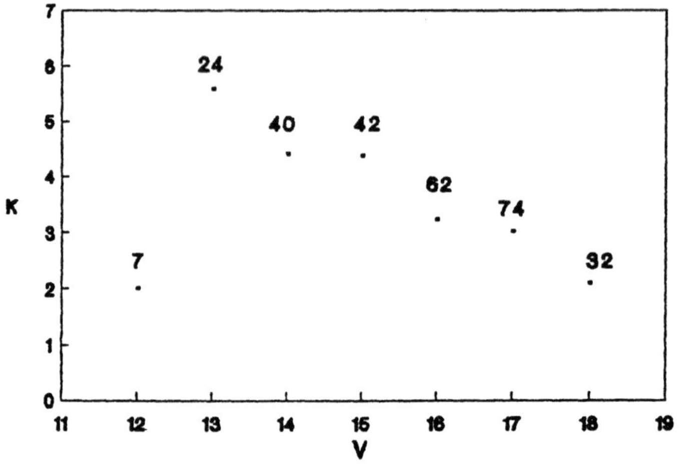

Figure 2.

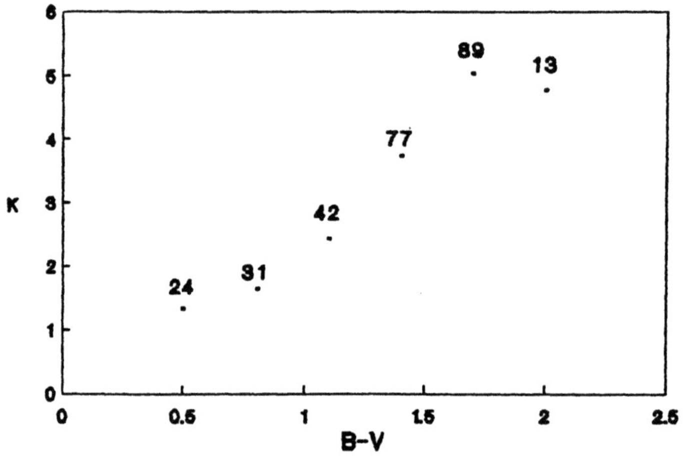

Figure 3.

duration of flares is on average shorter, and so the probability of their discovery is smaller. This correlation was found by Haro and Chavira (1955).

Taking into account Fig. 3, it must be assumed that the dependance of flare activity on colour is stronger than it follows from Fig. 3. A more detailed discussion of this problem is very desirable.

References

Chavushian, H.S., Garibjanian, A.T.: 1975, Afz, 11, 565

Haro, G., Chavira, E.: 1955, Bol. Obs. Tonantzintla, No. 12, 3

Jones, B.F.: 1981, AJ, 86, 290

Mirzoyan, A.L., Mnatsakanian, M.A.: 1990, in *Flare Stars in Star Clusters, Associations and the Solar Vicinity*, L.V. Mirzoyan, B.R. Pettersen and M.K. Tsvetkov (eds), Kluwer Academic Publishers. Dordrecht, p. 77

Mirzoyan, L.V., Hambarian, V.V., Garibjanian, A.T., Mirzoyan, A.L.: 1988, Afz, 29, 531

ON SPECTRA OF SOME PLEIADES FLARE STARS

L.V. Mirzoyan, V.V. Hambarian and A.T.Garibjanian
Byurakan Astrophysical Observatory, Armenia 378433, USSR

1. Introduction

The discovery of young flare stars in stellar associations and clusters brought Haro (1968) and Ambartsumian (1969, 1971) to the conclusion that the flare stars represent one of the early stages in the evolution of red dwarf stars. The results of extensive photographic observations of flare stars in the nearest associations and comparatively young clusters have confirmed this principle conclusion (1984). In this paper we discuss some results of low-dispersion spectral observations of flare stars in the Pleiades cluster.

2. Spectral Observations of Flare Stars

The spectra of 17 Pleiades flare stars were observed with the 6-m telescope of the Special Astrophysical Observatory in North Caucasus. The 1000-channel television scanner spectrometer was used. The effective resolution of the spectra obtained is ≈ 5Å. The majority cover the spectral region 3600 - 6800 Å.

In the spectra of 14 flare stars intense hydrogen Balmer and CaII H & K emission lines are observed. Molecular bands, TiO in particular, are very strong. In one spectrum (PFS 243 PFS = Pleiades Flare Star; Haro *et al.* 1982) emission lines are absent but molecular bands are observed. Only in two spectra (PFS 70 and 191) neither Hα emission nor molecular bands are present. All three of these stars are excluded from further discussion. Probably they are not flare stars (Mirzoyan *et al.* 1990) since the number of flares observed for each is only one (Haro *et al.* 1982).

3. General Properties of the Spectra

In their general shape the spectra are very similar to those of UV Cet flare stars of the solar vicinity. This is evident when a comparison is made with the flare star spectra of Petterson and Hawley (1989). It is important to note that Stauffer (1980) has in fact studied 34 pre-main-sequence stars in the Pleiades. Comparison with Haro em et al. (1982) shows that 27 are known flare stars. The other 7 stars are probably also flare stars Mirzoyan *et al.* (1990). This seems to be new evidence confirming the evolutionary status of flare stars.

4. Spectral Indices, (R-I) Colours and Spectral Classes

Here we have used the TiO spectral indices, introduced by Stauffer (1980, 1982). They are determined for the TiO-bands near the wavelengths 5400, 5900, 6100 and 6500Å from the relations between integral fluxes at the bottom of the band and in the neighbouring "continuum". These indices, which are designated D54, D59, D61 and D65, define the colour index (R-I) and therefore define the spectral class of each star using the tabulation given by Joy and Abt (1974).

Our TiO spectral indices Mirzoyan *et al.* (1990) differ slightly from those of Stauffer. Therefore we have calibrated the relationship between our TiO indices and (R-I) using our data on the UV Cet type flare stars. Using these relationships we carried out the spectral classification of flare stars. The results of our study are given in Table 1, where the following data are presented; PFS number, V, k, the number of observed flares, TiO spectral indices, derived colour index (R-I) and the resulting spectral class. The differences between colour indices and spectral classes, determined by different spectral indices are not large: about 0.05 and one spectral subclass Stauffer (1980). Thus, in agreement with Stauffer the TiO spectral indices are sensitive

Table 1. TiO Spectral Indices, R-I Colour Indices, Spectral Classes and Absolute Visual Magnitudes of the Pleiades Flare Stars

Star (PFS)	V	k	TiO Indices D54	D59	D61	D65	R-I (Av)	Sp (Av)	M_V TiO	m-M=5.54	ΔM_V
36	17.17	9	0.251	0.627	0.205	0.292	1.36	M4.5	12.3	11.5	0.8
49	18.28	4	0.232	0.605	0.185	0.258	1.32	M4.3	12.1	12.6	-0.5
73	17.99	5	0.193	0.616	0.160	0.231	1.28	M4.1	11.7	12.3	-0.6
79	17.60	5	0.223	0.663	0.154	0.258	1.35	M4.5	12.0	11.9	0.1
124	17.67	4	0.289	0.652	0.206	0.381	1.44	M4.9	12.7	12.0	0.7
135	17.93	2	0.223	0.615	0.187	0.279	1.33	M4.4	12.1	12.3	-0.2
184	17.72	3	0.159	0.638	0.138	0.244	1.28	M4.1	11.6	12.1	-0.5
263	16.91	8	0.237	0.570	0.088	0.210	1.20	M3.8	11.5	11.2	0.3
275N	15.73	-	0.223	0.610	0.207	0.189	1.30	M4.2	11.9	10.1	1.8
275S	15.73	-	0.205	0.556	0.173	0.108	1.19	M3.7	11.4	10.1	1.3
313	17.09	8	0.247	0.591	0.166	0.257	1.29	M4.2	12.0	11.4	0.6
394	17.54	1	0.231	0.611	0.159	0.257	1.30	M4.2	11.9	11.9	0.0
441	17.90	5	0.281	0.634	0.276	0.398	1.48	M5.1	13.0	12.2	0.8
465	18.35	3	0.205	0.627	0.212	0.334	1.37	M4.6	12.3	12.7	-0.4

PFS 275 is a double star, components of which have approximately the same brightness, The V-magnitudes of the components are calculated from the integral V-magnitude. The number of flares *k* is given for it, as a single star (Stauffer, 1984).

indicators of stellar surface temperatures.

5. Flare Star Luminosities

Pettersen and Hawley (1989) have shown that the TiO spectral indices of UV Cet flare stars of solar vicinity are sensitive to their luminosities. True, the TiO spectral indices were determined by them in another way. But the TiO spectral indices determined by Stauffer's method are also sensitive to stellar luminosity. Mirzoyan *et al.* (1990) have shown that, for emission red dwarfs of the solar vicinity, there exists a linear relationship between the TiO spectral indices determined by Stauffer and Hartmann (1986) and their absolute magnitudes, taken from Gliese (1969).

Flare stars in star clusters and associations and the UV Cet type flare stars of the solar vicinity constitute a common class possessing a united property, *i.e.* flare activity, and the observed differences between them are consequences of their age differences Mirzoyan and Hambarian (1988). Consequently, for flare stars in systems one can expect the same relationship between the TiO spectral indices and absolute magnitude. In what follows we use this assumption.

In the last two columns of Table 1 luminosities of the Pleiades flare stars are studied are given which were determined in two ways, *viz. a)* using the relationship between spectral indices and absolute maguitude and *b)* assuming that they are all members of the Pleiades ($m - M = 5.54$, $A_V = 0.12$, Crawford and Perry (1976)). The last column of Table 1 shows that the differences ΔM_V between absolute visual magnitudes determined by the two independent methods are significant only for the double star PFS 275. For other stars they are small and all these stars can be considered as members of the Pleiades. As to PFS 275 one can assume that it is not a member. According to Jones (1981) the probabilility for it to be a member is 0.93. This problem probably needs further study.

Table 1 confirms our assumption on the similarity of flare stars in systems and in the solar vicinity. This gives us the possibility to use the TiO spectral indices of flare stars for estimating their absolute magnitudes. The data presented in Table 1, on the whole, show that the TiO spectral indices of flare stars are sensitive not only to stellar surface temperature but also to luminosity.

6. Conclusion

The data on the spectral properties of the Pleiades flare stars discussed here confirm the common physical nature of the flare stars in star clusters and associations and the UV Cet flare stars of the solar vicinity.

References

Ambartsumian, V.A.: 1969, in *Stars, Nebulae, Galaxies,* V.V.Sobolev (ed.), Acad. Sci. Armenian SSR, Yerevan, p. 283

Ambartsumian, V.A., Mirzoyan, L.V.: 1971, *New Directions and New Frontiers in Variable Star Research,* Veroff. Bamberg, Bd. 9, Nr. 100, p. 98

Crawford, D.L., Perry, C.L.: 1976, AJ, 81, 419

Gliese, W.: 1969, *Catalogue of nearby stars,* Veroff. Astron. Rechen-Inst. Heidelberg, No. 22

Jones, B.F.: 1981, AJ, 86, 290

Haro, G.: 1968, in *Stars and Stellar Systems,* Vol. 7, Chicago Univ. Press, p. 141

Haro, G., Chavira, E., Gonzalez,G.: 1982, Bol. Inst. Tonantzintla, 3, 3

Joy, A.H., Abt, H.A.: 1974, ApJS, 28, 1

Mirzoyan, L.V.: 1984, *Vistas in Astronomy,* 27, 77

Mirzoyan, L.V., Hambarian, V.V.: 1988, Afz, 28, 375

Mirzoyan, L.V., Hambarian, V.V., Garibjanian, A.T.: 1990, Afz, in press

Pettersen, B.R., Hawley, S.L.: 1989, A&A, 217, 187

Stauffer, J.R.: 1980, AJ, 85, 1341

Stauffer, J.R.: 1982, AJ, 87, 899

Stauffer, J.R.: 1984, ApJ, 280, 189

Stauffer, J.R., Hartmann, L.W.: 1986, ApJS, 61, 531

ELECTRON DENSITIES AND TEMPERATURES IN SOLAR AND STELLAR ATMOSPHERES

B.C. Monsignori-Fossi[1], F. Tribioli[1] and M. Landini[2]

[1]Osservatorio Astrofisico di Arcetri
Largo E.Fermi 5, 50125 Firenze, Italy

[2]Dipartimento di Astronomia e Scienza dello Spazio,
Universita' di Firenze, Largo E.Fermi 5, 50125 Firenze, Italy

1. Introduction

The densities and temperatures in the transition region of the Sun and late-type stars may be derived using the measured intensities of UV lines (1200 - 2000Å). The allowed lines (NVλ1240Å, SiIVλ1393Å, CIVλ1550Å) permit us to infer the temperature of the emitting plasma. Intersystem lines may be used to determine electron densities. We use a selection of these lines (OIII]λ1666Å, SiIII]λ1892Å, CIII]λ1909Å) measured in some solar active and quiet regions by Skylab and in two active late type stars (λ And and UX Ari) that have been observed simultaneously by the IUE and EXOSAT satellites. The deduced temperature and density give an indication of the pressure which is assumed to hold up to the coronal level. Moreover, the observed X-ray and UV emissions are well reproduced assuming a cool loop-like structure plus a high temperature isothermal region.

2. Analysis of ultraviolet observations

The total emitted line intensity in an optically thin spectral line is given by:

$$I = (4\pi D^2)^{-1} \int_V G(n_e,T)\, n_e^2\, dV \qquad (\mathrm{erg\,cm^{-2}\,sec^{-1}})$$

where n_e is the electron density (cm^{-3}), dV is the emitting volume and $G(n_e,T)$ is the line emissivity per unit emission measure. The intensity ratio of two allowed transition lines of different ions is only a function of temperature. The intensity ratio of an allowed line to an intersystem line of different ions but with similar $G(T)$, is mainly a function of density.

In IUE low dispersion spectra it is possible to resolve the intersystem lines of SiIII(1892Å), CIII(1908.7Å), OIII(1666Å) blended with AlII. An allowed line with temperature distribution similar to these lines is the SiIV resonance doublet. One may use the intensity ratios: CIII/SiIII, SiIII/SiIV and CIII/SiIV to deduce T and n_e (Table 1).

Table 1

		$n_e(cm^{-3})$	T (K)
λ And	Nov 11, 1985	$2.5\,10^{11} \pm 2.0\,10^{11}$	$6.9\,10^4 \pm 2.0\,10^4$
λ And	Nov 06, 1983	$5.2\,10^{11} \pm 2.7\,10^{11}$	$6.5\,10^4 \pm 1.5\,10^4$
UX Ari	Sep 26, 1985	$2.9\,10^{11} \pm 2.5\,10^{11}$	$9.0\,10^4 \pm 2.8\,10^4$
Active Region		$6.0\,10^{10} \pm 2.0\,10^{10}$	$5.6\,10^4 \pm 1.0\,10^4$
Quiet Sun		$3.0\,10^{10} \pm 1.0\,10^{10}$	$5.4\,10^4 \pm 1.0\,10^4$

3. Analysis of IUE and EXOSAT observations

Simultaneous observations were obtained with EXOSAT LE and ME experiments and with

Table 2

	λ And (1)	UX Ari	λ And (2)
Norm	$7.2\,10^7 \pm 7\,10^6$	$7.7\,10^8 \pm 3\,10^8$	$4.5\,10^7 \pm 8\,10^6$
p_o	$2.4 \pm 1.$	1.2 ± 0.9	1.9 ± 0.8
$\log T_o$	4.35 - 4.85	4.3 - 4.9	4.3 - 4.8
$\log T_c$	5.4 ± 0.1	5.8 ± 0.1	5.9 ± 0.5
β	1.96 ± 0.1	-0.8 ± 0.6	1.5 ± 0.2
EM	$2.2\,10^{53} \pm 3\,10^{52}$	$8.8\,10^{53} \pm 5\,10^{52}$	$2.2\,10^{53} \pm 3\,10^{52}$
$\log T_i$	7.18 ± 0.05	7.34 ± 0.05	7.2 ± 0.05
χ^2_{red}	2.1	0.65	0.9

IUE for λ And on November 11, 1985 and for UX Ari on September 26, 1985. It is not possible to reproduce both the UV and X-ray observations with a single temperature model and a multi-temperature analysis must be performed.

Using the spectral code of Landini and Monsignori-Fossi (1990) and assuming solar abundances, a parametrized form of the differential emission measure ($DEM = n_e^2\,dV$) proposed by Monsignori-Fossi and Landini (1988) is used to reproduce all the observations.

$$DEM = f(T) = Norm\, p_o^2\, exp(T_o/T)\,(T/T_c)^{-\beta}\,[T/(1-(T/T_c)^{2.5}]^{-0.5}$$

An isothermal component is added to reproduce the ME data. The χ^2 test gives the parameters ($Norm$, p_o, T_o, T_c, β) of the DEM distribution and the temperature (T_i) and EM of the isothermal emitting plasma (Table 2). For λ And the χ^2 test does not give acceptable results (column 1). The EXOSAT LE filters are not reproduced. The Al/Pa filter has maximum efficiency in the spectral range 170 - 210Åwhere a lot of FeIX - XII lines are present. The 3-LEX and 4-LEX filters have maximum efficiency around 100Å, where FeVIII - X lines are prominent. A lower abundance of Fe ($\log[Fe_{\lambda And}/Fe_{Sun}]$ = -0.7) (Helfer and Wallerstein, 1968) allows a better reproduction of LE data (column 2).

Table 3

λ And	UX Ari
$L = 5\,10^8$ cm	$L \leq 2\,10^{10}$ cm
ff = 0.5	ff$\geq$ 1
$p_o = 1.9$ dyn cm^{-2}	$p_o = 1.2$ dyn cm^{-2}
$\log T_c = 6.0$	$\log T_c = 5.8$

4. Results

We have examined the simultaneous observations of IUE and EXOSAT of two close binary system λ And and UX Ari.

- Densities and temperatures in the transition zone are deduced from IUE observations.
- The emission observed by IUE and EXOSAT may be explained with the presence of a high temperature plasma plus a cool loop-like structure.

- In order to have a good reproduction of all observations of λ And, according to Helfer and Wallerstein (1968), a Fe abundance five times lower than for the Sun is assumed.
- The length, filling factor and base pressure (Table 3) for the cool loop structures may be deduced from the DEM model, assuming that in the energy balance a positive input is present everywhere in the model.

References

Landini, M., Monsignori-Fossi, B.C.: 1990, A&AS, 82, 229

Monsignori-Fossi, B.C., Landini, M: 1988, in *Activity in Cool Star Envelopes,* O.Havnes Ed., p. 237

Helfer, H.L., Wallerstein, G.: 1968, ApJS, 16, 1

(l to r) M. Giampapa spells it out but P. Maltby looks unconvinced!

ON THE AR Lac ACTIVE REGIONS

I. Pagano[1], M. Rodonó[1,2], J.E. Neff[3]

[1]Institute of Astronomy, University of Catania, v.le A.Doria, 6, I-95125 Catania, Italy
[2]Catania Astrophysical Observatory, v.le A.Doria, 6, I-95125 Catania. Italy
[3]Goddard Space Flight Center, Greenbelt MD20771, USA

1. Introduction

AR Lac (G2IV + K0III/IV) is an RS CVn-type eclipsing binary with two components of nearly equal mass, 1.35 $M_\odot$, and radii of 2.77 $R_\odot$ and 1.52 $R_\odot$ for the cooler secondary and the hotter primary component, respectively. The system has a semi-major axis of 9.1 $R_\odot$, an inclination of 87°, and an orbital period of $\approx$ 1.98 day. The optical light curve shows variations due to a varying wave-like distortion. The ultraviolet lines are in strong emission, with rapid variations both in intensity and in profile. The system is also a radio variable star and an X-ray emitter. For further information on this system see Rodonó *et al.* (1986, 1987).

2. Photometric Data and Photospheric Spots

Due to the unfavorable photometric period, almost exactly two days, it is difficult to acquire a reasonably well covered light curve from only one place in a short time interval. Because of this, to evaluate the wave-like distortion, we have useful data only for the periods: Aug - Nov 1983, Jul - Dec 1988, and Jul - Oct 1989. The 1983 and 1989 photometry was obtained with the 91 cm and 61 cm Cassegrain reflectors at Catania Astrophysical Observatory. The 1988 UBV light curves were obtained with the 25 cm APT at Mt. Hopkins. The 1989 V light curve is shown in Fig. 1. The phases are reckoned from the ephemeris by Rodonó *et al.* (1986), *viz.* HJD = 2444977.0216 + 1.983170 E. By taking into account the ellipticity and other effects by a Fourier series expansion of the outside-of-eclipse light curve, we extracted the photometric wave. Then, we applied an interactive spot model based on the analytical method outlined by Friedemann and Gurtler (1975) and already used by Rodonó *et al.* (1986). For the model parameters we adopted the values listed in Table 1. The resulting latitudes, longitudes and radii for the two spots are listed in Table 2.

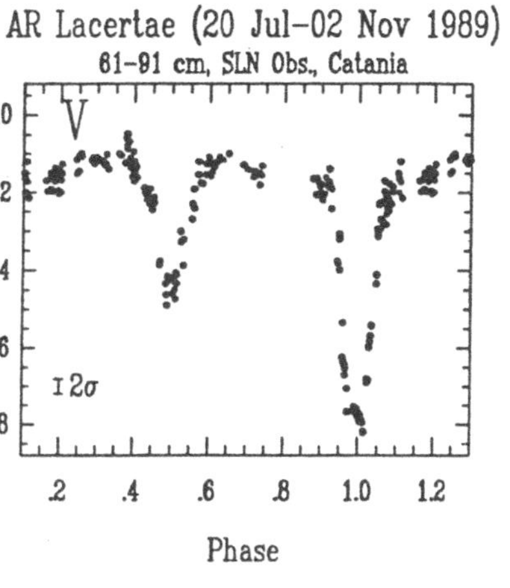

Figure 1. The V light curve of AR Lac in 1989.

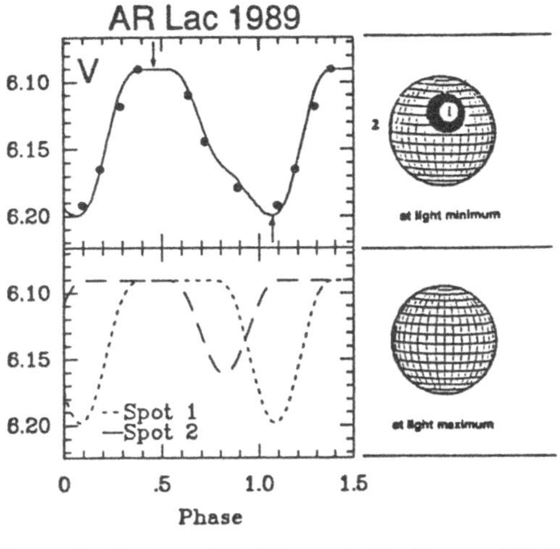

Figure 2. Spot model of the out-of-eclipse rectified curve observed in 1989.

Table 1. Parameters used in the spot models

T_{phot}	T_{spot}	μ	L_K/L_G	i
4700 K	3500 K	0.78	0.71	87°

μ is the limb darkening coefficient and
L_K/L_G, is the ratio between the two stars' luminosities

Table 2. Spot model results

	1983		1988		1989	
Spot Parameter	1	2	1	2	1	2
Longitude (°)	180	85	90	332	30	320
Latitude (°)	59	50	35	26	25	5
Radius (°)	13	9	6	15	23	18

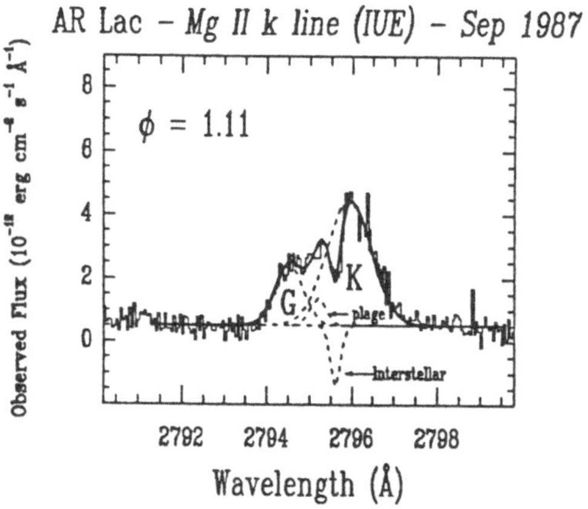

Figure 3. Typical multi-component gaussian fit of the high-dispersion MgII k line. In addition to the stellar components (G & K) a good fit required one plage on the K star and one interstellar absorption feature.

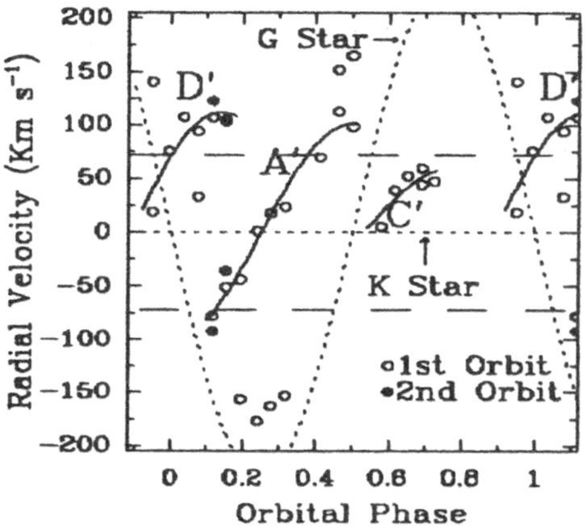

Figure 4. Radial velocity curves of the G star of AR Lac and of three bright features (A', C', D') detected on the MgII k line profiles of the K star, in the velocity rest frame of the K star.

The 1989 outside-of-eclipse light curve, corrected for ellipticity effects, together with the fitted modelling curve, is shown in Fig. 2. In the same figure the contributions of each spot and the spot configurations at light curve minimum and maximum are shown.

3. Ultraviolet Spectra and Chromospheric Structures

To map the active regions on the chromosphere of AR Lac we applied the Spectral Imaging Technique (see Walter *et al.* 1987; Neff *et al.* 1989) to 34 high resolution spectra of the Mg II k line obtained during two consecutive full orbital cycles in Sep 1987 and to an additional 23 spectra obtained during one single cycle in Aug - Sep 1989. Every observed profile was fitted with a symmetric component centered at the stellar velocity plus additional components to account for residual emissions (see Fig. 3). The primary constraints in this procedure are the positions of the uniform stellar components, while secondary constraints are the rough constancy of the stellar emission line widths and smooth point-to-point variation of the stellar flux. The velocity relative to the K star of the discrete emission components, deduced from the 1989 spectra, are

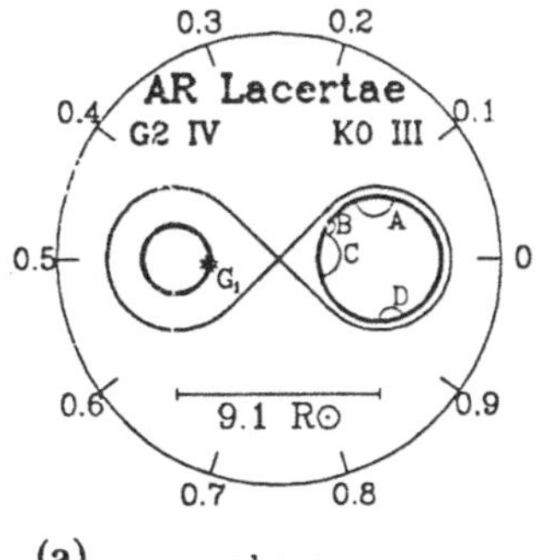

(a)

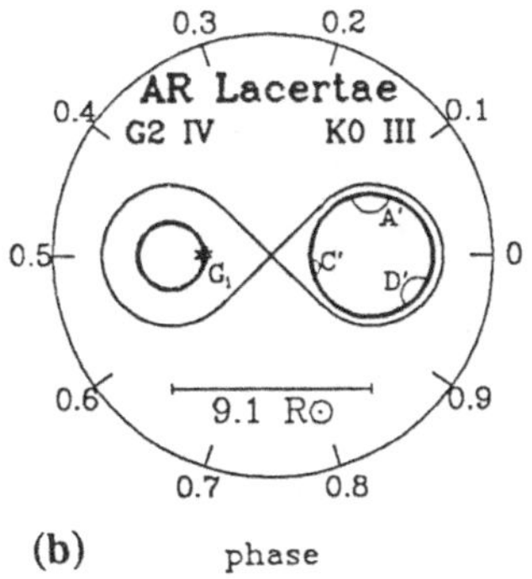

(b)

Figure 5. The chromospheric active regions on AR Lac in 1987 (*a*) and in 1989 (*b*) as derived from spectral imaging of Mg II k line profiles.

shown in Fig. 4. These points can be well fitted by three distinct sinusoidal rotation curves, each representing an individual plage.

From the phases and amplitudes of these sinusoidal rotation curves we can deduce the longitude and the product between the height above the photosphere and the cosine of the latitude, respectively. When the amplitude is greater than the stellar $v_e \sin i$, we can infer only the minimum height by assuming the plage to be on the equator. If the amplitude is less than $v_e \sin i$, we can deduce the minimum latitude by assuming the plage to lie on the photosphere. The chromospheric configurations inferred for 1987 and 1989 are shown in Fig. 5 and listed in Table 3. Note that only for one plage can we infer the minimum latitude.

4. Spot and Plage Evolution

The migrations in longitude and latitude of spots and plages are shown in Fig. 6 and Fig. 7, respectively. The 1983 and 1985 data were taken from (the literature (Walter *et al.* 1987; Neff *et al.* 1989). There is firm evidence of migration both in longitude and in latitude, with a

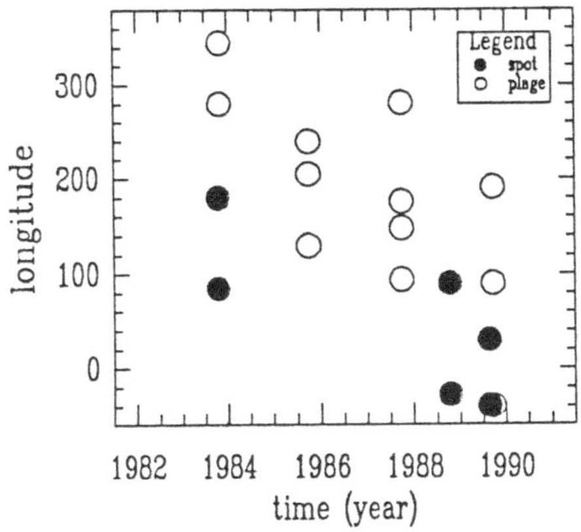

Figure 6. Spot and plage longitudes, inferred from spot modelling and spectral imaging, respectively, are shown versus time.

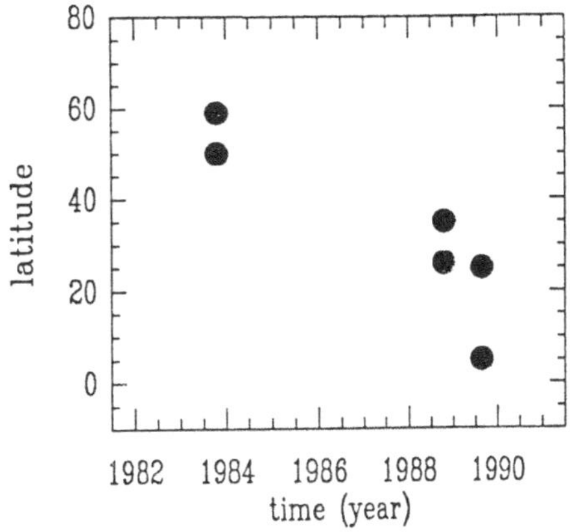

Figure 7. Spot latitudes, as derived from spot models, are plotted versus time.

Table 3. Plage parameters

	1987				1989		
	A	B	C	D	A'	C'	D'
H_{min}	$1.46 \pm .10$	$1.44 \pm .15$	$1.08 \pm .08$	$1.86 \pm .10$	$1.40 \pm .10$	—	$1.56 \pm .08$
L (deg)	94 ± 4	148 ± 1	176 ± 7	281 ± 4	90 ± 4	191 ± 11	320 ± 7
ΔL (deg)	~ 36	~ 15	~ 41	~ 23	~ 35	≤ 15	~ 32
μ	0.10	0.02	0.12	0.04	0.09	≤ 0.02	0.08
l_{min} (deg)	—	—	—	—	—	$33.6 \pm .3$	—

L is the longitude, l, the latitude, H, in units of R/R_*, the height above the photosphere, ΔL, the extension in longitude and μ, the covering factor.

mean rate of about $-22 \pm 6\,\mathrm{deg\,yr^{-1}}$ and $-14 \pm 10\,\mathrm{deg\,yr^{-1}}$ for the spot and plage longitudes, respectively, and $-6 \pm 2\,\mathrm{deg\,yr^{-1}}$ for the spot latitude, resulting from least square fits to the data from 1983 - 1989. Consistent with previous results, a very low spot differential rotation ($\simeq 10^{-5}\,\mathrm{rad\,s^{-1}\,deg^{-1}}$) results.

Unfortunately, only sparse photometry is available for 1985 and 1987. From 1983 data no spatial correlation between photospheric and chromospheric active regions is apparent, while from 1989 data at least one spatial coincidence is clearly seen. The 1983 spectral results are seriously limited by poor phase coverage (only eight spectra were obtained), so that this negative result for 1983 is not conclusive. Taking into account previous positive and negative results on spatial correlation between spots and plages (Rodonó *et al.* 1980; Baliunas *et al.* 1982; Rodonó *et al.* 1987; Butler *et al.* 1987; Andrews *et al.* 1988), we may conclude that, although well documented positive examples exist, further studies are needed.

References

Andrews, A.D., *et al.*: 1988, A&A, 204, 177

Baliunas, S.L., Dupree, A.K.: 1982, ApJ, 252, 668

Butler, C.J., *et al.*: 1987, A&A, 174, 139

Friedemann, C., Gurtler, J.: 1975, Astron. Nachr. 296, 125

Neff, J.E., Walter, F.M., Rodonó, M., Linsky, J.L.: 1989, A&A, 215, 79

Rodonó, M., Romeo, G., Strazzulla, G.: 1980, in *2nd Europ. IUE Conf.*, ESA SP-157, 55

Rodonó, M., *et al.*: 1986, A&A, 165, 135

Rodonó, M., *et al.*: 1987, A&A, 176, 267

Walter, F.M., Neff, J.E., Gibson, D.M., Linsky, J.L., Rodonó, M., Gary, D.E., Butler, C.J.: 1987, A&A, 186, 241

A SEARCH FOR ROTATIONAL MODULATION OF THE LITHIUM LINE IN SPOTTED STARS

R. Pallavicini[1], G. Cutispoto[2], S. Randich[3],
[1]Osservatorio Astrofisico di Arcetri. Florence, Italy
[2]Osservatorio Astrofisico di Catania, Italy
[3]Dipartimento di Astronomia, University of Florence, Italy

1. Introduction

The LiIλ6708Å line has been observed to be a factor $\approx$20 - 40 times stronger in sunspots, and about a factor 2 weaker in plages, with respect to the undisturbed solar photosphere (Giampapa, 1984). Giampapa has suggested that also in cool stars surface activity should significantly affect the strength of the Li line: in particular, because of the larger abundance of neutral Lithium, a stronger line is expected in starspots.

If this is the case, the strong Li feature often observed in RS CVn binaries and other chromospherically active stars could be explained by the enhancement of the Lithium line in large cool spots (Pallavicini et al. 1987). On the other hand, if the strength of the Lithium line is significantly affected by stellar surface activity, rotational modulation of the Li line in antiphase with the photometric variations (*i.e.* the Li line should be stronger at the minimum of the photometric light curve) should be seen.

We therefore searched for rotational modulation of the LiI line in four spotted stars, using nearly simultaneous spectroscopic and photometric observations obtained at ESO. In this paper we report on this search showing that no rotational modulation of the Li line was observed for any of our stars, in spite of the fact that they showed photometric variations that could be attributed to the presence of large starspots.

2. The Observations

2.1 The programme stars

We monitored nearly simultaneously in LiIλ6708Å and in broad-band UBV(RI)$_c$ filters the

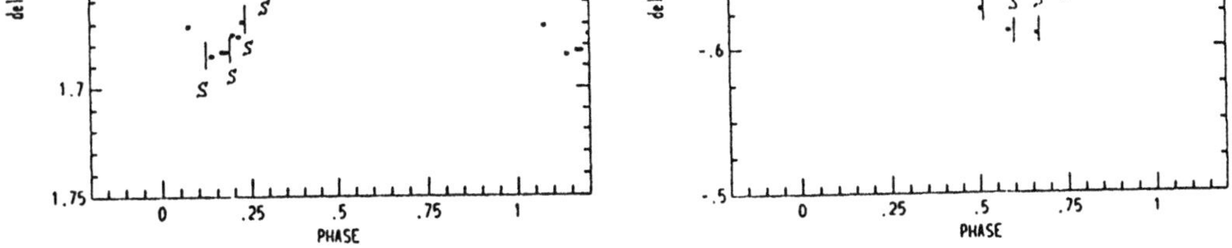

Figure 1. Derived light-curves for AB Dor and IL Hya. The phases at which spectroscopic data were taken are marked by *S*.

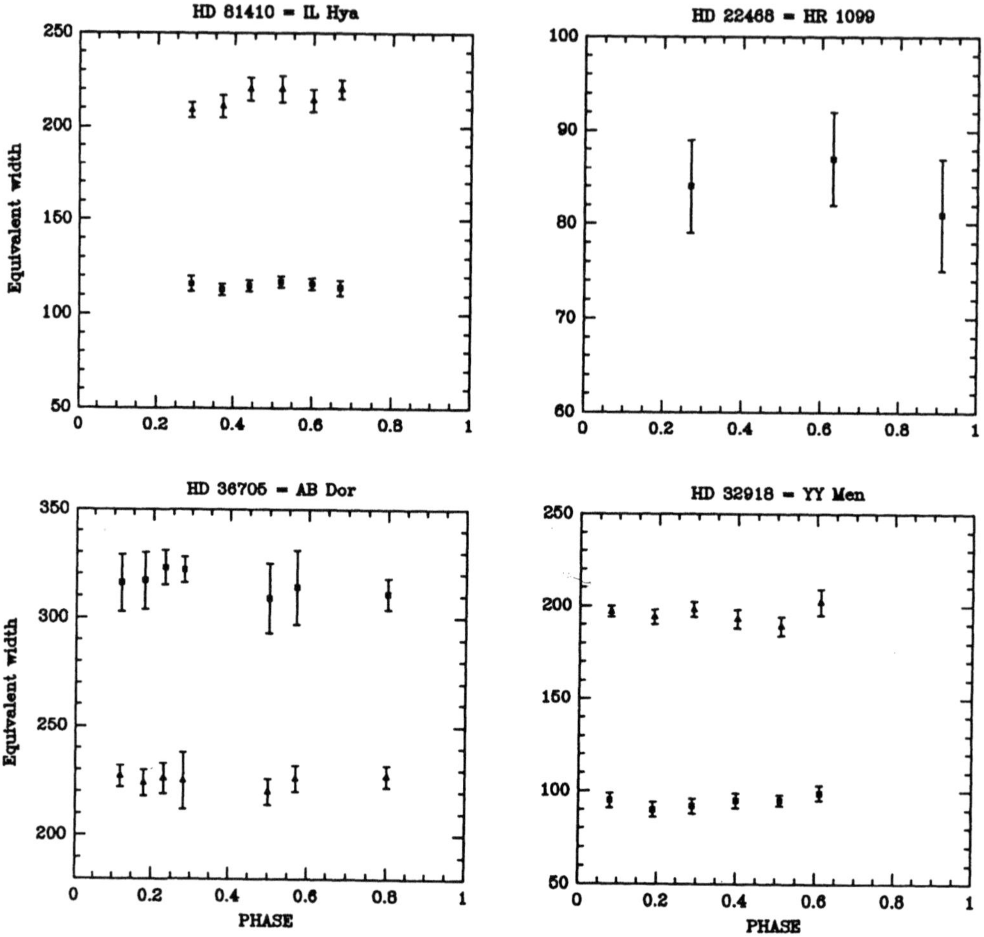

Figure 2. Equivalent width of the LiI (*filled square*) and CaI (*filled triangles*) *vs.* phase.

following stars: HR1099, YY Men, AB Dor and IL Hya. These are well-known "spotted" stars as indicated by their photometric variations; moreover, from previous observations (see Pallavicini, Randich and Giampapa elsewhere in this volume), they were also known to have a strong Li line.

2.2 Photometry.

Photometric observations were carried out during a 2-week period; we used the 50 cm ESO telescope equipped with a single-channel photometer and standard $UBV(RI)_c$ filters. The derived light-curves in the V-band (the curves for AB Dor and IL Hya are shown in Figure 1) show, for all stars, a clear photometric variation (with amplitude of $\approx 0.05 - 0.1$ magnitudes). The ephemerides derived from the photometric variations were used to determine the phases at which the spectroscopic data were obtained (marked by S in the figure).

2.3 Spectroscopy

Spectroscopic observations in the LiIλ6708Å line were obtained on 6 consecutive nights (14-20 Dec 1987) immediately following the photometric observations. The instrument used was the Coudé Echelle Spectrometer fed by the 1.4 m CAT telescope. The short-camera and a cooled

CCD detector were used at a resolving power R = 50 000. The phase coverage of the spectroscopic observations was good for AB Dor, YY Men and IL Hya and included both the minimum and the maximum of the photometric light curve. For HR1099 the spectroscopic observations were more sparse but still sufficient to reveal phase-dependent variations if they were present.

The Li line is strong in all stars, with equivalent widths ranging from 85 mÅ for HR1099 to 315 mÅ for AB Dor. We converted the equivalent widths to Lithium abundances using the curves of growth of Pallavicini et al. (1987), after correction for the contribution of the nearby FeIλ6707.4Å line to the Li blend: we obtained, especially for AB Dor ($\log n(\mathrm{Li}) = 3.4$), Lithium abundances that were anomalously large with respect to typical values for late-type stars of similar spectral type.

3 Results

Figure 2 shows the measured equivalent widths of the blend LiIλ6707.8Å + FeIλ6707.4Å as a function of phase. For comparison, we also plot the equivalent width of the CaIλ6718Å line as a function of phase. The error bars are $\pm 1\,\sigma$ standard deviations as derived from repeated measurements on the same spectra. No significant variations of the equivalent widths with phase are apparent for both the LiI + FeIλ6708Å blend and for the CaIλ6718Å line, in contrast to previous results by Robinson et al. (1986).

4 Conclusions

No significant rotational modulation of the equivalent width of the LiI blend was detected for any of the programme stars. The derived upper limits ($\Delta W/W < 5$ - 10%) are substantially smaller than what had been suggested previously on the basis of the solar analogy. If we assume that the enhancement of the equivalent width of the LiI line in starspots is about the same as for the Sun, the above upper limits imply a spot coverage factor of only a few percent, much smaller than that derived from the photometric variations (15 - 25%). It is clear therefore that the enhancement of the Lithium line in the spots of these stars, if present, is certainly lower than for the Sun.

In any event, the absence of rotational modulation of the Li line at the same time when strong photometric variations were observed, indicates that large cool spots are not a viable explanation for the presence of a strong Li feature in RS CVn binaries and other chromospherically active stars.

References

Giampapa, M.S.: 1984, ApJ, 277, 235

Pallavicini, R., Cerruti-Sola, M., Duncan, D.K.: 1987, A&A, 174, 116

Robinson, R.D., Thompson, K., Innis. J.L.: 1986, PASA, 6, 4

FK Comae - A DECREASE OF THE PERIOD?

K.P. Panov, A. Antov
Department of Astronomy and National Astronomical Observatory
Bulgarian Academy of Sciences,
1784 Sofia, Lenin Blvd 72, Bulgaria

1. Introduction

FK Comae (= HD 117555, G2III, V = 8.2) is a peculiar star, recognized as the prototype of a small group of stars (FK Com, UZ Lib, HD 199178) by Bopp and Rucinski (1981). FK Com is characterized by rapid rotation ($v\sin i \approx 75$ - $120\,\mathrm{km\,s^{-1}}$), first noticed by Merrill (1948), and the presence of strong CaII H & K emission, as well as a broad and variable Hα emission (Bidelman, 1954, Herbig, 1958, Ramsey, Nations and Barden, 1981, Walter and Basri, 1982). IUE observations in the ultraviolet (Bopp and Stencel, 1981, Bianchi *et al.* 1985) show strong chromospheric and transition region emission lines similar to those found in RS CVn stars, and indicating a very active chromosphere. The similarity with the RS CVn stars is also seen in the photometric behaviour of FK Com. Chugainov (1966) discovered the light variability of this star with an amplitude of ≈ 0.1 mag and period of 2.412 day. Further studies by Chugainov (1976a,b) refined the value of the period to 2.400 day. The light variation is generally believed to result from rotational modulation due to star spots and the rotational period seems to be consistent with the large $v\sin i$.

These early studies of FK Com revealed an interesting paradox. If it is a single star, evolved from an A-type star on the main sequence without substantial loss of angular momentum, then its rotational velocity on the main sequence would have been faster than the break-up velocity. The only way out of this problem would be the presence of close companion and a tidal interaction. However, extensive spectroscopic studies by Bopp (1982) failed to reveal any radial velocity variation. This result has been confirmed by McCarthy and Ramsey (1982) who could not detect any RV variations greater than $5\,\mathrm{km\,s^{-1}}$, which places an upper limit for the mass of the companion at 0.025 solar masses. Bopp and Stencel (1981) suggested that the primary and the secondary stars have coalesced, following a scenario studied by Webbink (1976). The point of this scenario is that the secondary may eventually be absorbed by the primary through the Lagrangian point of the system. In this way a single, rapidly rotating star would be created. An alternative model has been proposed by Walter and Basri (1982). In their model the light variations and the Hα emission result from a hot spot on the surface of the G-type star, which is formed by the impact of accreting material, originating from an undetectible low mass companion.

Since the early photometric work of Chugainov many other photometric studies have been carried out (Rucinski, 1981, Morris and Milone, 1983, Dorren *et al.* 1984, Holzman and Nations, 1984, Bianchi *et al.* 1985, Cellino *et al.* 1986, Huovelin *et al.* 1987). The main result of these studies is the finding of strongly variable amplitude of the light variations. The amplitude in V changes from 0.05 mag to 0.20 mag. The colour behaviour is somewhat inconclusive (see Dorren *et al.* 1984), but most researchers seem to agree that the star is redder at the luminosity minimum.

Flares of FK Com have been reported both in Hα and in continuum (Chugainov, 1976, Rucinski, 1981, Ramsey and Nations, 1982, Walter and Basri, 1982, Morris and Milone, 1983, Holzman and Nations, 1984). An important point is the stability of the period. Indeed, several authors already noticed possible changes in the phase of the minimum light and Morris and Milone (1983) proposed a refined value of the period of FK Com, *viz.* 2.39960 d. However, follow-up

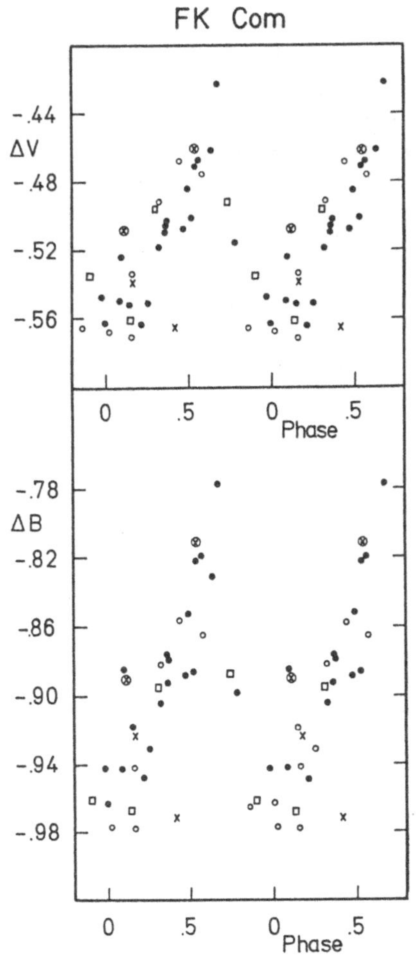

Figure 1. Differential observations of HD 117567 - FK Com. × = 1984, • = 1985, ⊗ = 1987, □ = 1990.

observations of Holzman and Nations (1984), Bianchi *et al.* (1985), and Cellino *et al.* (1986) supported the Chugainov's period of 2.400 d. Until May-June, 1984, Chugainov's ephemeris for FK Com seems to be correct, indicating stability of the rotational period for at least 15 years. Occasional discrepancies of observations from the above ephemeris have been attributed to the changes of the amplitude of the light curve. In this paper we present observations of FK Com which indicate strongly a decrease of the period during the last years.

2. Observations and Discussion

Photoelectric UBV observations of FK Com were carried out with the photon-counting photometer, attached to the 60 cm telescope of the Rozhen Bulgarian National Astronomical Observatory. The comparison star was HD 117567. The observations were obtained with an integration of 10 s and nightly mean data were obtained from typically 3-5 individual observations. The data were collected on 2 nights in 1984, 18 nights in 1985, 2 nights in 1987, 7 nights in 1988, and

4 nights in 1990. The observations from the largest sample of 1985 show a systematic decrease in phase, when plotted with the 2.400 d period, indicating that the actual period should be shorter. After taking into account the phasing of all our data, we derived a period of 2.36666d. Fig. 1 shows the light curves in δV and δB (in the instrumental system) with the ephemeris $JD_{minlight} = 2446194.379 + 2.36666E$.

The magnitude differences are in the sense HD 117567 - FK Com. The decrease of the period is about 1.4%. It is not possible to determine from our observations the exact time of the period change nor the way it occurred. The two points from 1984 (see Fig. 1) deviate from the light curve but they agree well with the old 2.4 d period. Since the photometry of Huovelin *et al.* (1987) from 1984 is consistent with the 2.400 d period, the change of the period should have occured in late 1984, or early 1985. Period changes could be expected from both the spot model and the mass-transfer model. In the spot model period changes could occur if the position of the spot changes either in longitude, or in latitude. In the second case a latitudinal dependence of rotation, *i.e.* differential rotation has to be assumed. Such phenomena are seen in some RS CVn stars. On the other hand, mass transfer from a companion would result in a continuous decrease of the period and this could even be abrupt, if the process of mass transfer is not uniform. Two difficulties arise with the mass transfer model. First, the lack of radial velocity variations. Second, a brightening of the star has to be expected around 1985, if an increased mass transfer took place. However, the mean brightness of FK Com did not change. As a working hypothesis, we could assume that the old spot disappeared around 1984-1985 and a new one was formed on a different latitude with corresponding (faster) rotation.

References

Bianchi, L., Grewing, M., Kappelmann, N.: 1985, A&Ap, 149, 41.

Bidelman, W.P.: 1954, ApJS, 1, 175.

Bopp, B.W.: 1982, SAO Special Rep. No 392, 207.

Bopp, B.W., Rucinski, S.M.: 1981, IAU Symp. 93, p. 177.

Bopp, B.W., Stencel, R.E.: 1981, ApJ, 247, L131.

Cellino, A., Scaltriti, F., Busso, M.: 1986, Ap&SS, 121, 265.

Chugainov, P.F.: 1966, IBVS, No 172.

Chugainov, P.F.: 1976a, Izv. Krymskoj Astr. Obs. 54, 89.

Chugainov, P.F.: 1976b, Izv. Krymskoj Astr. Obs. 55, 85.

Dorren, J.D., Guinan, E.F., McCook, G.P.: 1984, PASP, 96, 250.

Herbig, G.W.: 1958, ApJ, 128, 259.

Holzman, J.A., Nations, H.L.: 1984, AJ, 89, 391.

Huovelin, J., Piirola, V., Vilhu, O., Efimov, Y.S., Shakhovskoy, N.M.: 1987, A&Ap, 176, 83.

McCarthy, J.K., Ramsey, L.W.: 1982, BAAS, 14, 780.

Merrill, P.W.: 1948, PASP, 60, 382.

Morris. S.L., Milone, E.F.: 1983, PASP, 95, 376.

Ramsey, L.W., Nations, H.L., Barden, S.C.: 1981, ApJ, 251, L101.

Ramsey, L.W., Nations, H.L.: 1982, SAO Special Rep. No 392, p. 225.

Rucinski, S.M.: 1981, A&Ap, 104, 260.

Walter, F.M., Basri, G.S.: 1982, ApJ, 260, 735.

Webbink, R.F.: 1976, ApJ, 209, 829.

ON VARIABILITY OF TZ Ori

E.S. Parsamian, A.S. Melkonian, K.G. Gasparian, G.B. Ohanian
Byurakan Astrophysical Observatory, Armenia, USSR

1. Introduction

The variable star TZ Ori was considered as a flare star for its flare-like variations in the lightcurve (Roslund, 1969). By analysing data covering 600 observations, it was found that the star has periodic light variations with a period of $P = 5\,\mathrm{hr}$ (Parsamian, 1986). This result was acheived using the mean stellar magnitudes of TZ Ori over each night. In that same paper an assumption was made, that there is a possibility for the existence of periodic light variations with the period shorter than 1 day.

2. Analysis of Photometric Data

In the present work data consisting 519 points over the time interval from JD = 2436907 - JD = 2439916, have permitted us to uncover periodic light variations of TZ Ori with the period $P = 11\,\mathrm{hr}$. For the periodogram analysis we use Deeming's method for unequally spaced data. (Deeming, 1975). Power spectrum analysis shows that there is only one real period (11 hr) in the light curve of TZ Ori ($\nu = 2.18\,\mathrm{day}^{-1}$), the remaining peaks in the power spectra are aliases.

All observational data were brought to one period, *i.e.* $P = 11\,\mathrm{hr}$ and the resulting light curve is shown in Fig. 1.

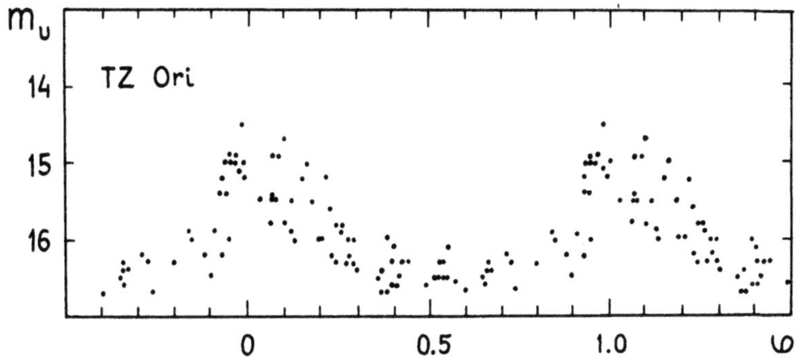

Figure 1.

Fig. 1 could be interpreted as suggesting that the light curve is the sum of many flares with similar shapes but with different amplitudes. The large scattering of the points figure can be explained by the following reasons:

1. by the different amplitudes of different flares.
2. by different durations of the flares.
3. by observing errors.

There are good reasons for the first assumption, because there is a clear limit in the light curve from the bigger values of magnitudes, and large scattering from the smaller magnitudes. It can be established that the star has a definite brightness at the minimum and flares with different amplitudes. The existence of periodic light variations of TZ Ori, with $P = 11\,\mathrm{hr}$ was also found by Natsvlishvili (1987) from the analysis of the times of flare maxima.

3. Spectral Observations

Spectral observations of TZ Ori were made with the 6 m telescope of the USSR SAO and with the 2.6m telescope of Byurakan Observatory in the spectral range 3800 - 7000Å(Fig. 2).

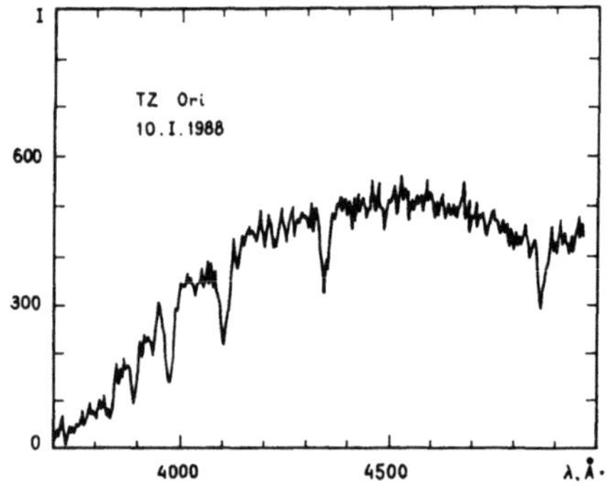

Figure 2.

The spectral class of TZ Ori was estimated as A by the Balmer lines of Hydrogen and CaII K. Equivalent widths of the spectral lines show changes which may be correlated with photometric variations. CaII K exhibits strong changes until to disappearance. According to both spectral and photometric observations, as well as from the angular distance from the center of the Orion association, TZ Ori is not a member of this association.

3. Conclusion

The variability intervals of TZ Ori in UBV are very roughly (Parsamian, 1986):

U	B	V
14.0 - 16.7	14.0 - 16.4	13.8 - 15.8

Besides variability discussed, TZ Ori has flare-like light variations, when, for instance, the star brightened by $\Delta U \sim 1.5$ in 8 min. To understand the phenomenon of TZ Ori and to clear up the correlation between the photometric and spectral changes more detailed simultaneous spectral and photoelectric high-resolution observations are necessary. By light curve and spectral type TZ Ori may be a RR Lyr type star with $P = 0.458$ day.

References

Deeming T.I.: 1975, Ap&SS, 36, 137

Natsvlishvili, R.Sh.: 1987, private communication

Parsamian, E.S., Pogosian, G.A.: 1986, Afz, 24, 239

Roslund, C.: 1969, Uppsala Astron. Obs. Meddelande, 169, 381

Li ABUNDANCES AND CHROMOSPHERIC ACTIVITY IN F, G AND K SOUTHERN STARS

L.Pasquini
European Southern Observatory, La Silla,
Casilla 19001, Santiago 19, Chile

1. Introduction

An accurate determination of Li abundance in stars of different spectral types, ages and luminosity classes is fundamental for our understanding of several astrophysical problems, ranging from the abundance of the primordial elements, through the chemical evolution of the Galaxy, to the study of the internal structure of late type stars.

A quite successful scheme suggests that in late-type stars subphotospheric convective motions transport material from the surface to the hotter interior, where Li is rapidly destroyed; as a consequence of this , Li abundance should be a tracer of stellar age (Herbig, 1965). Early studies of G stars confirmed this hypothesis, and Skumanich (1972) was able to show that Li abundance, chromospheric activity and stellar rotation are well correlated with stellar age and follow similar decay laws. The availability of recent, high quality data has shown that Li depletion is more complex than the scheme proposed above. In particular Spite and Spite (1982) found significant amounts of Li in halo and old disk stars and Duncan (1981) and Pallavicini et al. (1987) found that, while *vsini* and chromospheric activity are well correlated, Li abundance presents a large scatter with respect to the other two quantities even within a restricted (F8-G5) spectral range.

I present new, high resolution, high S/N Li observations of 15 southern stars, for the majority of which CaII K absolute chromospheric fluxes have also been determined. Adding to this sample other data taken from the literature (homogeneous with the new ones), I briefly discuss the relationship between chromospheric activity and Li abundances in late-type dwarfs. A possible super-Li-rich object is separately discussed.

2. Observations, Li abundance and CaII chromospheric flux determination

Thc observations have been obtained at ESO, La Silla using the CES spectrograph and CCD detector fed by the CAT 1.4 m telescope. The resolving power was set to R = 60000. The raw data were flat fielded by dividing them by the spectrum of a quartz lamp and normalised by using a second order polynomial. The real S/N is limited by the flat fielding process, but always exceeds 100.

The CaII K data were also obtained at La Silla and are extensively discussed in Pasquini and Pallavicini (1990, submitted to A&A); the procedure used to obtain absolute chromospheric fluxes is that developed by Pasquini et al. (1988).

In order to derive Li abundances, stellar effective temperatures have been determined using the (B-V) - T_{eff} relation given by Popper (1980); the Li curves of growth used are those published by Pallavicini et al. 1987 based on Bell et al. (1976) atmospheric models. The adopted stellar parameters, Li abundances and CaII chromospheric fluxes are given in Table 1; the abundances are given on a scale where $\log N(H) = 12.00$.

In order to have a more comprehensive picture of the activity-Li abundance relation, I have enlarged the original sample using Lithium data taken from the literature. Most of the objects are from Pallavicini et al. (1987) and Soderblom (1985) and a few additional ones are from Andersen et al. (1984) and Cayrel de Strobel and Cayrel (1989). For all the stars in the sample, chromospheric absolute fluxes were obtained using spectra of comparable quality and the same calibration procedure. All Li observations were taken with the CAT + CES and stars in common

Table 1

Star Id.	(B-V)	(T_{eff})	log EW	log N_{Li}	log F'_k
HD199260	0.50	6250	1.93	3.02	6.40
HD202008	0.65	5800	1.67	2.19	6.05
HD 4308	0.65	5800	0.60	0.98	5.45
HD120690	0.69	5650	0.85	1.12	—
HD114613	0.70	5650	2.00	2.61	5.54
HD128620	0.71	5600	0.85	1.07	5.45
HD203244	0.72	5500	1.36	1.51	6.09
HD 36435	0.78	5300	1.36	1.30	6.05
HD112758	0.79	5250	0.85	0.15	—
HD 13445	0.82	5200	≤0.00	≤-0.19	6.02
HD114946	0.87	5150	0.78	0.54	—
HD128621	0.88	5000	≤0.3	≤-0.1	5.15
HD 52689	0.90	5000	≤0.3	≤-0.1	6.03
HD156026	1.16	4350	≤0.6	≤-0.5	5.45
HD155555A	0.78:	5300:	2.46:	3.6:	—
HD155555B	0.88:	5000:	2.55:	3.7:	—

between different authors reveal an excellent agreement in the measured Li EW. The curves of growth and temperature scales used by Soderblom (1985), Pallavicini et al. (1987) and in this work are the same; I have therefore retained the abundance values given by the above authors, obtaining a total of 42 stars in the ≈F7-≈K3 spectral range. In Figure 1 Li abundance *vs.* CaII K line chromospheric fluxes are plotted. The size of the points is proportional to the stellar effective temperature in the range $4350 \leq T_{eff} \leq 6400$ K.

Although the data points in Figure 1 appear extremely scattered, it is possible to identify some patterns, depending on the stellar T_{eff}. Active K stars (*i.e.* stars with $T_{eff} < \approx 5400$ K) with very similar F'_k values show a large spread in Li abundances. This suggests that for these objects the timescale for the decay of chromospheric activity is longer than the Li depletion time. Note that the only K dwarf in the sample showing a very strong Li line is HD17925, which is supposed to be only $\approx 10^7$ years old (Cayrel de Strobel and Cayrel 1989).

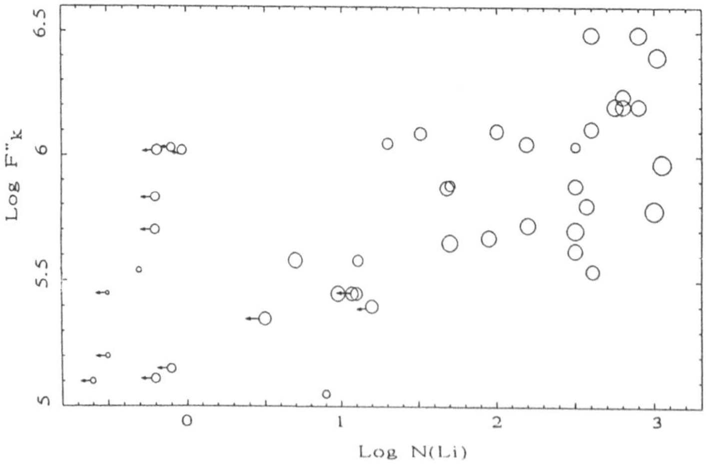

In Figure 2 $\log N(\mathrm{Li})$ is plotted *vs.* T_{eff} and the symbol size is proportional to the log of the chromospheric flux. Fixing our attention on the more active stars, they are quite well separated from the inactive ones (with some exceptions, see later) up to $T_{eff} \approx 5300$ K, while cooler objects are totally mixed.

The number of F stars ($T_{eff} > 6000$ K) comprising the sample is too small to allow firm conclusions, but we can tentatively say that for these objects, Li abundance and chromospheric activity appear uncorrelated, in the sense that low activity hot stars tend to preserve their initial Li content. The timescale for the decay of chromospheric activity should therefore be ***shorter*** than the Li depletion timescale.

Finally, the two timescales seem to be better correlated for solar-type stars. However, also for this, group the relationship is very scattered; in particular there are objects showing a strong Li content but low chromospheric activity, which are presumably old. Some examples were reported already by Duncan (1981) and Pallavicini et al. 1987. In my sample a new object of this class is HD114613 (see Table).

It has been suggested that the physical parameter causing Li preservation could be a reduced stellar metallicity (Spite and Spite 1982), but the effects of other possible causes, like stellar rotational history or stellar evolutionary status have not yet been fully investigated. Too few well studied cases are known to attempt any statement.

3. HD155555: a super Li-rich star?

One of the objects in the sample, the spectroscopic binary HD155555, needs particular discussion. This object has been considered for a long time an RS CVn variable. RS CVn's are close binary systems formed by a G dwarf and a K subgiant and with orbital periods of ≈ 15 days or shorter; they show photometric variations presumably due to surface inhomogeneities and are very active (Hall 1976). It is also commonly believed that RS CVn's are old stars, in which the cool subgiant is already evolved out of the main sequence. On the basis of X-ray and optical observations of the optical companion of HD155555, LDS 587B, Pasquini et al. (1989) suggested that this spectroscopic binary is instead formed by two young main sequence stars. I have obtained two CES spectra of HD155555 in the Li region at different phases. One of the two is shown in Figure 3 (*continuous line*); the two line systems are well separated and the LiIλ6708Å appears in both components to be far stronger than the Ca + Fe blend at λ6717Å. In order to get an estimate of Li equivalent widths and abundances I have constructed "synthetic spectra" using templates having color indices similar to those expected from the two components (Bennet et al. 1962) properly shifted in wavelength to match the HD155555 spectra.

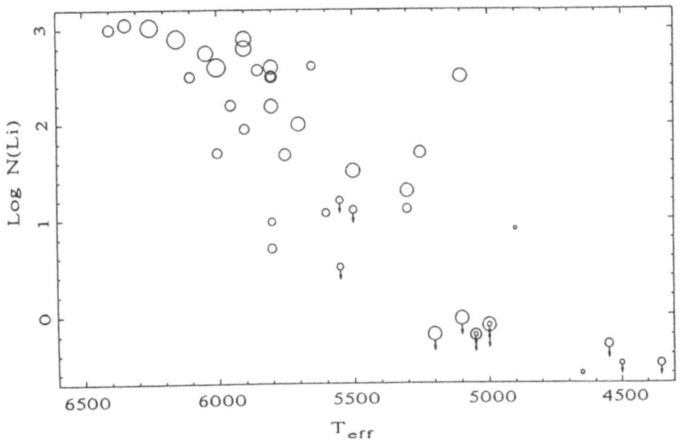

Since the luminosity class of the two components is not well established and the contribution of the two components must be weighted according to their luminosities, I have performed the whole analysis adopting several spectral type and luminosity class combinations. After measuring the Li equivalent widths in the HD155555 spectra, corrections have been applied by subtracting the contribution of blends and of the other component as estimated from the synthetic profiles. Regardless to the assumptions made, the resulting Li equivalent width is very similar to those given in the Table. Although the derived abundances must be considered as preliminary, we are certainly dealing with an extremely Li-rich object, with a Li content very close to, or higher than the initial or protostellar abundance ($\log N(\mathrm{Li}) = 3.0$ - 3.3) as inferred from Population I stars and Type I carbonaceous chondrites (Duncan 1981).

References

Bell, R.A., Eriksson, K., Gustafsson, B., Nordlund, A.: 1976, A&AS, 23, 37

Bennett, N.W.W., Evans, D.S., Laing, J.D.: 1962, Roy. Obs. Bull. 61, 92

Cayrel de Strobel, G., Cayrel, R.: 1989, A&A, 218, L9

Duncan, D.K.: 1981, ApJ, 248, 651

Hall, D.S.: 1976, IAU Coll. No. 29, Part 1, 287

Herbig, G.H.: 1965, ApJ, 141, 558

Pallavicini, R., Cerruti-Sola, M., Duncan, D.K.: 1987, A&A, 174, 116

Pasquini, L., Pallavicini, R., Pakull, M.: 1988, A&A, 191, 266

Pasquini, L., Schmitt, J., Harnden, F.R., Tozzi, G.P., Krautter, J.: 1989, A&A, 218, 187

Popper, D.M.: 1980, ARA&A, 18, 115

Skumanich, A.: 1972, ApJ, 110, 205

Soderblom, D.R.: 1985, PASP, 97, 54

Spite, F., Spite, M.: 1982, A&A, 115, 357

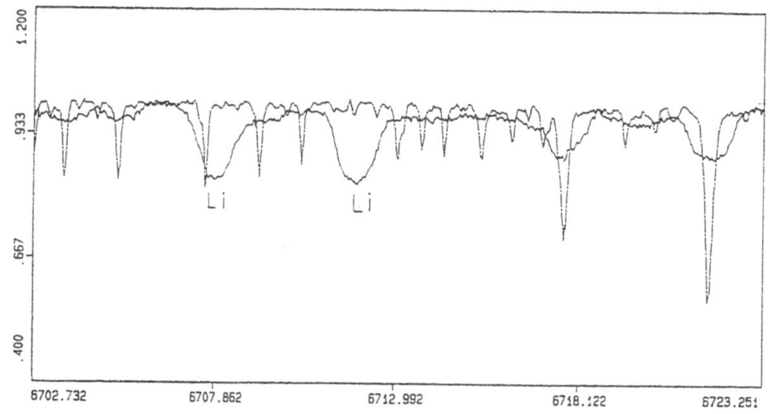

X-RAYS FROM BOTH COMPONENTS OF THE FLARE STAR BINARY Gliese 867

A.M.T. Pollock[1,3], G. Tagliaferri[1], R. Pallavicini[2]
[1]EXOSAT Observatory, ESTEC, Noordwijk, The Netherlands
[2]Osservatorio Astrofisico di Arcetri, Florence, Italy
[3]Now at Computer and Scientific Co. Ltd., Sheffield, England

1. Introduction

Gliese 867 is a visual binary whose components are separated by ≈ 25 arcsec. The primary, Gl 867 A (= FK Aqr) is a double-lined spectroscopic binary of period 4.083 day and combined spectrum dM2e. Gl 867 B is 2.4 mag fainter and is classified as dM3e. Both components of Gl 867 are known to flare in the optical and both also show quasi-periodic variations that are attributed to the presence of cool spots over their surfaces. The photometric period of Gl 867 A is 4.39 day. A possible period of 1.95 day has been reported for Gl 867 B.

Gl 867 A+B was first observed in X-rays using the IPC on Einstein. The unresolved Einstein observations were analysed by Agrawal (1988) who attributed the entire X-ray emission to component A. He also interpreted the 40% decrease observed over a 3-day interval to the X-ray analogue of the BY Dra variability found in the optical for Gl 867.

In this paper we present observations of Gl 867 obtained with EXOSAT. By applying a likelihood method we show that both components of the system were contributing to the observed X-ray emission with an intensity ratio A:B of about 3:1. We also show that X-ray flares with energies of the order of 10^{32} erg originated from both stars.

In the light of the EXOSAT results, we have reanalysed the Einstein observations and we

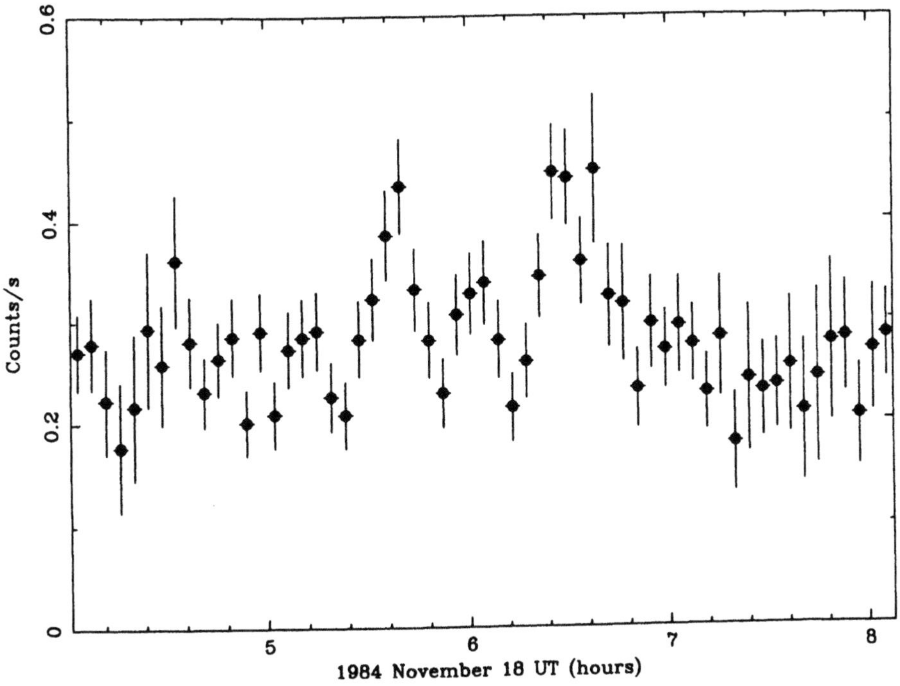

Figure 1. EXOSAT LE light curve for the combined emission of Gl 867 A + B.

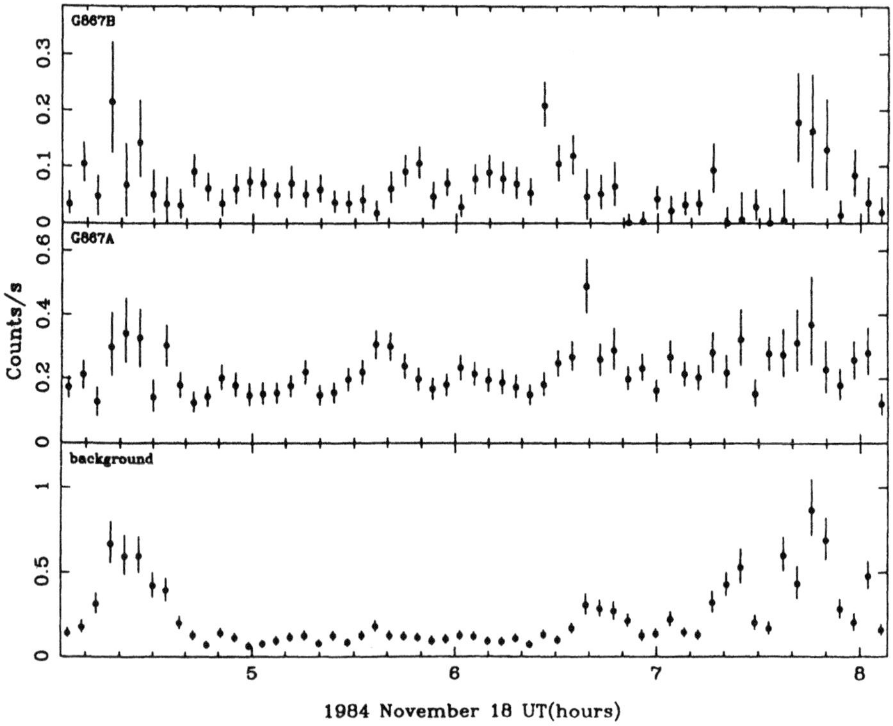

Figure 2. EXOSAT LE light curves resolved into the separate contributions of Gl 867 A, Gl 867 B and the background by a maximum likelihood analysis.

have found, quite unexpectedly, that during those observations Gl 867 B was rougly as bright as Gl 867 A. We also find that the variations observed in the Einstein data were due to changes in both components, similar to the flares observed by EXOSAT, rather than to rotational modulation of component A.

2. The EXOSAT observations

Gl 867 was observed with the Low Energy (LE) instrument aboard the EXOSAT Observatory on 18 Nov 1984 from 04:02 - 08:06 UT. Fig. 1 shows the background subtracted EXOSAT light curve for the combined X-ray emission from the system. Significant variability is present with peaks at $\approx$ 05:40 and 06:30 UT.

Since the separation of the two components of the binary is comparable to the spatial resolution of the EXOSAT LE, it is not a trivial task to separate them. In order to do so, we have used a likelihood method similar to that described by Pollock (1987; also Pallavicini *et al.* 1990 for an application of the same method to IPC observations of the triple system Castor).

We have first applied the likelihood method to the time-integrated EXOSAT observation by considering 1- and 2-source models. The increase in likelihood for the latter model clearly shows that two sources are required to model the data. The derived positions for the two sources agree with the expected positions of Gl 867 A and B. The derived intensity ratio A:B is 3.5, which compares favourably with the value expected from the ratio of the bolometric luminosities.

We have then subdivided the EXOSAT observation in 250 sec bins and we have repeated the likelihood analysis for the time-resolved data. The results are given in Fig. 2 which shows the

X-ray light curves for the two components separately and for the background. Both sources were clearly variable during the observation. The peak at $\approx$ 05:40 UT visible in the integrated light curve was entirely due to Gl 867 A, while the peak at $\approx$ 06:30 UT was the combined effect of flares occurring on both A and B. The energy released by each of the three flares was $\approx 10^{32}$ erg.

3. The Einstein observations

Gl 867 A + B was observed by the Einstein IPC on three consecutive days 11 - 13 May 1980. In each case the source was observed for about half an hour, while the intervals between successive observations were 36 hr and 25 hr, respectively. Since the $\approx$ 1 arcmin spatial resolution of the IPC was poorer than that of the LE instrument ($\approx$ 20 arcsec), these data represent a greater challenge to the likelihood method's power to separate overlapping sources. However, by assuming that two sources were present in the field at a fixed distance of 25 arcsec, we can use the likelihood method to determine the best positions and count rates for the two objects.

We have carried out the maximum likelihood calculations for the three Einstein observations added together, as well as for each of them separately. In both cases, the calculations were made for two standard energy bands, 0.2 - 0.8 keV and 0.8 - 4 keV. A third much weaker source is required at high energies only. There is no obvious optical counterpart for this weak source.

The positions derived for the two strongest sources in the field agree with those of Gl 867 A and B. However, we find, quite unexpectedly, that at the time of the Einstein observations the two sources were of roughly the same intensity. A ratio A:B of 3.5 like that observed with EXOSAT is excluded at greater than 5σ in both energy bands.

The results for the individual IPC observations clearly show that both components varied over the 3-day interval. There is no evidence that the gradual decrease observed in the total X-ray luminosity of the system was due to rotational modulation of component A, in contrast to Agrawal's suggestion.

4. Conclusions

The two components of Gl 867 are both strong X-ray sources as demonstrated by maximum-likelihood analysis of EXOSAT data as well as of previous Einstein observations. Both components appear highly variable in X-rays, as they are known to be at optical wavelengths. X-ray flares with energies of $\approx 10^{32}$ erg were observed during the continuous EXOSAT observation. The variations from day-to-day observed by Einstein were most likely due to flares on the two components rather than to rotational modulation of Gl 867 A. One of the two components of the binary, although we don't know which, changed its mean X-ray luminosity by a factor of roughly 3 between the 1980 Einstein observations and the observation made by EXOSAT in 1984.

References

Agrawal, P.C.: 1988, A&A, 204, 235

Pallavicini, R., Tagliaferri, G., Pollock, A.M.T., Schmitt, J.H.M.M., and Rosso, C.: 1990, A&A, 227, 483

Pollock, A.M.T.: 1987, A&A, 320, 283

CHARACTERIZATION OF LONG-TERM X-RAY VARIABILITY IN A SAMPLE OF LATE-TYPE STARS

S. Sciortino and G. Micela
Osservatorio Astronomico di Palermo
Piazza del Parlamento 1, 90134 Palermo, Italy

1. Introduction

The advent of imaging X-ray observatories and of the photon-counting imaging detectors with their impressive increase in sensitivity has drastically improved the capability to explore the $\log N$ - $\log S$ relation in the low flux region, increasing by ≈ 100 times the number of detected X-ray sources, many of which have been observed several times. The Einstein IPC (Imaging Proportional Counter; Gorenstein *et al.* 1981) has observed more than once about 1500 - 2000 celestial objects detected at least once. Hence the IPC observations contain a large amount of information on long-term X-ray variability, but this information requires proper methodology for being extracted and studied.

Motivated by these considerations we have developed an original method, especially suited to search for long-term variability in low-statistic X-ray sources, observed with photon-counting detectors having imaging capabilities. This method makes usage both of source counts and background counts and takes into account exposure time and source position in the detector field-of-view (Sciortino and Micela 1990).

2. The Method

For the sake of clarity we describe the application of our method to IPC data resulting from the final standard data processing (*cf.* Harnden *et al.* 1984; Harris *et al.* 1990). The final data processing has determined X-ray source count rates on the basis of two detection methods the "Map" and the "Local" detection algorithms. In the "Map" algorithm the background is determined from a reference map calibrated using the longest exposures of the Deep Survey (Giacconi *et al.* 1979, Primini *et al.* 1990) and is known with high accuracy and its error can be neglected. In this case the number of counts, C, collected in the detection cell is

$$C = \phi_{PRF}\, R_S T_e + B \tag{1}$$

where ϕ_{PRF} accounts for the width of the point-response function, R_S is the source rate, $T_e = T\phi_{MS}\,\phi_{vign}$ is the effective exposure time corrected for the vignetting and mirror scattering effects, and B are the background counts falling in the detection cell. In the " Local" algorithm the background is evaluated locally and is know with lesser accuracy than in the Map algorithm. This is accomplished measuring at the same time C and the number of counts, F, collected in a larger cell concentric to the detection cell. In this geometry we have the further relation:

$$F = \beta\, R_S T_e + \eta\, B \tag{2}$$

where $\beta\,(\approx 1)$ is the fraction of source counts falling in the larger cell and η is the ratio of large cell area and detection cell area. In real applications, one wants to infer conclusions on temporal variations of source rates, after having tested the occurrence of temporal variations of counts collected in the detection cell. For this purpose one has to distinguish between the "Map" and the "Local" rates.

Let consider only "Map" rate. According to the Poisson statistics, the probability to observe $(C_1, C_2, ...C_N)$ counts in N distinct observations of the same objects is:

$$P_{1,2,..,N} = \prod_{i=1}^{N} \frac{(\phi_{PRF}\, T_{e,i}\, R_S + B_i)^{C_i}}{C_i!}\, e^{-(\phi_{PRF}\, T_{e,i}\, R_S + B_i)} \quad (3)$$

The value $\tilde{R}_S$ maximizing this equation, *i.e.* the maximum likelihood estimate of R_S, is:

$$\tilde{R}_S = (\textstyle\sum_{i=1}^{N} C_i - \sum_{i=1}^{N} B_i)/(\phi_{PRF} \sum_{i=1}^{N} T_{e,i}) \quad (4)$$

In this case the expression for the fractional variability is:

$$\Delta R_i/\tilde{R}_S = (R_{S,i} - \tilde{R}_S)/\tilde{R}_S = (C_i - \tilde{C}_i)/\tilde{C}_i \quad (5)$$

with $\tilde{C}_i$ being the expected number of counts for $R_{S,i} = \tilde{R}_s$. Where C_i results to be compatible, at the chosen confidence level, with $\tilde{C}_i$ one can determine the upper limit to the amplitude of undetected variations.

In the most general case in which we have both N_L "Local" and N_M "Map" measurements the expression for $\tilde{R}_S$ is a bit more complicated. The expected number of source counts is $\tilde{S}_i = \tilde{R}_S T_{e,i}$ and it should be compared with the actual number of source count S_i - given by

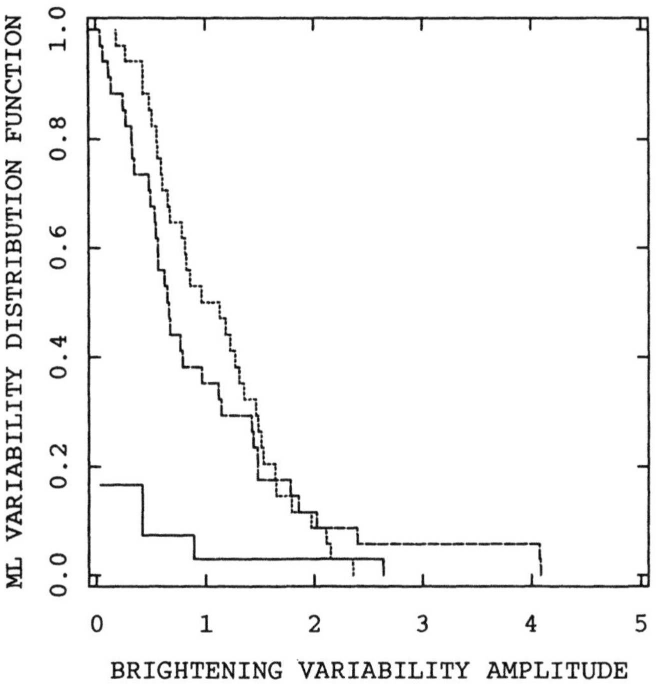

Figure 1. Maximum likelihood (ML) brightening variability distribution function for late-type Hyades stars (*solid line*) compared with the ML brightening variability threshold distribution for late-type Pleiades (*short dashed line*) and the Hyades (*long-dashed line*). While none of the Pleiades stars have been recognized as variable according to the adopted 3σ confidence level, the sensitivity to variability attainable with present data does not exclude that the Pleides stars could undergoes long-term variations of amplitude similar (and even greater) to that detected in the Hyades stars.

Table 1. Composition of Hyades and Pleiades Samples[1]

Spectral Type	Hyades Stars	Hyades Observations	Pleiades Stars	Pleiades Observations
dF	1	3	1	3
dG	7(2)	17(3)	5	16
dK	4	9	5	13
dM	2	4	1	2
gK	1(1)	2(1)	-	-

[1] In parentheses we indicate the number of stars and of observations with detected variability.

eqn (2) - to test, at the given chosen confidence level, the "null hypothesis" that $\tilde{S}_i$ is compatible with S_i. This can be accomplished noting that F and C are integers, so that a direct sum of the (implicit) probability distribution of S over the permitted range of C and F pairs can be performed (for more details see Sciortino and Micela 1990). This technique allows us to establish, at the chosen confidence level, whether the source counts S_i undergo a significant fluctuation with respect to the expected number of counts $\tilde{S}_i$. The amplitude of detected variability is easily computed according to equation (2) or (1) for the "Local" and the "Map" cases, respectively.

The adoption of the technique to build maximum likelihood integral distributions in the presence of upper bound (Avni *et al.* 1980; Schmitt 1985) allows us to use all available information to construct the Maximum Likelihood amplitude variability integral distribution for a given class of sources.

This distribution taken together with the distribution of time separations between pairs of observations of same objects allows us to statistically characterize and compare the variability properties of well defined classes of sources, such as the frequency of variations of a given amplitude, the sensitivity of available observations, and the range of time scale searched for variability occurrence.

3. The case of Hyades and Pleiades late-type stars

We use data taken with the IPC on board the Einstein Observatory, reduced with the standard *Rev-1* processing (Harnden *et al.* 1984); the final processing results are organized in a easy retrievable way in the Stellar Einstein X-ray database resident on SUN^{TM} at Center for Astrophysics (Cambridge, Ma), and on VAX^{TM} at Osservatorio Astronomico di Palermo (Sciortino *et al.* 1988; Harnden *et al.* 1990).

The Hyades and the Pleiades stars are from the complete Einstein Observatory surveys of these clusters (Micela *et al.* 1988; 1990). The Hyades and Pleiades samples consist of the 15 multiply observed stars with $L_X > 3\,10^{29}$, and of the 12 multiply observed stars with $L_X > 3\,10^{28}$, respectively. The subdivision according to spectral type is summarized in Table 1. We note that 5 Hyades stars and 8 Pleiades stars have been observed three times or more. Exposure times for the IPC images containing the Hyades and the Pleiades stars considered here range between 10^3 - $1.5\,10^4$ sec. Most of the observations of the Pleiades stars cover the time separation of few days. On the contrary the Hyades observations are distributed on a wider range of time separations.

Applying our technique to these two samples of data we have been able to detect significant (at confidence level $> 99.87\%$ ($3\,\sigma$)) brightening variability in $\approx 20\%$ of the Hyades stars (*cf. solid line* in Fig. 1). No significant variability has been detected in the Pleiades stars. However the sensitivity to variations in this last case (*short-dashed line* in Fig. 1) is significantly lower than for the Hyades (*long-dashed line* in Fig. 1); indeed variability amplitudes similar to those detected in the Hyades could have occurred in the Pleiades and pass undetected due to the lower count statistic of these more distant stars.

We conclude noting that the new data gathered with the pointed phase of ROSAT PSPC and ROSAT Wide Field Camera observations will be an elective area of application for the proposed methodology.

References

Avni, Y., Soltan, A., Tananbaum, H., Zamorani, G.: 1980, ApJ, 238, 800

Giacconi, R., *et al.*: 1979, ApJ, 234, L1

Gorenstein, P., Harnden, Jr., F.R., Fabricant, D.G.: 1981, IEEE Trans. Nucl. Sci., NS-28, 869

Harnden, Jr., F.R., Sciortino, S., Micela, G., Harris, D.E., Schwarz, J.: 1984, SAO Special Report No. 393

Harnden, F.R., Jr., Sciortino, S., Micela, G., Maggio, A., Vaiana, G.S., Schmitt, J.H.M.M., Rosner, R.: 1990, in *Imaging X-ray Astronomy,* ed. M. Elvis, Cambridge University Press

Harris, D.E. *et al.*: 1990, *The Einstein Observatory Catalog of IPC X-ray Sources,* in press

Micela, G., Sciortino, S., Vaiana, G.S., Schmitt, J.H.M.M., Stern, R.A., Harnden, F.R., Jr., Rosner, R.: 1988, ApJ, 325, 798

Micela, G., Sciortino, S., Vaiana, G.S., Schmitt, J.H.M.M., Harnden, F.R., Jr., Rosner, R.: 1990, ApJ, 348, 557

Primini, F.A., Murray, S.S., Huchra, J., Schild, R., Burg., R., Giacconi, R.: 1990, ApJ, in press

Schmitt, J.H.H.M.: 1985, Ap.J., 293, 176

Sciortino, S., Harnden, Jr. F.R., Maggio, A., Micela, G., Vaina, G.S., Rosner, R., Schmitt, J.H.H.M.: 1988, in *Astronomy from Large Database,* ESO Conference and Workshop Proceedings No. 28, p. 483

Sciortino, S., Micela, G. 1990: Ap.J., submitted

RADIAL VELOCITY MEASURES OF SUSPECTED CHROMOSPHERICALLY ACTIVE AND RSCVn-LIKE STARS

J. Sperauskas, V. Dadonas
Vilnius University Observatory, Ciurlionio 29
232009 Vilnius, Lithuania

1. Introduction

The upsurge of interest in chromospherically active stars in recent years has produced a huge amount of observational material on them. As a result, a number of catalogues have been compiled with information on the photometric, spectroscopic, orbital and physical properties of these interesting stars. The latest catalogue called *Chromospherically Active Binary Stars* (Strassmeier *et al.* 1988) includes the basic results of surveys on RS CVn, BY Dra and other binaries which show strong CaII H & K emission in their spectra and a candidate list of additional stars having similar characteristics but which are not definitely known binaries.

Stellar chromospheric activity includes all of the features of solar activity, but much more extreme and/or on a much large scale, and it is found in both single and binary stars (Hall, 1987). The latter fact causes tremendous interest; binarity is relevant in producing rapid rotation in stars which otherwise would be slow rotators (Stauffer and Hartmann, 1987), most of them are examples of very nearly synchronous rotation (Hall, 1976: *etc.*). Therefore radial velocity measures are desirable to establish the properties of typical chromoshperically active stars, to determine the states of suspected or to indentify the new members of those systems.

The present work has mainly been triggered by the aim of indentifying spectroscopic binaries among stars for which spectroscopic systems have been discovered or confirmed. Most of these objects are suspected chromospherically active and RS CVn-like stars, mainly from the Strassmeier *et al.* (1988) catalogue. New data on duplicity and $v \sin i$ are presented.

Table 1. Basic data for program stars

Name	Spectral Type	P_{phot} (days)	P_{orb} (days)	SB	Class
HD 1326	dM2.5			SB2?	flare
HD 9746	gK1	76.0			CA, RS CVn?
HD 25893	G8IV	7.35		SB1	CA
HD 26913	G5IV	6.8	7.2		CA
HD 27536	G8IV	309.6			CA, RS CVn?
HD 28591	K1III	21.45	21.30	SB1	CA
HD 29104	A-F/G5III-II			SB2	CA?
HD 30957	KO			SB2	CA?
HD 31993	K2III	13.0?		SB2?	CA?
HD 32357	KOIII	80.17/45.2?	80.17	SB1	RS CVn
HD 33798	G5	9.8			CA?
HD 62721	K4III				CA?
HD 86856	dK8				CA?
BD+25°2511	G9V				CA

SB - binary nature: SB1 = single-line spectrum, SB2 = double-line spectrum
CA - chromospherically active star: ? - doubtful value or suspected type

Table 2. Mean radial velocities and other results from our measures

Name	No. Obs.	v_{rad} km s^{-1}	Int. error	Ext. error	$v \sin i$ km s^{-1}	Error	Binary
HD 1326	2	11.28	0.43	0.03			single
HD 9746A	4	-7.54	1.00	1.54			SB2
B	5	-42.89	0.15	0.11	9.3	0.3	
HD 25893	4	26.53	0.20	0.27	4.5	1.0	SB1
HD 26913	4	-8.29	0.23	0.24	6.8	0.9	single
HD 27536	4	6.65	0.20	0.14			SB1?
HD 28591	7	15.31	0.22	6.44	25.8	0.8	SB1
HD 29104	4	14.03	0.20	0.12	4.3	1.2	SB1?
HD 30957A	6	-23.64	0.44	2.78	3.9?		SB2
B	6	26.71	0.25	4.90	6.3	1.4	
HD 31993	7	14.05	0.30	0.57	21.7	1.3	SB1?
HD 32357	6	7.33	0.14	3.11	11.3	0.3	SB1
HD 33798A	6	21.48	0.28	0.25	27.5	1.1	SB2
B	3	26.00	1.10	0.40			
HD 62721	5	74.21	0.14	0.29	3.9	0.4	single
HD 86856	4	30.02	0.25	0.19			SB1
BD+25°2511	24	-9.63	0.23	12.50			SB2

2. Observations and results

Radial velocity observations were obtained during the period February 10-29, 1990, on Maidanak (Uzbekistan). The 1 m telescope was used with the photoelectric radial velocity scanner CORAVEL (Tokovinin, 1987). Typical errors for the cross-correlation technique are estimated to be $\sim 0.6\,\mathrm{km\,s^{-1}}$.

Basic data for our program stars are listed in Table 1. The results of our observations are given in Table 2, where we present the mean radial velocities for all stars, internal and external errors, and the projected rotational velocities, $v \sin i$, derived from the widths of the cross-correlation dips.

To summarize, six of the fourteen stars from our program are undoubted binaries, *i.e.* HD 28591, 32357, 29746, 30957, 33798 and BD +25°2511, and the last four of these are spectroscopic binaries. In five cases (HD 25893, 27536, 29104, 31993, and 86856) more radial velocity measures are needed; we can only note that our data show a very slight and/or obscure variation, if in fact they vary at all.

References

Hall, D.S.: 1976. in IAU Coll. 29, *Multiple Periodic Variable Stars,* ed. W.S. Fitch (Dordrecht: Reidel), p. 287

Hall, D.S.: 1987, in Proc. 10th IAU Europ. Reg. Astron. Meet., ed. P. Harmanec, 5, 77

Stauffer, J., Hartmann. L.: 1987, ApJ, 318, 337

Strassmeier, K.G., Hall, D.S., Zeilik, M., Nelson, E., Eker, Z., Fekel, F.C.: 1988, A&AS, 72, 291

Tokovinin, A.: 1987, AZh, 64, 196

OPTICAL PROPERTIES OF COOL STARS DISCOVERED BY EXOSAT

G. Tagliaferri[1], G. Cutispoto[2], P. Giommi[1], R. Pallavicini[4], L. Pasquini[3], M. Rodonó[2]

[1]EXOSAT Observatory, ESTEC, Noordwijk, The Netherlands
[2]Osservatorio Astrofisico di Catania, Catania, Italy
[3]Osservatorio Astrofisico di Arcetri, Florence, Italy
[4]European Southern Observatory, La Silla, Chile

1. Introduction

X-ray observations are a powerful tool for studying stellar surface activity, its variation with stellar evolution and the determination of the stellar contribution to the diffuse soft X-ray background (*e.g.* Rosner et al. 1985; Pallavicini 1989). Most of our knowledge on the properties of stellar X-ray emission is based on the results of pointed X-ray obervations and the analysis of optically selected samples (*e.g.* Rosner et al. 1981, Schmitt et al. 1985, Bookbinder 1985, Sciortino et al. 1990). Recently, a different approach has been taken by studying in the optical band samples of sources serendipitously detected by X-ray satellites (Caillault et al. 1986; Favata et al. 1988; Fleming 1988, Fleming et al. 1988, 1989). This approach has the advantage of eliminating the unknown bias associated with observations of historically "interesting" objects and allows the investigation of the high tail of the stellar X-ray luminosity function. Typically, serendipitous stellar X-ray sources are the most active among stars.

The Low Energy Telescope and CMA detector on board the EXOSAT satellite were very good X-ray instruments for detecting serendipitous sources in the 0.05 - 2 keV energy band, and about one hundred late-type stars have been detected in this way (Giommi et al. 1988). This sample includes many variable stars. To date, three new CVs (Beuermann et al. 1987, 1989; Osborne et al. 1988), one suspected RS CVn (Pravdo et al. 1985), a flaring PMS star (Tagliaferri et al. 1988) and a few flaring M dwarfs have been detected. Among the latter is the faint M dwarf VB 8 (= Gl 644C, $M_V \sim 17.7$), probably the coolest star (Sp = dM7) so far detected at X-ray wavelengths (Tagliaferri et al. 1989).

2. Results

Cool stars comprise $\sim 25\%$ and $\sim 40\%$ of the sources discovered in the Einstein and EXOSAT X-ray surveys, respectively (Gioia et al. 1989, Giommi et al. 1988). In order to study the nature of X-ray selected cool stars in the EXOSAT sample, we started a campaign of optical observations. In March 1990 we obtained, at the European Southern Observatory at La Silla, Chile, moderate-resolution spectra both in the Hα and CaII H & K regions, and high-resolution spectra of the CaII H & K lines and of the LiI (6707Å) line. UBVRI photometric observations were also obtained. The preliminary results from these observations are very encouraging. The Hα line was detected in emission or filled-in in more than 50% of our stars (here we exclude the dMe stars which by definition have Hα in emission and constitute $\sim 40\%$ of our sample), providing a sure indication of extremely high chromoshperic activity. Two stars showed variable Hα with emission and filled-in profiles on different nights. This can be interpreted as the result of rotational modulation. Strong Ca II H & K emission was also detected in almost all our sources.

Among PMS, the Post T-Tauri stars (PTTS) are particularly interesting. PTTS were first proposed as a class of PMS by Herbig (1978), who noticed that the T-Tauri phase occupies only a small fraction (≈ 5 - 10%) of the contraction time of a 1 $M_\odot$ star towards the main sequence. As a consequence, many more PTTS should exist than classical T-Tauri stars. Unfortunately, it is quite difficult to detect PTTS by conventional methods (*e.g.* Hα and infrared surveys),

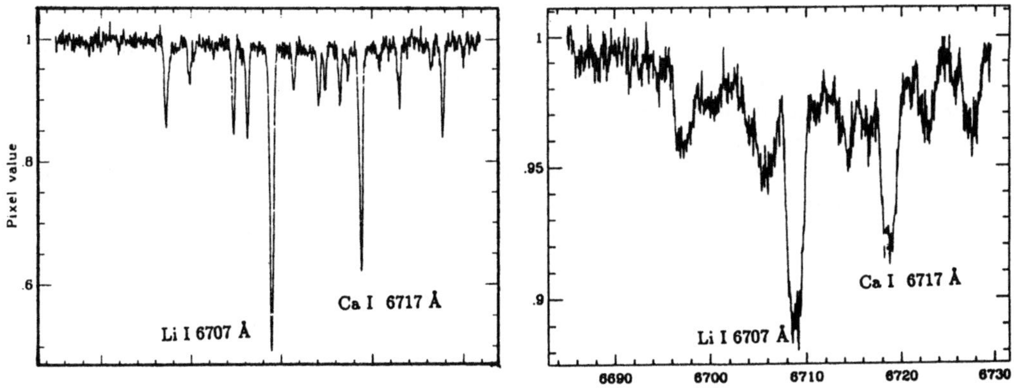

Figure 1. High resolution spectra of two EXOSAT serendipitous stars in the LiIλ6707Å line region. Note that the Li line is stronger than the nearby CaI line. The line broadening due to the star rotation is remarkable for the star in the bottom panel.

because these stars lack the extreme spectroscopic signatures of classical T-Tauri stars. So far, only a handful of *bona fide* PTTS have been identified (see reviews by Herbig (1978) and Haro (1983)). However PTTS are expected to be vigorous X-ray coronal sources and X-ray surveys may prove to be the best way to identify them. One PTTS (HD560B) has already been detected in the EXOSAT sample (Tagliaferri et al. 1988). Some of these elusive objects, whose study can provide crucial information on the evolution of stellar coronal activity during the pre-main sequence lifetime, could be present in our sample. To search for PTTS we took high resolution spectra of the λ6707Å Lithium line region. The line was clearly detected in 13 stars out of the 16 observed. All these stars are of spectral type later than G. In particular for 8 sources this line was stronger than the nearby CaIλ6717Å line (*c.f.* Figure 1). This is a clear indication of stellar youth. The presence of the Lithium line *per se* is not sufficient, however, to classify a star as a young object, since this line has recently been detected in some RS CVn binaries (Pallavicini et al. 1990), which are evolved systems. This provides an interesting challange for stellar evolution models.

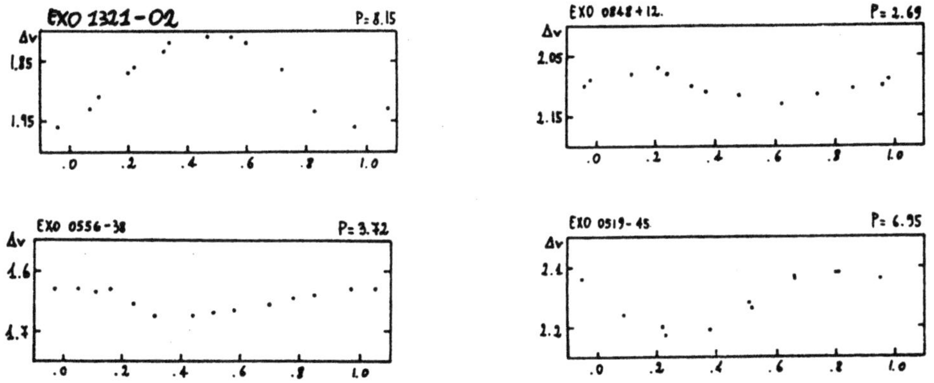

Figure 2. ΔV folded light curves of a few EXOSAT serendipitous stars observed in March 1990. Periods are in days.

The high resolution spectra in the Lithium region also revealed that many stars in our sample are very fast rotators, with a $v \sin i$ between 15 and 50 km sec^{-1} (these are preliminary estimates from comparison with spectra of template stars obtained with the same telescope configuration).

From the photometric observations we also found that many stars (13 out of the 20 observed) of our sample are variable with periods between one and few days with Δ mag $\simeq$ 0.2 - 0.05 (*e.g.* Figure 2). Previous photometric observations of eight southern stars of the EXOSAT sample gave similar results (Cutispoto et al. 1990).

Our optical data show that a large fraction of the EXOSAT serendipitous sources are very active stars, a number of which are especially interesting X-ray emitting objects, such as RS CVn systems, cataclysmic variables, pre-main sequence stars. Further study of this sample will provide important information on stellar coronal activity and its evolution.

References

Beuermann, K., Thomas, H., Giommi, P., Tagliaferri, G.: 1987, A&A, 175, L75

Beuermann, K., Thomas, H., Giommi, P., Tagliaferri, G., Schwope, A.: 1989, A&A, 219, L7

Bookbinder, J.A.: 1985, *PhD Thesis,* Harvard University, Cambridge, Mass. USA

Caillault, J.-P., Helfand, D.J., Nousek, J.A., Takao, L.O.: 1986, ApJ, 304, 318

Cutispoto, G., Tagliaferri, G., Giommi, P., Gouiffes, C., Pallavicini, R., Pasquini, L., Rodonó, M.: 1990, A&AS, in press

Favata, F., Rosner, R., Sciortino, S., Vaiana, G.S.: 1988, ApJ, 324, 1010

Fleming, T.A.: 1988, *PhD Thesis,* University of Arizona

Fleming, T.A., Gioia, I.M., Maccacaro, T.: 1989, ApJ, 340, 1011

Fleming, T.A., Liebert, J., Gioia, I.M., Maccacaro, T.: 1988, ApJ, 331, 958

Gioia, I.M., Maccacaro, T., Schild, R.E., Wolter, A., Stocke, J.T., Morris, S.L., Henry, J.P.: 1989, ApJS, 72, 567

Giommi, P., Tagliaferri, G., Angelini, L.: 1988, in *X-ray Astronomy with EXOSAT,* eds. R. Pallavicini and N.E. White, Mem. Soc. Astron. Ital. 59, 33

Haro, G.: 1983, Rev. Mexicana, 7, 183

Herbig, G.H.: 1978, in *Problems of Physics and the Evolution of the Universe,* ed. L.V. Mirzoyan. Yerevan: Armenia Academy of Science, p. 171

Osborne, J., Giommi, P., Angelini, L., Tagliaferri, G., Stella, L.: 1988, ApJ, 328, L45

Pallavicini, R.: 1989, A&A Rev. 1, 177

Pallavicini, R., Randich, S., Giampapa, M.: 1990, these proceedings

Pravdo, S.H., White, N.E., Giommi, P.: 1985, MNRAS 215, p11

Rosner, R., Avni, Y., Bookbinder, J., Giacconi, R., Golub, L., Harnden, F.R. Jr., Maxson, C.W., Topka, K., Vaiana, G.S.: 1981, ApJ, 249, L5

Rosner, R., Golub, L., Vaiana, G.S.: 1985, ARA&A, 23, 413

Schmitt, J.H., Golub, L., Harnden, F.R. Jr., Maxson, C.W., Rosner, R., Vaiana, G.S.: 1985, ApJ, 290, 307

Sciortino, S., Vaiana, G.S., Morosi, C., Ramella, A., Harnden, F.R. Jr., Rosner, R., Schmitt. J.H.,:1990, ApJ, in press

Tagliaferri, G., Giommi, P., Angelini, L., Osborne, J.P., Pallavicini, R.: 1988, ApJ, 331, L113

Tagliaferri, G., Doyle, J.G., Giommi, P.: 1989, A&A, 231, 131

SUBJECT INDEX

AUTHOR INDEX

OBJECT INDEX

Lecture Notes in Physics

For information about Vols. 1–365
please contact your bookseller or Springer-Verlag

Vol. 366: M.-J. Goupil, J.-P Zahn (Eds.), Rotation and Mixing in Stellar Interiors. Proceedings, 1989. XIII, 183 pages. 1990.

Vol. 367: Y Osaki, H. Shibahashi (Eds.), Progress of Seismology of the Sun and Stars. Proceedings, 1989. XIII, 467 pages. 1990.

Vol. 368: L. Garrido (Ed.), Statistical Mechanics of Neural Networks. Proceedings. VI, 477 pages. 1990.

Vol. 369: A. Cassatella, R. Viotti (Eds.), Physics of Classical Novae. Proceedings, 1989. XII, 462 pages. 1990.

Vol. 370: H.-D. Doebner, J.-D. Hennig (Eds.), Quantum Groups. Proceedings, 1989. X, 434 pages. 1990.

Vol. 371: K.W. Morton (Ed.), Twelfth International Conference on Numerical Methods in Fluid Dynamics. Proceedings, 1990. XIV, 562 pages. 1990.

Vol. 372: F. Dobran, Theory of Structured Multiphase Mixtures. IX, 223 pages. 1991.

Vol. 373: C. B. de Loore (Ed.), Late Stages of Stellar Evolution. Computational Methods in Astrophysical Hydrodynamics. Proceedings, 1989. VIII, 390 pages. 1991.

Vol. 374: L. Ting, R. Klein, Viscous Vortical Flows. V, 222 pages. 1991.

Vol. 375: C. Bartocci, U. Bruzzo, R. Cianci (Eds.), Differential Geometric Methods in Theoretical Physics. Proceedings, 1990. XIX, 401 pages. 1991.

Vol. 376: D. Berényi, G. Hock (Eds.), High-Energy Ion-Atom Collisions. Proceedings, 1990. IX, 364 pages. 1991.

Vol. 377: W. J. Duschl, S.J. Wagner, M. Camenzind (Eds.), Variability of Active Galaxies. Proceedings, 1990. XII, 312 pages. 1991.

Vol. 378: C. Bendjaballah, 0. Hirota, S. Reynaud (Eds.), Quantum Aspects of Optical Communications. Proceedings 1990. VII, 389 pages. 1991.

Vol. 379: J. D. Hennig, W. Lücke, J. Tolar (Eds.), Differential Geometry, Group Representations, and Quantization. XI, 280 pages. 1991.

Vol. 380: I. Tuominen, D. Moss, G. Rüdiger (Eds.), The Sun and Cool Stars: activity, magnetism, dynamos. Proceedings, 1990. X, 530 pages. 1991.

Vol. 381: J. Casas-Vazquez, D. Jou (Eds.), Rheological Modelling: Thermodynamical and Statistical Approaches. Proceedings, 1990. VII, 378 pages. 1991.

Vol. 382: V.V. Dodonov, V. I. Man'ko (Eds.), Group Theoretical Methods in Physics. Proceedings, 1990. XVII, 601 pages. 1991.

Vol. 384: M. D. Smooke (Ed.), Reduced Kinetic Mechanisms and Asymptotic Approximations for Methane-Air Flames. V, 245 pages. 1991.

Vol. 385: A. Treves, G. C. Perola, L. Stella (Eds.), Iron Line Diagnostics in X-Ray Sources. Proceedings, Como, Italy 1990. IX, 312 pages. 1991.

Vol. 386: G. Pétré, A. Sanfeld (Eds.), Capillarity Today. Proceedings, Belgium 1990. XI, 384 pages. 1991.

Vol. 387: Y. Uchida, R. C. Canfield, T. Watanabe, E. Hiei (Eds.), Flare Physics in Solar Activity Maximum 22. Proceedings, 1990. X, 360 pages. 1991.

Vol. 388: D. Gough, J. Toomre (Eds.), Challenges to Theories of the Structure of Moderate-Mass Stars. Proceedings, 1990. VII, 414 pages. 1991.

Vol. 389: J. C. Miller, R. F. Haglund (Eds.), Laser Ablation-Mechanisms and Applications. Proceedings. IX, 362 pages, 1991.

Vol. 390: J. Heidmann, M. J. Klein (Eds.), Bioastronomy - The Search for Extraterrestrial Life. Proceedings, 1990. XVII, 413 pages. 1991.

Vol. 391: A. Zdziarski, M. Sikora (Eds.), Ralativistic Hadrons in Cosmic Compact Objects. Proceedings, 1990. XII, 182 pages. 1991.

Vol. 392: J.-D. Fournier, P.-L. Sulem (Eds.), Large-Scale Structures in Nonlinear Physics. Proceedings. VIII, 353 pages. 1991.

Vol. 393: M. Remoissenet, M.Peyrard (Eds.), Nonlinear Coherent Structures in Physics and Biology. Proceedings. XII, 398 pages. 1991.

Vol. 394: M. R. J. Hoch, R. H. Lemmer (Eds.), Low Temperature Physics. Proceedings. XXX,XXX pages. 1991.

Vol. 395: H. E. Trease, M. J. Fritts, W. P. Crowley (Eds.), Advances in the Free-Lagrange Method. Proceedings, 1990. XI, 327 pages. 1991.

Vol. 396: H. Mitter, H. Gausterer (Eds.), Recent Aspects of Quantum Fields. Proceedings. XIII, 332 pages. 1991.

Vol. 397: P. B. Byrne, D. J. Mullan (Eds.), Surface Inhomogeneities on Late-Type Stars. Proceedings. XVI, 355 pages. 1992.

New Series m: Monographs

Vol. m 1: H. Hora, Plasmas at High Temperature and Density. VIII, 442 pages. 1991.

Vol. m 2: P. Busch, P. J. Lahti, P. Mittelstaedt, The Quantum Theory of Measurement. XIII, 165 pages. 1991.

Vol. m 3: A. Heck, J. M. Perdang (Eds.), Applying Fractals in Astronomy. IX, 210 pages. 1991.

Vol. m 4: R. K. Zeytounian, Mécanique des fluides fondamentale. XV, 615 pages, 1991.

Vol. m 5: R. K. Zeytounian, Meteorological Fluid Dynamics. XI, 346 pages. 1991.

Vol. m 6: N. M. J. Woodhouse, Special Relativity. VIII, 86 pages. 1992.

Vol. m 7: G. Morandi, The Role of Topology in Classical and Quantum Physics. XIII, 239 pages. 1992.

GPSR Compliance

The European Union's (EU) General Product Safety Regulation (GPSR) is a set of rules that requires consumer products to be safe and our obligations to ensure this.

If you have any concerns about our products, you can contact us on ProductSafety@springernature.com

In case Publisher is established outside the EU, the EU authorized representative is:

Springer Nature Customer Service Center GmbH
Europaplatz 3
69115 Heidelberg, Germany

Batch number: 10838062

Printed by Printforce, the Netherlands